Tropical Dry Forests in the Americas

Ecology, Conservation, and Management

Tropical Dry Forests in the Americas

Ecology, Conservation, and Management

Edited by
Arturo Sánchez-Azofeifa
Jennifer S. Powers
Geraldo W. Fernandes
Mauricio Quesada

CRC Press
Taylor & Francis Group
Boca Raton London New York

CRC Press is an imprint of the
Taylor & Francis Group, an **informa** business

CRC Press
Taylor & Francis Group
6000 Broken Sound Parkway NW, Suite 300
Boca Raton, FL 33487-2742

Printed on acid-free paper
Version Date: 20130809

International Standard Book Number-13: 978-1-4665-1200-9 (Hardback)

Visit the Taylor & Francis Web site at
http://www.taylorandfrancis.com

and the CRC Press Web site at
http://www.crcpress.com

Contents

Preface

This book is the result of several years of scientific research and networking, building interdisciplinary collaboration, and sharpening the appreciation of natural scientists for the human occupants and their activities in the dry forests of the Neotropics. Tropical dry forests are largely neglected ecosystems that rarely receive the scientific or conservation attention they deserve. They do not have the same popular appeal as tropical rain forests. Yet, tropical dry forests have tremendous biodiversity and are occupied by millions of people. Even the driest of the dry forest regions, in northeastern Brazil, is home to some 30 million people who largely live off agriculture and ranching.

Under the conditions of global climate change, the already marginal and variable climatic conditions of dry forests and the agriculture that is practiced in these regions make for highly vulnerable populations and ecosystems. Many such regions already show signs of desertification. On the other hand, these dry regions have a great production potential for suitably adapted production systems because they do not suffer from the high pest and disease incidence of the more humid regions. Dry forest plants have developed defenses against drought and herbivory that involve secondary metabolites such as waxes, resins, and aromatic biochemicals, which are useful in industry and medicine. The traditional ecological and phytochemical knowledge of dry forest inhabitants is considerable, and their culture of adaptation deserves study as societies need to understand adaptation to adverse and changing climate.

All the aforementioned reasons make a project on the natural and human dimensions of dry forests in the American tropics an important subject for Inter-American Institute for Global Change Research funding. The Tropi-Dry network has provided ecological insight and knowledge on the vulnerability of socio-ecological systems, the cultural adaptations of such systems, and external threats from globalization. The close link between the functioning of dry forests, including their use, and climatic and ecological gradients makes them ideal for exploring the richness of ecosystems and cultures. Research on dry forest regions also serves to highlight the risks under which other ecosystems find themselves under the conditions of climate change.

In the Americas we find tropical dry forests in southern Mexico, Central America, Venezuela, and Brazil, and the Tropi-Dry network is engaged in research in all of these. This has permitted the comparison of climate trends at different latitudes and provided a comprehensive picture of climate change on the continent. The principal change detected in climatic conditions has been the length of the rainy season: in some regions it has lengthened, whereas in others it has become shorter. This not only affects the overall productivity of ecosystems but also causes changes in phenology, which, in turn, change the relationships among plants, animals, and insects. Rather than recording a simple productivity trend, it is therefore necessary to explore the complex interrelationships among plants, animals, insects, and human populations.

Not only obvious activities such as apiculture but also the vegetation–landscape relationships that will modify risks and opportunities for societies in the long run are affected by these changes. One characteristic of the highly diverse dry forest regions is a strong expression of ecological niches. For instance, pollinators may be insects and birds or bats, all of which will be affected differently by human activities, development, and climate change.

Tropical dry forests are ecosystems that harbor great biodiversity that has evolved critical adaptation mechanisms to extreme climatic conditions. The resilience to stress of even belowground assemblages of organisms shows a highly adapted and diverse ecology, whose understanding and utilization will be important as we search for resilience of (agro-) ecosystems under climate change. Conserving this resource is important. At the same time, human populations have been part of tropical dry forest systems for thousands of years. Their roles and rights must be part of this conservation effort. This book is a unique effort to bring together the many facets of function, use, heritage, and future potential of these forests. It presents an important and exciting synthesis of many years of work across countries, disciplines, and cultures.

Holm Tiessen
Director, Inter-American Institute for Global Change Research (IAI)
Sao Jose dos Campos, Sao Paulo, Brazil

Contributors

María de Jesús Aguilar-Aguilar
Centro de Investigaciones en Ecosistemas
Universidad Nacional Autonoma de Mexico
Morelia, Michoacan, Mexico

Esteban Alvarez
Medellin Botanical Gardens
Medellin, Colombia

Mariana Álvarez-Añorve
Centro de Investigaciones en Ecosistemas
Universidad Nacional Autonoma de Mexico
Morelia, Michoacan, Mexico

Felisa C. Anaya
Departmento de Saúde Mental e Coletiva
Universidade Estadual de Montes Claros
Montes Claros, Minas Gerais, Brazil

Carla I. Aranguren
Centro de Ecología
Instituto Venezolano de Investigaciones Científicas Altos de Pipe
Caracas, Venezuela

Luis Ávila-Cabadilla
Centro de Investigaciones en Ecosistemas
Universidad Nacional Autonoma de Mexico
Morelia, Michoacan, Mexico

Francisco Balvino-Olvera
Centro de Investigaciones en Ecosistemas
Universidad Nacional Autonoma de Mexico
Morelia, Michoacan, Mexico

Rômulo S. Barbosa
Departmento de Política e Ciências Sociais
Universidade Estadual de Montes Claros
Montes Claros, Minas Gerais, Brazil

Brad Boyle
Department of Ecology and Evolutionary Biology
University of Arizona
Tucson, Arizona

Diego Brandão
Departamento de Biologia Geral
Universidade Estadual de Montes Claros
Montes Claros, Minas Gerais, Brazil

Alberto Burquez
Instituto de Ecología
Universidad Nacional Autónoma de Mexico
Unidad Hermosillo, Sonora

Julio Calvo-Alvarado
Escuela de Ingenieria Forestal
Instituto Tecnologico de Costa Rica
Cartago, Costa Rica

Dorian Carvajal-Vanegas
Escuela de Ingenieria Forestal
Instituto Tecnologico de Costa Rica
Cartago, Costa Rica

Alicia Castillo
Centro de Investigaciones en Ecosistemas
Universidad Nacional Autonoma de Mexico
Morelia, Michoacan, Mexico

Marcel Coelho
Ecologia Evolutiva & Biodiversidade
Universidade Federal de Minas Gerais
Belo Horizonte, Minas Gerias, Brazil

José Miguel Contreras-Sánchez
Centro de Investigaciones en Ecosistemas
Universidad Nacional Autonoma de Mexico
Morelia, Michoacan, Mexico

Joselândio Correa-Santos
Departamento de Biologia Geral
Universidade Estadual de Montes Claros
Montes Claros, Minas Gerais, Brazil

Fernanda Covacevich
Laboratory of Soil Microbiology
Agronomy INTA
Balcarce, Buenos Aires, Argentina

Jacob Cristobal-Perez
Centro de Investigaciones en Ecosistemas
Universidad Nacional Autonoma de Mexico
Morelia, Michoacan, Mexico

Luiz Alberto Dolabela-Falcão
Departamento de Biologia Geral
Universidade Estadual de Montes Claros
Montes Claros, Minas Gerais, Brazil

Maria das Dores Magalhães-Veloso
Departamento de Biologia Geral
Universidade Estadual de Montes Claros
Montes Claros, Minas Gerais, Brazil

Emmanuel Duarte-Almada
Ecologia Evolutiva & Biodiversidade
Universidade Federal de Minas Gerais
Belo Horizonte, Minas Gerias, Brazil

Brian J. Enquist
Department of Ecology and Evolutionary Biology
University of Arizona
Tucson, Arizona

Marcos Esdras-Leite
Departamento de Geografia
Universidade Estadual de Montes Claros
Montes Claros, Minas Gerais, Brazil

Mário Marcos do Espírito-Santo
Departamento de Biologia Geral
Universidade Estadual de Montes Claros
Montes Claros, Minas Gerais, Brazil

Geraldo Wilson Fernandes
Ecologia Evolutiva & Biodiversidade
Universidade Federal de Minas Gerais
Belo Horizonte, Minas Gerias, Brazil

Paola Fernandez
Centro de Investigaciones en Ecosistemas
Universidad Nacional Autonoma de Mexico
Morelia, Michoacan, Mexico

Fernando Fernández-Méndez
Forestry Science Department
Universidad del Tolima
Tolima, Colombia

Yule Roberta Ferreira-Nunes
Departamento de Biologia Geral
Universidade Estadual de Montes Claros
Montes Claros, Minas Gerais, Brazil

Henrique M.A.C. Fonseca
Department of Biology
Universidade de Aveiro
Aveiro, Portugal

Flávia Fonseca-Pezzini
Ecologia Evolutiva & Biodiversidade
Universidade Federal de Minas Gerais
Belo Horizonte, Minas Gerias, Brazil

Maria Gabriela Gei
College of Biological Sciences
University of Minnesota
Minneapolis, Minnesota

Claudia Galicia
Centro de Investigaciones en Ecosistemas
Universidad Nacional Autonoma de Mexico
Morelia, Michoacan, Mexico

José A. González-Carcacía
Centro de Ecología
Instituto Venezolano de Investigaciones Científicas Altos de Pipe
Caracas, Venezuela

Bruno T. Goto
Department of Botany, Ecology, and Zoology
Federal University of Rio Grande do Norte
Lagoa Nova, Natal, Brazil

Michael Hesketh
Earth and Atmospheric Sciences Department
University of Alberta
Edmonton, Alberta, Canada

Branco Hilje
Escuela de Ingenieria Forestal
Instituto Tecnologico de Costa Rica
Cartago, Costa Rica

Yingduan Huang
Earth and Atmospheric Sciences Department
University of Alberta
Edmonton, Alberta, Canada

Catherine M. Hulshof
Department of Ecology and Evolutionary Biology
University of Arizona
Tucson, Arizona

Claudia Jacobi
Departamento de Biologia Geral
Universidade Federal de Minas Gerais
Belo Horizonte, Minas Gerias, Brazil

César D. Jiménez-Rodríguez
Escuela de Ingenieria Forestal
Instituto Tecnologico de Costa Rica
Cartago, Costa Rica

Juan Manuel Lobato-García
Centro de Investigaciones en Ecosistemas
Universidad Nacional Autonoma de Mexico
Morelia, Michoacan, Mexico

Martha Lopezaraiza-Mikel
Centro de Investigaciones en Ecosistemas
Universidad Nacional Autonoma de Mexico
Morelia, Michoacan, Mexico

Sergio Lopez-Valencia
Centro de Investigaciones en Ecosistemas
Universidad Nacional Autonoma de Mexico
Morelia, Michoacan, Mexico

Ricardo Louro-Berbara
Soil Department
Federal Rural University of Rio de Janeiro
Rio de Janeiro, Brazil

Alfredo Lozano
Forestry Science Department
Universidad del Tolima
Tolima, Colombia

Bruno Madeira
Departamento de Biologia Geral
Universidade Estadual de Montes Claros
Montes Claros, Minas Gerais, Brazil

Henrique Maia-Valério
Departamento de Biologia Geral
Universidade Estadual de Montes Claros
Montes Claros, Minas Gerais, Brazil

Tatianne Marques
Departamento de Entomologia
Universidade Federal de Viçosa
Viçosa-MG, Brazil

Silvana Martén-Rodríguez
Instituto de Ecologia A. C.
Xalapa, Veracruz, Mexico

Angelina Martínez-Yrízar
Instituto de Ecología
Universidad Nacional Autónoma de Mexico
Unidad Hermosillo, Sonora, Mexico

Hugo N. de Matos-Brandão
Departamento de Biologia Geral
Universidade Estadual de Montes Claros
Montes Claros, Minas Gerais, Brazil

Omar Melo
Forestry Science Department
Universidad del Tolima
Tolima, Colombia

Francisco Monge
Escuela de Ingenieria Forestal
Instituto Tecnologico de Costa Rica
Cartago, Costa Rica

João Gabriel Mota-Souza
Departamento de Biologia Geral
Universidade Estadual de Montes Claros
Montes Claros, Minas Gerais, Brazil

Jafet Nassar
Centro de Ecología
Instituto Venezolano de Investigaciones Científicas Altos de Pipe
Caracas, Venezuela

Frederico de Siqueira Neves
Ecologia Evolutiva & Biodiversidade
Universidade Federal de Minas Gerais
Belo Horizonte, Minas Gerias, Brazil

Camila P. Nobre
Soil Department
Federal Rural University of Rio de Janeiro
Rio de Janeiro, Brazil

Yumi Oki
Ecologia Evolutiva & Biodiversidade
Universidade Federal de Minas Gerais
Belo Horizonte, Minas Gerias, Brazil

Lemuel Olívio-Leite
Departamento de Biologia Geral
Universidade Estadual de Montes Claros
Montes Claros, Minas Gerais, Brazil

Sergio Ricardo Olvera-García
Centro de Investigaciones en Ecosistemas
Universidad Nacional Autonoma de Mexico
Morelia, Michoacan, Mexico

Uriel Perez
Engineering Department
Universidad del Tolima
Tolima, Colombia

Alexander Pfaff
Public Policy, Economics and Environment
Duke University
Durham, North Carolina

Carlos Portillo-Quintero
Earth and Atmospheric Sciences Department
University of Alberta,
Edmonton, Alberta, Canada

Jennifer S. Powers
College of Biological Sciences
University of Minnesota
Minneapolis, Minnesota

Mauricio Quesada
Centro de Investigaciones en Ecosistemas
Universidad Nacional Autonoma de Mexico
Morelia, Michoacan, Mexico

André Vieira Quitino
Ecologia Evolutiva & Biodiversidade
Universidade Federal de Minas Gerais
Belo Horizonte, Minas Gerias, Brazil

Manoel Reinaldo-Leite
Departamento de Geografia
Universidade Estadual de Montes Claros
Montes Claros, Minas Gerais, Brazil

Benoit Rivard
Earth and Atmospheric Sciences Department
University of Alberta,
Edmonton, Alberta, Canada

Juan Robalino
Centro Agronómico Tropical de Investigación y Enseñanza
Turrialba, Costa Rica

Giovana Rodrigues da Luz
Departamento de Biologia Geral
Universidade Estadual de Montes Claros
Montes Claros, Minas Gerais, Brazil

Francisco Rodriguez
Escuela de Ciencias y Letras
Instituto Tecnológico de Costa Rica, Sede San Carlos
San Carlos, Costa Rica

Jon Paul Rodríguez
Centro de Ecología
Instituto Venezolano de Investigaciones Científicas Altos de Pipe
Caracas, Venezuela

Fernando Rosas
Centro de Investigaciones Biológicas
Universidad Autónoma del Estado de Hidalgo
Pachuca, Hidalgo, Mexico

Victor Rosas-Guerrero
Centro de Investigaciones en Ecosistemas
Universidad Nacional Autonoma de Mexico
Morelia, Michoacan, Mexico

Arturo Sánchez-Azofeifa
Earth and Atmospheric Sciences Department
University of Alberta,
Edmonton, Alberta, Canada

Gumersindo Sánchez-Montoya
Centro de Investigaciones en Ecosistemas
Universidad Nacional Autonoma de Mexico
Morelia, Michoacan, Mexico

Rubens Manoel dos Santos
Departamento de Ciências Florestais
Universidade Federal de Lavras
Lavras, Minas Gerais, Brazil

Carlos Magno Santos-Clemente
Departamento de Geografia
Universidade Estadual de Montes Claros
Montes Claros, Minas Gerais, Brazil

Roberto Sáyago
Centro de Investigaciones en Ecosistemas
Universidad Nacional Autonoma de Mexico
Morelia, Michoacan, Mexico

Diellen Librelon da Silva
Departamento de Biologia Geral
Universidade Estadual de Montes Claros
Montes Claros, Minas Gerais, Brazil

Jhonathan O. Silva
Ecologia Evolutiva & Biodiversidade
Universidade Federal de Minas Gerais
Belo Horizonte, Minas Gerias, Brazil

Saimo Rebleth de Souza
Departamento de Biologia Geral
Universidade Estadual de Montes Claros
Montes Claros, Minas Gerais, Brazil

Kathryn E. Stoner
Department of Biological and Health Sciences
Texas A&M University-Kingsville
Kingsville, Texas

Natalia Valdespino-Vázquez
Centro de Investigaciones en Ecosistemas
Universidad Nacional Autonoma de Mexico
Morelia, Michoacan, Mexico

Arely Vázquez-Ramírez
Centro de Investigaciones en Ecosistemas
Universidad Nacional Autonoma de Mexico
Morelia, Michoacan, Mexico

Soraya del Carmen Villalobos
Centro de Ecología
Instituto Venezolano de Investigaciones Científicas Altos de Pipe
Caracas, Venezuela

Payri Yamarte-Loreto
Earth and Atmospheric Sciences Department
University of Alberta,
Edmonton, Alberta, Canada

Magno Augusto Zazá-Borges
Departamento de Biologia Geral
Universidade Estadual de Montes Claros
Montes Claros, Minas Gerais, Brazil

Andréa Zhouri
Dept. de Ciências Sociais e Antropologia
Universidade Federal de Minas Gerais
Belo Horizonte, Minas Gerias, Brazil

1

Tropical Dry Forests in the Americas: The Tropi-Dry Endeavor

Arturo Sánchez-Azofeifa, Julio Calvo-Alvarado, Mário Marcos do Espírito-Santo, Geraldo Wilson Fernandes, Jeniffer S. Powers, and Mauricio Quesada

CONTENTS

1.1 Introduction

Tropical dry forests (TDFs) are considered among the most endangered ecosystems worldwide, and are even more endangered than tropical rainforests. When evaluated at a continental level, degradation and deforestation processes for TDFs in the Americas are no exception to this rule. With almost 60% of their total extent currently extinct, and the remaining forests experiencing high levels of forest fragmentation, new efforts are urgently required for understanding land use/cover change processes. Kalácska et al. (2005d) clearly presented an argument to increase efforts that aimed at understanding drivers of change and ecological processes in TDFs. The authors compared papers on TDFs with papers on rainforest ecosystems that were published between 1945 and 2005, and identified a ratio of 1 TDF paper for every 300 rainforest ecosystem papers published. This startling number underscores our limited understanding of this endangered ecosystem and also how much work is needed to support comprehensive and holistic management and conservation policies (Sánchez-Azofeifa et al. 2005a).

Conservation efforts related to TDFs are further hindered by this ecosystem's tremendous agricultural potential and resultant tendency to attract human settlement. Portillo and Sánchez-Azofeifa (2010) emphasized that deforestation drivers and forces of land-cover change cannot be universally applied to all neotropical TDFs. The authors demonstrated a sharp differentiation between human influences on insular forests where tourism and urban expansion are the dominant socioeconomic forces and human influences on continental regions where expansion of agriculture is the dominant driver of deforestation and forest fragmentation. This is not a surprise, especially with tourism being one of the main drivers of land-cover change in islands and premium-value coastal areas in Latin America (e.g., Mexico and Costa Rica). In addition, there are essential differences between tourism-driven land-use change and agricultural land-use change; hence, they cannot be easily compared.

Besides the deforestation processes mentioned earlier, another key element that is associated with the socioeconomic forces that drive land conversion is the emergence of secondary TDFs across the Latin-American landscape. Much of the current TDF landscape is comprised of unique mosaics of primary TDFs, secondary TDFs, and agricultural land. A combination of different forces is making these landscapes extremely dynamic. Secondary TDFs hold a fragile position. If preserved, these fragmented forests can eventually become primary TDFs; but their fragmented state makes them more accessible and easily deforested, leaving them quite vulnerable and resulting in a bifurcated prediction path for secondary TDFs. In Mexico and Brazil, strong deforestation forces are dominant; whereas in Costa Rica, secondary forests are becoming more dominant in the landscape. In other words, the trajectory for TDF deforestation, fragmentation, and conservation cannot be generalized to a single explanation or prediction.

It is fundamentally clear that our knowledge of TDFs is disjointed and incomplete. Integrated analyses of land-cover change, ecological processes, and human dimensions, and the manner in which these elements are related both to each other and to primary and secondary forests serve as critical building blocks for comprehensive decision making.

1.2 Tropi-Dry's Study Areas

The work presented in this book is the summary of almost 14 years of collaborative work in key TDFs across the Americas (Figure 1.1). Key research sites are located in Mexico (Figure 1.2), Costa Rica (Figure 1.3),

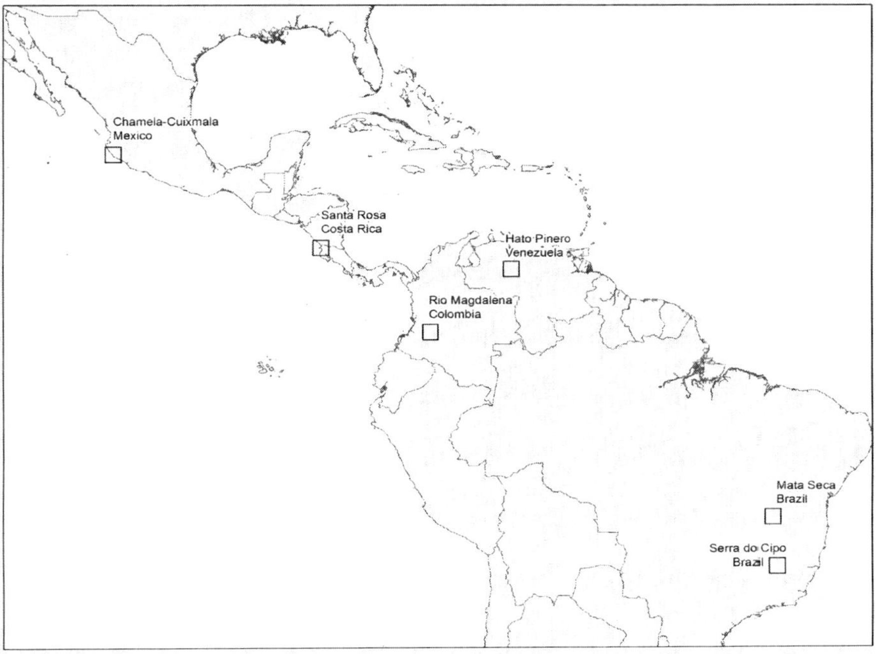

FIGURE 1.1
Location of Tropi-Dry comparative sites across the Americas.

Brazil (Figure 1.4), Venezuela (Figure 17.1), and Colombia (Figure 3.6). Key Tropi-Dry's research sites are summarized next:

1.2.1 Mexico

One of the most important and protected areas of TDFs in Mexico is the Chamela–Cuixmala Biosphere Reserve (Figure 1.2) (Quesada et al. 2009; Sánchez-Azofeifa et al. 2009c). This biosphere reserve primarily consists of mature forests, has an extension of 13,200 ha, and is located along the central western coast of Mexico in the state of Jalisco, where the predominant vegetation is that of TDFs (Lott 2002). This region presents an annual average temperature of 24.6°C with a seasonal pattern of rainfall (García-Oliva et al. 1995). Average annual precipitation based on 30 years (1977–2006) is 763 ± 258 (SD) mm and occurs mainly between June and October.

This region is rich in species, with more than 1149 known vascular plant species and a high incidence of endemisms of more than 125 species (Lott et al. 1987). This pattern of endemism is evident in many other groups of organisms, as indicated by several studies of insect groups. It is important

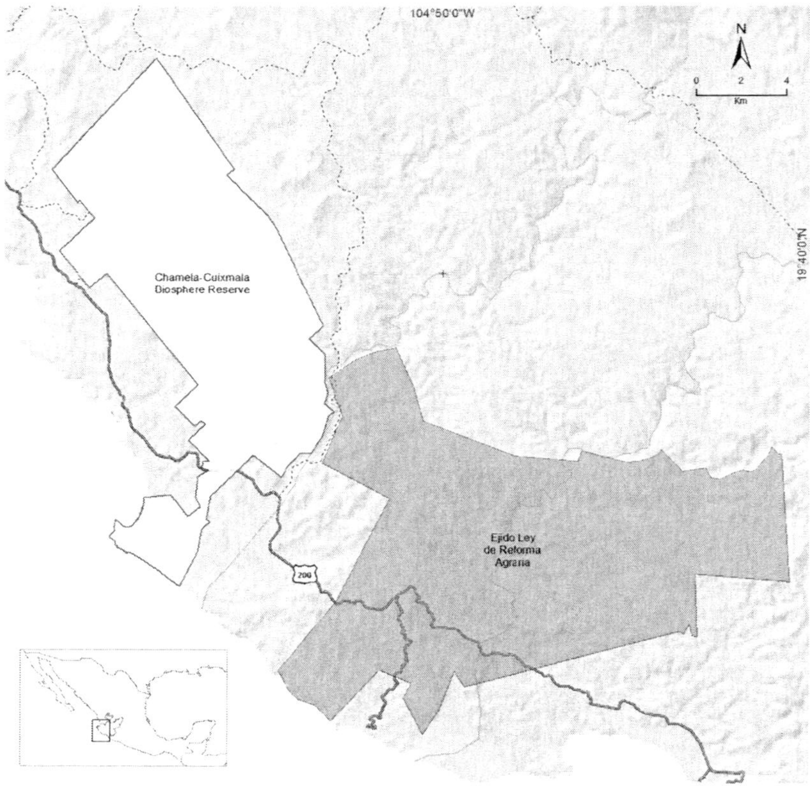

FIGURE 1.2
Location of Mexico's Tropi-Dry research site at the Chamela–Cuixmala Biosphere Reserve, state of Jalisco. The Ejido Ley de Reforma Agraria is Tropi-Dry's social science research site.

to highlight that the occurrence of endemism is significantly lower in the deciduous forests of Central America. In a geographic comparison of the flora of Mesoamerica, Gentry (1995) showed that the highest concentration of endemic species and TDF diversity occurs along the western coast of Mexico and that the flora of southern Central America possesses a heterogeneity of abundant species. Therefore, this region of Mexico is considered the apex of TDF diversity in the neotropics.

The area surrounding the Chamela–Cuixmala Biosphere Reserve is covered by a mosaic of pasture, secondary and primary TDFs with different land-use histories. It is located to the south of Puerto Vallarta and to the north of Manzanillo (Mexico) along the Pacific coast of the state of Jalisco and is bordered on the west by the Pacific Ocean. A remote sensing study of land-cover change indicates that the percentage of area covered by TDFs of the Chamela–Cuixmala Biosphere Reserve is greater than the percentage of the

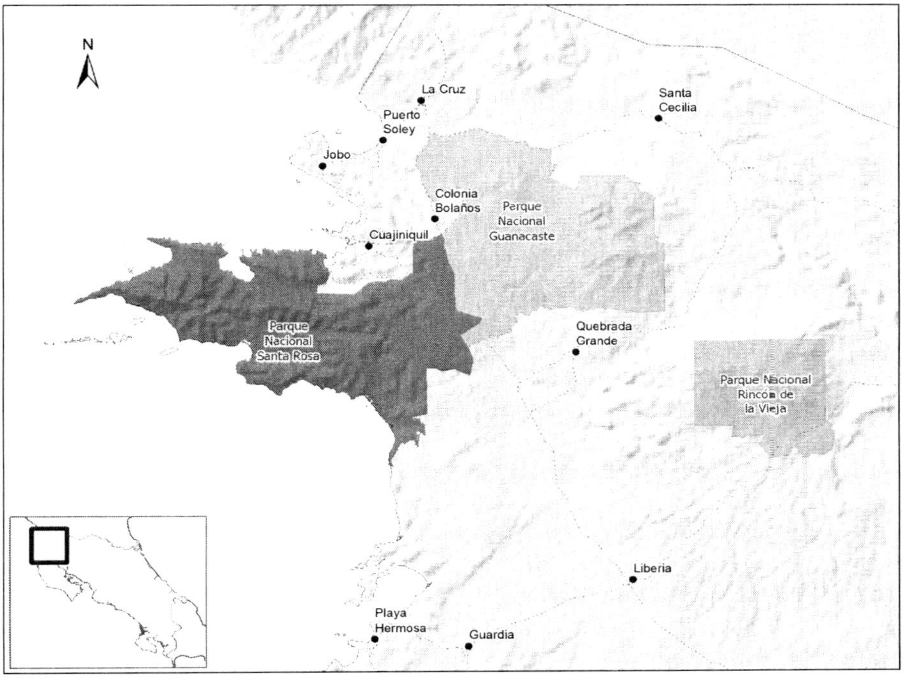

FIGURE 1.3
Location of Costa Rica's Tropi-Dry research site at the Santa Rosa National Park, Guanacaste. Local communities such as Liberia, Quebrada Grande, and La Cruz are Tropi-Dry's social science research sites.

cover originally estimated by Trejo and Dirzo (2000) for the entire country (Sánchez-Azofeifa et al. 2009a). In addition, around 70%–80% of TDFs surrounding the Biosphere Reserve is still maintained. Much of the reserve is surrounded by human settlements of organized local peasant communities (*Ejidos*). The *Ejidos* surrounding the Biosphere Reserve have a forest cover that ranges from 81.2% to 98.0% but much of which is vegetation under secondary succession. The study area is fully described by Avila Cabadilla et al. (2009).

1.2.2 Costa Rica

The study was established in Área de Conservación Guanacaste, specifically in Santa Rosa National Park (SRNP), Costa Rica. In 1972, the SRNP was founded and was Costa Rica's first national park, in the northwest province of Guanacaste, for protecting one of the most important dry forest remnants in Central America. The park is a home to Costa Rica's most famous monument, the Hacienda Santa Rosa, which is also known as La Casona. The

Casona marks the site of Costa Rica's "Epic" War of 1856, a 14-minute battle in which a Costa-Rican civilian militia defeated an invasion of mercenaries sponsored by Tennessee-born William Walker.

Before becoming a protected parkland, SRNP was a part of a large Hacienda-system cattle ranch for almost 200 years. Logging activity, mainly of Swietenia macrophylla (mahogany), took place in the early twentieth century (Burnham 1997). The topography of the study area is relatively flat, with an average slope of 7%. The elevation ranges from 325 masl (meters above sea level) in the northwest to sea level in the southeast. The soils are classified as Usthortent under the U.S. Department of Agriculture (USDA) soil taxonomy system. The mean annual temperature is 26.6°C, and the mean annual precipitation is 1390.8 mm a^{-1} (period from January 2005 to December 2009). The five-month dry period extends from December to April, and the monthly water availability during the wet period exceeds 100 mm.

The SRNP, with its 50,000 ha of coastline, forest, and savannah, protects numerous species of flora and fauna, including 240 species of trees and shrubs, 253 bird species, 10,000 insect species (including 3,200 butterfly species), 100 amphibian and reptile species, and 115 mammal species, of which about 50 species are bats. The park's actual vegetation is a mixture of successional stages due to the historic land use of the region, which included cattle raising, agriculture, and selective logging (Kalácska et al. 2004b; Quesada and Stoner 2004). As a result, SRNP has heterogeneous forests that exhibit a high diversity of habitats.

For decades before the establishment of national park, the entire region encompassing Santa Rosa and the northwest region of Guanacaste suffered due to intense deforestation. One of the primary forces behind forest clearance involved the development of pasturelands when international beef prices were high (Quesada and Stoner 2004). Nevertheless, from 1986 to 2005, the province of Guanacaste experienced the highest rate of forest regrowth. Data analysis (Calvo-Alvarado et al. 2009) showed that by 2005, only 20% of the total forest cover in the region was protected in conservation units. Data also reveal that during the same period, 80% of the mature and secondary forests were located on private lands. Taken together, these factors imply that forest restoration in this region has increased significantly due to three essential reasons: (1) a fall in the price of exported meat in the 1980s, causing marginal pasturelands to be abandoned; (2) the creation of several national parks expropriating many large ranches in the region; and (3) increased tourism-diversified employment opportunities, reducing the demand for agricultural jobs. These changes were augmented by the introduction of the Forest Law in 1996, which implemented a permit system that restricted timber extraction, forest-cover land-use change on private land, and a program of Payments for Environmental Services, known as *PSA*, for forest conservation.

1.2.3 Brazil

1.2.3.1 Parque Estadual da Mata Seca, Minas Gerais

The Mata Seca State Park (MSSP) is a conservation unit (CU) of restricted use (equivalent to the International Union for Conservation of Nature [IUCN] categories Ia, II, and III) that is located in Manga, Minas Gerais, Brazil (Figure 1.4). The park was created in the year 2000 after the expropriation of four farms, and it currently has an area of 15,466.44 ha under the

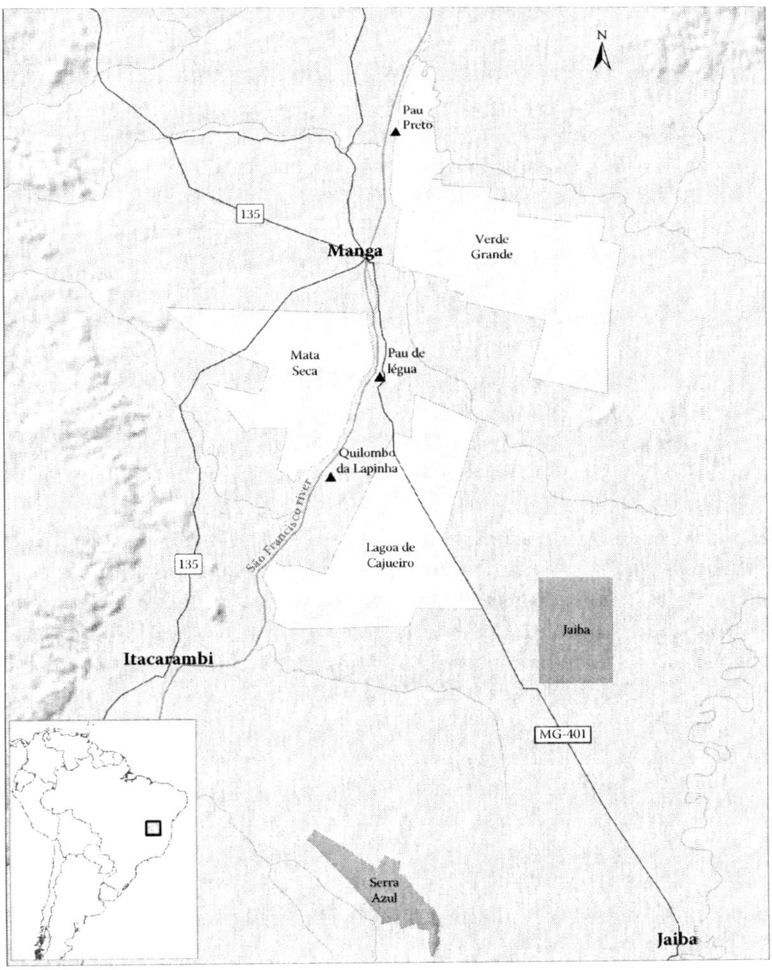

FIGURE 1.4
Location of Brazil's Tropi-Dry research site at the Mata Seca State Park, Manga District, Minas Gerais. Manga and Quilombi da Lapinha are Tropi-Dry's main social science research sites.

responsibility of the Forestry Institute of Minas Gerais State (IEF-MG). The MSSP is situated in a wide transitional area between the Cerrado (savanna) and the Caatinga (scrub and dry forest) biomes in the valley of the São Francisco River (Madeira et al. 2009). In Köppen's classification, the climate is Aw with a pronounced dry season in winter. The annual average temperature is 24°C (Antunes 1994), with an annual average rainfall of 818 ± 242 mm (Madeira et al. 2009) and a monthly average of less than 60 mm during the dry season, which lasts from May to September (Espírito-Santo et al. 2008).

In Minas Gerais, the MSSP is the only CU with dry forests on non-karst soils (IEF 2000) to the west of the São Francisco River. The vegetation is composed of distinct formations and is dominated by deciduous species that lose between 90% and 95% of their leaves during the dry season (Pezzini et al. 2008). Before the establishment of the park, farming was one of the main economic activities, with extensive cattle ranching, and the cultivation of corn, beans, and tomatoes in two 80-ha. However, large areas of the original primary forests were either left intact or suffered only occasional, low-intensity selective logging. Two main regimes were applied: clear-cutting and furrowing for agriculture, and a more general clearing for pasture development (cattle raising), with each disposition requiring distinct management techniques. In 2000, approximately 1525 ha of the MSSP were covered with abandoned pastures and crop fields at an early regeneration stage; whereas the remaining area of the park was a mosaic of primary and secondary dry forests and riparian forests (IEF 2000).

The MSSP is part of the System of Protected Areas (SPA) of the Jaíba Project, which is the largest irrigated perimeter in Latin America (approximately 100,000 ha) that uses water from the São Francisco River. Thus, the park was created as part of the Environmental Compensation Program that had been initiated to balance the deforestation caused by the expansion of irrigation projects at the end of the 1990s. The SPA comprise five CUs of restricted use (three state parks and two biological reserves [Figure 1.4]) and two CUs of sustainable use (Areas of Environmental Protection; equivalent to IUCN category V), totaling approximately 167,000 ha of government-protected areas. However, the creation of the SPA of the Jaíba Project did not account for the numerous traditional populations that inhabit this region (see Chapter 22). Several CUs, including the MSSP (Figure 1.4), enclosed the ancestral territories of these populations within their boundaries, which spurred resistance to the Jaíba Project, threatened the livelihoods of the traditional populations, and generally diminished the efficacy of the CUs.

1.2.3.2 Serra do Cipó National Park, Minas Gerais

The TDFs in Serra do Cipó, which is in the municipality of Santana do Riacho, Minas Gerais, southeastern Brazil, is mostly represented by enclaves of limestone that support tree-sized vegetation with a distinctive floristic composition. The Serra do Cipó is located in the southern portion of the Espinhaço Range and is dominated by Cerrado and Rupestrian Fields vegetation. The climate is mesothermic and is characterized by dry winters and rainy summers, with an annual average rainfall of 1500 mm and an annual average temperature that ranges from 17.4°C to 19.8°C (Madeira and Fernandes 1999). The Serra do Cipó is located in a high-diversity region, and is part of the Espinhaço Range Biosphere Reserve. The dry forests are part of the Cerrado ecosystem biome that is located at approximately 800 masl. These forests are subject to severe pressures imposed by logging, urban development, cattle grazing, and fire.

The patches of the dry forests consist of several sizes and under several successional stages. Intermediate and late-stage forests form the most frequent successional stages in the Serra do Cipó. Fabaceae, Apocynaceae, and Malvaceae are the main plant families in these dry forests. The species *Myracrodruon urundeuva*, *Rauwolfia sellowii*, *Inga platyptera*, *Anadenanthera colubrina*, and *Bauhinia brevipes* are the most predominant plant taxa. Due to the marked isolation, this TDF has a unique floristic composition (for details, see Coelho et al. 2012).

1.3 Book Sections

Given the context just mentioned, the Tropi-Dry network serves as a catalyzer for generating and promoting research in TDFs across the Americas. This edited volume represents a comprehensive overview of the work conducted at Tropi-Dry field sites (Figure 1.1; Table 1.1) across the Americas. Tropi-Dry operates in three countries: Mexico (Figure 1.2), Costa Rica (Figure 1.3), and Brazil (Figure 1.4), with emerging field sites under development in Colombia. The fundamental goal of Tropi-Dry is to produce an integrative overview of the human and biophysical dimensions of TDFs that can be used as an objective platform for policy and decision making. This book reflects that spirit.

The book is divided into three sections. Section I presents an evaluation of critical hot spots of change for TDFs: Mexico (Figure 1.2), Brazil (Figure 1.3), and Colombia. Chapters by Quesada et al. (Chapter 2), Fernández et al. (Chapter 3), Coelho et al. (Chapter 4), and Santo et al. (Chapter 5) provide a presentation of the challenges to TDFs and Tropi-Dry's efforts to balance conservation and socioeconomic concerns in countries with high rates of TDF deforestation. Section II presents reviews of major Tropi-Dry components. Hesketh et al. (Chapter 6) provide a review of the current application of remote sensing in TDFs, and Lopezariaza et al. (Chapter 7) provide a synthesis of the phenology processes that are being developed at Tropi-Dry's long-term monitoring plots. Hulshof et al. (Chapter 8)

TABLE 1.1

Vegetation Structure Attributes (Mean ± 1 SD) of Woody Species in the Successional Plots Studied in Chamela–Cuixmala Biosphere Reserve (CCBR), Mexico; Santa Rosa National Park, Costa Rica; Unidad Productiva Socialista Agropecuaria-Piñero (UPSAP), Venezuela; and Mata Seca State Park (MSSP), Brazil

Successional Stage Plots	No. of Individuals	No. of Families	No. of Species	CHCI
Mexico				
Early	89.3 (74.5)	6.7 (0.6)	15.0 (5.3)	0.16 (0.3)
Intermediate	201.0 (27.0)	20.7 (0.6)	43.0 (2.7)	161.6 (93.7)
Late	212.0 (27.7)	21.7 (4.6)	45.3 (4.5)	257.1 (109.4)
Costa Rica				
Early	92.3 (28.0)	9.3 (4.2)	12.7 (5.7)	1.6 (2.4)
Intermediate	91.7 (9.8)	15.3 (2.1)	27.7 (6.7)	50.9 (39.7)
Late	119.7 (25.8)	16.0 (3.6)	24.7 (4.5)	74.1 (28.1)
Venezuela				
Early	193.0 (7.1)	19.0 (4.2)	27.0 (9.9)	64.1 (18.6)
Intermediate	176.3 (50.1)	12.3 (0.6)	16.3 (3.1)	44.3 (31.5)
Late	131.0 (22.7)	19.3 (3.1)	30.3 (2.9)	152.3 (50.4)
Brazil				
Early	98.3 (22.3)	7.3 (1.5)	13.7 (2.5)	5.2 (2.4)
Intermediate	90.7 (17.8)	12.7 (0.6)	21.0 (3.5)	28.7 (8.4)
Late	118.0 (14.7)	14.0 (1.0)	24.7 (4.9)	95.4 (58.1)

Source: Nassar, J.M., et al., *Manual of Methods: Human, Ecological and Biophysical Dimensions of Tropical Dry Forests*, Ediciones IVIC (Instituto Venezolano de Investigaciones Científicas), Caracas, Venezuela, 135, 2008; Espírito-Santo, M.M., et al., *Cienc. Hoje*, 288, 74–76, 2011; Quesada, M., et al., Human impacts on pollination, reproduction and breeding systems in tropical forest plants, in *Seasonally Dry Tropical Forests*, eds. R. Dirzo, H. Mooney, and G. Ceballos, 173–194, Island Press, Washington, DC, 2011.

provide complementary material to Chapters 4 and 5. Work on nutrient cycling (Gei and Powers, Chapter 9), edge effects (Portillo et al., Chapter 10), and a comprehensive synthesis on birds and bats in TDFs (Nassar et al., Chapter 11) constitutes this section. Section III of the book provides specific case studies from Tropi-Dry sites that can be used to develop new comparative studies across emerging sites. Neves et al. (Chapter 12) provide an overview of one of Tropi-Dry's main works on herbivory, whereas Berbara et al. (Chapter 13) study the links between mycorrhizal diversity and forest succession. Jimenez and Calvo (Chapter 14) provide an innovative overview of rainfall interception as a function of succession, information that is critical to the development of comprehensive hydrological models. This chapter is complemented by Huang et al. (Chapter 15), who explore temporal changes in Leaf Area Index (LAI) in Mexico in relation to changes in remote sensing observations. Rodrigues et al. (Chapter 16), Villalobos et al. (Chapter 17), Nunes et al. (Chapter 18), and Carvajal et al. (Chapter 19) provide important information on secondary growth processes at Tropi-Dry sites in Brazil, Venezuela, and Costa Rica. Finally, the section closes with a chapter by Oki et al. (Chapter 20), who explore the potential response of TDF trees and lianas to elevated atmospheric CO_2 levels and provide a discussion of how these endangered ecosystems could respond in the years to come as climate changes become the norm. Section IV, the final portion of the book, considers research on the human dimensions of TDF conservation and deforestation. Castillo et al. (Chapter 21) open the section with a historical overview of the pressures forcing land-cover change at all research sites documented in this book. Anaya et al. (Chapter 22) provide an evaluation of the sociopolitical forces driving the deforestation of TDFs in Minas Gerais. Pfaff and Robalino (Chapter 23) provide a discussion of how econometrics can be used to estimate the effectiveness of policies that are aimed at controlling tropical deforestation. Finally, the book closes with an element that is generally overlooked in tropical research: the use of forests by the people who live in them. The role of traditional ecological knowledge, a resource that is quickly diminishing, is explored by Almada et al. (Chapter 24) in their study of a rural community in Minas Gerais.

1.4 Tropi-Dry and the Way Forward

A complete understanding of the regeneration and restoration of TDFs requires a detailed and integrated knowledge of functional traits, ecological processes, and evolutionary relatedness of the species within plant communities. Several studies have proposed that TDFs recover relatively quickly after disturbance, but an exhaustive evaluation of the recovery of the species' richness, structure, function, and phylogenetic constraints has not been performed for this highly threatened ecosystem. Little is known about how phylogenetically neutral or niche forces structure the regeneration of plant communities (Webb et al. 2002). The first point of view predicts that neutral processes would randomly structure plant communities, whereas a second view states that evolved ecological differences between lineages are responsible for the structuring of communities. Explicit evaluations that connect the analysis of a phylogenetic community structure to assembly processes are relatively rare (Kraft et al. 2008). However, the potential of functional traits to reveal processes that structure communities is increasingly recognized. Only a few studies have shown that a complete functional recovery of TDFs takes longer than the time period inferred from floristic or structural data. Alvarez-Añorve et al. (2012) demonstrated that plant functional traits changed along succession from those that maximize photo protection and heat dissipation in early succession, where temperature is an environmental constraint, to those that enhance light acquisition in late succession, where light may be a limiting factor.

In addition, other ecological processes related to species interactions can also affect the recovery of TDF communities. Plant–herbivore and plant–pathogen interactions can regulate, through density-dependent factors, species richness and the structure of plant populations (Connell 1978). Other important ecological processes that regulate the incorporation of seeds to the seed bank, and the subsequent regeneration of plant communities, are related to plant–pollinator and plant–disperser interactions. However, these interactions in relation to ecological succession remain highly unexplored in TDFs. Native pollinators contribute to the provisioning of environmental services by being actively involved in the process of fruit and seed production in plants used by humans as well as in the plants involved in the regeneration and maintenance of natural forests. Wind and vertebrates play key roles in primary seed dispersal in TDFs. Given this fact, changes in environmental and frugivorous animal communities during succession will undoubtedly affect areas undergoing succession and, ultimately, the regeneration of the larger TDF ecosystem. Therefore, understanding the role of species interactions and environmental changes in ecological succession is key for the management and conservation of natural forests and nearby croplands. In sum, a multilevel functional approach should be incorporated in the analysis of TDF recovery and regeneration.

Basic science research in TDFs obviously requires the implementation of monitoring efforts using state-of-the-art technologies. Advanced remote sensing approaches in conjunction with ground-based observations of biophysical properties should be standardized. The integration of these remote sensing observations with emerging technologies such as distributed wireless sensor networks can help provide a considerable amount of information, which is quickly and reliably acquired, for forests that are undergoing anthropogenic and climatic stress. Advanced monitoring systems can help in promoting a better understanding of how changes in climatic regimes will affect phenology, structure, and composition of TDFs. This information, in turn, can be used to gain a deeper understanding of the capacity of TDFs to sequester carbon and produce water.

Knowledge derived from basic science and advanced technology is not of much use if it cannot be integrated into the domain of decision making and conservation policy. As of today, we are still unclear about what works and what does not work in terms of building sound economic incentives for the conservation of TDFs. Because these ecosystems are in the best agricultural soils and in the proximity of areas with great potential for tourism development, any cost/benefit conservation analysis should consider both the value of regular economic activities and the economic value of ecosystem services that TDFs provide to local communities. Unfortunately, our knowledge of the ecosystem services provided by TDFs is extremely limited, particularly with regard to hydrological systems. Water, due to its scarcity, is the most vital service to be considered when assessing ecosystem services. In addition, we need to better understand how ecosystem services, under climate change conditions, will be affected and whether those changes represent significant economic variances that can be mitigated by regional and national conservation efforts.

1.5 Concluding Remarks

To better understand the mechanisms and processes governing the dry forest at a global scale, we should simulate and scale up the knowledge gained by Tropi-Dry in old world dry forests. Although it is important to understand their extent, fragmentation, fragility, and resilience, the fundamental ecological understanding of ecosystem processes and their relationship to people's livelihoods is a paramount endeavor. There is a fundamental need to consolidate Tropi-Dry's vision and perspectives in other regions of Latin America; this could be the first step for a true global effort to preserve, protect, and conserve what is left of one of the most endangered tropical ecosystems.

Acknowledgments

Tropi-Dry's main work is carried out with the aid of a grant from the Inter-American Institute for Global Change Research (IAI) CRN 2-021, which is supported by the U.S. National Science Foundation (Grant GEO-0452325). The support of Dr. Holm Tiessen, Ione Anderson, and Dr. Gerhard Breulman has been crucial since the beginning of this Tropi-Dry Initiative, and it is deeply appreciated. The authors are grateful for the support rendered by the following organizations and individuals:

Canada:

The authors acknowledge the support from the National Science and Engineering Research Council of Canada (NSERC) via their Discovery Grant Program. We also thank the Vice President for Research (Dr. Renee Elio and Dr. Lorne Babiuk), the Department of Earth and Atmospheric Sciences (Dr. Martin Sharp and Mary-Jane Turnell), and the Office of the Dean of Science at the University of Alberta for their financial support to Tropi-Dry. Mei Mei Chong, Alena Lundel, Abdel Houchaimi, and Doug Calvert provided logistical support.

The authors acknowledge the great support provided by Donnette Thayer over the years.

The Americas:

The authors acknowledge the support from the UNESCO-L'ORÉAL Co-Sponsored Fellowship for Young Women in Life Sciences, the U.S. National Science Foundation's CAREER Award Program (DEB-1053237), and a National Science Foundation Graduate Diversity Fellowship. Logistical support was provided by Heberto Ferreira and Alberto Valencia.

The authors are grateful to Jennifer Powers and Arturo Sánchez-Azofeifa for inviting us to contribute to this chapter as well as to an anonymous reviewer for providing insightful comments and suggestions. Catherine Hulshof was supported by a National Science Foundation Graduate Diversity Fellowship. Angelina Martinez-Yrizar and Alberto Burquez Montijo contributed to this work during a sabbatical year at the University of Arizona supported by a Dirección General de Asuntos del Personal Académico–Subdirección de Formación Académica, Universidad Nacional Autónoma de Mexico Fellowship.

Mexico:

Financial support was provided by the Consejo Nacional de Ciencia y Tecnología, México (CONACYT 2009-C01-131008; SEMARNAT-CONACyT 2002-C01-0597 and 2002-C01-0544; CONACYT 2010-155016; and the Dirección General de Asuntos del Personal Académico at the Universidad Nacional Autónoma de México [Grant Nos IN304308 and IN201011]).

Costa Rica:

Financial support in Costa Rica was provided by the Vicerectoria de Investigación y Extension of the Instituto Tecnológico de Costa Rica through the following projects (1) Human, Ecological, and Biophysical Dimensions of TDFs (5402-1401-9001) and (2) Monitoring of Ecological Processes in TDFs: Remote Sensing Applied to Landscape and Global Change Scales (5402-1401-1012). Logistical and administrative support was provided by FUNDATEC and Centro de Investigaciones Integración Bosque Industría. The authors appreciate the collaboration of the research project "Long-Term Research in Environmental Biology: Streams of the Area de Conservación Guanacaste" conducted by the Stroud Water Research Center (U.S. National Science Foundation DEB 05-16516). We thank the Area de Conservación Guanacaste, especially Roger Blanco and Maria Marta Chavarría. We also appreciate the field and laboratory support from our technical assistants: Oscar Arias, Juan Carlos Solano, Branko Hilje, Dorian Carvajal, Cristian Baltonado, Ana Julieta Calvo, and

César Jiménez, and many student assistants from the School of Forestry at the Institute Tecnologico de Costa Rica.

Brazil:

Financial support was provided by Fundação de Amparo à Pesquisa de Minas Gerais (FAPEMIG) (CRA 2288/07, CRA-APQ-04738-10, and CRA-APQ 00001-11), the Conselho Nacional de Pesquisa e Desenvolvimento (CNPq) (563304/2010-3), the Coordenação de Aperfeiçoamento de Pessoal de Nível Superior (CAPES), and PPBio semi-arido. The authors also want to thank the Instituto Estadual de Florestas (IEF) of the state of Minas Gerais for their logistical support.

We also thank all the communities' residents who were studied and who were willing to share their knowledge, especially Dona Zita, Dadiane, Dewey, Tiao, Julius, John Melos, John Paraúna, Ana Miguel, Thomas, Mislene, Didico, Gerard, John Ciqueira, Washingtom, D. Lourdes, Mario D. Liquinha, Agnaldo, and Lucia. The authors appreciate the support from E. C. Araújo, L. Monteiro, and P. Matos for field assistance We are grateful to the Laboratory of Plant Systematics from the Universidade Federal de MinasGerais for help with the species identification. Much of the work related to Brazil would not have been possible without the help from many students from the Laboratório de Ecologia e Propagação Vegetal Universidade Estadual de Montes Claros (UNIMONTES) for their help with field and laboratory work. The authors also thank Mariana Rodrigues Santos of Universidade Federal de Viçosa (UFV) for the cactus identification and professor Rubens Manoel dos Santos of Universidade Federal de Lavras (UFLA) for identification of the other tree species. Finally, we are very grateful to the "Vazanteiros" in Movement, Center for Alternative Agriculture, and Pastoral Land Commission of the north of Minas Gerais state for their invaluable collaboration in this study.

2

Tropical Dry Forest Ecological Succession in Mexico: Synthesis of a Long-Term Study

Mauricio Quesada, Mariana Álvarez-Añorve, Luis Ávila-Cabadilla, Alicia Castillo, Martha Lopezaraiza-Mikel, Silvana Martén-Rodríguez, Victor Rosas-Guerrero, Roberto Sáyago, Gumersindo Sánchez-Montoya, José Miguel Contreras-Sánchez, Francisco Balvino-Olvera, Sergio Ricardo Olvera-García, Sergio Lopez-Valencia, and Natalia Valdespino-Vázquez

CONTENTS

2.1 Introduction

The current status and rates of conversion of mature tropical forests indicate that these habitats will eventually disappear, leaving behind a complex landscape matrix of agricultural fields and forest patches under different levels of succession. Tropical dry forests (TDFs) are not an exception, and the current management derived from human activities will clearly result in the complete loss of this habitat worldwide (Quesada and Stoner 2004; Miles et al. 2006; Quesada et al. 2009). The TDFs have been extensively transformed and occupied by urban and agricultural areas at significantly higher rates than tropical rainforests (Murphy and Lugo 1986a). Therefore, understanding tropical succession in the context of ecological and human dimensions represents one of the major challenges to promoting and developing conservation and management programs for this threatened ecosystem.

Few studies have analyzed ecological succession in TDFs, indicating contrasting results; in some cases, a relatively faster structural recovery is found for this system than for other tropical systems (i.e., Ceccon et al. 2002; Ruiz et al. 2005; Vieira and Scariot 2006), but in others, a slower process is found in TDFs than in wet forests in terms of plant growth and other developmental features (Ewel 1977; Murphy and Lugo 1986a). However, the interpretation of succession is not clear because the recovery of species richness and the composition are dependent on structural change (Sheil 2001; Ceccon et al. 2002; Pascarella et al. 2004; Ruiz et al. 2005; Toledo and Salick 2006; Quesada et al. 2009; Alvarez-Añorve et al. 2012). In addition, some studies claim that TDFs are relatively simple and small in structure and composition, and that they recover predominantly through coppicing after disturbance (Ewel 1977; Murphy and Lugo 1986a; Chazdon et al. 2007). However, Quesada et al. (2009, 2011) challenge this view, as the predominant mode of reproduction in TDFs is through a wide variety of sexual systems in which seeds are mainly produced via outcrossing. If coppicing or asexual reproduction were the main drivers of regeneration, changes in species composition along secondary succession would not be expected. Several studies that analyze a chronosequence have found differences in species composition in TDFs. Therefore, the regeneration of TDFs is expected to be slow and very susceptible to human disturbance because the growth rate of many tree species is slow, reproduction is highly seasonal, and most plants are mainly outcrossed and dependent on animal pollination (Bawa 1974, 1990; Frankie 1974; Murphy and Lugo 1986a; Hamrick and Murawski 1990; Bullock 1995; Jaimes and Ramirez 1999; Cascante et al. 2002; Fuchs et al. 2003; Quesada et al. 2001, 2004, 2009). Another important aspect to consider in the process of regeneration is the functional recovery of the community, which identifies groups of plants and animals that exhibit similar responses to environmental conditions and have similar effects on dominant ecosystem processes that are associated with successional stages (Gitay and Noble 1997; Lebrija-Trejos et al. 2010; Alvarez-Añorve et al. 2012; Avila-Cabadilla et al. 2012). Only a few studies have simultaneously evaluated TDF succession in floristic, structural, and functional terms.

Alvarez-Añorve et al. (2012) found that plant functional traits along succession change from those that maximize heat dissipation in early successional stages to those that enhance light acquisition and water use in late successional stages. This study suggests that the functional recovery of TDFs could take longer than inferred when the process is evaluated from just a floristic and/or structural perspective, but more studies from other tropical regions are required to corroborate patterns of functional succession. In addition, the variation in vertebrate guild assemblages is associated with the variation in landscape habitat attributes under different successional stages.

TABLE 2.1

Tropi-Dry Plot Abbreviations and Age of Abandonment in 2004 and 2009

Successional Stage	Plots	Age in 2004 (Years of Abandonment)	Age in 2009 (Years of Abandonment)
Initial (pastures)	P1-P3	0	5
Early	E1-E3	3–5	8–10
Intermediate	I1-I3	8–12	13–17
Late	L1-L3	>50	>55

Avila-Cabadilla et al. (2012) found that nectarivore bats tend to be associated with TDF patches, whereas frugivore bats are associated with riparian forests. This probably reflects the prevalence of species that produce nectar resources for bats in dry forests, and of species which produce fruits that are eaten and dispersed by bats in riparian forests. In conclusion, the main mechanisms of succession and regeneration of TDFs still remain unexplored, and more efforts are required to understand the ecological processes of these important ecosystems.

The main goal of this synthesis is to understand the successional process underlying the natural regeneration of TDFs for the development of conservation strategies in the Chamela–Cuixmala Biosphere Reserve region, which is located along the central Pacific coast of Mexico (Figure 1.2). Little is known about the regeneration process of these forests, and our study is one of the few that provides basic and applied ecological information on the succession in TDFs in Mexico. To characterize TDF successional patterns, we assessed the successional changes in vegetation attributes (i.e., structure, species composition), ecosystem functioning (i.e., functional traits, herbivory, phenology), and fauna diversity and abundance in a highly diverse Mexican TDF. For this purpose, we performed a five-year study in a chronosequence that represented four TDF successional stages: initial, early, intermediate, and late (Table 2.1). We also conducted socio-ecological research to understand the changes in land-use history and their effects on succession and forest regeneration of TDFs (see Chapter 21 for an in-depth comparative study). We emphasize the need to integrate ecological knowledge with the human dimension as a tool that supports sound conservation, management, and understanding of TDFs.

2.2 Ecosystem Structure and Composition

Between 2004 and 2009, vegetation structure and species composition of the plots were compared among the different successional stages. Changes in the chronosequence were compared within plots, between successional stages, and over the course of the five-year study (Table 2.1). The study design and methods are detailed in Chapter 1.

2.2.1 Species Richness

In general, species richness increased with successional age (Figure 2.1), and intermediate stage plots showed similar species richness than did late successional plots. This parameter differed significantly among successional stages during both years (Kruskal–Wallis $x^2_{(2004)} = 9.46$, df = 3, $p = 0.024$; $x^2_{(2009)} = 8.08$, df = 3, $p = 0.044$) (Figure 2.1). In 2004, pastures and early stages had a significantly lower species richness than intermediate and late stages. These results indicate that management practices used in the chronosequence plots left many species of late successional stages standing in the intermediate plots. In 2009, early, intermediate, and late successional stages were similar to each other; only pastures differed from all the other three successional stages. During the same, pastures (5 years old by 2009) also showed high intra-stage variations, suggesting higher stochasticity in early stages. The lowest species richness occurred at site P3, which is dominated by trees and shrubs of the genus *Mimosa* (Leguminosae) and that is surrounded by other pastures. In contrast, the highest species richness occurred at site P2, which is surrounded by secondary forests that could facilitate the regeneration of several species. In general, from 2004 to 2009, the highest increase in species richness occurred in pastures, whereas there was no significant change in intermediate and late successional plots. Plots that are 8 years old and older present similar species richness, indicating a relatively rapid regeneration and recovery. This suggests an important role

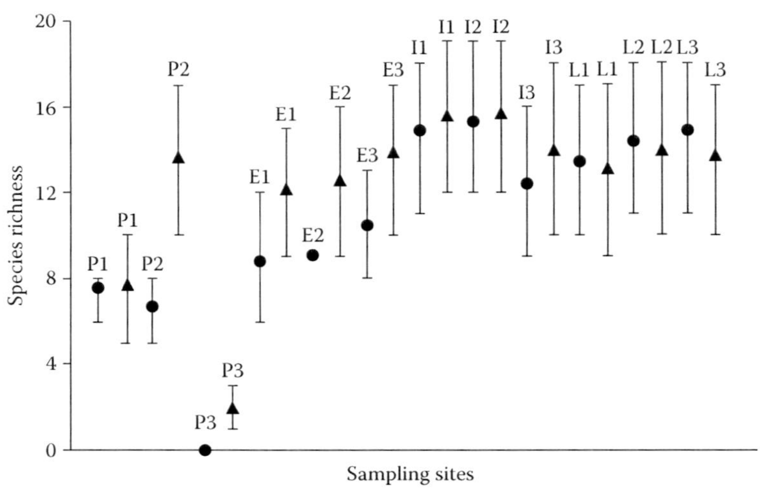

FIGURE 2.1
Species richness of Tropi-Dry plots under different successional stages for years 2004 (circles) and 2009 (triangles). Species richness was rarified at 23 individuals. P1-P3 = pastures, E1-E3 = early successional plots, I1-I3 = intermediate successional plots, and L1-L3 = late successional plots.

of the surrounding landscape attributes of the vegetation matrix in the successional process of this system.

2.2.2 Species Density

Species density differed significantly among successional stages and increased with successional age (ANOVA 2004 $F_{(3,8)} = 19.4$, $p = 0.0004$; ANOVA 2009 $F_{(3,8)} = 12.4$, $p = 0.002$) (Figure 2.2). In 2004, similar to species richness, pastures and early stages were significantly different from intermediate and late stages. In 2009, only pastures differed from other successional stages. The density parameter appears to be useful in differentiating successional stages as well as in predicting temporal dynamics from the chronosequence due to the following reasons: (1) Density showed a gradual increase along succession. (2) There were significant differences among successional stages. (3) Early successional plots of the year 2009 (8–10 years old) showed similar values to intermediate successional plots of the year 2004 (8–12 years old). (4) Pastures of the year 2009 (5 years old) showed similar values to early successional plots of the year 2004 (3–5 years old).

2.2.3 Species Composition

Species composition was analyzed by means of a nonmetric multidimensional scaling (NMDS) method that was based on a Bray–Curtis dissimilitude matrix (Figure 2.3). The scores of the axis that explained most of the variation (synthetic variable) were used to compare the successional stages

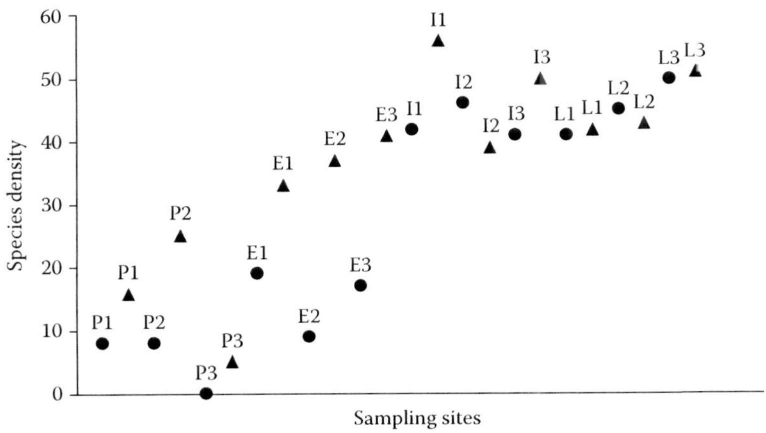

FIGURE 2.2
Species density of Tropi-Dry plots under different successional stages for years 2004 (circles) and 2009 (triangles). P1-P3 = pastures, E1-E3 = early successional plots, I1-I3 = intermediate successional plots, and L1-L3 = late successional plots.

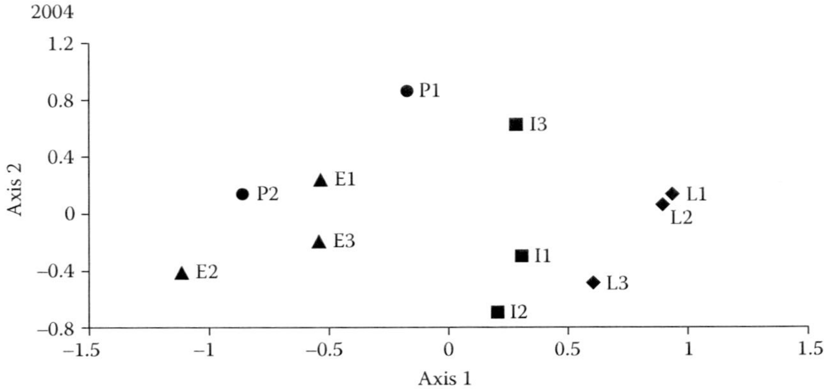

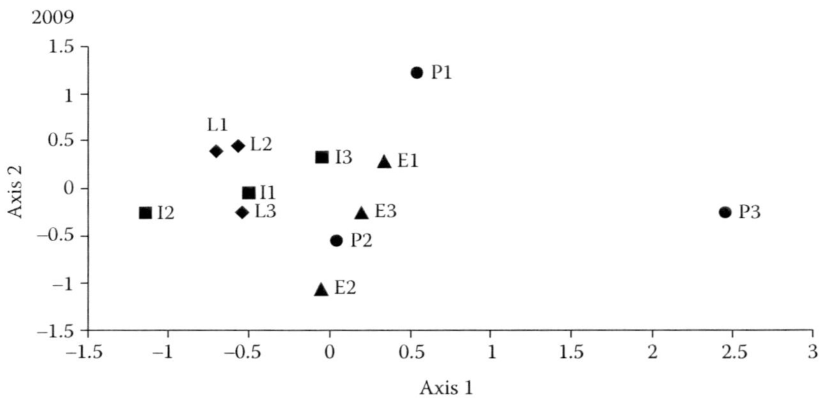

FIGURE 2.3
Nonmetric multidimensional scaling (NMDS) of the species composition under different successional stages for years 2004 and 2009. Plots are presented according to their successional stage: circles = pastures, triangles = early, squares = intermediate, and diamonds = late.

in compositional terms. In 2004, pastures and early successional plots significantly differed from intermediate and late successional stages (Pillai test $F_{(3,7)} = 3.25$, $p = 0.032$). In contrast, in 2009, species composition did not differ among the successional stages, suggesting that this parameter became similar to late successional stages in a short period of time (Pillai test $F_{(3,7)} = 1.51$, $p = 0.237$). However, pastures showed great variations in this parameter, and these variations could be masking real differences in the species composition of this successional stage. The great variations among the pastures of the year 2009 (5 years old) again suggest higher stochasticity in the assembly of early successional communities. Stochasticity can be influenced by landscape attributes, as pastures that are separated from the rest of the sites in the analysis are surrounded by pastures; whereas P2, the pasture that is closest to early successional plots in the analysis, is surrounded by secondary forests (Figure 2.3). Intermediate successional plots of 2009 (13–17 years old) appear more grouped than those in 2004 (Figure 2.3), suggesting a rapid homogenization of plot species composition.

When we analyzed the Bray–Curtis dissimilitude values between the plots, we observed a greater dissimilitude between pasture plots than between plots of other successional stages (Figure 2.4). This suggests that pastures present higher beta diversity than older successional stages. Intra-successional-stage beta diversity should decrease with a decrease along succession in dissimilitude between the plots of a given successional stage. This trend is

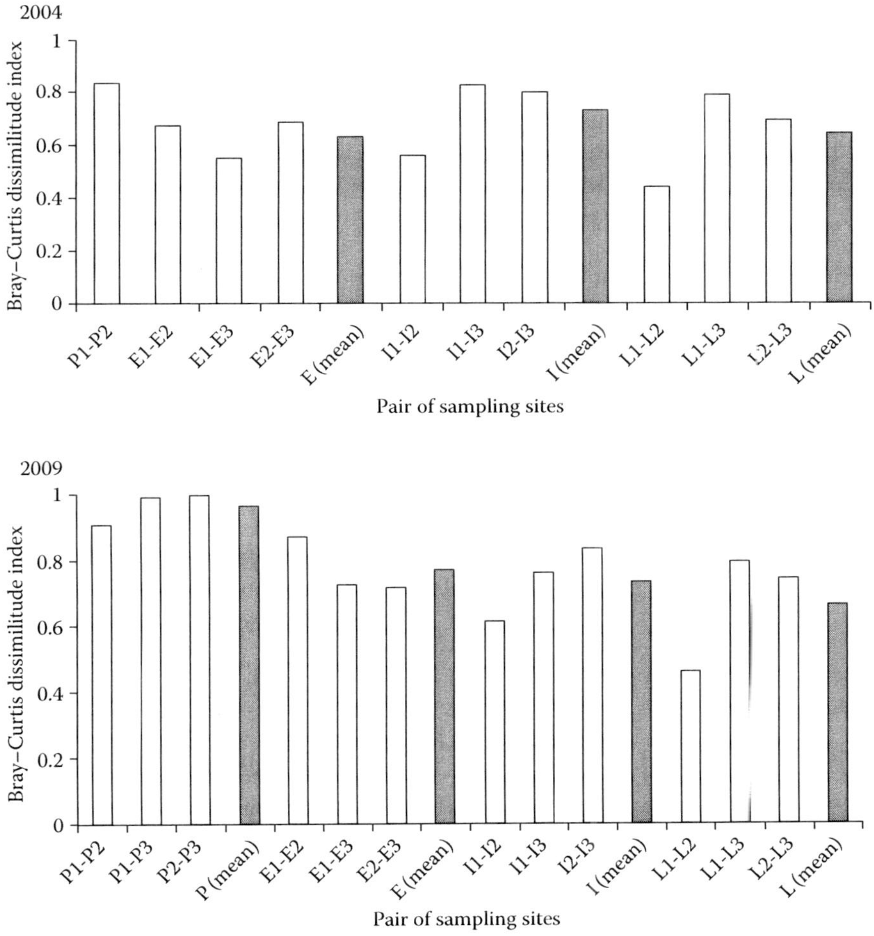

FIGURE 2.4
Bray–Curtis dissimilitude values between Tropi-Dry plots (open bars) and mean Bray–Curtis dissimilitude values among all the plots of a given successional stage (solid bars) in two different years (2004 and 2009). P1-P3 = pastures, E1-E3 = early successional plots, I1-I3 = intermediate successional plots, and L1-L3 = late successional plots.

consistent in both years (2004 and 2009). The higher beta diversity in pastures reinforces the idea of greater stochasticity in the assembly of early successional communities, resembling what has been found in other successional models (Leishman and Murray 2001), where stochastic and niche processes dominate early successional stages and only niche processes dominate late successional stages. In these models, both processes have an impact on the community's assembly in the early stages, but deterministic forces dominate the late stages.

2.2.4 Vegetation Structure

Vegetation structure was analyzed through principal component analysis (PCA) (Figure 2.5), which allowed for the ordination of plots in terms of four variables (basal area, number of stems, number of individuals, and number of species). Principal components (i.e., PC1, PC2) that explained most of the variation were used to compare different successional stages through ANOVAs and *post hoc* Tukey's mean separation tests. In 2004, intermediate and late successional plots appeared grouped and separated from the rest of the plots, indicating that structural traits are similar among themselves but different from pastures and early successional stages. In 2009, in contrast, late successional plots appeared grouped along PC1, but there was no clear separation among successional stages along this axis. Intermediate successional plots appeared separated from the rest of the plots along PC2. Most variations in PC2 were accounted for by the number of individuals and the number of species, and parameters were highest in intermediate successional stages. Accordingly, in 2004, ANOVA tests showed significant differences in structural traits among successional stages ($F_{(3,8)} = 9.835$, $p = 0.005$). Tukey's test showed that intermediate and late successional plots were similar among themselves but different from pastures, whereas pastures were different from all the other successional stages. In 2009, ANOVAs showed no differences among successional stages in PC1 ($F_{(3,8)} = 1.22$, $p = 0.365$), but significant differences among the stages were detected in PC2 ($F_{(3,8)} = 4.33$, $p = 0.043$). Tukey's test indicated that, along this axis, intermediate successional plots were similar to late successional plots but different from pastures and early successional plots (which were also similar to each other).

From 2004 to 2009, when the percentage of change in structural variables was analyzed, again, the greatest changes occurred in pastures and early successional plots. The early successional plot E1, however, showed a small increase compared with other early successional sites, probably as a consequence of the high mortality of woody individuals that was caused by an invasion of *Ipomoea* lianas. This mortality reduced the net increase in structural traits. The intermediate successional plot I2 also experienced *Ipomoea* invasion and high mortality, which is reflected in a net reduction in the values of structural traits. In contrast, site I3, which was the youngest intermediate plot, showed the greatest increase during

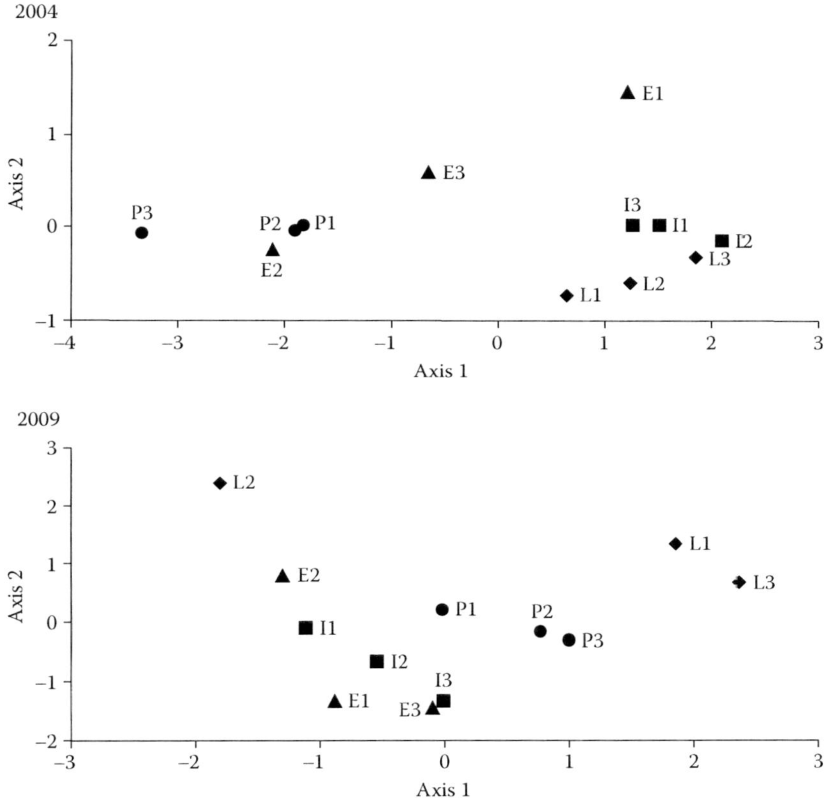

FIGURE 2.5
Principal component analysis (PCA) of the vegetation structure of four successional stages for years 2004 and 2009 in the region of Chamela, Jalisco, Mexico. In 2004, PC1 accounted for 86% of the variation; in 2009, PC1 accounted for 40%, and PC2 accounted for 34% of the variation. Plots are presented according to their successional stage: circles = pastures (P1-P3), triangles = early (E1-E3), squares = intermediate (I1-I3), and diamonds = late (L1-L3).

this successional stage. These cases constitute an example of site-specific effects that can determine particular successional trajectories in different sites of the same region.

In general, late successional plots showed low rates of change for most variables, and most of them showed decreases in the number of individuals and species, indicating mortality of woody individuals. Meteorological data of the Chamela Biological Station over the last 20 years indicate an increase in the number of dry days per year, which could be related to an increase in vegetation mortality (unpublished data). The main structural changes occurred in basal area and stem number, mainly due to increases in stem number and not increases in the number of individuals (as this parameter decreased in most cases). This idea was analyzed through a hierarchical partitioning analysis, a statistical technique that determines how much of the variation in a given variable (basal area in this case) is explained by other correlated variables (number of stems, number of individuals, etc.). This analysis showed that 93% of the changes in basal area from 2004 to 2009 were explained by changes in stem number (54%), followed by changes in the number of individuals (44%).

The analysis of the ecological succession of a chronosequence of TDFs in Mexico showed that during the first year of the study (2004), pastures and early successional stages were similar to each other but different from later successional stages in terms of species richness, density, and composition. In terms of vegetation structure, early successional stages were similar to intermediate and late stages, as all these stages had an equivalent total basal area. However, this result reflects the high number of small stems derived from the resprouting and recruitment of juveniles in early successional stages. Thus, a casual interpretation of results for total basal area could lead to an underestimation of the time required for structural recovery. Based on what has been stated earlier, the contribution of distinct stem diametric classes to the total basal area should be evaluated. This would enable a more accurate assessment of the vegetation structure at finer scales.

Five years later, in 2009, the four successional stages still differed in most of the variables evaluated, however in a more complex way. Pastures (5 years old in 2009) maintained their distinctness in terms of all the variables analyzed, although they were more similar to early successional stages in terms of species density and structure. Early stages of succession became more similar to later stages in terms of species richness and composition, whereas intermediate and late stages generally remained similar to each other (see results above, Figures 2.1 and 2.3). The increasing similarity in species composition with the advancement of succession is likely the result of a strong effect of deterministic factors. In contrast, stochastic factors, which appear to be important in explaining intersite variations in pastures, become less relevant in later stages of succession. It is of particular importance to state that the composition and structure of intermediate and late successional stages are similar from the beginning to the end of the chronosequence study, indicating that management practices that occurred in the intermediate stages maintained many species and trees in disturbed areas.

Although it appears that species composition is similar between intermediate and late successional stages, less frequent and rare species are not likely to be similar in both stages. For instance, other studies have demonstrated that beta diversity is high in mature TDFs in the region (Lott and Atkinson 2002; Balvanera et al. 2002), creating a heterogeneous landscape. In addition, the most diverse plant community of Chamela–Cuixmala, which is found in the canopy stratum (lianas, orchids, and bromeliads), is likely to be one of the communities that is most affected by disturbance, although this idea has not been assessed in the context of succession. However, a study of epiphyte-host networks in the Chamela–Cuixmala region in Mexico showed that the assembly of these commensalistic interactions is determined by the host-species abundance, species spatial overlap, host size, and wood density (Sáyago et al. 2013). Only the host plant communities of late successional stages are capable of supporting the unique, diverse canopy plant community of Chamela–Cuixmala. Further work on the succession of TDFs should include an analysis of all canopy-level strata. Meanwhile, it is important to interpret results from vegetation analyses with caution and to critically evaluate apparent similarities among successional stages that may reflect different underlying processes or causes.

2.3 Phenology

Phenological data were collected during four years (2006–2010) for a total of 695 individuals who belonged to 90 species. Results for this part of the study are presented in detail in Chapter 7 by Lopezaraiza et al.; next, we present a summary of the major findings of this study.

2.3.1 Leafing

All successional sites showed a distinct and consistent leafing pattern both within and across years. The months with the highest proportion of individuals with no leaves were April, May, and June. During the drier months, the proportion of individuals and species with no leaves was higher in the late successional sites. During the greener months, there was no difference among successional stages in the proportion of individuals or species with 50%–100% leaves. In the late successional sites, a higher proportion of individuals had no leaves for five or more months in a year; whereas at early and intermediate sites, a higher proportion of individuals maintained their leaves for six or more months per year. Thus, in general, individuals keep their leaves for longer at the early and intermediate sites.

2.3.2 Flowering and Fruiting

General patterns of flowering and fruiting at the community level show peaks at different times of the year. This may be due to differences in community composition and species abundance among plots, as well as due to the differences in the local physical environment. Almost every month, at least one species is flowering or fruiting, with large variation among sites. The most consistent flowering peak across sites and years occurs at the end of the dry season and at the beginning of the rainy season, around the months of June and July. There are other flowering peaks at different times for various sites at the end of the rainy season and at the beginning of the dry season.

At the early and intermediate successional sites, some individuals flowered for three to five months in a year; whereas at the late successional sites, individuals were only recorded flowering for one or two months in a year. In contrast, at the early successional sites, individuals were recorded with fruits for approximately three months in a year; whereas at the intermediate and late sites, some individuals were recorded with fruits for 4 to 10 months in a year. Thus, individuals tend to flower longer at early successional stages, whereas they ripen or bear fruits longer at late successional stages.

2.4 Successional Changes in Vertebrate Communities

In this study, we analyzed mainly bird and bat communities along successional stages.

The chapter by Nassar et al. (Chapter 11) and the studies by Avila-Cabadilla et al. (2009, 2012) describe the main findings of this section in detail.

2.4.1 Birds

In total, we captured 2,775 individuals of 84 different species. The most abundant species were *Vireo flavoviridis* (12.3%), *Passerina lechanclerii* (11.29%), *Cyanocompsa parellina* (11.03%), and *Columbina passerina* (9.4%). Fifty-one species were considered rare, because they occurred at low abundance (<0.5% of total captures). When comparing bird species richness and diversity among successional stages, the stages did not differ significantly in terms of rarefied species richness (Kruskal–Wallis tests $x^2 = 3.77, p = 0.28$) and rarefied Shannon diversity indices ($x^2 = 1.33, p = 0.72$). However, bird species composition differed between late successional plots and the remaining successional stages when analyzed through NMDS. Species composition among plots that were evaluated through the Morisita index showed a gradient of similarity along successional stages, where pastures were more similar to early successional sites, and intermediate sites were more similar to late successional sites. A comparison of plots through rank-abundance curves showed some species that occurred exclusively in particular successional stages. These species could potentially be considered indicator species. In general, our results suggest that the mosaic of secondary forests characterizing this region plays an important role in the maintenance of bird species biodiversity.

2.4.2 Bats

We documented the changes in the structure of bat assemblages among the secondary successional stages of Chamela–Cuixmala TDF over 42 nights of sampling and captured 606 phyllostomid bats belonging to 16 species.

In general, the late stage had the highest species richness, sustaining all 16 species, followed by the intermediate site with nine species, and the pasture with four species. Species found within any successional stage were a combination of species found at the previous stage as well as additional species. Bat diversity and abundance did not differ significantly among early, intermediate, and late stages. However, nectarivores were more abundant in early stages than in late stages, likely as a consequence of differences in food availability. Our results suggest that areas of forest that are recognized as late successional are the most important reservoirs of species richness.

We also evaluated variations in the occurrence of phyllostomid bat assemblages in different successional stages and variations in relation to habitat attributes at local (vegetation structure complexity) and landscape levels (percentage of forest cover, mean patch area, and diversity of patch types). We found that frugivore abundance was mainly explained by variations in the amount of riparian vegetation, whereas nectarivore abundance was mainly explained by variations in the amount of dry forest vegetation. These results reflect that fruit resources for bats mainly occur in the riparian habitat, whereas bat floral resources are mainly found in the dry forest habitat. We conclude that the preservation of the riparian vegetation is crucial for the conservation of bat diversity and the important ecological interactions in which bats are involved in TDF-transformed landscapes.

2.5 Human Dimensions

Humans have inhabited the coast of Jalisco for thousands of years (Mountjoy 2008). Before the Spaniards' arrival, indigenous communities had low impacts on the ecosystems of the region and even during early colonial times, the region remained relatively undeveloped (Rodríguez 1991). Eventually, the colonists directed the conversion of indigenous people's lands into extensive Haciendas that were dedicated to cattle ranching and agriculture, which led to land disputes that lasted over centuries. The coast of Jalisco was characterized by the presence of extensive Haciendas. Today, the Chamela–Cuixmala Biosphere Reserve is surrounded by Ejidos (see Chapter 21). During the 1950s, the federal law of Mexico promoted the colonization of the Pacific coast. This policy sparked TDF transformation in the Jalisco coastal region and land-ownership conflicts that have lasted until the present day. For many decades, government policies considered forested areas as useless lands, promoting the destruction and conversion of TDFs into pastures for cattle ranching and agricultural fields. Over time, governmental policies did not produce the expected results. On the contrary, TDFs have been cleared and fragmented, and local families have not accomplished dignified livelihoods. For example, in the Ejido Ley Federal de la Reforma Agraria, 70% of Ejido's young men migrate to the Americas in search of job opportunities. These migrants send a monthly stipend to their families, and this has become the most important source of income along the coast of Jalisco.

When we query the local population on their perceptions of policy outcomes, the opinion of local Ejidatarios is that government policies only favor those who are in better economic positions or investors who are interested in developing the coast for tourism. The current Coast of Jalisco Land Use Planning Program—a program that enhances social development and biological conservation in the region—prohibits clearing land for agriculture and favors private investment (Castillo et al. 2009). Most Ejidatarios disagree with these policies. Our analysis also reveals that, although the conservation of TDFs is clearly necessary to preserve biodiversity and ecosystem services, many local people are in conflict with conservation policies. Local and federal governments have favored land concessions to national and foreign investors to develop large-scale tourism projects that promote socioeconomic injustice and environmental damage. People recognize the need to create policies that regulate large-scale exploitation of TDFs and coastal resources by a few stakeholders and to preserve TDF habitats and ecosystem services.

2.6 Conclusions and Recommendations

Our study showed that secondary TDF succession is not a simple, unidirectional linear sequence of change in functional groups/species composition. In general, the analysis of the ecological succession of a chronosequence of TDFs in Mexico showed that, after five years, pastures maintained their distinctness in terms of all composition and structure variables analyzed, although they were more similar to early successional stages in terms of species density and structure. Early stages of succession became similar to later stages in terms of species richness and composition, whereas intermediate and late stages remained similar to each other. The increasing similarity in species composition with the advancement of succession is likely the result of strong deterministic factors. The composition and structure of intermediate and late successional stages are similar from the beginning to the end of the chronosequence study, indicating that management practices in the intermediate stages maintain many species and trees in disturbed areas. Although it seems that species composition is similar between intermediate and late successional stages, the high beta diversity found in the plant communities of the TDFs of Mexico indicates that less frequent and rare species may not be similar in both stages, and a more heterogeneous landscape should be found at a larger scale. Further research should identify the main parameters involved in TDF succession to develop models of recruitment dynamics of key dry forest plant species that facilitate the process of succession following natural or human-induced disturbances.

One component of TDFs for which an assessment of succession has not been conducted is found in the canopy stratum (lianas, orchids, and bromeliads), which encompasses the most diverse plant community of the Chamela–Cuixmala region. Changes in vegetation attributes in this forest stratum are likely to differ greatly with succession. Specifically, a decrease in species composition in early and intermediate successional stages would be expected due to a reduction in host species abundance, host size, and wood density (Sáyago et al. 2013). Apparently, only the host plant communities of late successional stages are capable of supporting the canopy plant community of the Chamela–Cuixmala TDFs. Further work on the succession of TDFs should include an analysis of all canopy-level strata. Meanwhile, it is important to interpret results from vegetation analyses with caution and to critically evaluate apparent similarities among successional stages that may reflect the different underlying processes or causes.

Another important aspect to be considered in the process of ecological succession is the functional response of the community, in which certain groups of plants and animals respond in a similar fashion to environmental conditions and are associated with particular successional stages. Preliminary data from our study in Mexico indicate that certain plant groups that share specific plant functional traits are more likely to be represented in a particular successional stage. For example, groups of plants in early successional stages that are more exposed to high temperatures and solar radiation tend to maximize heat dissipation more than plants from late successional stages, which tend to enhance more light acquisition and water use. Therefore, the functional recovery of TDFs might be more complex than inferred by just analyzing floristic and/or structural components of the community. In addition, the variations in animal community assemblages also seem to be associated with different successional stages. Our studies show that bird and bat guilds tend to be more associated with certain habitats under certain successional stages. More studies from other tropical regions are required to corroborate patterns of functional succession. In conclusion, the main mechanisms of succession and regeneration of TDFs still remain unexplored, and more efforts are required to understand the ecological processes of these important ecosystems.

Another important process to consider in ecological succession is related to plant phenology. Phenological differences encountered among successional stages might be related to biophysical parameters in which early successional stages should show high canopy openness (high light transmission) and low Leaf Area Index as a consequence of their low vegetation density and plant cover. This, in turn, should be reflected in low water availability, high light availability, and high temperature. Some of these environmental events have been proposed to trigger phenological patterns in plants. Leaf flushing is a phenological process that is directly associated to primary productivity.

The timing of leaf flushing is similar in all successional stages, but early and intermediate stages appear to retain leaves longer than do late successional stages. In addition, early successional stages also maintain flowers all year round with higher peaks of fruiting as well. It seems that plants in early successional stages are capable of using high light environments with high water efficiency and are very effective at temperature regulation. Extended patterns of flowering phenology in early successional stages are likely to maintain pollinators, particularly bees and butterflies, for long periods of time. Future research should be designed to study how intraspecific variation in the frequency, duration, amplitude, and synchrony of flowering may affect the reproductive success and genetic structure of plant communities in relation to the regeneration capacity of the different successional stages.

Finally, our study showed that human settlements have transformed a part of the landscape surrounding the Biosphere Reserve, but a high percentage of mature and successional forest is still found in fragments and continuous patches, which are owned by the communal land ownership of Ejidos. Although this Biosphere Reserve is still surrounded by an almost continuous forest, this important protected area of Mexico may turn into an island in less than 50 years at the current rate of deforestation, especially if trends related to Ejidos lands, private ranches, and urban developments far from the Biosphere Reserve are maintained (Sánchez-Azofeifa et al. 2009a). Ejidos encompass 70% of the territory located in the vicinity of the Biosphere Reserve, and the remaining land is divided between tourist developments and private properties. Ejidos are also characterized by poverty, high levels of illiteracy, migration, and limited access to secure employment. Agricultural production is the main economic activity, but it cannot secure peasant families' livelihoods. Although peasants recognize the services provided by TDFs, such as cooler climate, shade, and plant and animal species for family consumption, they are proud of their pasture fields and economic activities that are aimed at creating jobs and reducing migration to the Americas. Consequently, the conservation and restoration of TDFs is not perceived as a necessary activity, as has been found in interviews carried out in this study. As a communal tenure forum, Ejidos have a social organization that helps the conduction of collective activities such as forest exploitation. Their organization is also positioned with the local social institutions with which a regional conservation and development plan can be formulated and implemented.

New strategies for the preservation and restoration of TDFs should promote conservation via payment for environmental services. Among the most important environmental services are (1) pollination of crops by wild pollinator species, a service that is particularly important for temporal crops grown in riparian habitats; (2) protection of watersheds and aquifers, a service that helps prevent natural disaster and ensures availability of one of the most essential and endangered natural resources—water; and (3) carbon sequestration by mature and regenerating forests, a service that provides a healthy environment for tourists and local communities, while contributing to a reduction in the impacts of global climate change. Consequently, to accomplish the goal of promoting conservation through environmental services, we propose the creation of a "Red de Areas Ejidales Protegidas" (Ejidos' protected areas network) in the region surrounding the Chamela–Cuixmala Biosphere Reserve (Sánchez-Azofeifa et al. 2009a). In this network, Ejidos will commit to protect land within their property, and the government and the Biosphere Reserve will commit to pay for environmental services and to provide technical assistance and training for alternative ecosystem management strategies. For this to be implemented, the government agency that is responsible, along with the Biosphere Reserve, should accept a leading role in the construction of a payment form that improves peasant families' livelihoods and secures the long-term maintenance of TDFs and associated habitats. Promotion of environmental educational programs and training activities for the Ejidos should be implemented to modify the way in which they perceive and value TDFs. Furthermore, a continuous communication exchange between the Ejidos and the Biosphere Reserve should be enforced. Research groups and institutions would also play an important role in terms of providing information for evaluating the services provided by the Ejidos' reserves. Schemes similar to those developed in countries such as Costa Rica (Sánchez-Azofeifa et al. 2007) may be used as an initial template. The implementation of economic schemes under programs for payment of environmental services will be a very dynamic way of pragmatically enforcing sustainable development, conservation, and management in tropical rural Mexico.

3

Status of Knowledge, Conservation, and Management of Tropical Dry Forest in the Magdalena River Valley, Colombia

Fernando Fernández-Méndez, Omar Melo, Esteban Alvarez, Uriel Perez, and Alfredo Lozano

CONTENTS

3.1 Introduction

More than 61 million km^2 (40%) of the planet's surface experiences dry seasonal climates (Miles et al. 2006; Portillo-Quintero and Sánchez-Azofeifa 2010). Colombia has a 1.14 million km^2 continental extension, 245,342 km^2 of which are dry ecosystems ranging from arid to semiarid and subhumid dry zones; such areas are considerably affected by anthropogenic factors (IAvH 1997; FAO 1988). It has been estimated that areas covered by natural forest in Colombia account for 1.5%–2% of its original potential surface (ca. 80,000 km^2) (Etter 1993). The most recent data state that 9,955 km^2 out of 76,581 km^2 of the large biome of tropical dry forest (TDF) in Colombia are covered by secondary vegetation (IDEAM et al. 2007). A recent analysis summarizing the characteristics of dry forest (DF) in Colombia has estimated the extent of Colombia's forested areas to be 154,628 km^2 in mosaics having differing degrees of fragmentation (Alvarez et al. in press).

The extent of existing TDF is frequently not taken into account when forest conservation efforts are being made (Miles et al. 2006; Prance 2006; Portillo-Quintero and Sánchez-Azofeifa 2010) and Colombia is no exception; in fact, this type of ecosystem is poorly represented in the Colombian portfolio of protected areas (IDEAM et al. 2002; Arange et al. 2003; Ruiz and Fandiño 2007). Protected DF relicts only exist in two of Colombia's national parks: the Tayrona Park in the Magdalena department and the Old Providence McBean Lagoon in Providencia (Linares and Fandiño 2009). Such ecosystems are poorly represented in regional and local categories in the system of protected areas (IAvH 2002); it is currently estimated that only 0.7% of areas coming within protection categories lie within TDF (IAvH et al. 2012).

TDFs comprise complex and fragile ecosystems harboring unique richness in terms of biodiversity. TDFs occur where evaporation exceeds rainfall, presenting one or two periods of drought that may last for four to six months per year (Janzen 1986; Murphy and Lugo 1995; Luttge 2008), leading to a deficit of water in soil and defoliation of vegetation as a response to hydric stress (Murphy and Lugo 1996; IAvH 2002; Lobo et al. 2003). In turn, this leads to greater structural and physiological diversity of life-forms (Mooney et al. 1996), for example, deciduous species becoming mixed with evergreen types, resulting in complex ecophysiological patterns (Burnham 1997).

TDFs usually have high anthropogenic pressure due to their soil fertility and suitability for large-scale agriculture through the use of irrigation systems to such a point that it is considered one of the most degraded ecosystems (Quesada and Stoner 2004). TDFs have low growth rates compared to humid tropical forests and present restricted reproduction cycles, increasing their fragility and making them more susceptible to being disturbed (Murphy and Lugo 1996). They are also characterized by high floristic endemism (43%–73%) (Kaláscka 2004).

Fewer studies have been reported for TDFs than for tropical rain forests and Andean forests (Bazzaz and Pickett 1980; Montagnini and Jordan 2005; Sánchez-Azofeifa et al. 2005; Prance 2006). The current state of knowledge regarding TDFs in Colombia is poor, given that few places have complete or partial ecological monitoring or plant inventories and only a few research groups have kept or made inventories (IAvH 1998). There is little information concerning TDF structure and biodiversity dynamics in Colombia. The consolidated state of coverage, distribution, structure, and diversity of the remaining dry forest is currently unknown Mendoza 1999; Bernate and Fernández 2002; IAvH 2002; Cabrera and Galindo 2006; Linares and Fandiño 2009).

In Colombia and elsewhere in the tropics, the agricultural frontier's advance, changes in soil use, deforestation, as well as natural processes have led to a progressive loss of vegetation, endangering natural resources and their diversity (FAO 1988). TDFs thus continue to be one of the most threatened ecosystems in the Neotropics (Janzen 1983); the success of efforts to restore them will greatly depend on knowledge regarding successional processes, as less information is currently available on TDFs than on temperate forest (Bazzas 1979) and/or tropical rain forest ecosystems (Bazzaz & Pickett 1980; Mulkey et al. 1996; Sánchez-Azofeifa et al. 2005; Prance 2006).

This chapter compiles information based on a series of scientific studies regarding the economic, social, and ecological aspects of TDFs, especially that in the inter-Andean Magdalena river valley in Colombia where a TDF life zone predominates. We discuss the forest's current situation within its socioeconomic context regarding conservation, use, management, and restoration research priorities.

3.2 Dry Forests in Colombia

Evaluation of TDF characteristics and state of conservation fundamentally depends on the definition of vegetation type adopted for such a purpose (Miles et al. 2006); several authors have highlighted the complexity of attempting such definition since there are no discrete limits for any type of vegetation (Furley et al. 1992).

Holdridge's approach (1947, 1967) was used for mapping Colombia's vegetation formations in 1977 zoning the country according to the territory's life zones and complementing them with a detailed description of the characteristics of vegetation of regions having a dry seasonal climate in land lying below 2000 masl (Espinal 1977) (Table 3.1 and Figure 3.1). Olson et al. (2001) prepared a more recent map of the world's regions and included Colombia's dry formations separated into three broad categories: (1) tropical and subtropical dry broadleaf forest; (2) desert and xeric shrubland; and (3) tropical and subtropical grassland, savannah, and shrubland. This map was then modified by the World Wildlife Fund (WWF 2008). Figure 3.1 shows that the three maps contained differences regarding the extent of dry forest on the Caribbean coast and in eastern Colombia. Nevertheless, as Espinal's (1977) map included detailed fieldwork, it has been considered the best approach to determine the extent of dry forest in Colombia. The TDF in Colombia will thus refer to all the formations presented in Table 3.1, characterized by occurring below 2000 masl. Based on such a classification system, potential TDF

TABLE 3.1

TDF and Other Arid Formations in Colombia

	Name	Acronym	Altitude (masl)	Province Humidity	Temperature (°C)	Rainfall (mm/year)	Area of Colombia (km²)	Percentage of Dry Forest Area	Percentage Area of Colombia
1	Tropical dry forest	TDF	500	Subhumid	28	1.500	114,791	74.2	10.1
2	Premontane dry forest	PMDF	<1,000	Subhumid	21	750	14,330	9.3	1.3
3	Low montane dry forest	LMDF	2,000	Subhumid	15	750	9,339	6.0	0.8
4	Thorny subtropical montane	TSTM	500	Semiarid	28	375	5,600	3.6	0.5
5	Subtropical desert scrub	STDS	500	Arid	28	190	4,313	2.8	0.4
6	Tropical very dry forest	TVDF	500	Semiarid	28	750	3,559	2.3	0.3
7	Subtropical dry forest	STDF	1,000	Subhumid	24	750	2,583	1.7	0.2
8	Premontane thorn forest	PMTF	1,500	Semiarid	21	325	111	0.1	0.0
	Total						154,628	100.0	13.5

Source: Espinal, S., Zonas de vida o formaciones vegetales de Colombia, En Memoria explicativa sobre el mapa ecológico, Instituto Geográfico Agustín Codazzi, Bogotá, http://library.wur.nl/isric/index2.html?url = http://library.wur.nl/WebQuery/isric/6600, 1977.

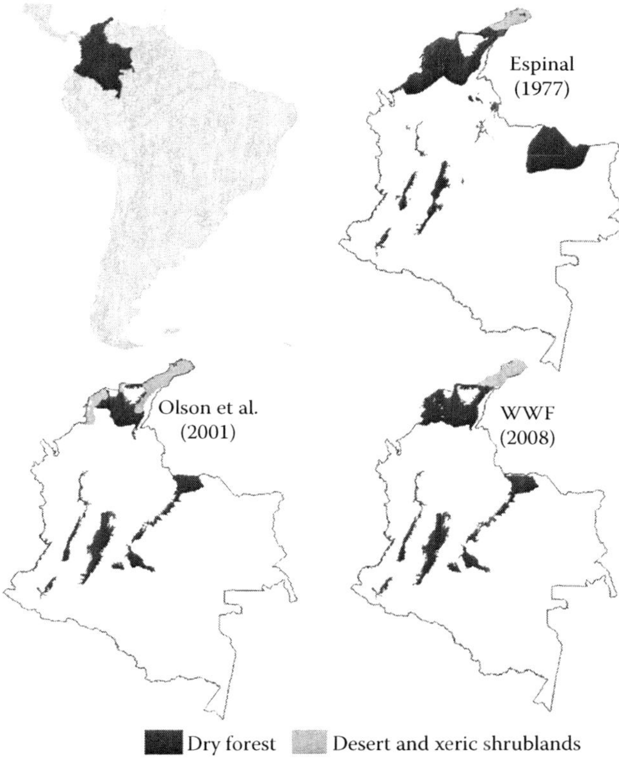

FIGURE 3.1
(See color insert.) Maps of dry forest distribution in Colombia drawn by several authors/organizations (Espinal 1977) grouped the dry forest into different life zones (TDF, PMDF, TVDF, and STDF), as well as desert and xeric scrubland (TSTM and STDS) (see Table 3.1 for an explanation of the acronyms).

in Colombia covers 145,000 km^2 (i.e., 12% of Colombia's continental territory), encompassing a relatively broad range of climatic conditions (Figure 3.2, Table 3.1).

Figure 3.3 shows the location of TDF in Colombia. The TDF in Colombia was originally distributed throughout the inter-Andean valleys of the Magdalena and Cauca rivers and on the Caribbean plains (Repizzo and Devia 2008); it also covered smaller areas in the dry enclaves alongside the Patía and Dagua Rivers and on the islands of San Andrés and Providencia (Linares and Fandiño 2009) located to the extreme north of Colombia (18° N and 81°W) off Nicaragua's Caribbean coast (not shown in Figure 3.3). The Caribbean coastal region has the greatest coverage today, followed by the dry region of the Magdalena river valley and then the Cauca river valley and canyon where there are still small isolated remnants; the latter is possibly the dry forest area most affected by human (agricultural) activity (Cavelier et al. 1996; Repizo and Devia 2008; Linares and Fandiño 2009).

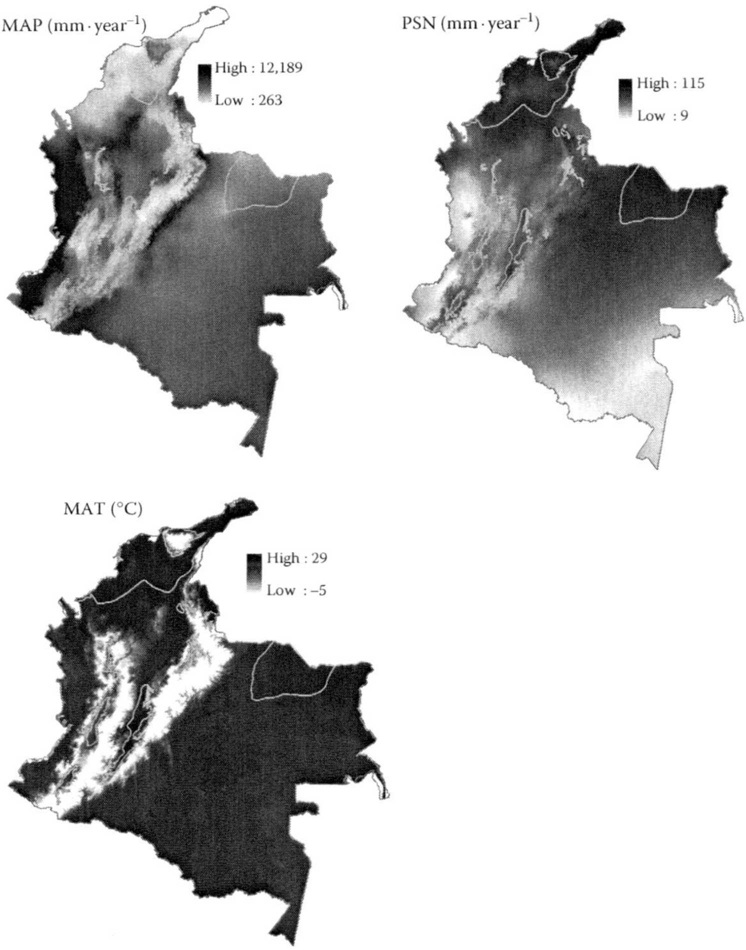

FIGURE 3.2
(**See color insert.**) Climatic characteristics of the dry formations in Colombia: the figure presents the ranges for Colombia in terms of mean annual precipitation (MAP); seasonality of precipitation (PSN), defined as the coefficient of variation of mean monthly precipitation; and mean annual temperature (MAT). TDF ranges are MAP = 21–29 mm·year^{-1}, PSN = 263–2300 mm·year^{-1}, and MAT = 55°C–115°C. (From Hijmans, R.J., et al., *Int. J. Climatol.*, 25, 1965–1978, 2005.)

The upper Magdalena river valley's dry forest covers around 20,000 km^2 where most forest has been replaced by pasture and agro-industrial crops. The most recent maps of the region date to the first decade of this century and show around 31 fragments covering 155.5 ha on average (IAvH 1998) accompanied by a high degree of fragmentation and irregularity (Lozano et al. 2011).

FIGURE 3.3
(See color insert.) Regions having dry forests (tropical, subtropical, and mountainous) in Colombia: (1) the Caribbean coast, (2) the Guajira, (3) the Catatumbo region, (4) Aguachica, (5) the Chicamocha canyon, (6) Arauca, (7) the Cauca valley and river canyon, (8) the upper Magdalena River valley, and (9) the dry enclave of the Patía and the Dagua Rivers.

3.3 Upper Magdalena River Valley's Dry Forest: Biophysical Aspects

3.3.1 Climatological Aspects

Colombia's mountainous relief (i.e., its orographic physiognomy) from 0 to 1000 masl in the Magdalena valley experiences two well-marked periods of rainfall (Figure 3.4), with average rainfall ranging from 831 to 2268 mm/year. The mean annual temperature is 26.8°C. July, August, and September have the highest temperatures (29.8°C, 29.8°C, and 29.1°C interannual average monthly temperatures, respectively) (Figure 3.4) (Santoro 2002).

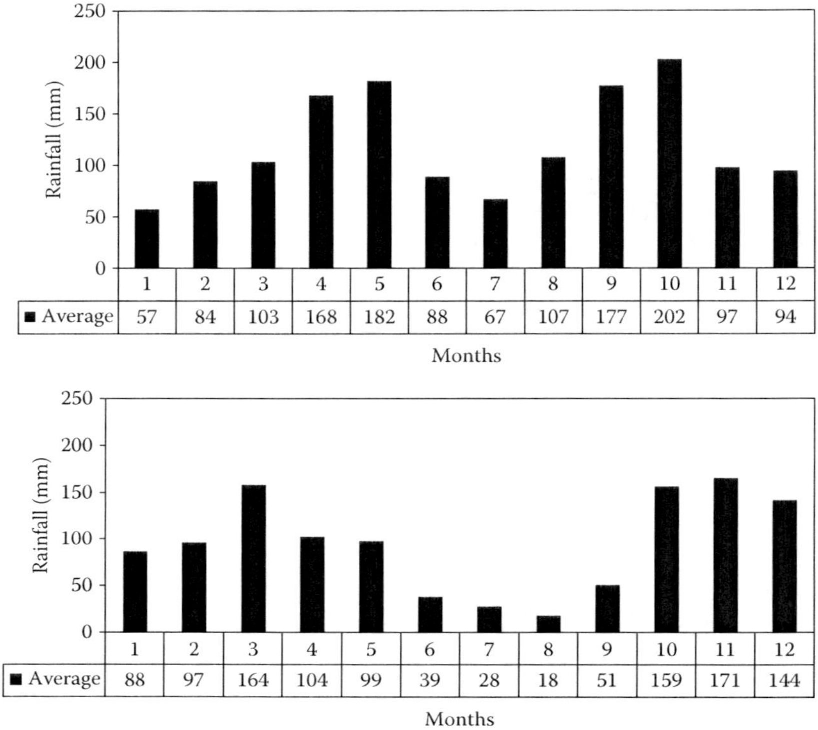

Months	1	2	3	4	5	6	7	8	9	10	11	12
■ Average	57	84	103	168	182	88	67	107	177	202	97	94

Months	1	2	3	4	5	6	7	8	9	10	11	12
■ Average	88	97	164	104	99	39	28	18	51	159	171	144

FIGURE 3.4
Interannual monthly rainfall for two areas in the inter-Andean Magdalena valley (1975–2001). (From Santoro, H., *Estudios de caracterización biofísica y socioeconómica de la ecorregión estratégica del valle del Alto Magdalena*, Componente Aguas, Ministerio de Ambiente, CORTOLIMA, CAM, Universidad del Tolima and Universidad Surcolombiana, Ibagué, Tolima, 2002.)

Dry periods during July, August, and September have 63.1%, 61.9%, and 65.1% relative humidity, respectively. Maximum relative humidity occurs in the north of the valley during the rainy season (83.1%) and exposure to hours of sunshine is constant, ranging from 120.7 h/month to 218.9 h/month. The annual solar intensity ranges from 1692.8 to 2288.1 h/year. The region has an average annual potential evapotranspiration of 1738 mm/year, suffering water deficiency from April to September (Santoro 2002).

3.3.2 Physiography and Soils

The region has six broad physical–graphical landscapes: fluvial–erosional mountainous relief, fluvial–erosional flattened surface, structural–erosional mountainous relief, hilly fluvial–erosional relief, piedmont alluvium and colluvium, and alluvial valley. The best soils are found in the extensive piedmont and alluvial valley landscape (54.41% of the area); this is where irrigated farming takes place. By contrast, the large structural–erosional mountainous landscape (18.13% of the total) has the least favorable and most deteriorated eroded conditions, whose potential land uses are practically only suitable for protection and conservation (Soto 2002).

Most of the area's soils are classified under the orders Entisols and Inceptisols and, to a lesser extent, Alfisols and Molisols, having a ustic moisture regimen in the large Ustorthent, Ustropept, Haplustalf, and Haplustoll groups. As for pH, most soils range from slightly acidic to slightly alkaline (pH 5.8–7.5). The soils in the area have low exchange capacity. Their organic material and organic carbon content are low to very low and phosphorus content is low to medium (Soto 2002). A warm, dry climate negatively influences the soil's natural fertility; it mostly limits areas lacking the availability of irrigation. Soils that are very susceptible to erosion predominate in the region (69%), having low to very low natural fertility (85%), and more than 60% are classified as conservation and reserve areas by ecological zoning (Perea 2001).

3.3.3 Soil Degradation

Vegetation cover mainly depends on the dry, hot climatic conditions determining high evapotranspiration indices, which cause water deficiency for plants and the disappearance of vegetation cover. Soil lacking vegetation cover may suffer severe impacts when heavy rains come; this has led to the establishment of five degradation categories (Table 3.2).

3.3.4 Managing the Region's Soils

Management categories have been established according to economic, social, and environmental factors defining areas for integral regional development (Table 3.3).

TABLE 3.2

Degradation Categories for the Inter-Andean Magdalena Valley's Soil

Description	Area (ha)	Percentage
No erosion	240.66	39.26
Light to moderate	93.17	15.20
Moderate to severe	76.48	12.47
Light, moderate, and severe	148.95	24.30
Moderate to severe	47.45	7.74

Source: Soto, O., *Estudios de caracterización biofísica y socioeconómica de la ecorregión estratégica del valle del Alto Magdalena,* Componente Fisiografía y Suelos, Ministerio de Ambiente, CORTOLIMA, CAM, Universidad del Tolima and Universidad Surcolombiana, Ibagué, Tolima, 2002.

TABLE 3.3

Management Categories Recommended for the Magdalena Valley's Inter-Andean Soils

Recommended Management Style	Area (ha)	Area (%)
Intensive agriculture	196.45	32.04
Semi-intensive agriculture	51.91	8.47
Semi-intensive livestock rearing	96.59	15.76
Agroforestry and silvopastoral	31.37	5.17
Recuperation and/or control of degradation	131.54	21.46
Productive protector-type forest management	103.37	16.86

Source: Soto, O., *Estudios de caracterización biofísica y socioeconómica de la ecorregión estratégica del valle del Alto Magdalena*, Componente Fisiografía y Suelos, Ministerio de Ambiente, CORTOLIMA, CAM, Universidad del Tolima and Universidad Surcolombiana, Ibagué, Tolima, 2002.

3.4 Socioeconomic Aspects

A population concentration viewpoint would suggest that the region is an urbanized zone; nevertheless, its agrarian lifestyle, customs, and economy would suggest that it is a rural area according to the work done by researchers from the University of Tolima (2002), Tolima department, Colombia; the institution leads the largest scale characterization projects that have been carried out in the upper Magdalena region.

The rural area's inhabitants depend on cisterns and rural aqueducts lacking prior treatment for human/animal consumption for their sources of water. This means that such a habitat is an environment that does not offer suitable conditions for living with dignity (i.e., in terms of social infrastructure), which substantially affects its inhabitants' quality of life. The rural area does not have modern wastewater or solid waste management systems; such a situation (together with a lack of educational programs and the promotion of a city-type culture) means that the fragile ecosystem becomes overloaded by negative impacts, leading to a loss of human settlements quality of life and a loss of ecosystem viability resulting from the constant contamination of surface water and soils by unsuitable practices adopted by the inhabitants in their attempts to resolve such basic shortages (Aldana and Oviedo 2002).

The public health-care system's lack of coordination limits the population's ability to enjoy a complete state of physical, mental, and social well-being. Likewise, preventative health care should prevail as opposed to a curative form of health-care attention, although the latter predominates due to a lack of preventative health-care plans plus the users' lack of pertinent knowledge.

The population aged 12 years or less have high global, chronic, and acute malnutrition indices (according to the figures established by the Pan American Health Organization and the World Health Organization); however, the region's precarious socioeconomic conditions limit the guarantee of food safety that the population has, this being reflected in the morbidity indices. Most diseases occurring in the ecoregion are related to intestinal infection and parasitosis, enteritis, acute infections of the respiratory system and dental problems associated with crop fumigation activities, inadequate treatment of water for consumption, and the population's customs and education. Rustic artesian filters for treating water for human consumption could be installed, thereby improving the health of the infant population affected by drinking poor-quality water (Aldana and Oviedo 2002).

Groups of low-scale traditional agricultural economy-based producers rely on simple and extended accumulation production systems having low or zero savings, making this population sector a net credit applicant. However, their financial conditions limit or impede their ability to apply for formal credit as they do not comply with established conditions and requirements. Due to the number of intermediation agents and the market's structure, farmers do not have an appropriate share in intermediation retention value per product unit, which is focused on brokers (intermediary agents), thereby minimizing primary producers' profit margins. The proliferation of populated centers (shanty towns) around large farms (haciendas) is creating belts of misery; home improvement plans must thus be promoted where the use of a type of strong thorny bamboo (*Guadúa angustifolia*) represents an alternative to the usual building materials (Aldana and Oviedo 2002).

Socioeconomic projects for recovering the ecoregion's sustainable human development are oriented toward a rational approach to the rural population in terms of its expectations regarding the family unit and the production unit and its personal development. The communities still conserve the traditional use of biodiversity regarding medicinal plants; an example is that 93% of the inhabitants of the flat region around Armero Guayabal control disease with plants (Rojas 1995). Studies concerning the plant use of local communities are scarce regarding dry forests. Work considering all vascular plants carried out by the ISA (2005) in the Anakarko indigenous area in the upper Magdalena region covering 4000 m^2 reported a use for 132 morphospecies (65% of the total). There are very few family vegetable/fruit gardens in this community due to its livestock tradition; however, 40 cultivated species can be found (30% of the total that they use). Phillips and Gentry (1993) proposed the categories of use and reported 48 species as having medicinal use, 40 being used in construction, 38 as having technological use, 31 being food related, and 9 as having commercial use and some species being included in several categories of use such as West Indian elm/bastard cedar (*Guazuma ulmifolia*), which has a wide range of uses.

The most common production systems on mechanizable flat soils are rice grown under irrigation in rotation with sorghum and cattle farming, followed by rainfed sorghum crops with cotton, corn, and/or dual-purpose sheep and cattle raising/ranching; rainfed sesame plants and fruit trees are also grown on such soils. Dual-purpose cattle raising and twice-yearly food and fruit crops and livestock are found on flat to rolling soils, as well as in areas of indigenous communities, which predominate in the center of the Magdalena valley. Attempts are being made to grow extensive permanent fruit crops (citric fruit, soursop, and mangoes) in business-oriented areas. Regarding livestock activities, poultry farming, pig farming, fish farming, and beekeeping are carried on both rural and business areas on a lesser scale. Plantains, yucca, and horticultural fruit-producing lines (tomatoes, beans, cucumbers, passion fruit, and citric fruit) are produced in rural economy areas in the south of the Huila department on flat and undulating soils (Aldana and Oviedo 2002).

The region's socioeconomic conditions (according to a 2002 report) have improved during the last 10 years; however, a new balance must be made for the region aimed at understanding the landscape's dynamics in the face of socioeconomic development. Any new approaches involving the overall economy pose a challenge regarding a change in how production is viewed in the region, mainly depending on production systems like rice and intensive livestock breeding. This necessitates the rethinking of conservation and management priorities, which leads to projecting and promoting the region faced with the vagaries and variations involved in climate change, to foresee long-term sustainable development alternatives.

3.5 Fauna

Some studies have been carried out highlighting the need for an in-depth evaluation of endangered populations, since a variety of species has been found in different taxonomic groups such as the yellow-striped poison frog (*Dendrobates truncatus*), *Iguana iguana*, the Magdalena River turtle (*Podonecmis lewyana*), the margay (a small, long-tailed ocelot—*Leopardus wiedii*), the long-tailed or Neotropical otter (*Lutra longicauda*), the New World rodent (*Agouti paca*), the grey titi or white-handed tamarind (*Saguinus leucopus*), the red brocket deer (*Mazama americana*), the white-tailed deer (*Odocoileus virginianus*), the central American agouti (*Dasyprocta punctata*), the brown-eared woolly opossum (*Caluromys lanatus*), the white-tailed hawk (*Buteo albicaudatus*), the roadside hawk (*Buteo magnirostris*), the yellow-headed caracara/falcon (*Milvago chimachima*), the American kestrel or sparrow hawk (*Falco sparverius*), the red-fronted conure/parrot (*Aratinga wagleri*), the spectacled parrotlet (*Forpus conspicillatus*), the common barn owl (*Tyto alba*), the tropical screech owl (*Otus choliba*), the black-throated mango/hummingbird (*Anthracothorax nigricollis*), the grey-lined hawk (*Buteo nitidus*), the orange-chinned parakeet (*Brotogeris jugularis*), the yellow-crowned Amazon parrot (*Amazona ochrocephala*), and the pale-bellied hermit/hummingbird (*Phaethornis anthophilus*) (Parga and Quevedo 2002).

Lozada and Martinez (2011) have reported 297 bird species, 35 of them having been determined as being directly associated with the forest. Another relevant study in the area (Jimenez et al. 2008) concerned ants, highlighting the importance of conserving forests to ensure the diversity of this group of insects.

Another study showed that using fauna represents a source of income by establishing zoo nurseries; this means that promising species that are susceptible to being exploited and are still present in the zone must be evaluated, ascertaining their environmental supply and potential demand. Species reported in the area as being potential material for zoo nurseries (bearing existent regulations in mind) are the red brocket deer (*M. americana*), the white-tailed deer (*O. virginianus*), the New World rodent (*A. paca*), the central American agouti (*D. punctata*), the eastern cottontail rabbit (*Sylvilagus floridanus*), the forest/Brazilian rabbit or tapiti (*Sylvilagus brasiliensis*), the poison dart frog (*Dendrobates* sp.), toads (*Bufos* sp.), the harlequin toad (*Atelopus* sp.), the tree frog (*Hyla* sp.), the rain frog (*Eleuterodactylus* sp.), the foam-nest frog (*Leptodactylus* sp.), the spectacled or common caiman (*Caiman crocodilus*), *I. iguana*, and the boa constrictor. These species cannot be hunted without a license for hunting such breeding-type animals; however, they represent the potential for using new approaches to conservation and exploiting many of them. This would tend to reduce pressure on them and lead to alternatives for the population after studies have been carried out aimed at understanding the dynamics of existing populations, their genetic state, and potential use (Rojas 2001; Parga and Quevedo 2002).

3.6 Vegetation

The floral composition reported for the region covers 52 arboreal species of trees belonging to 29 botanical families (Rojas et al. 2011); vegetation diversity in general covers 104 species for the northern part of the Tolima department and 234 for the Caribbean region, 52 of which are trees and shrubs (Mendoza 1999). A total of 121 arboreal species have been reported for the southern part of the Tolima department (Huertas 2001; Bernante and Fernández 2002; Mahecha 2002).

From the community point of view and overall distribution of abundance, the forest relicts fit log and lognormal series, so-called communities in a state of intermediate succession, mostly having a high predominance of one to three species (Bernate and Fernández 2002); 33.86% of the reported species belong to the Mimosaceae (7.22%), Myrtaceae (7.22%), Euphorbiaceae (6.11%), Flacourtiaceae (5.55%), Rubiaceae (3.88%), and Rutaceae families (3.88%), and the remaining 66.14% constitute 41 families (Quiroga and Roa 2002).

The two species having the greatest ecological importance in forest communities (mainly in the southern part) are the rough-leaf tree (*Curatella americana*) and the golden trumpet tree (*Tabebuia chysantha*). The greatest taxonomic diversity recorded in the region in 0.25 ha plots has been reported in a forest located in the Anakarko Indian reservation, having 34 species. The highest forest uniformity for dominant species such as Cape ash (*Machaerium capote*) and yellow Spanish plum (*Spondias mombin*) has been 66.8%, followed by 58.33% for the flowering cashew or tigerwood tree (*Astronium graveolens*) (Quiroga and Roa 2002). The golden trumpet tree (*T. chysantha*) and the West Indian elm or bastard cedar tree (*G. ulmifolia*) represent a typical association in transitional areas and at the edge of forests with open grassy areas; beta diversity is high, emphasizing the endemism demonstrated by the percentage of unique species and the percentage of unique species having unit abundance, reaching 47% and 17%, respectively, out of the 121 species found in the south of the region (Bernate and Fernández 2002).

The profile diagrams show structures having 20–25 m canopies in areas having steep riparian slopes (not greater than 10 m), in associations dominated by rough-leaf tree (*C. americana*) on flatter ground. The different forest strata are frequently represented by species such as the wax myrtle (*Myrcia* sp.), the holy wood tree (*Bursera graveolens*), the potbellied ceiba tree (tropical canopy tree) (*Pseudobombax septenatum*), the West Indian birch tree, gumbo-limbo or bald cypress (*Bursera simaruba*), and the tigerwood tree or flowering cashew tree (*A. graveolens*). Trees having a high canopy and presence in advanced successions are mainly the West Indian birch tree, the gumbo-limbo or bald cypress tree (*Bursera simaruba*), the tigerwood tree or flowering cashew tree (*A. graveolens*), the shade tree (*Pseudosamanea guachapele*), the cedar (*Cedrela angustifolia*), and the flowering cashew or tigerwood tree (*A. graveolens*) (Figure 3.5). Regarding total structure and diameter distribution, 58.7% of individuals in different areas of the valley had 5–9 cm outside bark

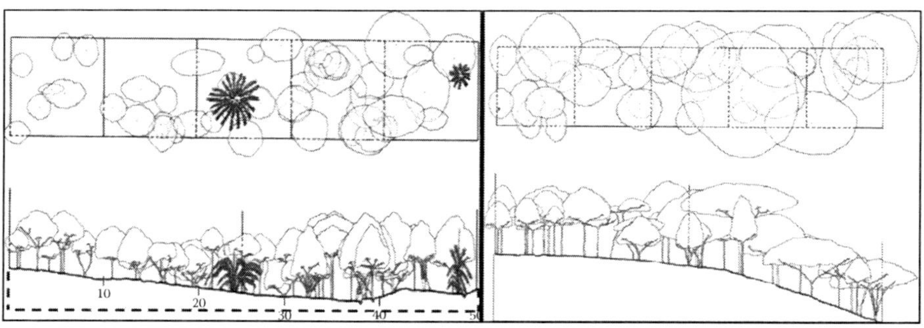

(a) (b)

FIGURE 3.5
(a) Vegetation profile for an early secondary forest, chaparral association (i.e., dominated by dense, spiny, evergreen shrubs); the dominant species is the sandpaper tree (*Curatella americana*). (b) Vegetation profile for a secondary forest in recuperation; the dominant species are immature fruit plant (*Casearia corymbosa*), golden trumpet tree (*Tabebuia chysantha*), West Indian elm or bastard cedar (*Guazuma ulmifolia*), salmwood (*Cordia alliodora*), and shade tree (*Pseudosamane guachapele*). (From Quiroga, J.A., and H.Y. Roa, Estructura de los fragmentos boscosos de la ecorregión de la Tatacoa y su área de influencia, en el Tolima, BSc Forestry Engineering thesis, Universidad del Tolima, Tolima, 2002).

diameter at breast height (DBH), showing that tall species in secondary forests in recuperation do not reach their maximum dimension, mostly due to constant domestic exploitation of wood (Quiroga and Roa 2002).

Regarding the dynamics of forests in the Tatacoa strategic ecoregion and their area of influence in Tolima, the major cause of mortality is trees dying from the roots upward in 75% of the forest fragments. Dead individuals exceeded those recruited for the study at 50% of the sites sampled; such forests were being strongly intervened in the area. In general, in the zone there was a noticeable tendency toward the shrinkage of forested areas and a lack of any probability of them becoming conserved; conserving such forest remnants has thus become a major priority to avoid further alteration in their status quo (Melo 2002).

Regarding priority areas for conservation and restoration, Quimbayo (2009) determined the state of TDF fragmentation in areas in the center of the Magdalena valley region, finding high beta diversity among the fragments for the fragmentation associated. Lozano (2005) carried out an in-depth study of a TDF relict's ecological patterns and described the typical composition of a conserved, dense riverside forest, which could represent a potential arrangement for restoring riverside areas, especially in piedmont areas, which would protect river channels and streams and provide environmental recreation-related and water supply-related services on all of the Magdalena river's tertiary tributaries.

Bermudez and Sierra (2006) studied one of the few reserve areas in the northern part of the region; they analyzed the arboreal coverage structure of Vallecitas (a forestry reserve owned by CORTOLIMA, one of the region's environmental entities), finding patterns similar to those previously described for the region. They recommended that this reserve should be protected as an area for research. Salmwood (*Cordia alliodora*) dominates natural succession in dry forests in the north of the Tolima department (Melo and Vargas 2003), thereby making it a potential source for managing natural regeneration aimed at restoration (given its great abundance in different states in recovery areas) (Henao and Moreno 2001). A lack of defined conservation and management areas has meant that successions in the northern area of the Tolima department have mostly been modeled for land use and fire (Rey 2008) and have not envisaged the forest reaching a state of advanced succession.

Dry forests are large-scale providers of goods and services for local communities. Albarracin et al. (2003) approached an environmental evaluation of the Tolima department's TDF ecosystem, specifically at the University of Tolima's North Regional University Center (NRUC). Their methodology involved estimating the availability/willingness to pay and established the following as direct use values for such ecosystems: wood and firewood production, research, education, species breeding sources, and biomass production. Indirect use values concerned the ecosystem as water regulator, sediment and nutrient retainer, food and firewood producer, research center, educational center, and biodiversity support. It was observed that greater knowledge of the forest was linked to greater willingness/availability to pay for its conservation and/or protection. Likewise, greater willingness/availability to pay for forest conservation and/or protection was related to a higher level of income and economic level; 88% of the people interviewed said that they knew about the TDF and valued it highly, especially because of its hydric regulation service.

3.7 Status of Dry Forest Conservation in the Upper Magdalena Valley and the Need for Research and Management

3.7.1 Status of Conservation

Figure 3.6 shows the extent of dry forest throughout the whole of Colombia and the upper Magdalena valley; it also indicates the state of forest coverage obtained from moderate-resolution imaging spectroradiometer images (Hansen et al. 2003a, 2003b), which is expressed as the percentage of vegetation in each 1 km × 1 km quadrant (available at http://glcf.umiacs.umd.edu/data/vcf/). It includes three layers: (1) percentage tree coverage, (2) herbs/shrubs, and (3) soil lacking vegetation. It shows tree coverage; this variable was chosen as a proxy for deforestation and/or disturbance since it displays forest coverage density (Buermann et al. 2008), showing that conserved forest areas are few and mainly relegated to riparian forest and high slopes.

3.7.2 Research

The real area of forest coverage in the life zone of the Colombian TDF still remains unknown, especially in heterogeneous landscapes dominated by an agricultural matrix. No studies of ecosystems' functionality have been carried out, bearing in mind how the forests' ecological processes influence the quality of the ecosystem as a whole. There are no records for species reported in threatened categories nor are there any follow-up records to enable the updating of their status and ensure that pertinent measures are taken.

The lack of knowledge regarding landscape dynamics hampers the relevance of new forest conservation and management approaches being appreciated. Determining TDF remnants and their current state and consolidating current knowledge regarding such forests has thus become an urgent priority. The use of remote sensors to ascertain coverage dynamics has enabled the understanding of forest degradation patterns regarding area, dynamics, productivity, and response to ongoing changes in the region's climatic patterns.

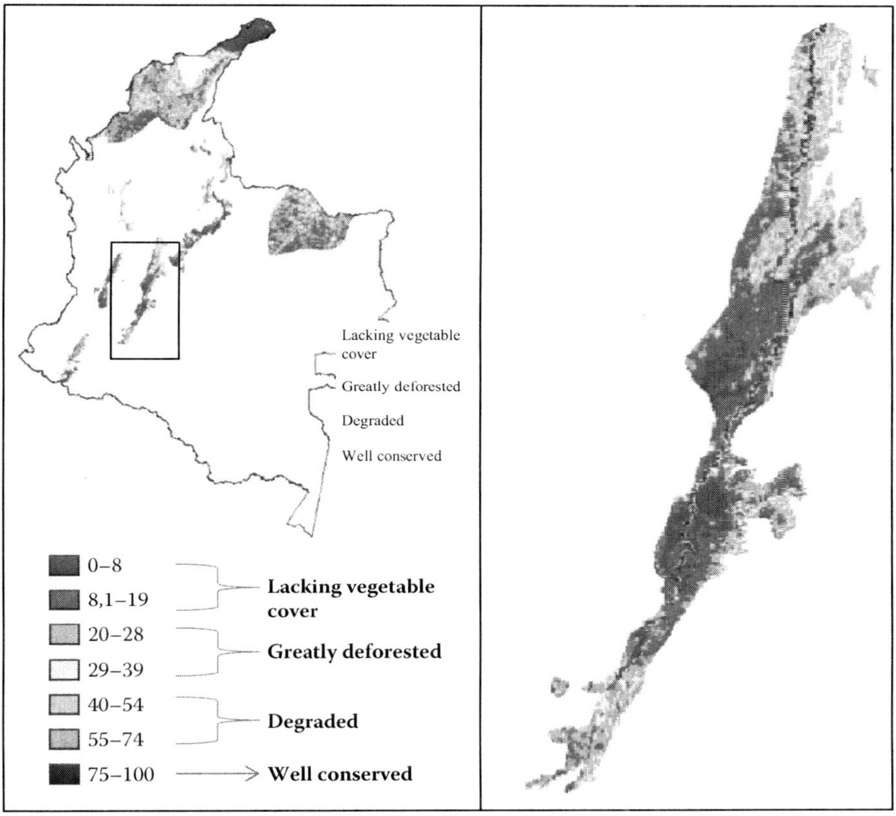

FIGURE 3.6
(See color insert.) The status of dry forest conservation in Colombia and the upper Magdalena Valley.

Knowledge regarding how forests can be restored must be based on an existing policy for conserving forest functionality; such a fund of knowledge (and local know-how) would result in maintaining ecosystem service provision and regulation (i.e., carbon conservation and capture, hydric regulation and conservation, biological diversity, and everything else that as yet has not been included in the awareness of those living in Colombia's dry region).

This would specifically mean that knowledge of reproduction of native tree species and forestry management is indispensable, especially regarding something that has any degree of being/becoming endangered. This would ensure that repopulation programs are established for each of them, in line with criteria such as inducing more regeneration and improving many species' genetic diversity. Academic research priorities must be established regarding participatory action with communities led by environmental and municipal entities. This would be aimed at recuperating and conserving forest remnant integrity, thereby providing many ecosystem services for the area's population and workers' associations. Studies regarding dynamics of forest remnants (i.e., phenology, ecosystem productivity, and evaluating ecosystem services) must be carried out soon.

3.7.3 Climate Change

Evaluating the structure of forest fragments located in the Tatacoa strategic ecoregion and their area of influence has shown that its critical state of deforestation could affect more than 90% of the total area. The natural vegetation is fragmented into small relicts (not exceeding 50 ha continuously); such fragments are isolated and their overall structure consists of early secondary successions, riverside vegetation, and (in very few cases) private or public reserve areas having mature forest characteristics (Melo 2002). Due to ongoing changes in temperature and precipitation regimes, obtaining accurate forecasts of the effects on the region's flora and fauna is a high priority.

3.7.4 Ecosystem Services

Albarracin et al. (2003) have shown that a lack of both knowledge and relating ecosystem services to externalities has meant that the forest has not been evaluated/assessed and/or protected, slowly leading to the ecosystem's deterioration. Assessment is a fundamental tool for changing the course of current tendencies in a region; values that are not easily detected by the community and decision makers are thereby been introduced into the economy. This provides the basis for knowledge aimed at changing perceptions and attitudes regarding landscape management as a whole, without productivity being lost in areas that today are a source of financial income for local economies.

3.7.5 Socioeconomic Dimension

The main factors directly affecting a region's inhabitants are quality of life, access to basic needs, and surplus, leading to increasing satisfaction. This means that the region's socioeconomic factors must be reevaluated in detail, together with a set of concrete tools, allowing the dynamics of the population and the region's economy to be ascertained for foreseeing what action should be taken for sustained management.

Orienting production support programs must be guided by production diversification in farm systems as a strategy for improving the flow of income from production units to face the risk and uncertainty involved in market prices. The family unit's food safety would thereby be protected, based on technological supply leading to a gradual transition from a type of agriculture involving the use of agrotoxic products to a more organic type of agriculture. This would thus contribute toward reducing production costs and simultaneously offer society clean products. An attempt must also be made to become involved in strategies contributing toward improving women's personal development and generating family unit income (Aldana and Oviedo 2002).

Including new activities in the economy that differ from traditional ones is not an easy task; however, the dry forest region provides a broad spectrum of new services that communities living in cities do require. Ecotourism could be an emerging activity triggering such socioeconomic change; this should be encouraged by local decision makers who require pertinent training for initiating joint processes regarding new forms of work in the region.

Environmental institutions must exercise greater control over the illegal trafficking of wildlife and forest products because many avifauna, mammalian fauna, and herpetofauna (reptiles and amphibians) species are seriously affected by man's activities, which (without taking existing regulations into account) involve hunting and selling products (Parga and Quevedo 2002). Environmental education-guided innovative production projects will lead to the area's inhabitants continuing to coexist in the region without coming to see conservation as an obstacle to their subsistence and economic development.

3.7.6 Management and Restoration

Considerable areas need to be restored in the Tolima department and throughout the whole of the Magdalena valley after becoming degraded by changes in soil use and overexploitation. Northern Tolima mainly has abundant dry forest relicts, having the potential to become connected by ecological restoration projects. These fragments' fragility is reflected by their irregularity and low compaction (Lozano et al. 2011). This makes their restoration a priority for the region; exploring new production systems such as fruit trees, forest plantations, and other permanent crops in agroforestry and forestry–pastoral systems that contribute toward increasing long-term income and improving landscape connectivity must be a priority to be tested in attempts at finding production alternatives for the region.

It is recommended that immediate action regarding restoration should involve the use of knowledge and the lists concerning the native species produced to date (Duran 1980; Henao and Moreno 2001; Mahecha 2002; Peña and Peña 2003), as well as the knowledge of the local communities that survive daily in the region under the adverse conditions presented by the environment. They have knowledge that should be used and validated for improving the use of natural resources in the area.

3.7.7 Payment for Environmental Services

Some scheme involving payment for environmental services should be studied; actors directly and indirectly benefitting from ecosystem services should be sought to enable institutions and inhabitants to receive such resources and ensure that the ecosystem does not continue to be degraded. This is the final objective in achieving conservation and production settings aimed at providing long-term durability and stability for both biodiversity and the society benefitting from it. Mechanisms such as the United Nations collaborative program on reducing emissions from deforestation and forest degradation (UN-REDD) project could be an option for exploring the possibility of developing some sort of pilot experience in the region. Adopting an initiative of this type is an arduous task due to all that it involves, but it is not imposable. Unquestionably, this is where all the efforts of academic, social, political, environmental, and state entities should be directed.

3.8 Summary and Conclusion

TDFs occupy a large part of the planet's surface and Colombia; however, the implications of excessive soil use and socioeconomic development in such regions still remain unknown. Dry forests have extreme climatic conditions; in spite of this, some people do live in them and exploit their implicit ecosystem services and even large-scale production projects are carried out there. The few remaining sections of forest involve dry ecosystems, although biodiversity is still conserved there (i.e., having great environmental value for society's subsistence). Knowledge regarding fauna and flora is not enough, above all the topics concerning their dynamics and functionality; research must be increased in the region that is aimed at constructing knowledge (know-how) to facilitate the planning of sustainable development coupled to a vision of mitigation and adaptation to current climate change. It must favor the restoration of wooded ecosystems on par with the socioeconomic development of the communities living in Colombia's TDF. Overall strategies regarding access to knowledge (and local know-how) must thus be rethought, as well as implementing new production-based approaches leading to a gradual change in the pertinent actors' perception of forest conservation, thereby forming an integral part of regional development.

4

Brazilian Tropical Dry Forest on Basalt and Limestone Outcrops: Status of Knowledge and Perspectives

Marcel Coelho, Geraldo Wilson Fernandes, and Arturo Sánchez-Azofeifa

CONTENTS

4.1 Introduction

South American tropical dry forests (TDFs) have a peculiar and well-studied biogeographical history. Because of low temperature and moisture changes at the time of Hemisphere Wisconsin-Würn (between 18,000 and 12,000 years before present (BP), the dry forests (caatinga) expanded across the Amazon, cerrado, and Chaco, resulting in what was defined as the "Pleistocene arc" (Prance 1973; Brown and Ab`Sáber 1979; Oliveira et al. 2005). At that time, tropical rain forests went through a contraction to isolated areas. Although in recent times during the expansion of tropical rain forests and the concurrent retraction of dry forests, some semideciduous and deciduous forests have remained surrounded by Amazonian, cerrado, and Chaco vegetation as witnesses of important preterit environmental changes (Meguro et al. 2007). Many studies have accumulated evidences of those climatic fluctuations with dry alternating to wet periods (Pennington et al. 2009). Furthermore, the current distributions of some species co-occurring on different Brazilian domains (Amazon, caatinga, cerrado, Chaco, and the Atlantic rain forest) reinforce the biogeographical history of seasonal formations and reveal biological links among deciduous and semideciduous forests across South America (Prado and Gibbs 1993; Santos et al. 2012).

In Brazil, TDFs can be easily separated into two categories: (1) TDFs on flat soils, and (2) TDFs of basalt and limestone (TDFBL) outcrops (Nascimento et al. 2004; Coelho et al. 2012). The TDFs of flat soils or arboreal caatinga are predominantly concentrated in northeast Brazil and comprise the largest area of continuous dry forests, located between two domains, caatinga and cerrado. The average of the canopy is high (ca. 12 m), with fertile soils and these TDFs are not associated with watercourses (Madeira et al. 2009). The main threats for TDFs on flat soils are agriculture and livestock (Espírito-Santo et al. 2009). Most of the studies on Brazilian TDFs were carried out on such forests (e.g., studies by Pezzini et al. [2008], Madeira et al. [2009], and Arruda et al. [2011]). On the other hand, TDFs on limestone outcrops generally comprise small fragments associated with the cerrado domain, in Central Brazil. These fragments are found across the Brazilian territory: in northern (Altamira/Itaituba–Pará), northeast (Ibiapaba–Ceará, Apodi–Rio Grande do Norte State, and Araripe–Pernambuco), southeast (Brasília–federal district, the Espinhaço Range, and the São Francisco River–Minas Gerais [MG], Pardo–MG, and Alto Ribeira–São Paulo), midwest (Alto Paraguai–Mato Grosso, Corumbá and Bodoquena–Mato Grosso do Sul), and south Brazil (Itacaré–Para until Serra Geral–Rio Grande do Sul state) (Meguro et al. 2007).

The floristic composition of this TDF is strongly related to the Amazon, the Atlantic rain forest, caatinga, cerrado, and pantanal (Chaco), but there is only a low similarity among them (Pérez-García et al. 2009). The average canopy height is 8 m (Meguro et al. 2007; Coelho et al. 2012). Limestone mining and livestock are the main threats to this ecosystem (Silva and Scariot 2003; Coelho et al. 2009). Contrary to TDFs on flat soils, studies of TDFs on limestone outcrops are still incipient and focused primarily on vegetation structure and composition (Pedralli 1997; Silva and Sacariot 2003, 2004; Nascimento et al. 2004; Almeida and Machado 2007; Felfili et al. 2007; Meguro et al. 2007; Santos et al. 2007). TDFs on basalt outcrops are almost extinct due to human activities but still found in southeastern Brazil and have similar topography and plant phenology; in spite of the soils being rich in phosphorous (P), potassium (K), calcium (Ca), and magnesium (Mg), these forests have some similarities with TDFs on limestone outcrops (Rodrigues and Araújo 1997; Oliveira-Filho et al. 1998; Werneck et al. 2000).

In this chapter, we identify the similarities and differences among Brazilian TDFBL outcrops by analyzing their plant structure and composition, reproductive patterns, phenology, and conservation status.

4.2 Plant Structure and Composition

Most studies conducted on TDFBL outcrops focused on several aspects related to the structure and composition of their flora (e.g., studies by Pedersoli and Martins [1972], Pedralli [1997], Rodrigues and Araújo [1997],

Werneck et al. [2000], Silva and Scariot [2003, 2004], Nascimento et al. [2004], Salis et al. [2004], Felfili et al. [2007], Meguro et al. [2007], Arruda et al. [2011], and Coelho et al. [2012]). The first records for such forests were made by the naturalist Johannes Eugenius Bülow Warming (1841–1924), who was known as Eugene Warming. In 1892, Warming described in the region surrounding Lagoa Santa–MG a forest with trees ranging between 15 and 20 m in height (see the study by Warming and Ferri [1973]). Pedralli (1997) described forests from different locations in MG with trees ranging between 16 and 20 m, shrubs between 3 and 10 m, and herbs between 1 and 1.5 m in height. Whereas some studies recorded that emergent trees achieved 13 m (Arruda et al. 2011) and 30 m in height (Oliveira-Filho et al. 1998), other studies specifically addressed the structure and composition of TDFs with recorded average heights of 8 m (Arruda et al. 2011), 6.3 m, and 7.7 m (Coelho et al. 2012). The basal area average of these TDFs varies widely: from 8.54 $m^2 \cdot ha^{-1}$ (Silva and Scariot 2003) to 46 $m^2 \cdot ha^{-1}$ (Meguro et al. 2007). Most studies, however, recorded intermediate values: 45.92 $m^2 \cdot ha^{-1}$ (Oliveira-Filho et al. 1998), 29.3 $m^2 \cdot ha^{-1}$ (Coelho et al. 2012), 20.54 $m^2 \cdot ha^{-1}$ (Oliveira-Filho et al. 1997), 19.36 $m^2 \cdot ha^{-1}$ (Nascimento et al. 2004), 18.6 $m^2 \cdot ha^{-1}$ (Silva and Scariot 2004), 17.8 $m^2 \cdot ha^{-1}$ (Coelho et al. 2012), 16.73 $m^2 \cdot ha^{-1}$ (Felfili et al. 2007), 16.17 $m^2 \cdot ha^{-1}$ (Werneck et al. 2000), and 9.9 $m^2 \cdot ha^{-1}$ (Silva and Scariot 2004).

Although some species stood out in many studies with high importance values (IVs) such as *Anadenanthera colubrina* (Oliveira-Filho et al. 1998; Werneck et al. 2000; Meguro et al. 2007; Coelho et al. 2012), *Myracrodruon urundeuva* (Oliveira-Filho et al. 1998; Nascimento et al. 2004; Silva and Scariot 2004; Felfili et al. 2007; Meguro et al. 2007; Arruda et al. 2011; Coelho et al. 2012), *Combretum duarteanum* (Nascimento et al. 2004; Arruda et al. 2011), *Pseudobombax tomentosum* (Nascimento et al. 2004; Silva and Scariot 2004; Felfili et al. 2007), and *Dilodendron bipinnatum* (Silva and Scariot 2004; Felfili et al. 2007), in most studies the species with the highest IVs were not the most abundant species. These included *Acacia polyphylla* (Oliveira-Filho et al. 1998), *Cupania vernalis* (Werneck et al. 2000), *Casearia sylvestris* (Werneck et al. 2000), *Tabebuia impetiginosa* (Silva and Scariot 2003), *Aspidosperma pyrifolium* (Silva and Scariot 2003), *Attalea phalerata* (Salis et al. 2004), *Aspidosperma australe* (Salis et al. 2004), *Copernicia alba* (Salis et al. 2004), *Acosmium cardenasii* (Salis et al. 2004), *Albizia niopoides* (Salis et al. 2004), *Ceiba pubiflora* (Salis et al. 2004), *Sebastiania discolor* (Salis et al. 2004), *Phyllostylon rhamnoides* (Salis et al. 2004), *Coutarea hexandra* (Salis et al. 2004), *Zizyphus oblongifolius* (Salis et al. 2004), *Tabebuia impetiginosa* (Silva and Scariot 2004), *Ficus calyptroceras* (Meguro et al. 2007), *Combretum leprosum* (Arruda et al. 2011), *Rauwolfia sellowii* (Coelho et al. 2012), *Inga platyptera* (Coelho et al. 2012), and *Bauhinia brevipes* (Coelho et al. 2012). In many studies, Fabaceae stood out as the most speciose family (Oliveira-Filho et al. 1998; Werneck et al. 2000; Nascimento et al. 2004; Salis et al. 2004; Silva and Scariot 2004; Felfili et al. 2007; Meguro et al. 2007;

Coelho et al. 2012). However, other families also stood out as speciose due to the presence of some key species like Moraceae (Oliveira-Filho et al. 1998), Myrtaceae, Rubiaceae (Werneck et al. 2000), Bombacaceae (Silva and Scariot 2003), Euphorbiaceae (Silva and Scariot 2003; Salis et al. 2004; Meguro et al. 2007), Anacardiaceae (Nascimento et al. 2004; Arruda et al. 2011), Bignoniaceae (Nascimento et al. 2004; Silva and Scariot 2004, Felfili et al. 2007; Meguro et al. 2007; Arruda et al. 2011), Apocynaceae (Silva and Scariot 2003, 2004; Salis et al. 2004; Felfili et al. 2007; Coelho et al. 2012), and Malvaceae (Coelho et al. 2012).

Although there are differences among the methods applied to evaluate the botanical structure and composition in TDFBL outcrops, some trends have emerged. We recorded 770 species distributed into 108 plant families. The most speciose family was Fabaceae with 109 species, followed by Euphorbiaceae (46), Malvaceae (36), Myrtaceae (33), Rubiaceae (32), Bignoniaceea (28), Apocynaceae (27), and Sapindaceae (25). The most abundant habitat type was tree, representing 59% of the species recorded, followed by liana (13.5%), shrub (13%), herb (12%), epiphyte (2%), *Rupicola* (<1%), and holoparasite (<1%) (Figure 4.1). Furthermore, only a few studies recorded the same species. Generally, 78% of the species were present only in one

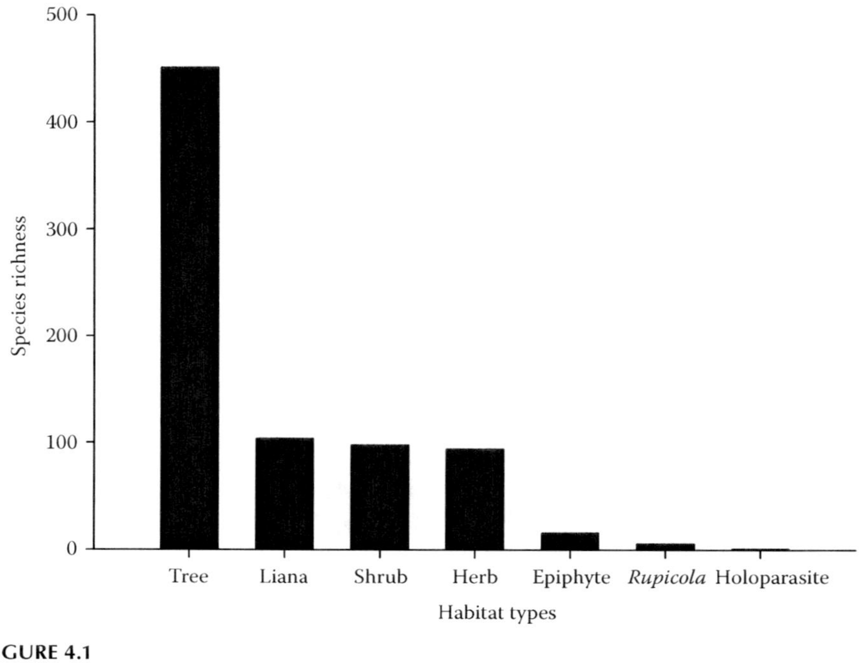

FIGURE 4.1
Habitat types of species from 14 studies carried out on Brazilian TDFs on basalt and limestone outcrops.

single study, whereas only 11% were present in two studies (Figure 4.2). The analysis of 12 studies reporting 17 sites of Brazilian TDFBL outcrops (Jaccard similarity indexes) clearly indicated a low similarity among sites pointing to a high beta diversity (Table 4.1) (see the study by Fernandes and Bezerra [1990]).

As the fragments of TDFBL outcrops are scattered in the Brazilian territory, it is important to track the biogeographical distribution of the species to better understand their common traits as well as their differences. The phytogeographic domains available for the species recorded in TDFBL outcrops were studied by Oliveira-Filho (2006) and Forzza et al. (2010). Few species have their occurrence restricted to a single Brazilian phytogeographic domain. Under a general analysis, summing every single ocurrence of the species in the Brazilian domains, cerrado was the most abundant domain followed by the Atlantic rain forest, caatinga, Amazon, pantanal, pampa, and Chaco (Figure 4.3). When the co-occurrence of the species in Brazilian phytogeographic domains were considered, most of the species (109) were simultaneously present in Amazon, caatinga, cerrado, and the Atlantic rain forest, followed by 74 species in cerrado and the Atlantic rain forest and 72 in caatinga, cerrado, and the Atlantic rain forest. Only 10 species out of 770 were exclusively characteristic of caatinga, which is a domain composed mainly of TDF and has a similar vegetation phenology as that of TDFBL outcrops.

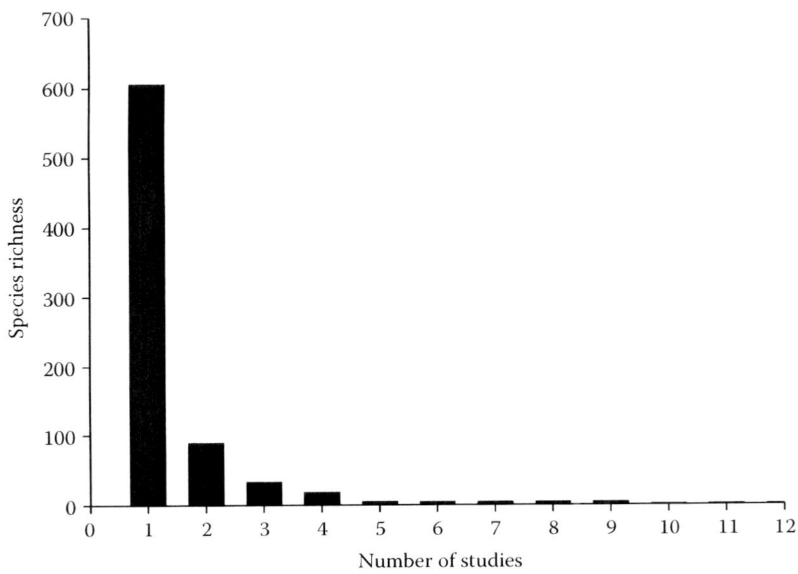

FIGURE 4.2
Co-occurrence of species recorded by 14 studies carried out on TDFs on basalt and limestone outcrops on different Brazilian phytogeographic domains.

TABLE 4.1

Jaccard Similarity Index among Species Recorded by 17 Surveys Available on 12 Studies from Brazilian TDFBL Outcrops

Studies	1	2	3	4	5	6	7	8	9	10	11	12	13	14	15	16	17
1	**1.000**	0.124	0.138	0.078	0.099	0.052	0.096	0.114	0.107	0.089	0.056	0.044	0.018	0.071	0.102	0.045	0.069
2	0.124	**1.000**	0.126	0.086	0.149	0.106	0.122	0.135	0.140	0.061	0.055	0.051	0.013	0.049	0.054	0.068	0.106
3	0.138	0.126	**1.000**	0.053	0.096	0.060	0.093	0.070	0.087	0.070	0.024	0.039	0.016	0.030	0.091	0.031	0.099
4	0.078	0.086	0.053	**1.000**	0.200	0.067	0.340	0.392	0.259	0.041	0.093	0.040	0.022	0.038	0.034	0.114	0.121
5	0.099	0.149	0.096	0.200	**1.000**	0.095	0.319	0.338	0.492	0.042	0.063	0.029	0.015	0.027	0.046	0.150	0.146
6	0.052	0.106	0.060	0.067	0.095	**1.000**	0.069	0.080	0.082	0.024	0.037	0.000	0.000	0.045	0.043	0.075	0.057
7	0.096	0.122	0.093	0.340	0.319	0.069	**1.000**	0.474	0.350	0.032	0.053	0.048	0.034	0.015	0.043	0.127	0.130
8	0.114	0.135	0.070	0.392	0.338	0.080	0.474	**1.000**	0.350	0.043	0.053	0.032	0.017	0.015	0.032	0.170	0.130
9	0.107	0.140	0.087	0.259	0.492	0.082	0.350	0.350	**1.000**	0.033	0.056	0.033	0.018	0.032	0.049	0.180	0.151
10	0.089	0.061	0.070	0.041	0.042	0.024	0.032	0.043	0.033	**1.000**	0.063	0.075	0.031	0.073	0.174	0.018	0.054
11	0.056	0.055	0.024	0.093	0.063	0.037	0.053	0.053	0.056	0.063	**1.000**	0.025	0.091	0.048	0.082	0.056	0.086
12	0.044	0.051	0.039	0.040	0.029	0.000	0.048	0.032	0.033	0.075	0.025	**1.000**	0.051	0.114	0.086	0.024	0.030
13	0.018	0.013	0.016	0.022	0.015	0.000	0.034	0.017	0.018	0.031	0.091	0.051	**1.000**	0.048	0.049	0.027	0.050
14	0.071	0.049	0.030	0.038	0.027	0.045	0.015	0.015	0.032	0.073	0.048	0.114	0.048	**1.000**	0.071	0.045	0.029
15	0.102	0.054	0.091	0.034	0.046	0.043	0.043	0.032	0.049	0.174	0.082	0.086	0.049	0.071	**1.000**	0.012	0.059
16	0.045	0.068	0.031	0.114	0.150	0.075	0.127	0.170	0.180	0.018	0.056	0.024	0.027	0.045	0.012	**1.000**	0.083
17	0.069	0.106	0.099	0.121	0.146	0.057	0.130	0.130	0.151	0.054	0.086	0.030	0.050	0.029	0.059	0.083	**1.000**

Note: 1 = Uberlândia–MG (Rodrigues and Araújo 1997); 2 = Santa Vitória–MG (Oliveira-Filho et al. 1998); 3 = Perdizes–MG (Werneck et al. 2000); 4 = São Domingos–GO (Silva and Scariot 2003); 5 = Monte Alegre–GO (Nascimento et al. 2004); 6 = Corumbá–MS (Salis et al. 2004); 7 = São Domingos–GO (Silva and Scariot 2004); 8 = São Domingos–GO (Silva and Scariot 2004); 9 = Iaciara–GO (Felfili et al. 2007); 10 = Santo Hipólito–MG (Meguro et al. 2007); 11 = Bocaiúva–MG (Meguro et al. 2007); 12 = Arcos–MG (Meguro et al. 2007); 13 = Pains–MG (Meguro et al. 2007); 14 = Vespasiano–MG (Meguro et al. 2007); 15 = Serra do Cipó–MG (Meguro et al. 2007); 16 = Capitão Enéas–MG (Arruda et al. 2011); 17 = Serra do Cipó (Coelho et al. 2012).

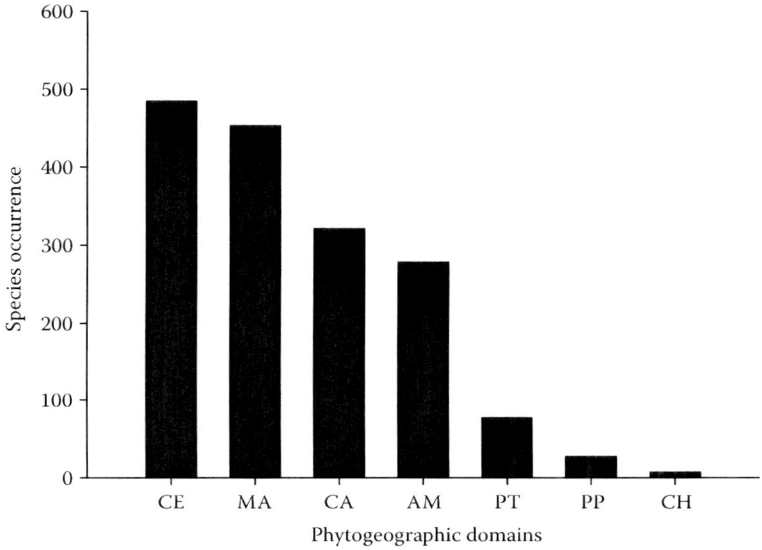

FIGURE 4.3
Brazilian phytogeographic domains of the species recorded by 14 studies carried out on TDFs on basalt and limestone outcrops from Brazil. Acronyms: CE, cerrado; MA, Atlantic rain forest; CA, caatinga; AM, Amazon; PT, pantanal; PP, pampa; and CH, Chaco.

In spite of the similarities, plant structure and composition varied enormously among the TDFBL fragments. The variation in structural aspects may reflect the many environmental factors acting on the widespread dry forest. The most important variables related to the structural aspects of vegetation are soil as well as topographical characteristics. TDFBL outcrops comprise a mosaic of different soils. Pedersoli and Martins (1972) provided the first information about the soil of TDFs on limestone outcrops. These authors classified five different soil types on which this dry forest was found: (1) calcareous and fertile soil requested for the development of luxurious vegetation (they highlighted that this soil is excellent for agriculture); (2) clay soil with some variations in the levels of silica, ferric oxide, and aluminum; (3) siliceous soil with a predominance of quartzite and sandstone but poor in organic elements; (4) red soil resulting from the decomposition of clay leaflets; and (5) limestone outcrops. Meguro et al. (2007) overlapped the distribution maps of carbonate karst regions and TDFs in Brazil and confirmed that most of the TDFs are associated with carbonate karst regions. Analyzing the soils of a TDF on limestone outcrops in Santo Hipólio–MG, Meguro et al. (2007) found a soil with the following characteristics: clayey, acid, pH between 6.0 and 6.5, with a good cation exchange capacity, high sum of bases, and high saturation degree. It is a eutrophic soil with very different chemical characteristics from the soils of lithic quartzitic areas that characterize the Espinhaço Range (see the study by Meguro et al. [2007] for granulometric composition and chemical properties).

In the case of basalt outcrops, which are restricted to an area between western MG and southern Goiás, the soils are eutrophic (base saturation >50%) and rich in P and K in addition to having high levels of Ca and Mg (Oliveira-Filho et al. 1998). Using the classification of the Soil Taxonomy System (Soil Survey Staff 1975), Oliveira-Filho et al. (1998) classified the soils into four categories: Hapludoll, Haplustoll, Ustropept, and Rhodustalf. Hapludolls and Haplustolls are Mollisol soils characterized by a mollic epipedon, a thick and dark surface horizon, rich organic matter, and moderate and high pH and base saturation. Ustropepts are Inceptisol soils or, in other words, young soils without a horizon definition. Rhodustalfs are Alfisol soils or, in other words, developed soils with an argilic horizon and moderate to high pH and base saturation (>35%) (see the study by Oliveira-Filho et al. [1998] for their granulometric composition and chemical properties). Therefore, species from TDFBL outcrops need to be adapted to fertile but simultaneously clayed, shallow, and stony soils with high runoff levels. The plants need to face high levels of lime and water stress. Some species can tolerate high lime levels in their tissues, which is an important advantage over other potential competitors. Other species can maintain the homeostasis of their metabolism through the process of exclusion of lime excess or the precipitation of oxalates and carbonates. Hence, species have different tolerances to the basic soil and to the presence of ions like HCO_3^- and their impacts on other important elements like P, manganese, and iron (see the study by Meguro et al. [2007]). However, almost all species shed their leaves throughout the dry season. Among the species recorded, the ones strictly related to TDFs on limestone formations as well as the ones with high IVs in many TDFs (e.g., *A. colubrina* and *M. urundeuva*) probably have special adaptations that provide them competitive advantages over others. Some tree species (*A. colubrina* and *M. urundeuva*) were considered emergent trees by some studies (Oliveira-Filho et al. 1998; Coelho et al. 2012). However, other variables can directly affect the composition of such forests, which was tested by Oliveira-Filho et al. (1998). Dry forests are more limited by physical constraints like water availability, topography, and soil physical features than soil chemical quality (Medina 1995). However, by considering topography, soil characteristics, and luminosity as explanatory variables for tree distribution at the same model in a TDF on basalt outcrops, Oliveira-Filho et al. (1998) found the strongest explanation given by luminosity (canopy gaps). This result was not expected considering the natural light penetration in TDFs in comparison with wet forests, even in the wet season, which is mostly the consequence of the richness of Fabaceae in the former. However, these results indicate a strong influence on the species distribution, mainly those species from early and late succession stages (Kalácska et al. 2004; Madeira et al. 2009; Coelho et al. 2012) (see the study by Oliveira-Filho et al. [1998] for more details).

In addition to the idiosyncratic patterns of each STFBL related to soil quality, topography, and light (canopy gaps), some studies have detected a low similarity among TDFBL fragments (Pedralli 1997; Pennington et al. 2000, 2006; Silva and Scariot 2003; Scariot and Sevilha 2005; Oliveira-Filho et al. 2006; Meguro et al. 2007; Arruda et al. 2011; Coelho et al. 2012) (see also the study by Pérez-Garcia et al. [2009]). Besides the reasons already mentioned, STFBL is strongly influenced by the surrounding environment. This influence appears clearer when we analyze the domains where the species come from (Figure 4.3). Most of the species come from cerrado, followed by the Atlantic rain forest, caatinga, Amazon, pantanal, pampa, and Chaco. As most of the sampling was conducted on the states of MG and Goiás, that is, inside cerrado's domain, it is expected that most species occur on other physiognomies of cerrado as well, followed by the geographically closer domains. Notwithstanding this, the presence of some witnesses species (e.g., *A. colubrina*, *Schinopsis brasiliensis*, and *Piptadenia viridiflora*) dates back to the Quaternary period, when dry and cool climates were responsible for contractions of rain forests and expansions of dry forests (see the studies by Prado and Gibbs [1993]; Oliveira-Filho et al. [2006]; Pennington et al. [2006]). For these reasons and for phenological similarities, many authors classify TDFBL outcrops as an extension of the caatinga domain, which approximate the caatinga with the Chaco (Andrade-Lima 1981; Pennington et al. 2000; Felfili 2003; Almeida and Machado 2007; Felfili et al. 2007). On the other hand, Fernandes and Bezerra (1990) pointed at a high dissimilarity among the deciduous dry forest enclaves in cerrado, caatinga, and the TDFs from Chaco. This high dissimilarity was also corroborated by our synthesis. As expected, the effect of TDFBL area reduction over time was responsible for the stronger influence of the surrounding vegetation. While fire is a recurrent phenomenon in the surrounding cerrado (see the study by Mistry [1998]), TDFBL species are not adapted to fire. Considering the isolated and small dimensions of TDFBL outcrops, fire can be an important factor of local extinctions and community relaxation, allowing the establishment of cerrado's species adapted to the presence of fire (Silva and Scariot 2003). This could justify the high TDFBL beta diversity and the significant presence of cerrado's species (Pérez-García et al. 2009). As the fragments are isolated on soil patches scattered throughout the Brazilian territory hosting small populations, reproductions among isolated populations from different fragments must keep the genetic variability of species under a local and regional scale (Caetano et al. 2008; Pennington et al. 2009).

4.3 Reproductive Patterns and Phenology

Despite the importance of reproductive patterns and phenology in isolated forest fragments, to our knowledge only two studies were conducted on TDFBL outcrops (Ragusa-Neto and Silva 2007; Coelho et al. 2013). Before conducting these studies, descriptions about the vegetation phenology of these forests were made by general observations based solely on the species present and from extrapolations from other studies conducted in the caatinga (Machado and Barros 1997; Machado and Lopes 2004). Some studies described the intensity of leaf fall and related it to the intensity of the dry season and the soil water retention capacity (Rizzini 1979; Andrade-Lima 1981; Pedralli 1997). Some of the results reported by Coelho et al. (2013) and Ragusa-Neto and Silva et al. (2007) are presented here.

Melittophily, phalaenophily, entomophily (not specialized insects), and chiropterophily were the prevailing pollination syndromes for most limestone TDF species. Entomophily and chiropterophily predominated in the studied TDFBL outcrops in terms of plant abundance. The pattern was consistent and hence did not change when the number of individuals was considered. The predominance of insect pollination syndrome is a common pattern in TDFs (see the studies by Machado and Lopes [2004] and Pezzini et al. [2008]). Other studies on TDFs in the cerrado and restinga (coastal dune formation) have also shown a predominance of pollination by insects (Silberbauer-Gottsberger and Gottsberger 1988; Ormond et al. 1993; Oliveira and Gibbs 2000). Bat pollination was higher in TDFs than that recorded for most studies conducted on other ecosystems such as the tropical rain forests, cerrado, and restinga. These results corroborate similar studies conducted on South American TDFs (Machado and Lopes 2004; Pezzini et al. 2008), which found a similar number of bat-pollinated plant species (Coelho et al. 2013).

Most species are animal dispersed followed by wind dispersion, as seen in the Brazilian cerrado (Almeida and Machado 2007; Meguro et al. 2007; Coelho et al. 2013). When the number of individuals is analyzed, the pattern is reversed. This reversion may be caused by a high abundance of typical TDF species, such as *A. colubrina* (Fabaceae) and *M. urundeuva* (Anarcadiaceae).

The vegetation phenology responded to climatic conditions since leaf production and fall were synchronized for all species. These responses to exogenous conditions are recurrent in the literature on plant phenology, especially in TDFs, which are characterized by a strong seasonality (Lieberman 1982). Unlike some authors, Rizzini (1979), Andrade-Lima (1981), Pedralli (1997), and Coelho et al. (2013) recorded complete deciduousness at a community level.

The reproductive phenophases were predominantly found throughout the wet season, although some isolated events were recorded in the dry season also. The general pattern for reproductive phenology is in accordance with other studies conducted in TDFs, in which forests with a predominance of zoochoric species concentrate their reproductive activities throughout the wet season as a strategy to maximize their fitness with the higher abundance of vectors (Ragusa-Neto and Silva 2007; Coelho et al. 2013).

4.4 Conservation Status and Natural Regeneration

Considering the wide distribution of TDFBL outcrops throughout the Brazilian territory (Meguro et al. 2007) and the different economic activities developed according to the region, these forests are under different anthropogenic pressures. As there is no specific study focusing on the conservation aspects of TDFBL outcrops, we are left with simple extrapolations from studies on TDFs on flat soils. As TDFBL outcrops and TDFs on flat soils co-occur in the same region, they might suffer similar anthropogenic pressures (see the study by Espírito Santo et al. [2009] for examples in northern MG and the Paranã River basin). However, we cannot rule out specificities exclusive of TDFBL outcrops. Similar to the TDFs on flat soils, TDFBL outcrops have suffered the impacts of the economic activities developed in the surrounding region. Hence, the main threats for these forests are cattle farms, charcoal production farms, logging, and fires (Werneck et al. 2000; Vieira and Scariot 2006; Vieira et al. 2006; Espírito-Santo et al. 2009). However, the impact imposed by grazing is smaller given the topography of TDFBL outcrops is not appropriate for this economic activity (Silva and Scariot 2004). On the other hand, TDFs on limestone outcrops are strongly explored by the cement industry. Such an economic activity is an irreversible activity that leaves a lifeless crater after resource depletion. When such extreme activities are conducted on TDFBL outcrops, it is almost impossible to implement a restoration of the forest considering that the substratum has been destroyed. However, when the human impact is not too extreme, what remains are fragments under different levels of succession. Surprisingly, we are not aware of any study conducted or attention given by the Brazilian authorities on the conservation of such endangered ecosystems.

Only three studies have been dedicated to analyze the ecological aspects of different succession stages in TDFBL outcrops (Werneck et al. 2000; Coelho et al. 2012, 2013) and the effect of surrounding activities on the edges of forests (Sampaio and Scariot 2011). Sampaio and Scariot (2011) conducted a study in a TDF on flat soil but argued that their results could be extrapolated to other TDFs facing high floristic similarities among different fragments at the region studied, including TDFs on limestone outcrops (São Domingos–Goiás). Coelho et al. (2012, 2013) focused on questions related to the structure, composition, and phenology of a TDF on limestone outcrops at two successional stages (intermediate and late), whereas Werneck et al. (2000) analyzed the floristic and structural changes during four consecutive years in a secondary TDF on basalt outcrops. Logging, fire, and cattle farms have impacted all forests. Despite the authors' recognition of the limitations of the results due to the impossibility of isolating the effects of topographic differences, some interesting trends were found (Werneck et al. 2000; Coelho et al. 2012, 2013). Edge effect is an important phenomenon in TDFs and was highly studied in wet forests (Ries et al. 2004). The edge effect can be defined as microclimatic (wind speed, temperature, moisture, and luminosity) and soil (physical and chemical) changes with consequences on plant mortality affecting the structure and composition of fragment edges (Tabarelli et al. 2004). However, even during the full leaf-growing season, TDFs have a discontinuous canopy with moderate light permittivity and additionally, as a consequence of the difference in richness, diversity, and structure between dry and wet forests, a negligible edge effect is expected on TDFs. As expected, Sampaio and Scariot (2011) did not find significant edge effects, with exceptions in seedling diversity and tree height, but the effect was negligible. If we consider the effects of gaps, topography, and soil quality on the distribution of trees in TDFBL outcrops, the conclusions found by the authors become even stronger (Oliveira et al. 1998).

Some studies have addressed the effects of succession on the structure, composition, and phenology of TDFBL outcrops' remnant fragments (Werneck et al. 2000; Coelho et al. 2012). When measuring structure parameters in a TDF on a limestone outcrop, Coelho et al. (2012) found greater values for basal area, density, tree height, and community indexes at the late succession stage in comparison with the intermediate stage of succession. Applying agglomerative hierarchical cluster analysis based on a matrix of Bray–Curtis coefficients, a significant difference in floristic composition between these two succession stages was detected, indicating that the successional process in these forests corresponds to the "relay floristic model" (Egler 1954). According to this model, TDFs would regenerate with a gradual substitution of pioneer species by late species. This pattern contradicts some recent studies in Brazilian TDFs on plain soils, which described the frequent occurrence of coppicing during succession in these ecosystems (but see the studies by Madeira et al. [2009]). Vieira et al. (2006), Vieira and Scariot (2006), and Sampaio et al. (2007) suggested that TDFs would conform to the "initial floristic composition model" (Egler 1954). In this case, late species would resprout after clear-cutting in TDFs and would be already present in early and intermediate successional stages. Hence, due to the presence of many species in TDFs that have the ability to resprout (e.g., *Astronium fraxinifolium*, *M. urundeuva*, and *Tabebuia impetiginosa*), the successional process will depend on the kind of impact that the forest has suffered as well as its intensity (Vieira and Scariot 2008). Reinforcing this argument, Werneck et al. (2000) have recorded the maintenance of community indexes (biodiversity and equability), an increase in basal area, a decrease in density, and the loss of pioneer tree species during succession.

On the same system, Coelho et al. (2013) did not record significant differences in successional stages for the following phenophases: green leaves, dead leaves, flower buds, flowers, and unripe fruits. On the other hand, the frequency of ripe fruits differed between successional stages. The absence of some functional differences among successional stages was found in other studies that focused on the structure and composition as well as the phenological features of TDFs (Kalacska et al. 2004; Madeira et al. 2009; Powers, Becknell et al. 2009). Pezzini et al. (2008) found significant differences in the reproductive phenology between initial and intermediate/late stages in a TDF on flat soils, but differences between intermediate and late stages of succession were not detected. It indicates that at the intermediate stage of succession, important ecosystem functions were already established. This study reinforced the importance of secondary forests for the maintenance of some ecosystem functions and services in TDFs.

Many studies have focused on the restoration principles of TDFs in Brazil (Vieira et al. 2006, 2008; Vieira and Scariot 2006, 2008; Sampaio et al. 2007). In spite of the studies being conducted on TDFs on flat soils, many resulting principles and conclusions can be applied to TDFs on limestone outcrops due to the co-occurrence of some species. However, these extrapolations must be applied with caution and consider the floristic, structural, and ecological process differences among TDFs. When the impact does not completely destroy a soil (e.g., mining), some restoration strategies can be applied in these forests. With the goal to recover and turn the restoration process faster, some principles discussed by Vieira and Scariot (2006) must be briefly considered. Resprouting after disturbances is probably the most important factor for the recovery of TDFs in Brazil (Vieira et al. 2006; Sampaio et al. 2007). However, other regeneration processes have been recorded as well. As mentioned earlier, Madeira et al. (2009) recorded a process based on the replacement of pioneer trees by trees from intermediate and late succession stages. Corroborating their results, Coelho et al. (2013) recorded a similar process in TDFs on limestone outcrops (Murphy and Lugo 1986b). In the case of TDFBL outcrops, shallow soils can expose tree roots to environmental elements (e.g., animal trampling, dissection, and fire), decreasing the resprouting capacity of some species that otherwise could resprout and recover the forest after a disturbance (Oliveira-Filho et al. 1998; Meguro et al. 2007). In addition to the main threats already mentioned, and as a consequence of the majority of the Brazilian TDFBL outcrops being surrounded by cerrado, which consists of vegetation adapted to the frequent presence of fire, this factor is considered as one of the most worrying threats to these ecosystems. The TDFBL outcrops are not adapted to fire, which can be an important factor of local extinctions in their isolated and small dimensions fragments. (Silva and Scariot 2003). This concern is augmented when we consider the fact that the Brazilian authorities and laws are exceedingly permissive, allowing the deforestation of the cerrado.

4.5 Final Remarks

Despite the similarities regarding phenology and topography, we have found important differences in structure and composition among Brazilian TDFBL outcrops according to the vegetation surrounding the outcrops. As TDFBL outcrops are located across different Brazilian biomes, submitted to different economic activities, under different conservation threats, and under different regional legislations, such vegetation formation must be a matter of much concern and be investigated in detail. TDFs on flat soils are considered part of the caatinga domain by some authors and part of the Atlantic rain forest domain by Brazilian legislation, thereby resulting in relevant conservation effects (Andrade-Lima 1981; IBGE 2008). However, this discussion is not clear yet in the case of TDFBL outcrops. According to our synthesis, TDFBL outcrops have a strong affinity to the surrounding cerrado domain. However, there are no specific legislations today to regulate the use as well as the protection of this important ecosystem. The wide distribution of TDFBL outcrops across the Brazilian territory and the proximity of different phytogeographic domains make the development of specific conservation policies difficult. More attention should be focused on TDFBL outcrops to enhance scientific knowledge, which is the most important requirement in the development of effective conservation strategies.

5

Tropical Dry Forests of Northern Minas Gerais, Brazil: Diversity, Conservation Status, and Natural Regeneration

Mário Marcos do Espírito-Santo, Lemuel Olívio-Leite, Frederico de Siqueira Neves, Yule Roberta Ferreira-Nunes, Magno Augusto Zazá-Borges, Luiz Alberto Dolabela-Falcão, Flávia Fonseca-Pezzini, Ricardo Louro-Berbara, Henrique Maia-Valério, Geraldo Wilson Fernandes, Manoel Reinaldo-Leite, Carlos Magno Santos-Clemente, and Marcos Esdras-Leite

CONTENTS

5.1 Introduction

Seasonally dry tropical forests (SDTFs) occupy 27,367,815 ha in Brazil, representing 3.21% of the country's territory (Sevilha et al. 2004). The delimitation of SDTFs used in this chapter follows that used by the Brazilian Institute for Geography and Statistics (IBGE 1992) and does not consider the entire caatinga biome as SDTFs, because most of the caatinga vegetation does not possess a forest structure. The Brazilian SDTFs have a naturally scattered distribution, occurring in several biomes but predominantly in the semiarid northeastern region (see Chapter 1, Figure 1.3). These ecosystems usually develop on nutrient-rich soils targeted for farming purposes, and the presence of valuable timber trees increases the logging pressure for SDTF deforestation (Espírito-Santo et al. 2009). They can also occur on limestone outcrops where mining activities pose severe threats to their integrity. Since 1993, Brazilian SDTFs have been considered to be one of the phytophysiognomies of the Atlantic rain forest biome (Federal Decree 750, 1993), a condition that was confirmed recently by the Federal Law 11428 (2006). For this reason, these forests are currently fully protected from deforestation (Espírito-Santo et al. 2011), provoking strong pressures from rural sectors and creating a long-standing dispute over SDTF classification and land use.

The region where such disputes are more exacerbated is the north of Minas Gerais state, in the southernmost distribution of Brazilian SDTFs (see Chapter 1, Figure 1.3), in a transition area between two large Brazilian biomes, the cerrado and the caatinga. This region concentrates 78% of the state's SDTFs and is economically poor, with social indicators (such as the human development index) standing below the state and country averages (Espírito-Santo et al. 2009). In this context, rural sectors claim that prohibiting SDTF deforestation will hinder economic development and increase poverty (Espírito-Santo et al. 2011). However, more than 50% of the original SDTF cover of the north of Minas Gerais is already lost (Scolforo and Carvalho 2006) and large areas consist of abandoned and/or degraded pasturelands (IBGE 2006). Extensive cattle raising is the main economic activity in this region, occupying 59% of the rural areas, usually according to the following land-use cycle: (1) the forest is completely cut for wood or charcoal and then the area is burned; (2) exotic grasses are planted for pasture formation; (3) extensive cattle ranching is started, with a regional density of one head per hectare; (4) pastures are annually managed with fire before the onset of the rainy season; and (5) soil fertility is exhausted, the pasture is abandoned, and a new area is cleared (Espírito-Santo et al. 2009).

We used satellite imagery (thematic mapper [TM] Landsat 5) analyses and geographic information system techniques to map vegetation cover in the entire area to the north of the Minas Gerais region, with an area of approximately 128,000 km^2 (see the study by Leite et al. [2011] for details). This process was conducted using images obtained for 1986, 1996, and 2006, and we compared SDTF cover and deforestation rates between 1986–1996 and 1996–2006. The current extent of SDTFs in this region is 17,000 km^2, and 2,200 km^2 (11.5%) were lost in 20 years (1986–2006) with an average of 0.6%/year, which is very similar to the values described by Miles et al. (2006) for SDTFs worldwide. Deforestation was more intense between 1986 and 1996, when 1630 km^2 were lost, compared to 1996–2006 (887 km^2). This temporal contrast in deforestation rates was caused by two different governmental policies: (1) tax incentives to economic development for the north of Minas Gerais between 1970 and 1990, including the establishment of the largest irrigated perimeter in Latin America (the Jaíba project, with 100,000 ha), which was fully installed in an area originally covered by SDTFs; (2) the inclusion of SDTFs in the protective "umbrella" of the Atlantic rain forest in 1993. Either way, legal and illegal deforestation rates in this region have been very high in the past decades, threatening with extinction many species that are not even known in this poorly studied ecosystem.

During the same period, SDTF natural regeneration also occurred in the north of Minas Gerais. In total, 550 km^2 that were under human use in 1986 were covered by SDTFs in different successional stages in 2006. Understanding forest recovery after anthropogenic disturbances is very important to support conservation programs based in the restoration of ecosystem functions (Guariguata and Ostertag 2001; Lebrija-Trejos et al. 2010b; Griscom and Ashton 2011). Current Brazilian environmental laws state that illegally cleared areas must be reforested, but techniques to scientifically orient this process are scarce, especially in SDTFs (Sánchez-Azofeifa et al. 2005a; Vieira and Scariot 2006; Sampaio et al. 2007; Quesada et al. 2009). It has been argued that natural regeneration is faster in SDTFs than in tropical wet forests mainly due to the higher frequency of coppicing in the former (Murphy and Lugo 1986a; Sampaio et al. 2007; Levésque et al. 2011). However, successional pathways depend on the type and intensity of previous land use (Madeira et al. 2009; Lebrija-Trejos et al. 2010b; Griscom and Ashton 2011) and also on the capacity of faunal recolonization and recovery of key ecological interactions such as pollination, seed dispersal, herbivory, and plant–fungi symbiotic relationships, among many others (Holl 1999; Avila-Cabadilla et al. 2009; Quesada et al. 2009; Neves et al. 2010a,b; Silva et al. 2012; Zangaro et al. 2012).

Since 2006, we have been investigating natural regeneration in SDTFs to the north of Minas Gerais. During this study, we also conducted a comprehensive survey of several groups of organisms, aiming to fill the gaps in biodiversity knowledge for this region, considered to be extremely important for conservation and determining priority for scientific investigation (Drummond et al. 2005). We used a chronosequence design to examine how forest structure and plant, animal, and arbuscular mycorrhizal fungi (AMF) communities change along a successional gradient. We discuss how such changes can affect ecosystem functioning and what the practical applications of this knowledge are for deforestation monitoring and law enforcement, elaboration of restoration and management strategies, and environmental policies.

5.2 Materials and Methods

5.2.1 Sampling Design

We used three areas at different stages of natural regeneration, which were selected by Madeira et al. (2009) based on the vertical (number of strata) and horizontal (tree density) structural characteristics of the forest. Time since the last disturbance and previous land use were determined through interviews with former farm managers and local dwellers. Thus, the early stage is mainly composed of sparse patches of woody vegetation, shrubs, and herbs, with a single vertical stratum formed by a discontinuous canopy approximately 4 m in height. This area was used as pasture for at least 20 years and was abandoned in 2000 (12 years old) with the creation of the Mata Seca State Park (MSSP). The intermediate stage has two vertical strata. The first stratum is composed of fast-growing trees, 10–12 m in height, forming a closed canopy, with a few emergent trees up to 15 m in height. The second stratum is composed of a dense understory with many lianas and juvenile trees. This area was used as pasture for an indefinite period and is under regeneration since the mid-1960s (45–50 years old). Pastures where both early and intermediate successional forest fragments now occur were managed similarly: after clear-cutting, the area was ploughed to plant exotic grasses and burned every two years right before the rainy season. The late stage also has two vertical strata. The first stratum is composed of tall trees, forming a closed canopy 18–20 m in height. The second stratum is a sparse understory with low light penetration and a low density of juvenile trees and lianas. The late stage is reported by a former farm manager to have the same structure since the 1970s and is probably a primary forest impacted by occasional selective logging (wood for fence construction). Free-ranging cattle at low density occurred in all stages until 2009.

Six fenced plots of 50 m × 20 m were delimited in January 2006 at early and intermediate stages, whereas eight plots were delimited in the late forest fragments, totaling 20 plots and 2 ha. All forest fragments were located under similar topographic, soil, and microclimatic characteristics, thus reducing the variation in physical conditions that could affect succession. The 20 plots were located along a 5 km transect encompassing these fragments, between 14°50′–14°51′ S and 43°57′–44°00′ W, inside the original area of a single farm. Plots from the same successional stage were located approximately 0.2–1.0 km from each other.

5.2.2 Forest Structure and Composition

In January 2006, we identified and measured the diameter at breast height (DBH) of all living trees with a DBH equal to or greater than 5 cm inside all plots. We also visually estimated the height of these individuals in each plot using a 2 m graduate stick as reference. Moreover, all independently growing liana stems with a DBH equal to or greater than 2 cm had their DBH measured and their height was estimated as the height of their host tree. Voucher specimens were deposited at the Montes Claros Herbarium at the State University of Montes Claros in Montes Claros, Brazil.

We computed the Holdridge complexity index (HCI) for the tree component only (Holdridge 1967; Holdridge et al. 1971) as a measure of community complexity. This index was calculated by using the following equation:

changed HCI = (height × density of stems × basal area × number of species)/1000. The original HCI considers only trees with DBH > 10 cm. Thus, we used a modified version of the index since we sampled trees with DBH ≥ 5 cm (Lugo et al. 1978). All structural variables were compared among stages using generalized linear models (GLMs) (Crawley 2002).

5.2.3 Soil and Arbuscular Mycorrhizal Fungi Sampling

Soil samples (0–10 and 10–20 cm depth) were collected during the wet season (February 2008) in each of the 20 plots (five replications per plot, totaling 1500 g); each was collected with five replications. Any individual sample (from each area and depth) was composed of nine subsamples of 0–10 and 10–20 cm depth. Soil physical and chemical traits were determined using standardized protocols (Embrapa 1979). To determine AMF abundance and richness, AMF spores produced in the trap cultures were extracted by wet sieving and sucrose density gradient centrifugation (Daniels and Skipper 1982). Quantification of spore number was carried out according to the studies by Lugo and Cabello (2002) and Schalamuk et al. (2006). Taxonomic identification was made at the Microbiology Laboratory-Estación Experimental Agropecuaria INTA, Balcarce, Argentina. About 70% of the spores fixed on slides could be identified at the genus level. Spore identification was based on current species descriptions and identification manuals (Schenck and Pérez 1990; INVAM 2012; Schüßler and Walker 2010). For further details on AMF extraction methods, see the study by Yang et al. (2010).

5.2.4 Faunal Sampling

We sampled five groups of insects and two groups of vertebrates considered to be good indicators of environmental disturbance (Leite et al. 2008; Neves et al. 2008, 2010a,b; Avila-Cabadilla et al. 2009) to assess the regeneration of SDTF fauna along succession. Among the insects, we sampled herbivorous insects, frugivorous butterflies, ants, dung beetles (see methodological details in Chapter 12 and the studies by Neves et al. [2010a,b]), and mosquitoes. Mosquitoes from the Culicidae family were sampled three times during dry seasons (August 2008, October 2010, and September 2011) and three times during wet seasons (March and December 2009 and January 2011). Sampling was conducted with modified Shannon traps for three hours at dusk (6 to 9 PM). These traps consist of a cloth hanging vertically from the top and two lateral compartments. The light of a lantern and the presence of a human collector serve as bait. The insects present inside the trap were captured using electric aspirators. One trap was installed at each plot, totaling three traps (nine hours per sampling period), and there were six temporal replications, totaling 54 hours of sampling effort. Among vertebrates, birds and bats were sampled using 15 and 10 mist nets per night

(12 m long and 2.6 m high), respectively. The overall sampling effort was 269,568 $h \cdot m^2$ for birds and 149,760 $h \cdot m^2$ for bats.

5.2.5 Data Analyses

All soil physical and chemical traits, species richness, and total abundance of each group of organisms were compared among successional stages using GLMs (Crawley 2002). Minimal adequate models were constructed with the removal of the nonsignificant terms of the full models. Models with statistically significant differences were subjected to contrast analyses, with the junction of nonsignificant categorical groups (amalgamation). The adjusted models were submitted to a residual analysis to evaluate the adequacy of the error distribution. Model construction and all analyses were conducted using the software R 2.6.2 (The R Development Core Team 2009). We also used principal components analysis (PCA) to determine whether the sampled plots were grouped according to the soil variables.

5.3 Results and Discussion

5.3.1 Soil Composition

We detected significant differences in soil composition among the three successional stages (Table 5.1), especially between the intermediate and the early/late plots. In general, all the forest stages at the MSSP have acid soils, within the range (pH from 5 to 9) usually observed for mineral soils in arid and humid regions (Buckman and Brady 1976). The carbon (C)/nitrogen (N) relationship is considered to be ideal (which ranges from 8/1 to 15/1) but close to the lower threshold of this category. In fact, the C/N ratio tends to be low in semiarid regions due to the limited water availability (Buckman and Brady 1976) (Table 5.1). This trait did not differ among successional stages.

The soils from all three stages can be considered as fertile, although nutrient levels, bases saturation (V), and cation exchange capacity (CEC) are significantly lower in the intermediate stage. Granulometry was also different among the stages (Table 5.1), except for sand levels. Silt levels were higher in the late stage, and clay levels were higher in the intermediate stage. Overall, we did not find a fertility gradient that matches the chronosequence in the studied SDTF. Indeed, this type of forest can develop on 13 different soil types in Brazil (Scariot and Sevilha 2005), and variation in soil types is marked inside the MSSP (Berbara, unpublished data) and for the entire region (Espírito-Santo, personal observation). Thus, it is possible that the lower soil fertility observed in the intermediate stage is a natural

TABLE 5.1

Soil Physical and Chemical Traits and Granulometry in Three Successional Stages in the Mata Seca State Park, Southeastern Brazil

	Physical and Chemical Properties		
Trait	**Stage**		
	Early	**Intermediate**	**Late**
pH (H_2O)	6.80 ± 0.20[a]	5.30 ± 0.10[b]	6.63 ± 0.29[a]
C ($g \cdot kg^{-1}$)	16.0 ± 1.00[ab]	14.6 ± 2.08[b]	18.4 ± 0.49[a]
Organic matter	27.5 ± 1.66[ab]	25.1 ± 3.61[b]	31.8 ± 0.87[a]
N ($g \cdot kg^{-1}$)	1.97 ± 0.15[ab]	1.73 ± 0.12[b]	2.17 ± 0.12[a]
C/N	8.14 ± 0.62[a]	8.39 ± 0.67[a]	8.52 ± 0.22[a]
Ca^{2+}	8.43 ± 0.32[a]	2.67 ± 1.01[b]	7.70 ± 0.87[a]
Mg^{2+}	2.37 ± 0.67[a]	1.13 ± 0.06[b]	2.00 ± 0.26[ab]
K^+	0.37 ± 0.08[a]	0.16 ± 0.01[b]	0.32 ± 0.06[a]
Na^+	0.01 ± 0.01	0.01 ± 0.00	0.01 ± 0.00
H^+	1.77 ± 0.55[a]	4.33 ± 0.32[b]	2.57 ± 0.40[a]
Al^{3+}	0	0.3 ± 0.2	0
CEC[a]	13.0 ± 0.86[a]	8.61 ± 0.98[b]	12.6 ± 1.50[a]
V (%)[b]	86.5 ± 3.40[a]	45.7 ± 7.28[b]	79.7 ± 0.86[a]
P ($mg \cdot kg^{-1}$)	7.0 ± 2.0[a]	5.0 ± 0.0[a]	5.33 ± 1.53[a]
Granulometry (%)			
Sand	53.4 ± 4.8[a]	47.7 ± 2.2[a]	43.2 ± 10.0[a]
Silt	19.9 ± 1.7[ab]	13.6 ± 1.2[b]	22.8 ± 5.8[a]
Clay	26.6 ± 4.1[a]	38.6 ± 1.1[b]	34.0 ± 5.2[ab]

Note: Different letters after each mean ± standard deviation indicate statistically significant differences between successional stages.

[a] Cation exchange capacity.

[b] V (bases saturation) = (Ca^{2+} + Mg^{2+} + K^+ + Na^+)/CEC.

characteristic of the soil type where the forest developed. On the other hand, such marked differences from the other stages may be a consequence of a more intense previous land use, but no precise information is available for the intermediate forest stage before 1972.

Moreover, the resemblance in soil characteristics between the early and late stages suggests that these areas have a similar soil type, and small, nonsignificant differences in some traits may be related to distinct topography and/or successional status. Indeed, the successional patterns for soil characteristics are not clear and some studies report an increase in soil nutrient levels during forest natural regeneration (Silver et al. 1996), although some other studies report otherwise (Uhl and Jordan 1984). Such variations are mainly attributed to differences in sampling methods, ecosystem type, and distinct kinds and intensities of disturbance (Guariguata and Ostertag 2001).

The aforementioned results can be visualized in the diagram obtained by PCA. In general, it is possible to observe segregation among the three successional stages in relation to soil traits (Figure 5.1). The first axis explained 52.3% of the variation and the second one explained 21.4%, making 73.7% of the total variation. The first axis formed a fertility gradient increasing toward the left side, separating the three successional stages, with the intermediate stage plots clustered on the right side, associated with the variables aluminum, hydrogen, and clay. The early and late stage plots, on the other hand, were clustered on the left side of the first axis, associated with the other variables. In the second axis, the plots of early and late stages were mainly grouped according to the variables C/N, phosphorus, and sand. The early stage plots were clustered below the second axis, associated with higher values of sand and phosphorus.

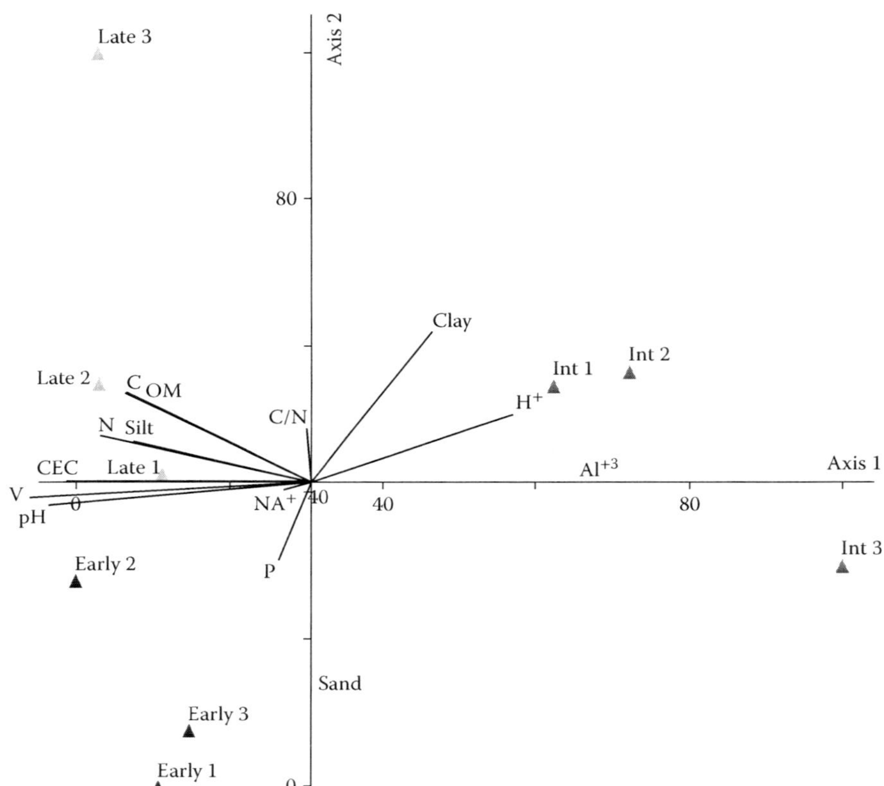

FIGURE 5.1
Ordination diagram of the principal component analysis (PCA) derived from soil chemical and physical variables to the early, intermediate (int), and late successional stages of a seasonally dry tropical forest (SDTF) in the Mata Seca State Park, Brazil. OM = organic matter.

5.3.2 Biodiversity Patterns

The studies conducted by the collaborative research network Tropi-Dry in the north of Minas Gerais comprise one of the few systematic, long-term biodiversity surveys for SDTFs in Brazil and are probably the first to investigate multiple groups of organisms ranging from fungi to plants, invertebrates, and vertebrates. The knowledge on Brazilian biodiversity is still poor, especially for microorganisms and terrestrial arthropods (Lewinsohn and Prado 2002). Brazil is considered a megadiverse country, and the estimated amount of species yet to be described in the region is enormous (Lewinsohn and Prado 2005). Furthermore, Brazilian SDTFs are frequently neglected in terms of public attention, research, and conservation efforts compared with moist ecosystems such as the Coastal Rain Forest and the Amazon (Espírito-Santo et al. 2006; Santos et al. 2011a). When determining priority areas for conservation in the Minas Gerais state (Drummond et al. 2005), the northern SDTFs were considered to be a "special" area, which is the highest category. The same study also included SDTFs as the top priority for scientific investigation, highlighting the relevance of our studies to conservation strategies at the state and national levels.

Practically all the sampled organisms were already identified at the species level with the exception of herbivorous insects, which were identified mostly at the family level. This group is composed of several different orders including leaf chewers and sap suckers; thus, a higher number of taxonomists must be involved in insect identification and many species are yet to be described. Considering all groups, we produced species lists that sum 621 species (see references in Table 5.2). For most groups, the recorded species richness is within the range observed for studies using similar sampling methods in SDTFs in the Neotropical region (e.g., studies by Pinheiro and Ortiz [1992]; Allen et al. [1998a]; Souza et al. [2003]; Kalácska et al. [2004b]; and Avila-Cabadilla et al. [2009]). The species encountered in our survey are typically associated with the cerrado and caatinga biomes, since the study site is situated in the transition between them, and only a few species are considered endemic to SDTFs. In the case of the plant community, many species recorded in our study site are also found in the semideciduous and evergreen Atlantic rain forest on the coast (see the studies by Oliveira-Filho et al. [2006] and Santos et al. [2012]), giving scientific support to the legal inclusion of SDTFs as part of this biome.

5.3.3 Successional Patterns

We detected significant differences in forest structure along the studied successional gradient, with a huge increase in HCI from early (0.6 ± 0.5) to intermediate (15.0 ± 8.3) to late (46.1 ± 25.7) stages (Madeira et al. 2009). All structural variables showed a gradual, significant increase as succession advanced (Figure 5.2). Accumulated tree species richness showed the same

TABLE 5.2

Successional Trends for Accumulated Species Richness and Composition for 10 Groups of Organisms Collected in the SDTFs of the Mata Seca State Park, Brazil, Using the Same Chronosequence Sampling Design

Group	Successional Stage				Richness[a]	Composition
	Early	Intermediate	Late	Total		
Mycorrhiza[b]	12	11	7	17	$E > I = L$	N/A
Trees[c]	36	48	59	97	$E < I = L$	$E \neq I \neq L$
Lianas[d]	2	24	14	27	$E < I > L$	$E \neq I \neq L$
Butterflies[e]	16	24	30	35	$E = I < L$	$E = I \neq L$
Herbivorous insects[f]	56	68	76	157	$E < I = L$	$E \neq I = L$
Ants[g]	60	72	75	95	$E < I < L$	$E \neq I = L$
Dung beetles[h]	13	32	27	38	$E < I > L$	$E \neq I \neq L$
Mosquitoes[i]	23	21	20	30	$E = I = L$	N/A
Birds[j]	73	70	50	100	$E < I > L$	$E \neq I = L$
Bats[k]	20	19	19	25	$E = I < L$	$E = I = L$
Total	**311**	**389**	**377**	**621**		

Note: See references for detailed statistical analyses of each group; for unpublished data, we used GLMs to compare average species richness per plot and nonparametric multidimensional scaling to compare species composition between stages. E, I, and L refer to early, intermediate, and late stages.

[a] Average species richness per plot.
[b] Santos (2010); the Shannon–Wiener diversity index was used for statistical comparisons.
[c] Madeira et al. (2009).
[d] Unpublished data.
[e] Madeira (2008).
[f] Neves (2009); only adult insects were considered.
[g] Marques (2011).
[h] Neves et al. (2010a).
[i] Santos (2011).
[j] Dornelas et al. (2012).
[k] Falcão (2010).

pattern (Table 5.2), but average species richness per plot differed between early and intermediate/late stages. No difference in average species richness was found between intermediate and late stages (see the study by Madeira et al. [2009]) (Table 5.2). Tree species composition differed significantly among the three stages and the early stage was completely dissimilar to the intermediate and late stages, which shared many of their most important species. In spite of this, tree functional groups differed among the three stages, which certainly indicates ecosystem function regeneration along succession (see Chapter 18). Considering that soil characteristics are more similar for plots at early and late stages, it seems that plot age is more important in determining tree richness and composition than soil type. It is possible that

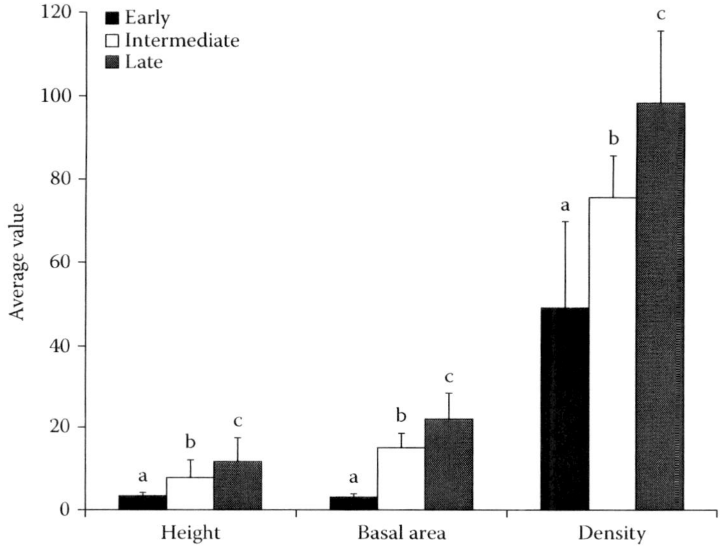

FIGURE 5.2
Forest structural characteristics (mean ± standard deviation [SD]) along a successional gradient in the Mata Seca State Park, Brazil: letters above the bars indicate statistically significant differences between means.

soil traits have a stronger effect on forest structure (which was very different between intermediate and late stages), but further studies considering forest stands from the same age growing on contrasting soil types are necessary to confirm this assertion.

Lianas were practically absent from early-stage plots (Table 5.2) (Madeira et al. 2009), probably because the trees at this stage are still short and do not provide adequate support for liana establishment. Contrary to the observed pattern of the tree community, all liana structural variables decreased from intermediate to late stages, and the same occurred for species richness (Table 5.2). This pattern is probably related to the fact that lianas are light demanding (Castellanos 1991; Teramura et al. 1991) and may not attain high diversity under the closed canopy of tall, late forests. On the other hand, the more open canopy of intermediate-stage forests usually constitutes a more suitable environment to liana colonization (Schnitzer 2005). Thus, lianas are a very important component of secondary dry forests, contributing substantially to wood area index in intermediate stages (Sánchez-Azofeifa et al. 2009b). Although some evidence indicates that they can arrest forest regeneration (Schnitzer and Bongers 2002), lianas also produce important food resources to herbivores, pollinators, and seed dispersers and positively affect animal diversity in disturbed areas.

The accumulated number of AMF species per stage showed a gradual decrease along the successional gradient, contrary to most of the other groups of organisms analyzed in the present study (Table 5.2). The early stage also had a higher AMF diversity index per plot compared with advanced stages (Shannon–Wiener index) (Santos 2010). AMF communities are key to plant establishment in harsh environments, since these organisms improve plant nutrient absorption and competitive ability (Jasper 1994; Yang et al. 2010). Thus, it is likely that AMF diversity is more important for pioneering, fast-growing plant species in early successional stages. Indeed, a recent study in three different Brazilian ecosystems showed that AMF colonization rates and spore density decreased with enhanced soil fertility and fine-root mass along a successional gradient (Zangaro et al. 2012). Thus, it is likely that AMF will show a negative correlation with plant species diversity along succession, but this relationship may be mediated by spatial variations in soil fertility, an issue that deserves further investigation.

We did not detect a clear, general successional pattern for animal richness in the studied chronosequence. In terms of total richness per stage (accumulated richness), butterflies, herbivorous insects, and ants showed a gradual increase from early to late stages and the opposite pattern was observed for mosquitoes and birds. For dung beetles species, richness was higher in intermediate stages, and bat richness remained practically unchanged along the successional gradient (Table 5.2). When we statistically compared average species richness per plot, successional trends were different: for herbivorous insects, the early stage showed a lower number of species than the intermediate/late stages, which was the same trend observed for trees. For butterflies and bats, richness was significantly higher in the late stage compared to both early and intermediate stages, which did not differ between each other. In the case of dung beetles and birds, richness was higher in the intermediate stage. A gradual increase in average richness from early to late stages was only observed for ants, and mosquito richness did not differ along the successional gradient.

Many successional models have been proposed mainly for plant communities (Egler 1954; Connell and Slatyer 1977; Connell 1978; Yodzis 1986; Tilman 1988), and animal diversity frequently tracks changes in forest structure and tree richness (Begon et al. 2006). In the present study, this was only the case for herbivorous insects, suggesting that other factors (e.g., diversity of other plant growth forms, such as herbs, epiphytes, and lianas; microclimatic factors; and soil traits) idiosyncratically affect the community regeneration pathway and speed for each animal group (see Chapters 11 and 12 for detailed discussions about insects and vertebrate patterns). Although there was little convergence in successional patterns of species richness between the animal groups analyzed here, a general trend can be depicted: early stages have a lower average richness than intermediate or late stages, clearly indicating that forest disturbance has a great impact on SDTF fauna and, consequently, on ecosystem functioning (seed dispersal, pollination, herbivory, nutrient cycling and forest productivity, etc.).

Although species richness is usually used as an indicator of ecosystem health (Costanza 1992), it is important to also analyze species composition and how functional diversity recovers along the successional process. In the present study, we assessed species composition and found differences among stages for all animal groups except bats (Table 5.2), especially between early and intermediate/late stages. Despite more detailed analyses of functional groups being necessary, species composition differed between only intermediate and late stages for butterflies and dung beetles. Given that species composition differed between these advanced stages for trees and lianas, this pattern suggest that most animal communities (both invertebrates and vertebrates) may recover faster than plant communities during SDTF succession. Although these results should be interpreted with caution, they highlight the importance of considering secondary tropical forests in biodiversity management and conservation strategies.

5.4 Conclusion

After six years of systematic studies, the Tropi-Dry collaborative research network made a huge contribution to the biodiversity knowledge of Brazilian SDTFs, including surveys and species lists for 10 different groups of organisms. Our results have a great impact on basic and applied research, since a more complete understanding of successional patterns is fundamental to support restoration strategies for degraded SDTF areas. The groups of organisms considered here are key to such endeavors, since they can be managed to improve forest productivity (AMF, lianas, herbivorous insects), nutrient cycling (AMF, ants, dung beetles), pollination (butterflies, birds, bats), seed dispersal (dung beetles, birds, bats), and biological control (birds and bats). In the case of mosquitoes, many species are important vectors of relevant diseases, such as yellow fever, dengue fever, and malaria. A better comprehension of how forest degradation and natural regeneration affect these organisms is extremely significant to public health programs and disease control.

Another important aspect of our studies is the possibility of using forest structural variables and plant and animal species compositions to differentiate among successional stages. Brazilian laws usually permit different land uses depending on the successional stage, being more restrictive for conversion as succession advances. Since Brazilian SDTFs are currently protected under the Atlantic Rain Forest Law, only early successional stages can be deforested, but environmental law-enforcement agencies frequently have difficulties in identifying these stages. Many of the groups we studied here can be surveyed with rapid assessment protocols and used as bioindicators of SDTF regeneration. In general, we believe Tropi-Dry helped by giving visibility to the neglected Brazilian SDTFs, which is important to sensitize public opinion and increase the government's efforts toward the sustainable use of this ecosystem.

6

A Review of Remote Sensing of Tropical Dry Forests

Michael Hesketh and Arturo Sánchez-Azofeifa

CONTENTS

6.1 Introduction

For more than 20 years, tropical dry forests (TDFs) have been recognized among the world's most threatened ecosystems (Murphy and Lugo 1986b; Janzen 1988a; Olson 2000). These forests account for 49% of the vegetated land cover in Mesoamerica and the Caribbean and 42% of all tropical forest vegetation worldwide (Murphy and Lugo 1995; Van Bloem et al. 2004) with a current estimated total global cover of 1,048,700 km^2 (Miles et al. 2006); they often are areas of intense human occupation and exploitation (Murphy and Lugo 1986a; Quesada and Stoner 2004; Sánchez-Azofeifa et al. 2005a). In spite of this, these forests have been the subject of only a fraction of the research devoted to tropical forests globally, with the majority of studies over the past 60 years having focused on tropical humid forests or rain forests (Sánchez-Azofeifa et al. 2005b). With increasing concern over the health and conservation status of TDFs (Stoner and Sánchez-Azofeifa 2009), there is a need for tools to better map and understand these important resources.

The benefit of satellite remote sensing to forest ecology is the potential for systematic, synoptic views of the earth at potentially large spatial scales and at regular intervals (Roughgarden et al. 1991; Nagendra 2001; Cohen and Goward 2004). Satellite data have been applied toward the mapping and monitoring of the distribution and change of plant species and land cover types, deforestation, fire and insect damage, and human impact on the environment. Additionally, these data have been used in extraction of biophysical characteristics (e.g., total aboveground biomass [TAGB], leaf area index [LAI] and woody area index, and canopy cover), which are key components in a variety of ecological models, as well as calculations of carbon balance and primary production (Lambin 1999; Nagendra 2001; Foody 2003; Kerr and Ostrovsky 2003; Castro-Esau et al. 2004). At smaller scales, the use of high-spectral-resolution data at the leaf and crown levels has allowed for the evaluation of the spectral elements of leaf properties and their variation between and among species, structural groups, locations, and seasons (Hesketh and Sánchez-Azofeifa 2012) in a manner that may be adapted to larger spatial scales and used to better understand ecosystems at the plot level and beyond (Asner and Martin 2008b).

The subject of this chapter is to explore the applications of remote sensing to forest studies and the use of these tools and techniques in better understanding TDFs. Following a discussion of the relationship between spectral reflectance and biophysical and structural properties of vegetation, the chapter reviews remote sensing challenges and findings in TDF research, which are broadly grouped under four main areas: (1) the use of satellite remote sensing data as an input to the classification of TDFs regionally and globally, with the aim of mapping their extent and distribution; (2) the estimation of TDF forest biomass and productivity; (3) the assessment of biodiversity through spectral evaluation of species and plant structural groups; and (4) the application of optical and light detection and ranging (LiDAR) remote sensing in the assessment of forest structure and successional stage.

6.2 Remote Sensing of Vegetation

6.2.1 Spectral Characteristics of Green Vegetation

Investigation into leaf optics dates back to the first half of the twentieth century (Shull 1929; Billings and Morris 1951; Gates et al. 1965; Loomis 1965), and a prime objective of leaf spectroscopy is to relate leaf optical properties to chemical and biophysical characteristics. Spectral reflectance at the leaf level is influenced by three characteristics: (1) the internal cellular structure of a leaf, (2) leaf pigment content, and (3) orientation relative to solar radiation (Turner et al. 1999). Although leaf morphology is highly variable by species and phenological stage, it tends to be characterized by a relatively open structure, with palisade and spongy mesophyll cells sandwiched between the upper and lower layers of the epidermis. The upper layer of the epidermis, called the cuticle, is a thin waxy coating that regulates the transmittance of radiation into a leaf's internal structure. Chloroplasts are found throughout the palisade and mesophyll cells, but tend to be concentrated toward the upper side of the leaf, provided that the leaf has a horizontal orientation relative to the sun (vertically oriented or erectophile leaves tend to have chloroplasts distributed along both edges [Jensen 2000]).

Radiation interacts with the leaf though absorption and scattering. The cell structures within the leaf are large with respect to the wavelengths of light, although the hairlike strands called grana within the chloroplasts are small enough (approximately 0.5 μm × 0.05 μm) to induce some scatter. Plants efficiently absorb the ultraviolet and visible wavelengths, although absorption dramatically decreases for the near- and mid-infrared parts of the spectrum (0.70–1.8 μm). The structural components influencing such decreases in the longer wavelengths are the large intercellular spaces in the spongy mesophyll cells (the site of oxygen [O_2] and carbon dioxide [CO_2] exchange), which result in high internal scattering of radiation in the near-infrared region, reducing absorption and increasing reflectance of these wavelengths. This lowered absorption in the higher energy part of the spectrum is a mechanism to control the thermal properties of the leaf, preventing overheating (Gamon et al. 2005). Water vapor saturating these intercellular spaces interacts with mid-infrared radiation, resulting in absorption peaks at 0.97, 1.19, 1.45, 1.94, and 2.7 μm and increased reflectance between them (Gates et al. 1965).

Leaf pigment content most directly affects the spectral response in the visible range of the spectrum (0.4–0.7 μm). Chlorophylls *a* and *b* absorb strongly in the blue and green wavelengths, but much less so in the green. The presence of other pigments within the leaf (carotenes and xanthophyll cycle pigments) with similar absorption characteristics broadens these absorption peaks. The ratio between this high absorption in the red region and high reflectance in the near-infrared region is exploited in a significant number of index-based approaches to vegetation monitoring (le Maire et al. 2004). As the leaf matures from initial flush to senescence the pigment levels shift, causing an alteration to the measured reflectance and the apparent color of the leaf as chlorophyll levels increase and decrease throughout the growing season, which are replaced by carotenes and anthrocyanin (Gates et al. 1965).

6.2.2 Satellite Analysis of Vegetation

Aside from the issues of appropriate spatial, spectral, and temporal resolutions associated with all satellite remote sensing analyses (Nagendra 2001), there are challenges associated with mapping and monitoring vegetation using airborne and satellite-derived spectral reflectance data. In particular, the potentially distorting effects of the atmosphere, topography, and canopy architecture and the influence of soil on spectral reflectance must be understood and accounted for as a preliminary to any analysis.

The effect of the atmosphere on reflectance is described as the difference between the actual top of copy reflectance and the measured top of atmosphere reflectance. For remotely sensed projects in which data are compared over either time or space, the effects of light scattering in the atmosphere due to dust and aerosols must be accounted for (Myneni et al. 1995b). Song et al. (2001) suggests several means of correction, which are summarized as follows: the first method is dark object subtraction (DOS), by which the effects of atmospheric scatter are registered as the brightening of the darkest objects in a scene. The DOS method uses this difference to reduce the brightness of the overall image relative to the difference between measured and assumed actual reflectances of these dark objects. The path radiance approach uses a similar set of relationships between the blue, red, and mid-infrared bands to approximate and correct for the effects of atmospheric aerosols. Finally, relative atmospheric correction takes advantage of the presence of pseudoinvariant features over a time series of images to account for atmospheric effects. These features, such as rock outcrops of built structures, may be used to adjust each image so that the reflectance of these features is standardized throughout the time series. This method does not require the estimation of any atmospheric optical property and can correct for systematic as well as atmospheric variance, but it is unsuitable in cases where extreme phenological or environmental changes have take place between the images or where images are spatially distributed. Similarly, topographic and forest structural characteristics can influence the measured reflectance. Distortions due to bidirectional reflectance from the canopy may cause "hot spots" in imagery (resulting from sun/sensor geometry, which places the sensor between the sun and the canopy, resulting in an artificially brightened image), and changes in canopy closure, gap spacing, and leaf clumping with the canopy can also induce variation in reflectance (Myneni et al. 1995b).

Because the radiance measured above a vegetated surface is a composite of both the vegetation itself and the background surface (typically soil), it is important to appreciate the contributions of both. Soil effects are most felt in areas of low canopy closure and low LAI, and spectral indexing methods may attempt to reduce sensitivity to its effects for studies in which soil registers as background noise, obscuring the object of interest (McDonald et al. 1998). Conversely, when surface parameters such as albedo are under study, it is important to accurately include the soil's contribution to total reflectance.

Table 6.1 details the air- and spaceborne optical sensors most commonly used in forest analysis. Although multispectral sensors still provide the majority of large-scale satellite data, hyperspectral data, characterized by many narrow, contiguous spectral bands, have become increasingly common in classification studies and the estimation of forest biophysical characteristics. Whereas the hyperspectral sensors listed are carried aboard aircraft or satellites, important primary research is also carried out at the leaf scale using small, portable field spectrometers.

TABLE 6.1

Current Electro-Optical Sensors and their Spatial and Spectral Resolutions

Sensor	Spatial Resolution (m)	Spectral Coverage (nm)	Number of Spectral Bands
NOAA AVHRR	1,000	580–11,500	5
MODIS	250–1,000	620–14,385	36
Landsat TM and ETM+	28.5	450–12,500	7
IKONOS	4	445–853	4
HYPERION	30	400–2,400	220
CASI	0.25–1.5	380–1,050	288
HyMap	3–10	450–2,500	126
AVIRIS	20	400–2,500	224

6.3 Mapping the Extent and Distribution of Tropical Dry Forests

The loss of TDF cover noted by Janzen (1988a) prompted his inclusion of this type of forest among the world's most threatened biomes. Taking into account the high degree of biodiversity and endemism found in TDFs (Gentry 1982a; Lott et al. 1987; Gillespie et al. 2000) and the concerns over forest degradation and fragmentation (Sánchez-Azofeifa et al. 2009), the importance of development of a clear inventory of TDF cover is clear. Previous assessments of ecosystems at risk have largely failed in addressing TDF status as being distinct from other tropical forest biomes (Miles et al. 2006), although areas containing high proportions of TDFs have been identified as biodiversity hot spots (Myers et al. 2000). Estimates of TDF cover and distribution not only are important tools for planning and conservation but also form part of an overall census of land cover. Inaccuracies in these forest maps, typically manifesting as an overestimation of tropical rain forests and the corresponding underestimation of TDFs (Portillo-Quintero and Sánchez-Azofeifa 2010), impact the accuracies of estimations of carbon stocks, which, in addition to their ecological importance, are essential for the implementation of conservation strategies based on payments for environmental services (Kalácska 2005).

In addition to methodological challenges that may vary based on the remote sensing platform and the analysis techniques employed in classifying TDF cover, two issues are consistent in the literature: (1) a lack of consensus on just what constitutes a TDF and (2) the problem of image acquisition, taking into account the inherent seasonality of TDFs.

There is general agreement that the basic characteristics of TDFs are relatively high temperatures, moderate but seasonal precipitation, and a forest canopy dominated by deciduous trees (Murphy and Lugo 1986a; Murphy and Lugo 1995). Holdridge (1967) defines tropical and subtropical dry forests as those in frost-free zones with a mean annual temperature >17°C, annual precipitation between 250 and 2000 mm, and a ratio of potential evapotranspiration to precipitation in the range of 1–2. Sánchez-Azofeifa et al. (2005b) amended this by specifying a mean temperature ≥25°C, precipitation between 700 and 2000 mm, and a minimum of three dry months per year of drought in which precipitation does not exceed 100 mm/month. Inconsistency among vegetation classification is of particular importance when comparing large geographic areas or in studies with varying methodologies (Blasco et al. 2000).

The second issue is the variability of vegetation spectral reflectance in response to phenological changes throughout the growing season. Since TDFs are intensely seasonal by definition, it is clear that spectral response follows similar annual patterns. A disregard for this variability may be responsible for the traditional misrepresentation of TDFs in satellite-based land cover assessments (Sánchez-Azofeifa et al. 2001). Kalácska et al. (2007b) and Portillo-Quintero and Sánchez-Azofeifa (2010) suggest the superiority of satellite data collected during the dry season for accurate classification of TDFs.

The only remotely sensed analysis of TDF cover at a global scale was carried out by Miles et al. (2006). Using a biogeographic classification scheme developed by Olson et al. (2001) and including tropical and subtropical grasslands, savannas, and shrublands as well as tropical and subtropical dry broadleaf forests as locations of potential TDF cover, they produced a potential cover map based on vegetation continuous fields data at a resolution of 500 m from a moderate-resolution imaging spectroradiometer (MODIS). Their findings point to 1,048,700 km^2 of total potential TDF cover globally, with the majority (66.7%) being located in the Americas. The authors further analyzed the degree of forest change during the period of 1980–2000 using 8-km-resolution data from an advanced very-high-resolution radiometer (AVHRR). During this period, deforestation was the greatest in Latin America, with an estimated 12% total decrease compared to an average estimated decrease of 2% throughout Asia and similarly low values across most of Africa. Also addressed was the risk to TDF areas from several potential threats: climate change, forest fragmentation, fire, conversion to agriculture, and increasing human population density. The study suggests that only 3.3% of the global forest cover is not subject to threat from one of these sources, with 31.7%–59.2% being subject to three or more and >95% being subject to at least two sources.

More recently, Portillo-Quintero and Sánchez-Azofeifa (2010) applied an approach based on spectral classification rather than the use of previously processed land cover products to generate a map of the extent and distribution of TDFs throughout the Neotropics. MODIS surface reflectance data were acquired during the dry season to improve the separability of TDFs from surrounding semideciduous and evergreen forests (Kalácska et al. 2007b) and mosaicked into a continuous image covering the American tropics. Using training sites collected from Landsat Thematic Mapper (TM) and Enhanced

Thematic Mapper Plus (ETM+) imagery over known vegetation types and validated using high-resolution imagery from Google Earth (Google Inc. 2012), the MODIS data were processed with the nonparametric decision tree classifier See5 (Rulequest Research 2008). The resulting map shows a total of 519,597 km^2 of TDF cover across the Americas, with the greatest cover found in Mexico (38%), Bolivia (25%), and Brazil (17%). Comparing this extent to the potential extent of the "tropical and subtropical broadleaf forest" defined by Olson et al. (2001), an average of 66% loss to anthropogenically attributed deforestation or conversion was noted across the study area.

Equally important as an accurate assessment of the extent and loss of TDFs derived from distribution maps is an understanding of the degree of continuity of the biome. Forest fragmentation poses a major risk to the health of both flora and fauna, with smaller fragments (<10 km^2) having higher rates of species extinction and a greater risk of conversion to other land covers (Laurance et al. 2002; Rodriguez et al. 2007a,b). Sánchez-Azofeifa et al. (2009) used a classification of TDFs based on data from National Aeronautics and Space Administration's 15 m Advanced Spaceborne Thermal Emission and Reflection Radiometer satellite to evaluate the degree of fragmentation in protected versus unprotected areas around Mexico's Chamela–Cuixmala Biosphere Reserve. They found that although both size and number of forest fragments remained relatively constant within a 15 km radius of the reserve, beyond that boundary the number of patches increased while the average patch size decreased considerably. Portillo-Quintero and Sánchez-Azofeifa (2010) examined the proportions of forest fragments falling into ≤2.5 km^2, 2.5–10 km^2, and ≥10 km^2 classes across the Neotropics and found generally high proportions of forest in the ≥10 km^2 class; the lowest proportions were found in Peru and Costa Rica. Both the aforementioned studies consider the impact of protected areas on TDF conservation and health. Sánchez-Azofeifa et al. (2009) noted the trend toward increasing fragmentation and decreasing patch size accompanying increasing distance from protected areas, noting the associated loss of connectivity between the residual fragments as a serious threat to forest biodiversity. In spite of this, Portillo-Quintero and Sánchez-Azofeifa (2010) report that only 4.5% of the TDF cover in the Neotropics is subject to protection, compared with a global estimate of 16%–18% of humid forests, savannas, and grasslands (Hoekstra et al. 2005).

6.4 Assessment of Forest Health (Biomass, Lear Area Index, and Productivity)

Remotely sensed analyses of forests have the potential to provide strong linkages between spectral reflectance and forest biophysical characteristics, which may then be used as proxy for physical inputs to models of ecosystem processes, biosphere–atmosphere transfer, and carbon exchange (Hall et al. 1995; Treitz and Howarth 1999). The goal of studies in this area has been to develop a methodology by which to accurately extract characteristics that are not commonly included in forest inventories, such as LAI, TAGB, and the fraction of absorbed photosynthetically active radiation (fAPAR), from remotely sensed data over large areas without further fieldwork. Although techniques involving neural networks have shown promise (Running et al. 1986; Carpenter 1997; Kalácska et al. 2005a), research in this field has focused primarily on two methods. Index-based approaches, based on empirical relationships between the ground-sampled measurements and their spectral properties, exploit differences in characteristic spectral regions of the reflected electromagnetic radiation (Running et al. 1986; Turner et al. 1999; Kalácska et al. 2004). Due to the relative simplicity of application, these methods have been by far the most common (McDonald et al. 1998).

Conversely, methods employing the inversion of physically-based radiative transfer models have been investigated with some success (Jacquemoud et al. 1996; Kuusk and Nilson 2000; le Maire et al. 2004). These models, such as those by Li and Strahler (1992, 1986), use reflectance as an input to derive biophysical variables such as LAI. Such models calculate canopy reflectance by incorporating nested models of the spectral properties of individual contributing factors. An example is the model by Kuusk and Nilson (2000); it incorporates the PROSPECT2 leaf optical model (Jacquemoud et al. 1996), the 6S atmospheric transfer model (Vermote et al. 1997), and the MCRM Markov-chain canopy reflectance model (Kuusk 1995), accounting for forest inventory and structural characteristics, as well as the effects of soil and ground bi-directional reflectance function. Raitianen (2005) found that, whereas this process only slightly overpredicted LAI values, such an approach is subject to error from a multitude of sources, most importantly the input parameter relating to stand characteristics, which must be taken from forest inventories (if available) or generalized. The ground truth LAI data are also subject to error, as they are generally modeled as well, taken from allometric equations. Even optical methods used for measuring LAI in situ operate on the often incorrect assumption of randomly dispersed clumping in the canopy. Privette et al. (1996) similarly note that although inverted physical models have the virtue of accounting for bidirectional effects, and requiring potentially less prior calibration than vegetation indices, they are computationally more demanding and require a priori knowledge of the vegetation characteristics of the study site, which may not be readily available.

Reliable estimates of forest biomass are essential for understanding the importance of forests in environmental processes and in regional and global carbon budgeting (Houghton et al. 2001; Foody 2003). The most direct method typically employed for biomass estimation is direct correlation with spectral reflectance or, more commonly, with a spectral vegetation index (SVI) such as the normalized difference vegetation index (NDVI), and then validation using either destructive forest sampling or comparison with the allometric equations derived from previous sampling (Castro-Esau et al. 2003). Although this approach has been employed with some success in temperate forests (Peterson et al. 1987; Curran et al. 1992; Danson and Curran 1993), the relationship between SVIs and biomass has been found to be generally poor in the tropics, with their overall denser forest cover, as indices tend to saturate at higher LAIs (Myneni et al. 1995a; Kalácska et al. 2005a). Foody et al. (2001) note only insignificant correlations ($p < 0.95$) between forest biomass and 230 permutations of six common ratio-based SVIs calculated from Landsat TM data collected over Borneo, confirming earlier results by Sader et al. (1989). An alternative to direct correlation with TDF spectral properties has been to exploit the relationship between forest age (or successional stage) and biomass. Issues in this approach, however, stem from a poor understanding of the age of succession and the site specificity of the equations used to derive biomass from stand age (Castro-Esau et al. 2003)

LAI, defined as the total one-sided surface area of all leaves in the canopy within a defined region (typically expressed in square meters per square meters) (Gong et al. 2003), is a key indicator of potential evapotranspiration and thus photosynthesis and stand productivity (Chason et al. 1991). As with forest biomass, LAI has often been estimated by the application of regressions between sampled LAI and SVIs, which show high sensitivity to changes in leaf area at low to moderate values, reaching an asymptote at index values of 3–5 (Chen and Cihlar 1996; White et al. 1997; Turner et al. 1999). Numerous ratio-based vegetation indices have been statistically related to LAI, typically exploiting the variation in the red and near-infrared reflectance of green plants. Turner et al. (1999) suggest that the relationships between vegetation indices from satellite multispectral data may be useful for retrieving LAI and note some issues that must be managed if these estimations are to be accurate. They remark on the impact of image-processing procedures and the importance of corrections for atmospheric effects, particularly when carrying out analyses between multiple sites or dates, even though they found that, although topographic corrections had a marked effect on raw vegetation index values, they had little to no effect on the strength of index–LAI relationships. The prime issue with the SVI approach is the tendency for SVIs to saturate at the higher LAI levels found in tropical systems (Birky 2001), attributed to the saturation of individual spectral bands when the forest reaches a certain level of green biomass (Kalácska et al. 2004). Turner et al. (1999) recognize the tendency for vegetation index values to reach an asymptote at LAIs >5 and note the importance of selecting vegetation indices that are appropriate for the cover type under evaluation. The goal in the development of these indices is to maximize the sensitivity to changes in the characteristic under study (like chlorophyll content) while minimizing the sensitivity to background effects (such as the influence of soil or the atmosphere) (Sims and Gamon 2002).

Kalácska et al. (2005c) tested the relationship between SVIs and LAI in the TDFs at two sites in Costa Rica and a third in Pacific Mexico. They found highly significant correlations between LAI and SVIs (calculated from 28.5 m Landsat ETM+ data) using nonlinear regression with a Lorentzian cumulative function. The best-fit spectral index varied slightly, with the modified simple ratio (MSR) (Chen 1996) providing the strongest result at two sites and the soil-adjusted vegetation index 2 (Qi et al. 1994) providing the best fit at the third. Kalácska et al. (2005a) also explored the use of Bayesian and neural network classifiers applied to Landsat ETM+ data as an alternative approach to estimating dry tropical LAI. Both Bayesian and neural network approaches were found to have a lower testing error than the SVI approach (48.7% and 56.9%, respectively, vs. 64.9%) when tested during the wet season at a TDF site in Costa Rica.

While estimations of both forest biomass and LAI stand as either proxies for forest productivity or inputs into further calculations, some research has been done on more directly measuring productivity by investigating the relationship between spectral reflectance and photosynthetic rates, in terms of both absorbed photosynthetically active radiation (APAR) and light use efficiency (LUE) at the leaf level. Gamon et al. (2005) broke down the gross photosynthetic rate into the product of LUE and APAR, adapted from an earlier work by Monteith and Moss (1977) on the components of net primary productivity. APAR can be evaluated with commonly used SVIs such as the simple ratio (SR) and NDVI (Gamon et al. 1995), and the authors found consistently strong relationships between SR and the measured fAPAR at the crown scale at a TDF site in Panama (NDVI was found to saturate over dense canopies). The LUE term was estimated using the hyperspectral photochemical reflectance index (PRI), which has been found to be sensitive to xanthophyll cycle activity (Gamon et al. 1992, 1997) and LUE at leaf (Gamon et al. 1992; Penuelas et al. 1995), canopy (Stylinski et al. 2002), and stand (Nichol et al. 2000; Rahman et al. 2001) scales. In the TDF context, leaf-scale PRI showed close correlation with measures of leaf fluorescence (a measure of radiation use efficiency [Genty et al. 1989]). The authors also noted a depression in PRI coincident with increased incident photosynthetic photon flux densities at the crown scale, supporting the relationship between PRI and LUE noted at the leaf scale and furthering the prospect of the evaluation of net photosynthesis via optical remote sensing.

6.5 Distinguishing between Species and Structural Groups

Although sensors with moderate spatial and spectral resolutions have been effective in identifying and classifying broad forest classes and estimating some forest properties, the high species densities found in both the humid and dry tropics (Myers et al. 2000; Zhang et al. 2006), as well as the general similarity in leaf reflectance among green vegetation (Portigal et al. 1997), make the identification of individual plant species difficult or impossible using these coarser resolution data. The issue of pixel resolution may be solved by the application of higher resolution data such as those from IKONOS or Digital Globe's Quickbird satellites, which are able to resolve individual tree crowns for analysis (Clark et al. 2005). Although these sensors have been used with moderate success in temperate forests (Nagendra and Gadgil 1999; Fuentes et al. 2001; Nagendra 2001; Ustin et al. 2004), the spectral resolution is insufficient in the context of the high species diversity found in TDFs. As such, the majority of studies have been conducted using leaf-level data collected using laboratory spectroradiometers.

Whereas the results of studies evaluating species discrimination using optical data in wet tropical environments are encouraging (Cochrane 2000; Clark et al. 2005; Gamon et al. 2006; Rivard et al. 2008), similar studies in TDF environments are slower in coming. Rather than work at the level of discriminating among individual species, some studies have concentrated on distinguishing the leaves of two principal structural groups in tropical canopies: trees and lianas (woody, self-supporting vines). This distinction is important as the impact of lianas at the leaf level bears heavily on the potential of automated species detection at the crown level and coarser resolutions. Both Castro-Esau et al. (2004) and Kalácska et al. (2007a) evaluated the spectral separability of these groups at sites in Panama. Both found that trees and lianas could be accurately distinguished using data collected from a dry forest site using a selection of supervised classifiers to process principal component- and wavelet-transformed data. Using the same procedures at a rain forest site, Castro-Esau et al. (2004) found they were less able to separate the two structural groups. To explain this difference in separability between trees and lianas in TDFs versus tropical rain forests, Sánchez-Azofeifa et al. (2009) proposed a liana syndrome, referring to a distinct set of plant traits exhibited by liana species in dry forest environments. They suggest that the evolutionary adaptations made by liana species to contend with increased water stress due to seasonal drought, such as delayed leaf loss at the end of the rainy season (Kalácska et al. 2005b) and increased leaf water content (Andrade et al. 2005; Schnitzer 2005), manifest spectrally as higher spectral transmittance, lower absorbance, and overall increased reflectance. The competitive advantage conferred by these adaptations in dry forests implies the potential that increasing liana cover will be an ongoing consideration for the automated evaluation of TDF biodiversity.

The basis for the differentiation between species or functional groups is that leaf (or canopy) biochemistry is unique for a given species, resulting in a chemical signature that may be used to identify that species and is expressed in that species' spectral reflectance (Peterson et al. 1988; Asner and Martin 2008a). Although the studies mentioned here have demonstrated strong correlations between taxonomy and leaf optical properties, it is clear that understanding the scope of variation in both leaf spectral properties and the biophysical traits that control them must be a priority. Castro-Esau et al. (2006) found sufficient difference between spectra of given species sampled at multiple sites in Costa Rica that accurate classification across sites was impossible. Martin et al. (2007) similarly found a strong environmentally attributable variation in pigment and optical characteristics among samples of *Metrosideros polymorpha* grown from seed sources spanning a wide environmental (soil–altitude) gradient. Asner et al. (2009) found in a study of 162 canopy species across a wide climatic gradient in Australia that, although biophysical variables were strongly related to leaf reflectance, variation in leaf chemical signatures varied far more in response to taxonomy and species richness than to changes in climate. They did, however, find the greatest chemical variation in lowland sites with moderate precipitation levels, which echoes the findings of Townsend et al. (2007), who found maximal nitrogen:phosphorous variation according to rainfall in highly seasonal sites in Costa Rica.

The impact of this chemical and spectral variability is particularly important in TDFs due to its strong seasonality and the accompanying variation in leaf properties. Although the impact of leaf phenology has been well noted (Kalácska et al. 2007a; Portillo-Quintero and Sánchez-Azofeifa 2010), spectral analysis has largely been limited to the tracking of NDVI (or similar spectral indices) throughout the growing season using spaceborne sensors (Schwartz and Reed 1999; Zhang et al. 2003). However, the importance of season on spectral response in the dry tropics has been demonstrated by Roberts et al. (1998), who documented spectral changes associated with leaf senescence in the Brazilian caatinga, and Hesketh and Sánchez-Azofeifa (2012), who found a tenfold decrease in classification accuracy when applying a single nonparametric classifier across both rainy and dry seasons. Underscoring the impact of seasonal and phenological cycles on forest monitoring, this leaf-level spectral variation has been found to be exaggerated when scaled up to the level of forest canopy (Zhang et al. 2006).

6.6 Forest Structure and Successional Stage

TDFs have long been areas of intense human activity (Miles et al. 2006), resulting in high levels of deforestation and fragmentation driven by fire, conversion for agriculture and habitation, and commercial logging (Colon and Lugo 2006; Calvo-Alvarado et al. 2009). These forests in stages of recovery from human-induced disturbances are termed secondary forests (Brown and Lugo 1990). Changes in the economic climate that drove forest degradation in many areas of the dry tropics have resulted in an increased rate of return of these forests on what was once cleared land (Calvo-Alvarado et al. 2009). These secondary forests represent an important element of the global capacity for carbon sequestration, but they are also a source of potential error in the estimation of carbon budgets as the capacity for carbon uptake is dependent on the forests' species composition and the age of the secondary growth (Uhl et al. 1988; Brown and Lugo 1990; Foody et al. 1996). In response to increasing interest in these young forests and the role they play in ecological and economic models (Feldpausch et al. 2004; Koning et al. 2005), a need for remote sensing tools to assess their structure and characteristics has been identified (Chambers et al. 2007).

The discrimination and mapping of forest structure in the tropics shares some of the challenges associated with the mapping of forest extents and TAGB addressed earlier, namely, the availability of quality, cloud-free satellite imagery over the often cloudy tropics, and the site specificity of the empirical relationships between stand age and reflectance. Nonetheless, the mapping of secondary forests and the age classes or successional stages they contain has been carried out with some success in both the wet and dry tropics. Whereas remote sensing analysis in the Brazilian Amazon using various data types and classification methods has shown the potential for differentiating primary and secondary forest classes (e.g., the study by Alves and Skole [1996]) as well as successional stages (e.g., the study by Kimes et al. [1999] and Lucas et al. [2000]), the increased variability in TDF sites has complicated similar studies in the seasonally dry tropics (Kalácska et al. 2005c).

The pronounced seasonality characteristic of TDFs presents an additional challenge in discriminating between age classes in these forests using optical data. As with studies mapping dry forest extent (Portillo-Quintero and Sánchez-Azofeifa 2010) and discriminating between plant functional groups (Castro-Esau et al. 2004; Kalácska et al. 2007a), Arroyo-Mora et al. (2005) used data from the dry season or the rainy-to-dry transitional seasons to map dry forest succession in Costa Rica using Landsat ETM+. Their work also addresses the issue of poor spectral correspondence with age-based definitions of successional classes. Rather, they define forest successional stages with respect to their structural elements (e.g., stem density, basal area (BA), and the number of canopy layers). Comparing the separability of these structural classes to that of stages based on time since abandonment (ca. 5–10 years, 19–22 years, 22–30 years, and primary growth) using cluster analysis and pattern recognition techniques, they found that the structural classes were consistently discernible while the age-based classes tended to overlap considerably, precluding accurate classification.

Hartter et al. (2008) also used dry-season data acquired from Landsat for successful discrimination of TDF successional stages. Rather than age-based definitions of secondary forest classes, they define early successional classes as those with a woody BA of $<15\ m^2 \cdot ha^{-1}$ and mid–late classes as those with a BA of $>30\ m^2 \cdot ha^{-1}$. Landsat TM data was acquired over two TDF sites in Mexico and validated with BA surveys at 28 field plots. The authors report an overall classification accuracy of 81% using a multistage classification approach. First, the land cover was segmented into forest, crops, and other categories using an SVI. Second, the forest class was further divided into early and mid–late successional classes using the SVI data plus the first three principal components calculated from the Landsat TM data and a texture layer (derived from the variance in spectral properties) as inputs. Despite the high accuracy of the classification, the authors echo the sentiment expressed in the previously cited studies that the variability imparted by shifting phenology is an essential element of the remote sensing analysis of TDFs.

Remotely sensed analysis can also contribute to an improved understanding of the ecological characteristics related to shifting forest succession. Using a 56 year chronosequence derived from orthorectified aerial photography and Landsat ETM+ data acquired between 1944 and 2000 over Providencia Island, Colombia, Ruiz et al. (2005) evaluated the variability in species richness and diversity according to six age classes (<6, 6–10, 11–16, 17–31, 32–56, and >56 years since abandonment) derived from the remote sensing investigation. Through a comparison of diversity metrics (Shannon's H, Simpson's D) calculated from field surveys carried out at plots located within the age classes identified using the chronosequence, they found that while species density reached a peak in the intermediate successional stages (32–56 years since abandonment), overall species richness increased linearly with stand age, reaching a maximum in stands over 56 years old. Although they acknowledge the limitation of this method compared to traditional chronosequence methodologies (interviews of residents, sometimes coupled with visual interpretation of air photographs) in identifying exact stand ages, the authors note the efficiency and effectiveness of the combination of Landsat data and orthophotographs in surveying and classifying a large geographic area, while maintaining the benefit of the long time series (more than 50 years in this example) available using archived aerial photographs.

Data outside the optical wavelengths have also been incorporated into the analysis of forest structure and succession. In a TDF site in Yucatan, Mexico, Southworth (2004) explored the incorporation of the thermal infrared data in Landsat TM band 6 to improve land cover classification, including discrimination between early–mid and mid–late successional stages. Using the relationship between surface temperature and successional stage, they developed a series of hybrid optical/thermal indices that they found to be allowed for a visually superior classification of primary and secondary forests, although the addition of the thermal band did not statistically improve the accuracy of the analysis.

An alternative with the potential to overcome some of the shortcomings of optical data, namely, the poor availability of cloud-free data, saturation of vegetation indices, and confusion between forest classes (Castro-Esau et al. 2003), is actively remotely sensed data such as LiDAR. LiDAR measures the strength of return of an emitted signal to estimate the distance between a target and a sensor (Lefsky et al. 2002). Waveform LiDAR, such as the LiDAR Vegetation Imaging System (LVIS), measures multiple returns, providing not only canopy height but also an estimation of the internal structure of the forest, with a three-dimensional picture of the canopy (Castillo-Nunez et al. 2011). Castillo et al. (2012) used LVIS data to discriminate and map successional stages in a site in Costa Rica. Using an unsupervised classification of three return levels, they generated a map of successional stages that corresponded well to a "literature-based" map extrapolated from the relationship between measured canopy height in 20 field plots and the estimated canopy height from the 100% return level. Further, they focused on the intermediate successional stage identified by the classification and identified three classes within this, better characterizing the areas of transition between the early–intermediate and intermediate–late successional stages.

6.7 Conclusion

In 2003, noting a paucity of research in TDFs relative to temperate wet tropical environments, Sánchez-Azofeifa et al. (2003) identified three principal research priorities for remote sensing in TDF environments: (1) the application of remote sensing tools and spectral analysis to the discrimination and characterization of secondary forests, citing the rapid regrowth and biomass accumulation of TDFs following abandonment and their then unquantified potential as carbon sinks; (2) the characterization of forest biophysical parameters, particularly LAI, using remotely sensed spectral proxies; and (3) the development of hyperspectral analysis techniques to characterize individual tree species based on their spectral reflectance. Subsequent work has demonstrated progress in all three of these research areas, but it has also highlighted new areas of importance in the remote sensing of TDFs. Of particular importance is investigation into the temporal and especially phenological characteristics of these forests. TDFs are intensely seasonable by definition, and the studies discussed in this chapter have cited this source of variation to be an obstacle to accurate spectral characterization of land cover, biophysical characteristics, and forest structure and composition. A second area of exploration is the integration of multiscale and multisensor data sources, taking advantage of overlapping spatial, spectral, and temporal characteristics to better address the challenges of remote analysis of TDFs.

The question of spectral variation in response to TDF phenology has been shown to impact the characterization of all aspects of these forests by optical remote sensing. Figure 6.1 summarizes the annual trajectory of NDVI (as a proxy for LAI) and the related utility of spectral data for TDF research. During the dry season, LAI (and accordingly NDVI) are low and the overall forest structure is less obscured by the leafy canopy. As such, discrimination of TDF boundaries (e.g., the study by Portillo-Quintero and Sánchez-Azofeifa [2010]) as well as structure and successional stage (e.g., the study by

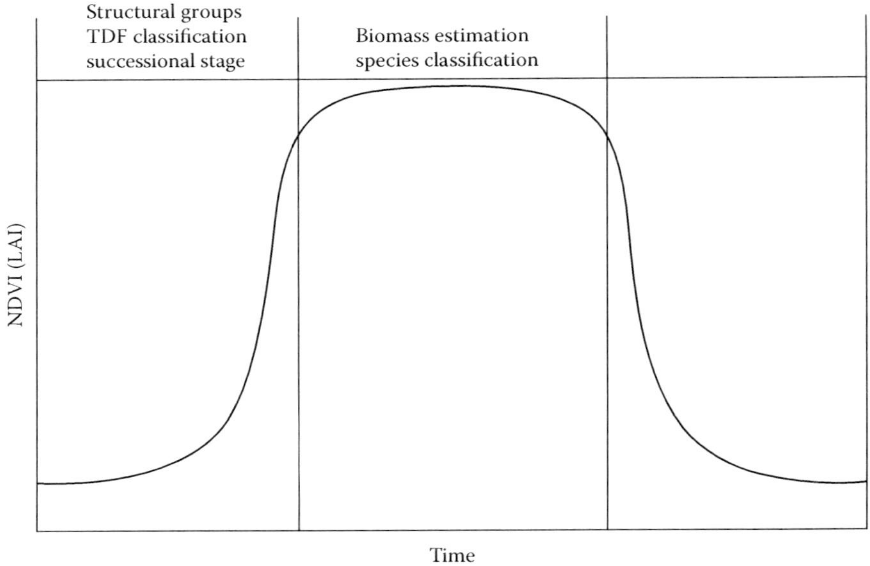

FIGURE 6.1
An idealized trajectory of normalized difference vegetation index (NDVI) values as a proxy for leaf area index (LAI) over the course of a year with a single growing season. Subjects of analysis associated with each phenophase are listed above the phenological curve.

Arroyo Mora et al. [2005], Kalácska et al. [2005c], Hartter et al. [2008]) have been most accurate during this phenophase. Conversely, LAI studies for forest biomass estimation may best be conducted during the rainy season when consistent NDVI values may be translated into more reliable biomass estimates. A complication of this method is the potential for spectral indices to reach an asymptote or saturate in response to high LAI (Turner et al. 1999). Figure 6.2 generalizes the potential for NDVI saturation as a function of stand age in both tropical dry and wet forest environments. Due to more rapid growth rates tropical wet forest canopies reach sufficient LAI values to saturate the index relatively early, whereas TDF canopies follow a slower successional trajectory and result in a larger temporal window within which NDVI remains sensitive to variations in LAI.

Classification at the species and structural group level seems to be optimally performed during different periods of the growing season. The potential for discrimination between tree and liana species, as explored by Castro-Esau et al. (2004) and Kalácska et al. (2007a), was highest during the dry season, likely due to the different adaptations to water stress of the two structural groups, as proposed by Sánchez-Azofeifa et al. (2009). Whereas aircraft and satellite sensors do not currently have the combination of high spatial and spectral resolutions required for species-level classification in the species-rich tropics (Asner and Martin 2008a), studies conducted at the leaf

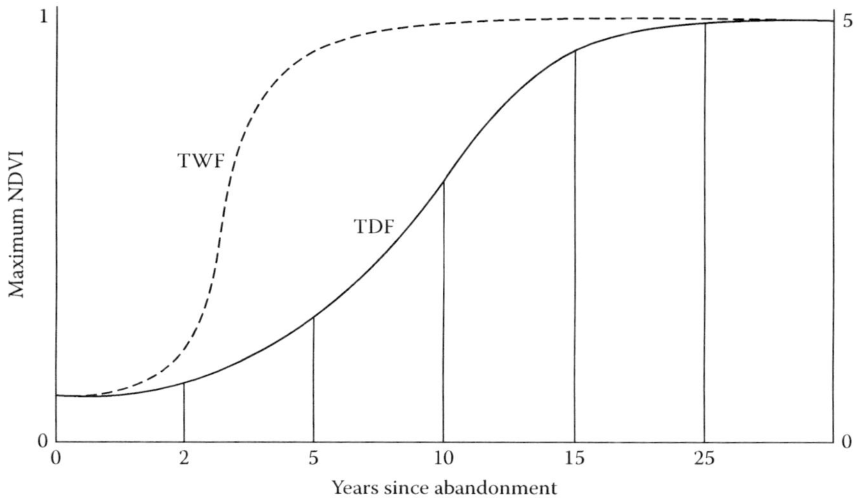

FIGURE 6.2
Idealized potential for NDVI saturation of tropical wet forest (TWF) (dotted line) and tropical dry forest (TDF) (solid line) plotted as a function of the number of years since abandonment. NDVI typically saturates at LAI values above 5.

level with hyperspectral instruments have demonstrated high accuracies in distinguishing between species at single sites within narrow temporal windows. Issues of temporal variability become an obstacle when these classifications are extended beyond the parameters of the original study, to the extent that at this stage the likelihood of automated classification of species using a library of consistent spectral signatures seems low (Castro-Esau et al. 2006; Hesketh and Sánchez-Azofeifa 2012). It is clear that virtually any remotely sensed analysis of TDFs is subject to phenologically induced spectral variability and that a clearer understanding of the nature and scope of this variation must be one of the priorities of future work.

A second area with the potential to advance our understanding of TDFs is the fusion of various data sources to provide a more detailed and accurate estimation of TDF characteristics. While the use of optical data at multiple spatial resolutions is discussed in this chapter (e.g., the study by Arroyo Mora et al. [2005]), the use of complementary data types has been underexplored in TDF studies. Synthetic aperture radar imagery has the benefit of cloud penetration, with the potential to fill imagery gaps in low temporal resolutions that result from high cloud cover over tropical areas, particularly during the wet season (Sánchez-Azofeifa et al. 2003). Radar data also provide complementary information content to optical data. While data in the visible and infrared spectral regions provide information on the chemical and structural characteristics of vegetation, radar data can supplement them with additional information on surface texture and dielectric properties, aiding in particular the separation of vegetation from bare soil and the estimation of moisture content (Held et al. 2003). Similarly, LiDAR data, which can provide detailed information on forest canopy structure (Skowronski et al. 2007), have shown to increase the accuracy of species classification when combined with an airborne hyperspectral data set over a temperate forest (Dalponte et al. 2008), although this fusion is as yet untried in TDFs.

A challenge to the application of these data fusions lies in the lack of infrastructure for the organization and integration of the various spectral and ecological data and metadata used in the analysis of forest characteristics and processes (Quesada et al. 2009). The establishment of this infrastructure has the potential to provide researchers with the opportunity to work at nested spatial, spectral, and temporal scales. For example, fAPAR data can be collected in near real time by ground-based sensor networks, linked to time-stamped daily phenological data from nearby sensor stations, which can be further linked via geographic-positioning-system coordinates to air- and satellite-borne remotely sensed products. This coupling of field data and remotely sensed imagery validates remotely sensed products but directly addresses the temporal dynamics of TDF ecosystems (Gamon et al. 2006).

Remote sensing analyses have become a critical component of ecological research (Kerr and Ostrovsky 2003), particularly in the assessment of remote regions and the exploration of patterns at various spatial and temporal scales. Continued research and development, particularly in the area of temporal and phenological variability, will be crucial in better understanding TDFs and providing linkages between forest biophysical and structural characteristics and the environmental factors that govern them.

7

Phenological Patterns of Tropical Dry Forests along Latitudinal and Successional Gradients in the Neotropics

Martha Lopezaraiza-Mikel, Mauricio Quesada, Mariana Álvarez-Añorve, Luis Ávila-Cabadilla, Silvana Martén-Rodríguez, Julio Calvo-Alvarado, Mário Marcos do Espírito-Santo, Geraldo Wilson Fernandes, Arturo Sánchez-Azofeifa, María de Jesús Aguilar-Aguilar, Francisco Balvino-Olvera, Diego Brandão, José Miguel Contreras-Sánchez, Joselândio Correa-Santos, Jacob Cristobal-Perez, Paola Fernandez, Branco Hilje, Claudia Jacobi, Flávia Fonseca-Pezzini, Fernando Rosas, Victor Rosas-Guerrero, Gumersindo Sánchez-Montoya, Roberto Sáyago, and Arely Vázquez-Ramírez

CONTENTS

7.1 Introduction

The study of periodic biological phenomena in relation to climatic changes over time is generally known as phenology (Stearns 1974). Phenology is a significant component of biological populations and communities because it affects several features of plant species; vegetation composition and structure; and ecological interactions between plants and animals such as pollination, herbivory, frugivory, and seed dispersal. Hence, the research of phenological patterns of plant growth and reproduction is critical to understand the processes related to productivity, forest succession, and the functioning of ecosystems (Newstrom et al. 1994).

Tropical forests are currently exposed to several threats from human activities, resulting in complex landscapes of agricultural fields, grasslands, and forest patches under different levels of succession (Sánchez-Azofeifa et al. 2005). After a forest disturbance, ecological succession (i.e., the process of species replacement over time) modifies not only species diversity and abundance (Quesada et al., Chapter 2) but also the phenological dynamics of the community. Inferring the functional relationship between successional status and phenological patterns of degraded forests is especially important for understanding the recovery rates and dynamics of heavily transformed tropical ecosystems.

Tropical dry forests (TDFs) have seasonal phenological patterns driven by cyclical regimes of precipitation and a marked dry season. Under such conditions, plant growth and reproduction are largely limited to the wet season. Seed germination, seedling establishment, and regeneration also respond to this marked climatic seasonality (Opler et al. 1976). TDFs hold remarkable biodiversity and have high endemism, and it is often assumed that TDFs can recover their functions and mature condition more quickly than wet tropical forests (Murphy and Lugo 1986a). However, although some studies indicate an apparent rapid recovery of species composition and structure, a functional recovery of the community in terms of physiological and morphological traits may take longer (Ceccon et al. 2002; Quesada et al. 2009; Alvarez-Añorve et al. 2012). Destruction and degradation of tropical forests have an effect on biotic and abiotic factors triggering phenological patterns of plant species with severe consequences for populations and communities.

Factors that determine plant phenology can be divided into proximate and ultimate causes. Proximate causes refer to short-term environmental events that may activate phenological patterns, whereas evolutionary processes underlie the ultimate causes responsible for such patterns. Short-term environmental events (proximate causes) include changes in the level of water stored by plants (Reich and Borchert 1984), seasonal variations in rainfall (Opler et al. 1976), changes in temperature (Williams-Linera 1997), photoperiod (van Schaik 1986), irradiance (Wright and van Schaik 1994), and sporadic climatic events (Sakai et al. 1999). Changes of environmental factors in disturbed habitats under succession indicate an increase in average temperature, a greater exposure to wind, a reduction in soil moisture, and an increase in evapotranspiration (Alvim 1960; Kapos 1989; Saunders et al. 1991; Wright 1996; Kapos et al. 1997; Wright et al. 1999; Laurance et al. 2002; Fuchs et al. 2003; Herrerias-Diego et al. 2006). Some of these abiotic factors are known to affect phenological patterns in plants, but only a few studies have evaluated the effect of abiotic factors on phenological patterns under disturbed conditions. For example, certain levels of drought or increases in temperature may trigger flowering (Alvim 1960; Wright 1999; Kudo et al. 2004). It has also been proposed that global warming is related to the earlier flowering of many species in temperate regions (Miller-Rushing and Primack 2008).

Changes in phenological patterns caused by variations in environmental proximate factors may also have immediate effects on the outcome of biological interactions that may have shaped those patterns. For instance, the disruption of flowering phenology caused by disturbance is likely to affect the behavior and visitation rates of pollinators, resulting in negative consequences for both the reproductive success of the plants and the ability of pollinators to obtain resources (Rolstad 1991; Saunders et al. 1991; Quesada et al. 2011). A shift in the flowering phenology of *Ceiba aesculifolia* in a Mexican TDF disrupted the pollination and reproduction of tree populations (Herrerias-Diego et al. 2006). Phenological changes will also disrupt interactions with long-distance pollinators (e.g., bats) and trap-liners (e.g., hawkmoths) that follow the flowering phenology of plants like Cactaceae and Agavaceae (Haber and Frankie 1989; Fleming et al. 1993; Frankie et al. 1997; Haber and Stevenson 2002).

Intraspecific variation in the intensity, duration, frequency, regularity, and synchrony of vegetative and reproductive phenophases may affect the growth, development, productivity, reproduction, establishment, and genetic structure of plant populations in disturbed habitats under succession (Newstrom et al. 1994; Doligez and Joly 1997; Herrerias-Diego et al. 2006). These variables may be affected by variations in environmental conditions associated with the successional process. Specifically, temperature, evapotranspiration, and irradiance are expected to be greater in early successional stages compared with mature forests, whereas water soil content should be lower in early successional stages. If these environmental cues are proximate factors that trigger the different phenophases, phenological variables will differ during succession.

Regarding leaf phenology, the amount of leaves is expected to increase in early successional stages compared to late successional stages (Baruah and Ramakrishnan 1989), due to higher irradiance. The duration of the leaf phenophase should be shorter at these stages due to lower soil humidity, a consequence of higher irradiance, temperature, and evapotranspiraton; therefore, leaf fall is expected to start earlier. The magnitude of the differences among successional stages should be greater in sites with lower annual precipitation. Given the greater production of leaves and the shorter leaf phenophase, plants in early successional stages should concentrate the production of flowers into short periods of greater intensity. Early successional species generally have higher growth rates (Baruah and Ramakrishnan 1989) and shorter life cycles (Baker 1974), which might make them more vulnerable to temporal variation in pollinator service and reproductive failure; therefore, continuous flowering in early succession might provide a way to ensure reproduction when pollinators are scarce (Kang and Bawa 2003). Alternatively, these plant species might use self-pollination, abiotic pollination, or generalist pollination. Fruiting patterns are expected to parallel flowering patterns, except when fruit production is pollen or resource limited (Kang and Bawa 2003). Anemochorous plants should be favored in early successional stages due to the lower availability of biotic dispersers (Quesada et al. 2009, 2011); therefore, the proportion of species that fruit during the dry season should be higher.

The study of plant reproduction, phenology, and ecological succession in the tropics dates back to the 1980s (Opler et al. 1980); nonetheless, few studies have surveyed the changes in phenological patterns across different phases of succession in tropical forests. Most studies have only evaluated floral phenology among successional stages in temperate regions (e.g., the studies by Kahmen and Poschlod [2004]; Vile et al. [2006]; Navas et al. [2010]), and only two studies have evaluated changes in leafing phenology (Baruah and Ramakrishnan 1989) and flowering phenology (Kang and Bawa 2003) in relation to ecological succession in subtropical dry and tropical wet forests, respectively. In this chapter, we use a replicated design to compare plant phenological patterns of leafing, flowering, and fruiting at three TDF sites in a latitudinal and successional gradient across four years. These sites are part of the Tropi-Dry research network. First, we assess the phenology of each community in relation to the precipitation data collected at each region and describe these patterns along a chronosequence of succession. Second, we analyze the flowering phenology of each species in the plant communities by determining the periodicity, frequency, and duration of flowering at each successional stage.

7.2 Methods and Analysis of Phenological Patterns

7.2.1 Study Sites

Three sites were selected in a latitudinal gradient: (1) the Chamela–Cuixmala Biosphere Reserve, Mexico (Chapter 1; Figure 1.2); (2) the Santa Rosa National Park, Costa Rica (Chapter 1; Figure 1.3); and (3) Mata Seca, Brazil (Chapter 1; Figure 1.4) (see Chapter 1 for general design and site descriptions). Mean annual temperatures for the three sites are 25°C, 28°C, and 24°C, respectively; in the same order, mean annual precipitations for the sites are 748, 800–2600, and 916 mm. The duration of the dry season at the sites are as follows: November–June in Mexico (eight months), mid-December–mid-May in

Costa Rica (six months), and May–November in Brazil (seven months). This climatic information is based on the studies by García-Oliva et al. (1995) for Mexico, Janzen (1983) for Costa Rica, and Madeira et al. (2009) for Brazil.

7.2.2 Data Collection and Analyses

Leafing, flowering, and fruiting phenological patterns were evaluated in three successional stages, early (3–5 years since abandonment), intermediate (8–12 years old), and late (more than 50 years old), with three plots surveyed at each successional stage (i.e., a total of nine plots). Monthly records of leaf, flower, and fruit production were collected in two transects of 3 m × 50 m laid at the edges of 20 m × 50 m plots. All individuals with a diameter at breast height (DBH) greater than 2.5 cm were marked and monitored for the following phenophases: green leaves, flowers, and fruits. For each individual, we estimated the intensity of each phenophase as the percentage of coverage of leaves, flowers, and fruits according to the following categories: I (0%), II (1%–25%), III (26%–50%), IV (51%–75%), and V (76%–100%). Data were collected from June 2006 to March 2010 in Mexico, from May 2007 to March 2010 in Costa Rica, and from May 2007 to February 2010 in Brazil.

We plotted climatic variables along with phenological variables for each month throughout the study period. Temperature and precipitation data were obtained from Chamela's Meteorological Station at the Chamela Biological Station, Universidad Nacional Autónoma de México; the Santa Rosa Meteorological Station at the Santa Rosa National Park; and the Mocambinho Meteorological Station, Minas Gerais, Brazil. Day-length data were obtained from the Astronomical Applications Department of the U.S. Naval Observatory (http://www.usno.navy.mil).

We described community-level patterns of each phenophase and compared these patterns among different successional stages of TDF, and among countries. To describe the mean monthly intensity of each phenophase, we averaged the score of all individuals at each plot, regardless of the species to which they belong, using the top value of the category in which they fell (i.e., 0, 25%, 50%, 75%, and 100%). To describe the frequency of the phenophases in the community, we calculated the number and proportion of species in categories IV and V for green leaves (indicating plants with mid to full leaf coverage) and in categories II–V for flowers and fruits (indicating the presence of flowering or fruiting) for each month of the study period. To compare the duration of the phenophases among different successional stages, we used the number of months per year that individuals and species present each phenophase (green leaves: categories IV and V; flowers and fruits: categories II–V). To calculate the length of a phenopase for a species, we used the mean number of months that the individuals of a species score in categories of mid to full leaf coverage (categories IV and V) and the maximum number of months that the individuals of the species were registered with flowers or fruits (categories II–V). For all variables describing the phenophases (intensity, proportion of species, and number of species), the means of the three plots of each successional stage and standard errors are plotted.

Flowering patterns were also analyzed at the species level in terms of frequency, regularity, and duration of the phenophase, according to definitions by Newstrom et al. (1994). For species classification, data from the three plots of each successional stage were pooled. Frequency is the number of cycles per year with respect to on/off phases. Species were classified as having "continual flowering," flowering year round with occasional brief gaps with no flowers; "subannual flowering," multiple irregular flowering phases per year; "annual flowering," one major flowering phase per year; and "supra-annual flowering," multiyear cycles of flowering. Regularity is the variability in the length of cycles and phases. This variable was classified as "regular," with predictable timing and duration of flowering, or "irregular," with unpredictable timing and duration of flowering. Duration is the length of time in each cycle or phase. We classified species according to the duration of flowering in a year as "brief flowering" (scoring categories II–IV in one month), "intermediate flowering" (two to four months), and "extended flowering" (five months or more). In this case, we jointly considered all individuals to calculate the flowering period of a species.

7.3 Results

The number of species and individuals that fell within the phenology transects tended to increase from early to late successional stages (Table 7.1 and Figure 7.11) and were significantly different between these stages for species (means: early 10, intermediate 26, late 31) and individuals (means: early 37, intermediate 84, late 92) in Mexico and for individuals in Brazil (means: early 60, intermediate 64, late 91).

7.3.1 Mexico

During the study years, the first rainfall of the season was in either June or July (Figure 7.1d). Total annual rainfall was 1066 mm in 2006–2007, 1026 mm in 2007–2008, 956 mm in 2008–2009, and 824 mm in 2009–2010 (10 month sampling).

7.3.1.1 Plot-Level Phenology: Leaves, Flowers, and Fruits

Leaf production started around July with the rainy season (Figures 7.1a). Leafing intensity reached high levels within one month of the start of the rainy season when most species were recorded in full leaf (Figures 7.1a and 7.2b). Leaf fall started in November and December, and mean intensity gradually decreased. Atypical rain events occurred in December 2009

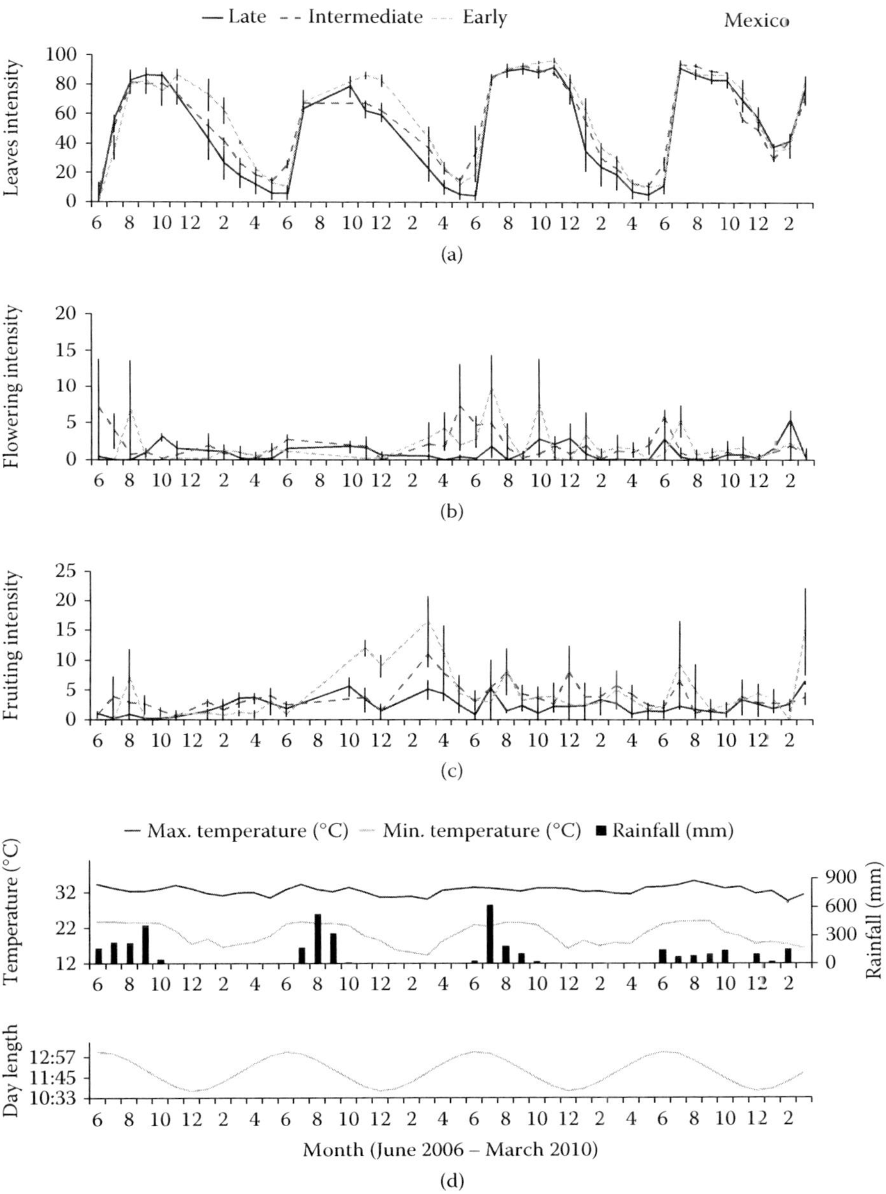

FIGURE 7.1
Mean monthly intensity (±standard error [±SE]) of individuals for the phenophases of (a) leaves, (b) flowers, and (c) fruits for plots of three different successional stages of a tropical dry forest (TDF) in Chamela, Mexico, for the period of study June 2006–March 2010. (d) Monthly rainfall, maximum and minimum temperatures, and day length are also shown. Intensity is measured as the percentage coverage of leaves, flowers, and fruits.

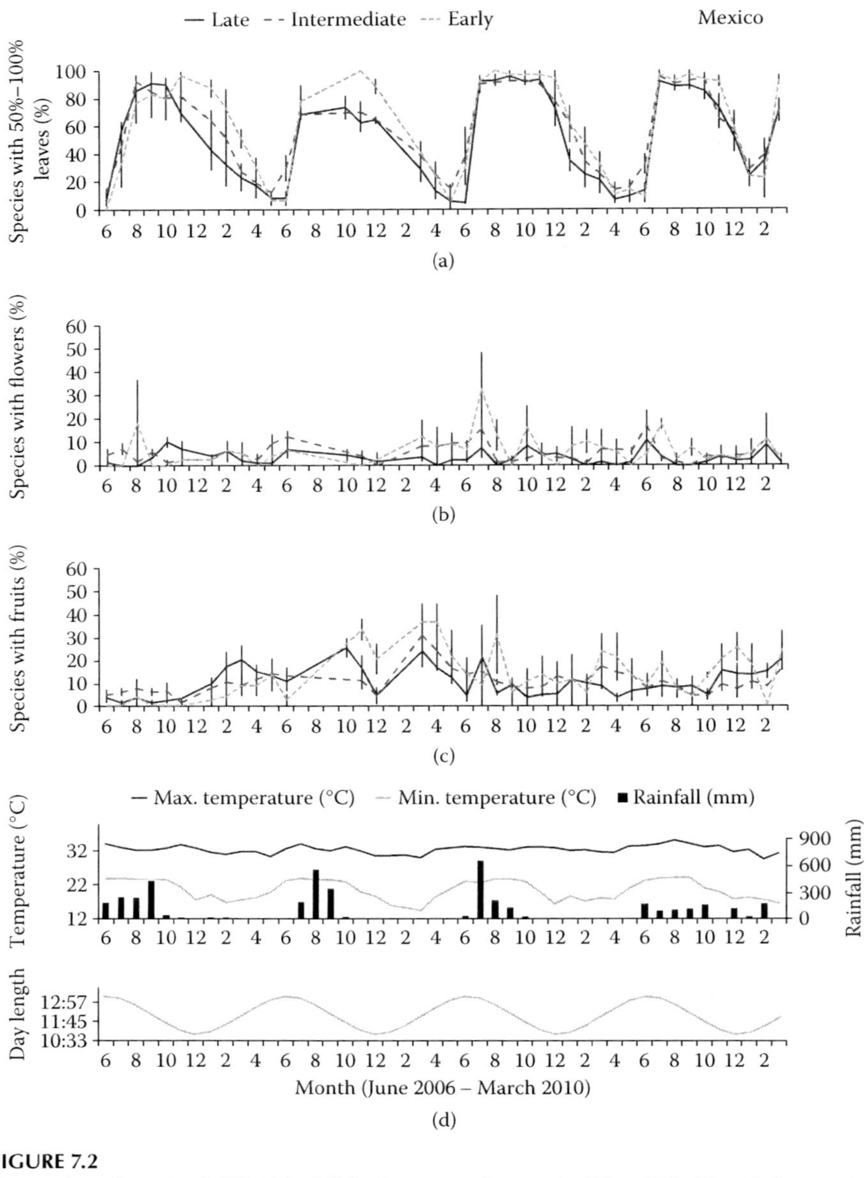

FIGURE 7.2
Proportion of species (±SE) (a) in full leaf coverage (categories IV and V), (b) with flowers (categories II–V), and (c) with fruits (categories II–V) for plots of three different successional stages of the TDF in Chamela, Mexico, for the period of study June 2006–March 2010. (d) Monthly rainfall, maximum and minimum temperatures, and day length are also shown.

through February 2010, which altered the frequency and regularity of leafing patterns, resulting in a leafing phase during the dry season. The lowest mean leafing intensity of the dry season was 30% in January 2009, compared to 4.5% in other years, and it was followed by leaf production that attained 83% by March 2009 (Figure 7.1b).

In early successional stages, yearly intensity tended to be higher, leafing intensity decreased more slowly, and the lowest levels of intensity in May were higher than they were in late stages (Figure 7.1a). Correspondingly, a higher proportion of species maintained their leaves during the dry season in early successional stages (Figure 7.2a). During the period of study, mean leafing intensity varied between 4.5% and 92% for late stages, 10% and 94% for intermediate stages, and 10% and 97% for early stages. A higher proportion of species had mid to full leaf coverage by June in intermediate stages than in early and late stages (Figure 7.2a), and this was even before the first rains were registered (Figures 7.2a and 7.2d, see June 2007). During the rainy season, the number of species with mid to full leaf coverage was generally highest in late successional stages and lowest in early stages (Figure 7.3a).

The overall duration of leafing tended to be higher in early successional stages, and a higher proportion of individuals and species had mid to full leaf coverage for extended periods of time (6–10 months: 62%–69% of individuals, 62%–67% of species for early stages; 51%–58% of individuals, 49%–58% of species for intermediate stages; and 39%–57% of individuals, 32%–49% of species for late stages). However, no evergreen species were observed (11–12 months in mid to full leaf). In contrast, a small proportion of individuals and species in later successional stages were in mid to full leaf coverage for 11–12 months of the year (1%–7% of individuals, 1.5%–5% of species).

Flowering intensity (Figure 7.1b) and frequency (Figures 7.2b and 7.3b) varied within the year, among years, and among successional sites, but there were records of individual species flowering every month. When the rainy season starts, there is a fairly synchronous and regular peaking of flowering across the three stages of succession (Figures 7.1b, 7.2b, and 7.3b; June–July 2008 and 2009). Another regular flowering peak that occurred in late successional stages is observed at the end of the rainy season and the start of the dry season (September–October and January–February, respectively). Frequency and regularity of flowering were affected by atypical rainfall at the end of 2009 to the start of 2010 (Figures 7.1b, 7.2b, and 7.3b; see flowering peak in February 2010). Some species that normally flower at the start of the rainy season flowered in February 2010: *Capparis indica, Casearia corymbosa, Croton roxanae, Croton suberosus, Lonchocarpus constrictus, Ruprechtia fusca*, and *Thouinia paucidentata*.

Intensity peaks and peaks in the monthly proportion of species in flower were higher in early and intermediate stages than in late stages of succession (Figures 7.1b and 7.2b). However, the absolute number of species in flower was comparable among successional stages. The annual proportion of individuals and species recorded as flowering in early and intermediate stages was also higher than in late stages (early: 22%–41% of individuals, 22%–56%

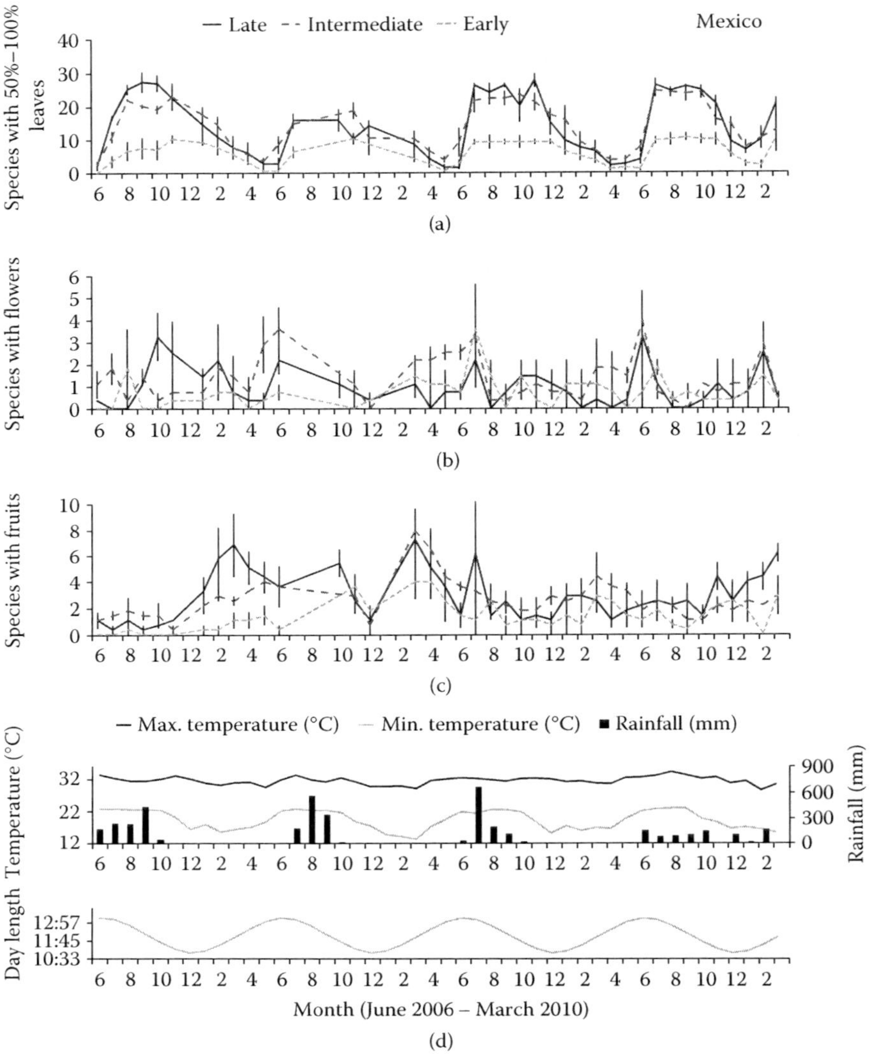

FIGURE 7.3
Number of species (±SE) (a) in full leaf (categories IV and V), (b) with flowers (categories II–V), and (c) with fruits (categories II–V) for plots of three different successional stages of the TDF in Chamela, Mexico, for the period of study June 2006–March 2010. (d) Monthly rainfall, maximum and minimum temperatures, and day length are also shown.

of species; intermediate: 23%–30% of individuals, 27%–31% of species; late: 5%–21% of individuals, 15%–24% of species). At the community level, flowering was constant in intermediate stages; only for 2 months in the whole study period, no species were found flowering at this successional stage (December 2007 and September 2009) (Figure 7.3b).

The duration of flowering was longer in early and intermediate successional stages. In late stages, individuals and species were observed flowering for one to two months, whereas in earlier stages some individuals (1%–5.5%) and species (1.5%–5.5%) flowered for up to five months. This suggests different reproductive strategies among species of different successional stages. Species whose individuals flowered longer in early and intermediate successional sites were *C. suberosus* (three months), *Opuntia excelsa* (five months), *Acacia farnesiana* (three months), *Cnidosculus spinosus* (five months), *Caesalpinia caladenia* (four months), and *Cordia alliodora* (three months).

Across successional stages, there were species in fruit throughout the year, but fruiting intensity was greater in early and intermediate successional stages (Figure 7.1c). Mean fruiting intensity varied little in late successional stages, whereas peaks in intensity were higher in early and intermediate stages (Figure 7.1c) (range of intensity: late 0%–6%, early 0%–17%, and intermediate 0%–11%). The mean number of species fruiting at any given time was variable within and between years and successional stages (Figure 7.3c), and there were fruiting peaks during the dry and the wet seasons. A very small proportion of individuals (2%–5%) and species (5%–12%) bore fruits for more than four months. Individuals and species that bore fruits for the longest periods occurred in intermediate and late successional stages.

7.3.1.2 Species-Level Flowering Patterns

Frequency: supra-annual (44%–50% of the species) flowering and annual (35%–50% of the species) flowering were observed in all successional stages (Figure 7.4a). Subannual flowering was found in 13%–16% of the species in intermediate and late successional stages. One species with continuous flowering occurred in early stages (*A. farnesiana*). Regularity: the proportion of species with irregular flowering decreased from 78% in early stages to 44% in late stages; consequently, the proportion of species with regular flowering was higher in late successional stages (Figure 7.4b). Duration: around 60% of the species at the three successional stages had a brief flowering period (Figure 7.4c). The proportion of species with extended flowering was lower for early successional stages than for intermediate and late stages. No species with intermediate flowering were observed in late successional stages.

7.3.2 Costa Rica

In the 35 month period that this study lasted, only one month was registered with no precipitation (Figure 7.5d). Total annual precipitation was 806 mm in 2007–2008, 3007 mm in 2008–2009, and 1425 mm in 2009–2010 (11 months).

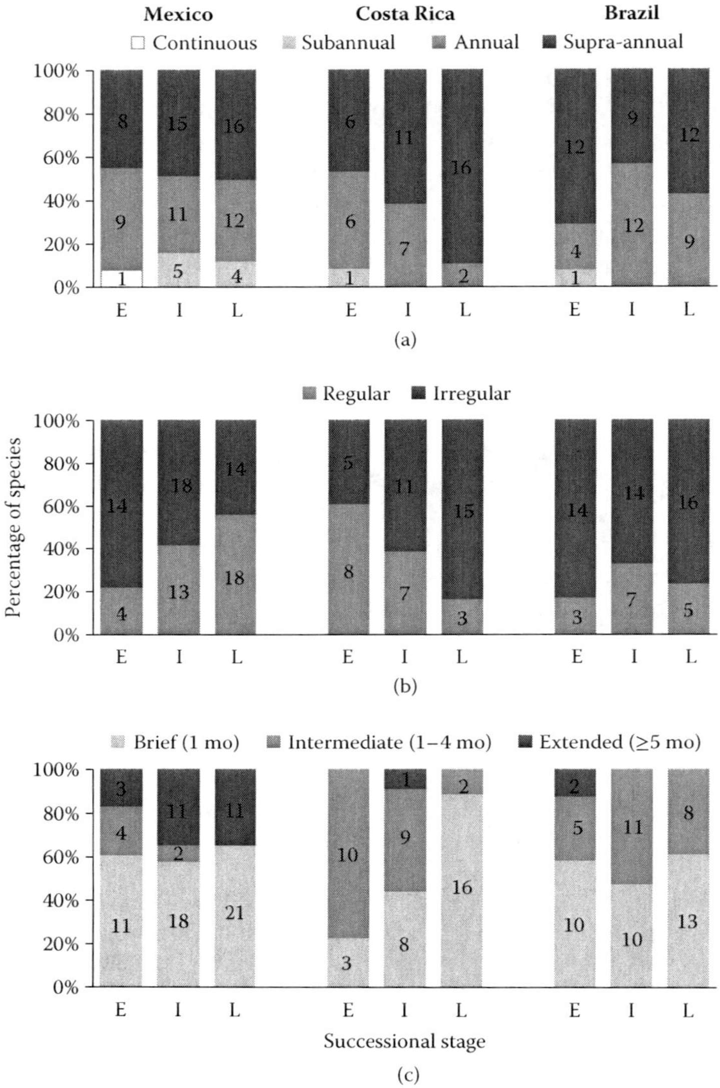

FIGURE 7.4
Proportion of species falling on the different categories of flowering frequency (a), regularity (b), and duration (c) at three different successional stages of the TDF in Chamela, Mexico (left); Santa Rosa, Costa Rica (central); and Mata Seca, Brazil, (right). The numbers inside the bars indicate the numbers of species in each category. In the figure, E = early successional stage; I = intermediate successional stage; and L = late successional stage.

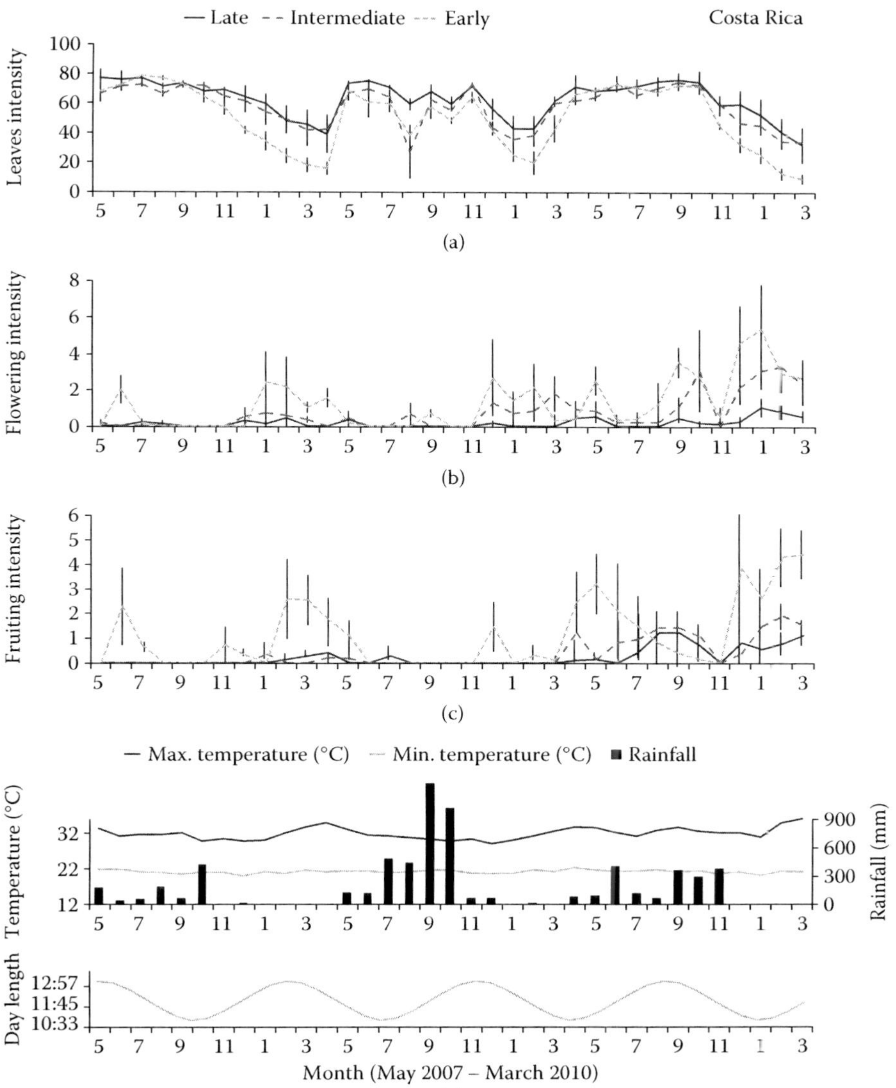

FIGURE 7.5
Mean monthly intensity (±SE) of individuals for the phenophases of (a) leaves, (b) flowers, and (c) fruits for the plots of three different successional stages of TDF in Santa Rosa, Costa Rica, for the period of study May 2007–March 2010. (d) Monthly rainfall, maximum and minimum temperatures, and day length are also shown. Intensity is measured as the percentage coverage of leaves, flowers, and fruits.

7.3.2.1 Plot-Level Phenology: Leaves, Flowers, and Fruits

Leafing intensity varied less across the year in Costa Rica than it did in Mexico and Brazil, with higher values of intensity being registered during the dry season and lower values during the rainy season (Figure 7.5a). Thus, seasonal cycles of leafing intensity were less marked across years in the Costa Rican TDF (Figure 7.5a). Leafing intensity started to decline in November–December and continued to drop during the dry season. Intermediate and late successional stages showed similar values of leafing intensity and number and proportion of species in mid to full leaf coverage (Figures 7.5a, 7.6a, and 7.7a). In contrast to the pattern observed in Mexico, early stages showed higher variation in leafing intensity than intermediate and late stages (late: 30%–71%; intermediate: 32%–67%; early: 9%–73%). Early successional stages also had a higher proportion of deciduous individuals and species (which lost their leaves during the dry season) and a lower proportion of species in mid to full leaf during the driest months (Figure 7.6a). The duration of the leaf phenophase within a cycle was greater for individuals and species in intermediate and late successional stages; a greater proportion of individuals (early: 25%–41%; intermediate: 46%–67%; late: 53%–73%) and species (early: 27%–56%; intermediate: 43%–70%; late: 45%–73%) maintained mid-to-full leaf coverage during 7–12 months.

As in Mexico, flowering intensity varied in time, and species could be found flowering almost every month of the year (Figures 7.5b and 7.7b). Flowering intensity and the proportion of species in flower were also higher in early and intermediate successional stages than in late stages (Figures 7.5b, 7.6b, and 7.7b). Flowering peaks are fairly synchronous among successional stages, with main peaks occurring during the dry season (Figures 7.5b, 7.6b, and 7.7b). Flowering peaks that occur during the rainy season are more evident in early successional stages. As in Mexico, the annual proportions of flowering individuals (early: 15%–35%; intermediate: 6%–17%; late: 3%–9%) and species (early: 27%–65%; intermediate: 10%–22%; late: 7%–20%) were higher in early stages. Longer flowering periods (three to five months) at the individual or the species level only occurred in early and intermediate stages. Species with more extended flowering periods in these sites include *Acacia collinssi* (three months), *Euphorbia schlechtendalii* (three months), *Semialarium mexicanum* (three months), *Byrsonima crassifolia* (four months), *Cordia guanacastensis* (five months), *Lippia oxyphyllaria* (three months), *Calycophyllum candidissimum* (three months), *C. alliodora* (three months), *Malvaviscus arborea* (five months), *Trichilia hirta* (three months), *Trophis racemosa* (four months), and *Gliricidia sepium* (three months).

Fruiting intensity varied between years and successional stages, but it was generally higher in 2009–2010. Early stages had peaks of fruiting intensity every year (Figure 7.5c). Species were recorded to fruit in both the dry and wet seasons, but in early successional stages the highest fruiting peaks tended to occur during the months with lower rainfall. Early stages also showed the greatest peaks of fruiting intensity, the greatest monthly percentage and

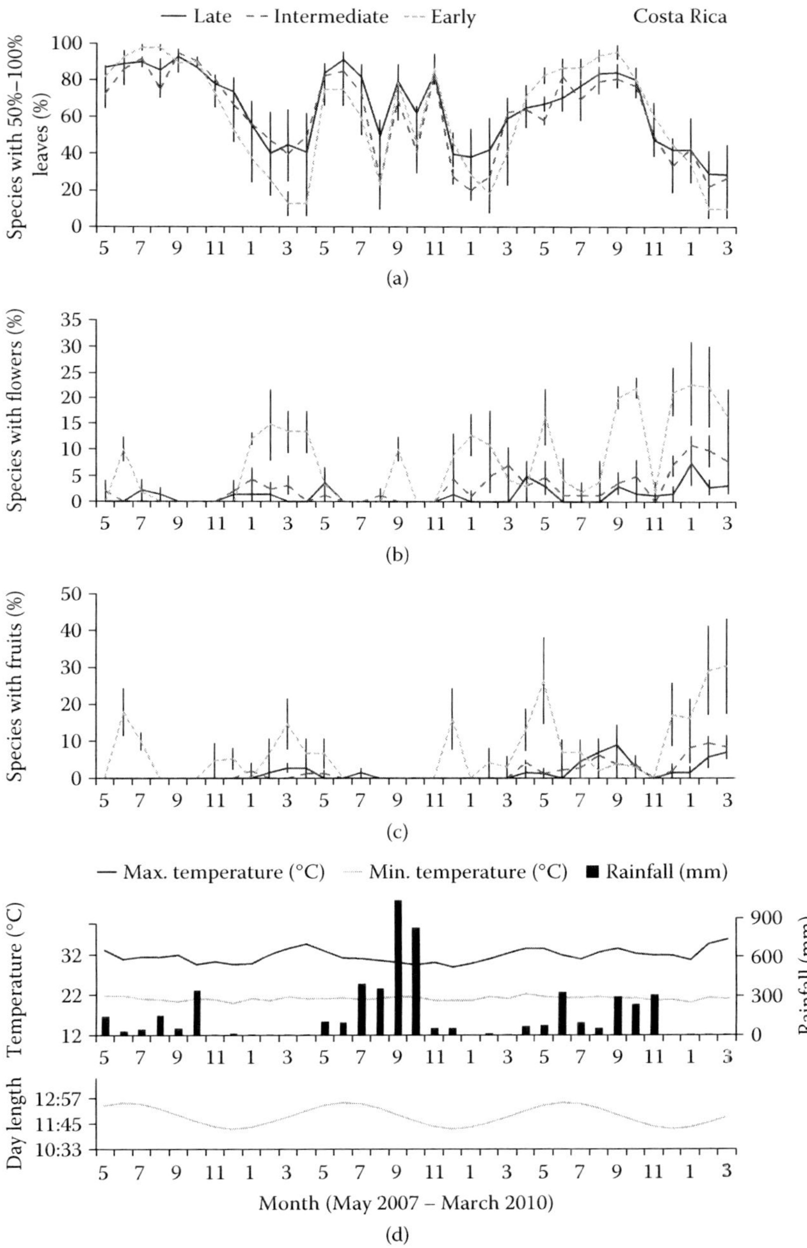

FIGURE 7.6
Proportion of species (±SE) (a) in full leaf (categories IV and V), (b) with flowers (categories II–V), and (c) with fruits (categories II–V) for plots of three different successional stages of TDF in Santa Rosa, Costa Rica, for the period of study May 2007–March 2010. (d) Monthly rainfall, maximum and minimum temperatures, and day length are shown.

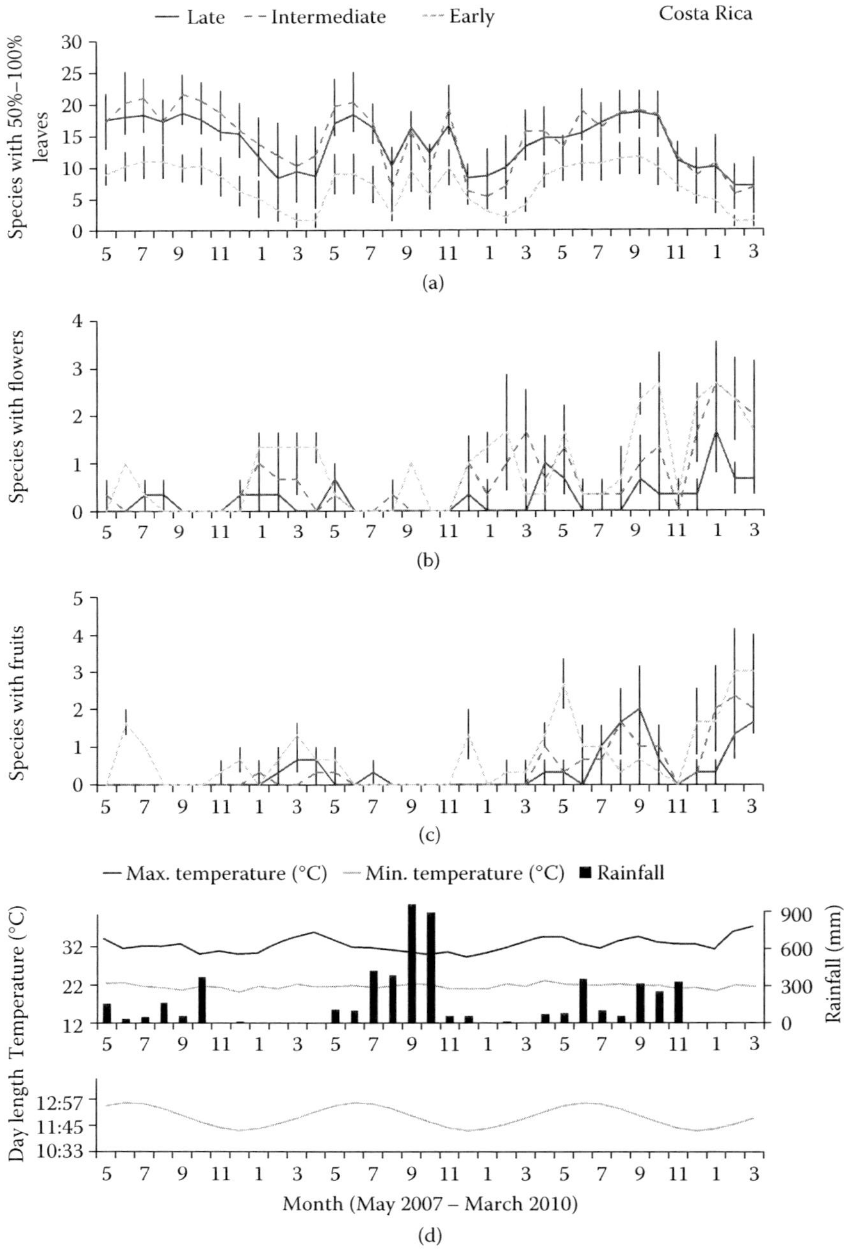

FIGURE 7.7
Number of species (±SE) (a) in full leaf (categories IV and V), (b) with flowers (categories II–V), and (c) with fruits (categories II–V) for plots of three different successional stages of TDF in Santa Rosa, Costa Rica, for the period of study May 2007–March 2010. (d) Monthly rainfall, maximum and minimum temperatures, and day length are shown.

number of species that fruited (Figures 7.5c, 7.6c, and 7.7c), and the highest annual proportion of individuals (early: 11%–21%; intermediate: 4%–15%; late: 1%–10%) and species (early: 23%–42%; intermediate: 5%–22%; late: 3%–22%) that fruited. Plants in early stages bore fruits for longer periods of time than in later stages (proportion of individuals that bore fruits for three to seven months: early 5%–8%, and intermediate and late 0%–3%; proportion of species: early 10%–22%, intermediate 0%–5%, and late 0%–5.5%) (in year two, no individuals bore fruits for more than two months in any stage).

7.3.2.2 Species-Level Flowering Patterns

Frequency: the proportion of supra-annual species increased from 46% in early stages to 61% and 89% in intermediate and late successional stages, respectively (Figure 7.4a). The remainder of the species was annual, except for one species (*S. mexicanum*) with subannual flowering in early stages. Regularity: the opposite pattern to the one found in Mexico was found in Costa Rica. The proportion of species with irregular flowering increased from 38% in early stages to 83% in late stages (Figure 7.4b). Duration: the proportion of species with brief flowering increased from 23% in early to 44% in intermediate and 89% in late stages (Figure 7.4c). The remainder of the species has intermediate periods of flowering, and one species (*M. arborea*) presented extended flowering in intermediate stages.

7.3.3 Brazil

Rainfall (Figure 7.8d) was 663 mm in 2007–2008, 1025 mm in 2008–2009 (data available for 11 months), and 563 mm in 2009–2010 (data for 10 months).

7.3.3.1 Plot-Level Phenology: Leaves, Flowers, and Fruits

Seasonal cycles of leafing intensity were well marked in the TDF sites in Brazil. The three successional stages had similar high and low levels of leafing intensity (late: 2%–76%; intermediate: 1%–72%; early: 1%–76%) in contrast to what happened in the TDFs in Mexico and Costa Rica. Leaf production started with the rainy season around October, with a tendency to occur at an earlier time in early stages of succession (Figures 7.8a, 7.9a, and 7.10a). The following species from early successional stages had leafing intensity values above 50% during October 2007, September 2008, and September 2009, coinciding with low precipitation values (Figure 7.9a and 7.9d): *Enterolobium contortisiliquum, Manihot anomala, Platypodium elegans, Zanthoxylum continifolium*, and *Zyziphus joazeiro*. During the rainy season, the proportion of species in full leaf tended to be lower in early stages than in late stages (Figure 7.9a).

In early successional stages, a greater proportion of individuals (early: 61%–73%; intermediate: 31%–47%; late: 41%–62%) and species (early: 54%–64%; intermediate: 23%–47%; late: 32%–60%) maintained mid-to-full leaf

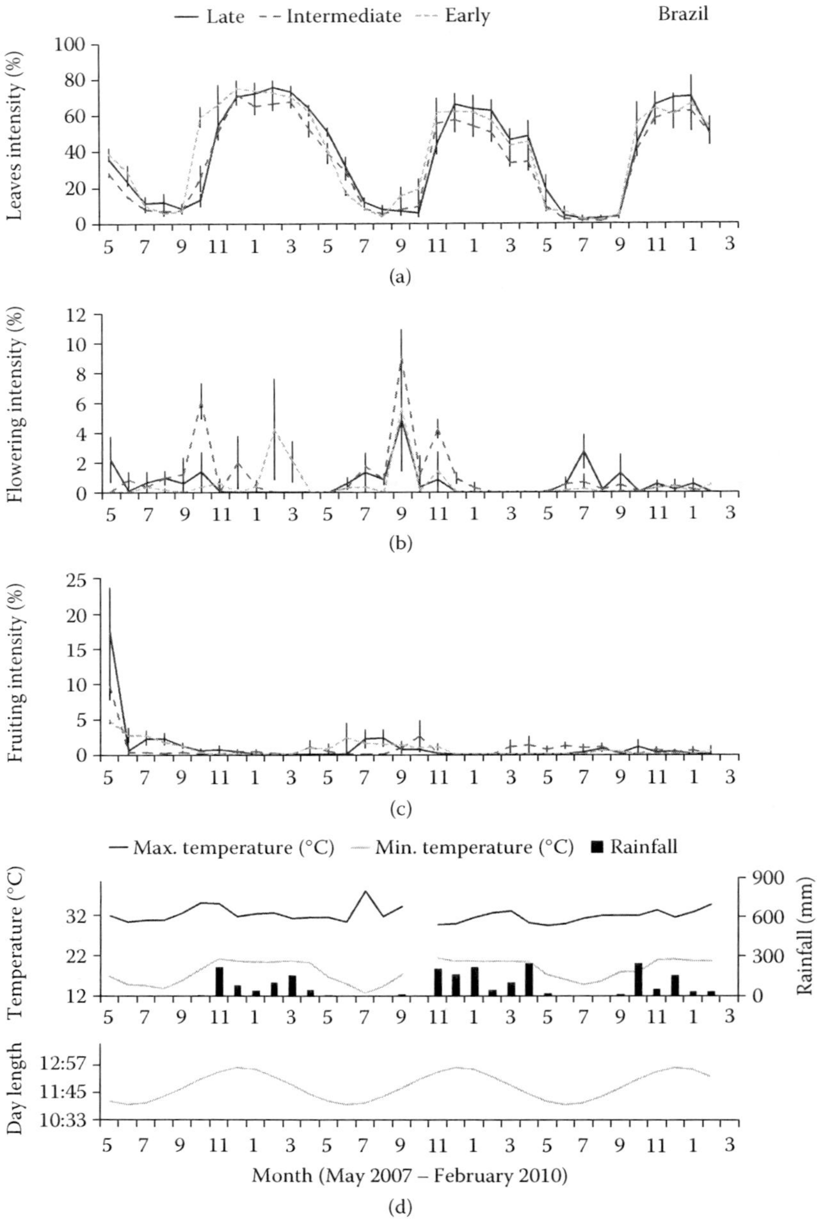

FIGURE 7.8
Mean monthly intensity (±SE) of individuals for the phenophases of (a) leaves, (b) flowers, and (c) fruits for plots of three different successional stages of TDF in Mata Seca, Brazil, for the period of study May 2007–February 2010. (d) Monthly rainfall, maximum and minimum temperatures, and day length are shown. Intensity is measured as the percentage of coverage of leaves, flowers, and fruits.

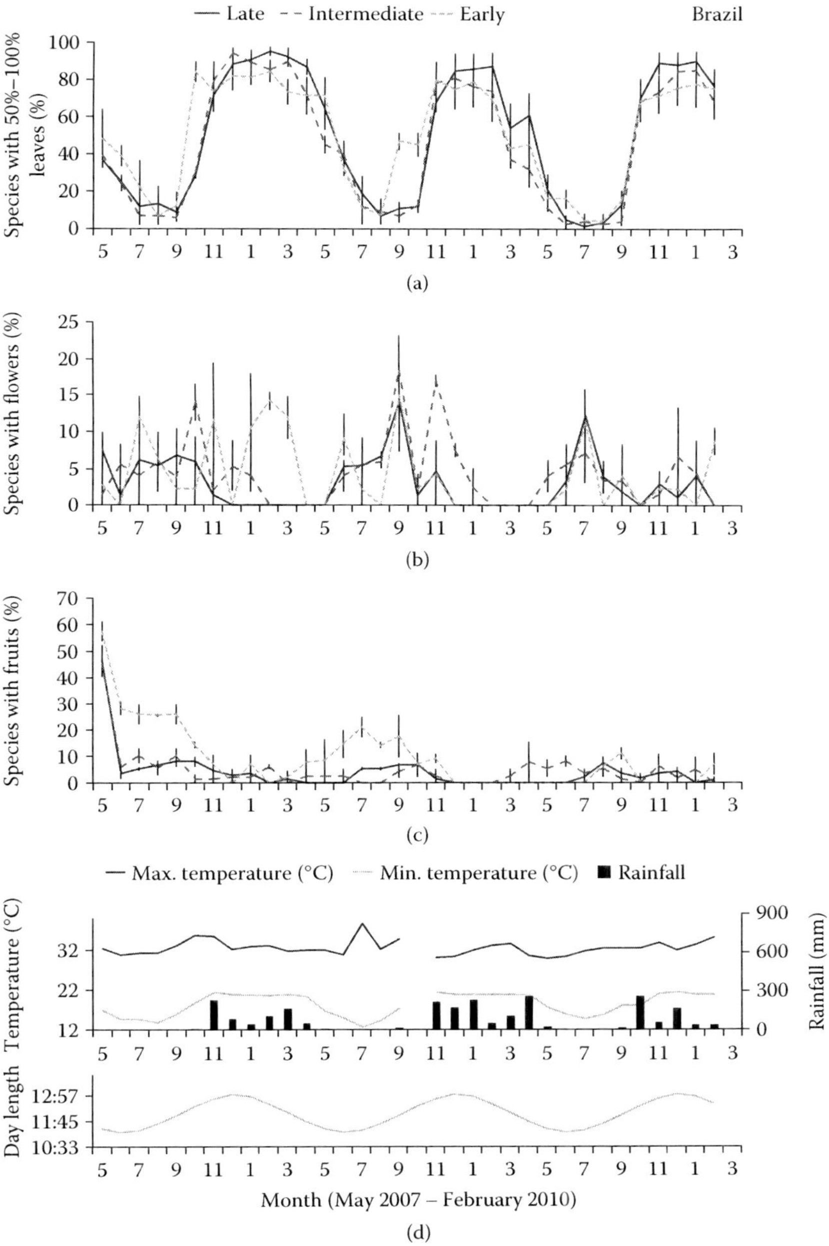

FIGURE 7.9
Proportion of species (±SE) (a) in full leaf (categories IV and V), (b) with flowers (categories II–V), and (c) with fruits (categories II–V) for plots of three different successional stages of TDF in Mata Seca, Brazil, for the period of study May 2007–February 2010. (d) Monthly rainfall, maximum and minimum temperatures, and day length are shown.

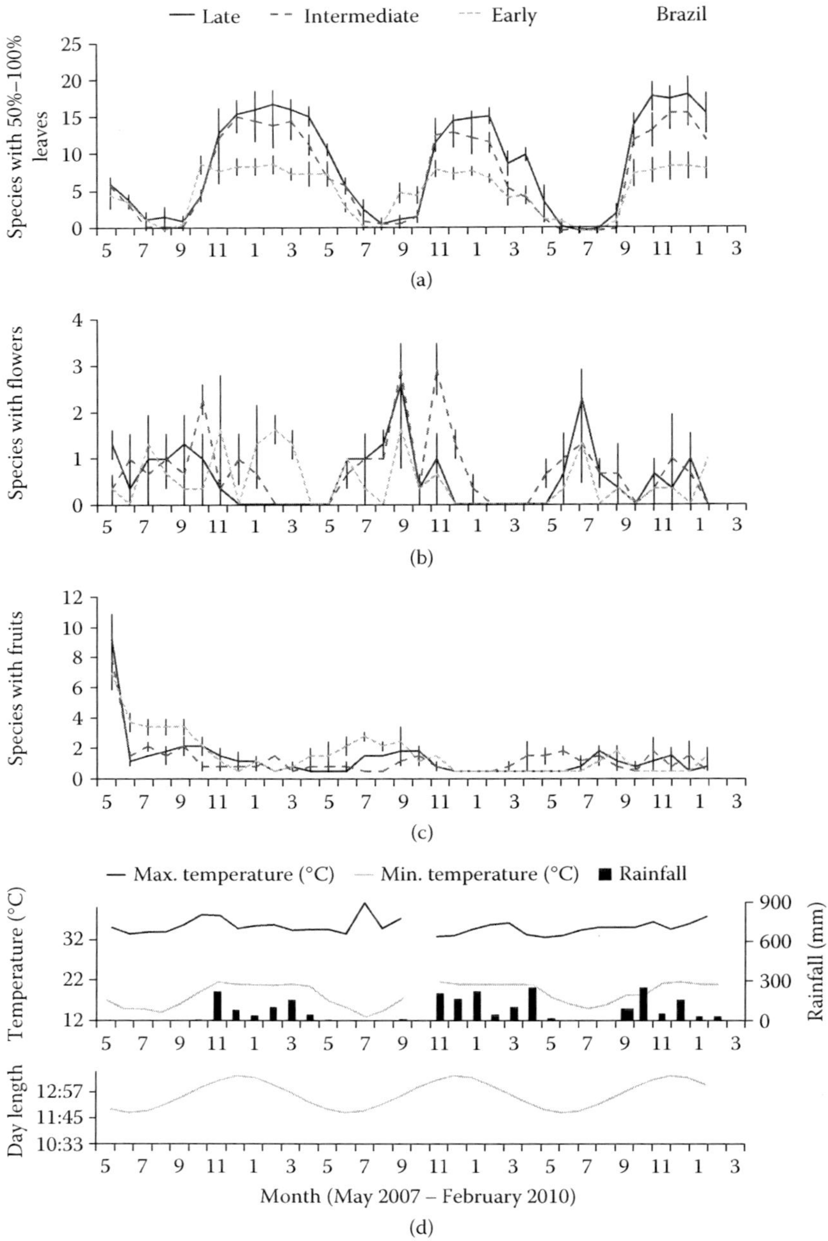

FIGURE 7.10
Number of species (±SE) (a) in full leaf (categories IV and V), (b) with flowers (categories II–V), and (c) with fruits (categories II–V) for plots of three different successional stages of TDF in Mata Seca, Brazil, for the period of study May 2007–February 2010. (d) Monthly rainfall, maximum and minimum temperatures, and day length are shown.

coverage for longer periods (6–12 months) than in later stages. The longer duration of the leaf phenophase in early successional stages is due to the earlier start in leaf production (Figure 7.9a). The start of leaf fall varied among years, and the pattern was similar across the three successional stages (Figure 7.8a).

As for the other sites, some species were recorded in flower most months, and flowering intensity varied within and among years (Figure 7.8b). Flowering was more regular across years and synchronous among successional stages during the dry season and the first half of the rainy season and more variable during the rainy season (Figures 7.8b, 7.9b, and 7.10b). The highest peaks in flowering intensity occurred in the intermediate stages of succession. Peaks in the monthly proportion and number of species in flower were of similar magnitude among successional stages (Figures 7.9b and 7.10b). A similar proportion of individuals (8%–29%) and species (17%–44%) flowered within years among successional stages. The duration of flowering was also similar among individuals and species of different successional stages; most flowered for one month and a few up to three months. In late stages, two species flowered for four months in the first year. The species with longest flowering were *Auxemma oncocalyx* (four months), *Cochlospermum vitifolium* (three months), *M. anomala* (three months), *Myracrodruon urundeuva* (four months), *Sapium* sp. (three months), *Senna spectabilis* (three months), and *Spondias tuberosa* (three months).

Mean fruiting intensity was highest during the first year (2007–2008) and during the dry season (Figure 7.8c). The proportion of fruiting species was highest for early successional stages in the dry season of the first two years of the study (Figure 7.9c). Furthermore, there was greater variability in fruiting intensity and frequency during the wet season across years and successional stages (Figures 7.8c and 7.9c). Synchrony and regularity in fruiting across the three successional stages occurs primarily at the end of the dry season in August, September, or October (Figure 7.10c). The highest proportion of fruiting individuals (45%–56%) and species (53%–63%) was recorded during the first year, and early successional stages consistently had the highest proportion of fruiting species across years (early: 63%, 35%, and 16%; intermediate: 59%, 13%, and 13%; and late: 53%, 9%, and 12% for years one, two, and three, respectively) and fruited for up to eight months in the first year.

7.3.3.2 Species-Level Flowering Patterns

Supra-annual flowering and annual flowering were the most common categories in the three successional stages, with the proportion of supra-annual species being higher for early stages (Figure 7.4a). One species with subannual flowering was observed in early stages (*Acacia polyphylla*). Most species presented irregular flowering in the three successional stages (67%–82%; Figure 7.4b). Most species had a brief flowering period (48%–62%; Figure 7.4c). Two species with extended flowering (*Calotropis procera* and *M. urundeuva*) were observed in early stages, and the remaining species in all the stages had intermediate flowering periods.

7.4 Discussion

7.4.1 Phenological Patterns of Tropical Dry Forests in a Latitudinal Gradient: A Synthesis

Leaf phenology patterns were similar across countries at different latitudes and across successional stages. In all sites, leaf flushing occurred immediately after rainfall accumulated for over 100 mm in the first month of the wet season and leaves persisted for several months until the precipitation dropped below 50 mm/month. Therefore, a precipitation threshold is necessary to maintain primary productivity and growth in TDFs. Rare, isolated rain events during the dry season may trigger leaf flushing, as shown in Mexico in 2010. Similarly, experimental irrigation studies in TDF trees have demonstrated that leaf production can be induced during the dry season for short periods of time (Borchert 1994; Hayden et al. 2010). However, if watering is not maintained these ephemeral leafing events do not lead to tree growth (Hayden et al. 2010).

Leafing intensity patterns were similar in the TDFs of Mexico and Brazil, whereas in Costa Rica leafing intensity was less variable and maintained higher levels over time. Similarities in the leafing intensities of Mexico and Brazil followed the average annual rainfall and seasonal precipitation episodes of these two countries. In Costa Rica, despite the seasonality of rainfall, greenness may be maintained for longer periods due to the higher annual precipitation.

Similar patterns of leaf phenology were described in previous studies in mature forests of Costa Rica and Mexico (Frankie et al. 1974; Bullock and Solis-Magallanes 1990), but our study is the first to describe phenological patterns in different successional stages in TDFs. Differences in leafing intensity patterns among successional stages were not consistent among TDFs. In Mexico, individuals retained their leaves longer in early successional stages during the first two years of the study; but this pattern disappeared during the last two years, after vegetation structure and species composition changed over time. A study in the TDF site of Mexico demonstrated that plants in early successional stages have adapted to high temperatures and high radiation loads by maximizing heat dissipation and minimizing water loss while maintaining photosynthetic performance (Alvarez-Añorve et al. 2012); furthermore, in species common to all successional sites leaf longevity is greater in early successional stages (Alvarez-Añorve et al. 2012). These functional adaptations may explain the leaf phenological patterns observed in Mexico and Brazil. In contrast, in early successional stages in Costa Rica individuals retained their leaves for a shorter period than in other stages. These results might be associated with the higher precipitation regime of Costa Rica, which is almost twice as high as the precipitation of the other two countries.

Flowering intensity patterns varied across countries, successional stages, and years. There were flowering peaks in the dry and the wet seasons in all sites and in all successional stages. Early and intermediate stages had higher levels of flowering intensity in all sites across years. Several studies of TDF trees have indicated that flowering intensity is greater in disturbed open sites—similar to early and intermediate successional stages—than in mature late successional forests (Cascante et al. 2002; Fuchs et al. 2003; Quesada et al. 2003; Herrerías-Diego et al. 2006; Rojas-Sandoval et al. 2008). Analyses of climatic environmental variables have indicated that disturbed habitats under succession have higher average temperatures, greater exposure to wind, lower soil moisture, and greater evapotranspiration than the understory of mature forests (Alvim 1960; Kapos 1989; Saunders et al. 1991; Wright 1996; Kapos 1997; Wright 1999; Laurance et al. 2002; Fuchs et al. 2003; Herrerias-Diego et al. 2006). However, little is known about the ecological and physiological processes that control flowering and fruiting phenology in tropical plants (Chapotin et al. 2003). Only a few studies have demonstrated that drought may trigger flowering (Alvim 1960; Wright et al. 1999) and that higher light conditions may prolong flowering in tropical plants (Marquis 1988). Therefore, these are fields of study that require more research to understand how environmental factors affect TDF regeneration processes.

The analysis of flowering frequency indicates that most species presented supra-annual or annual flowering at the three sites (Figure 7.4a). Overall, supra-annual flowering was found in a large percentage of the species (at least 43%), which is greater than that found in the wet tropical forest of La Selva Biological Station (10%–14%) (Newstrom et al. 1994; Bawa et al. 2003) or the TDF of Comelco in Costa Rica (11%) (Frankie et al. 1974). Every year, a large proportion of individuals and species did not flower. The high representation of supra-annual species may be due to the presence of juveniles in the samples, which might be more pronounced at early successional stages. Instead, the monthly censuses conducted in this study may have missed the flowering events of species with brief flowering periods of less than three weeks. Alternatively, water availability of TDFs may limit resources for individuals to reproduce annually, particularly for younger reproductive individuals. The representation of species with supra-annual flowering was lower at early successional sites only in Costa Rica, as would be expected due to the selection for rapid growth and early reproduction in early successional species (Baker 1974; Bazzaz 1979; Bawa et al. 2003). Consistent with this hypothesis, early successional stages showed continuous flowering in Mexico and subannual flowering in Costa Rica and Brazil. A higher abundance of species with continuous and subannual flowering in early and intermediate successional communities may contribute to the higher flowering intensity in these stages, described in the aforementioned results.

The analysis of flowering regularity showed that irregular flowering was common in all successional stages in Brazil, in late stages in Costa Rica, and in early stages in Mexico (Figure 7.4b). Irregular flowering should be more common in early successional stages because short-lived early successional species (*r*-selected species) should have been selected to flower or fruit at any time of the year (Kang and Bawa 2003). This hypothesis was true for Mexico, where 78% of the species had irregular flowering and the proportion decreased toward the late successional stage. However, the opposite pattern was observed in Costa Rica and no differences among successional stages were observed in Brazil where irregular flowering was predominant. Interestingly, the ratio of the number of regular to the number of irregular species was similar at the intermediate stages of all countries (Figure 7.4b), suggesting that flowering frequency strategies converge at these sites, possibly because they share species from early and late successional stages.

Flowering duration varied across countries (Figure 7.4c). In Mexico and Brazil, most species (between 48% and 66%) flowered for a brief period of time across all successional stages. Species with extended flowering were mainly found in the intermediate and late stages in Mexico. In Costa Rica, 77% of species showed intermediate flowering at the early stages and the proportion decreased toward the late successional stages. At the individual level, the duration of flowering was greater in earlier successional stages at all sites. In Mexico and Costa Rica, individuals in late stages flowered only for one or two months, whereas in early and intermediate stages some individuals flowered for up to five months. In Brazil, longer flowering periods were also more common in early and intermediate sites than in late sites. Therefore, the representation of species in the categories of frequency, regularity, and duration of flowering phenology of the communities was not always consistent with the higher flowering intensity observed in the early and intermediate stages. A high abundance of species with long and/or continuous flowering may contribute to the intensity patterns observed.

Variations in flowering intensity and phenological patterns associated with succession should also reflect variations in pollinator communities. Extended, continuous, and intense flowering may be associated with generalist pollination in early succession to ensure reproduction. Pollination systems are unexplored in this chapter, but a plant–pollination network study in Mexico (Lopezaraiza and Quesada, unpublished data) indicates that the abundance and species richness of pollinators is higher in early succession and is explained by a higher floral abundance at this stage. The network analysis shows that the pollinator community is generalist at early and intermediate successional stages, and more specialized at late stages, and consequently pollinator assemblages are more similar among plant species at early stages than at late successional stages. Variations in the synchrony and intensity of flowering associated with pollination at different successional stages will affect the reproduction, genetic structure, and mating patterns of tropical plant populations in disturbed and mature TDFs (Quesada et al. 2004, 2011).

Timing of fruit and seed production is key to understanding the dispersal, regeneration, and establishment of natural populations along succession. The phenological intensity and timing of flowering patterns will directly affect fruiting and seed dispersal patterns. Fruiting intensity varied across countries, successional stages, and years. If reproduction was successful, the higher intensity of flowering at early and intermediate stages should be reflected in fruiting patterns and fruiting peaks should lag behind flowering peaks. Early and intermediate stages did have higher fruiting intensities than late successional stages in Mexico, and in Costa Rica early stages had the highest intensity. By contrast, fruiting intensities in Brazil were similar for all successional stages. Fruiting peaks did not necessarily follow flowering peaks. However, a higher proportion of species fruited during the dry season across countries and successional stages (Figures 7.2c, 7.6c, and 7.9c), and fruiting peaks were greater for the early successional stages. Frankie et al. (1974) found for the Comelco TDF in Costa Rica that most species fruit during the dry season regardless of their flowering season and dispersal syndrome, although anemochory (42% of species) is slightly more common than endozoochory (36%) during this season. In TDFs, a large proportion of species are wind dispersed and depend on high temperature and low relative humidity for abscission and dispersal (Greene et al. 2008). Variations in such environmental conditions in disturbed habitats will change fruit maturation and seed dispersal patterns in successional environments.

Seed dispersal is an integral part of successional and regenerative processes in tropical ecosystems (Hardwick et al. 2004). Successful colonization by particular species within early successional areas depends not only on many microclimatic factors (Maluf de Souza and Ferreria Batista 2004; Holl 1999) but also on species' ability to disperse. If no nuclear trees (isolated trees that can attract animal dispersers) were found in disturbed areas, early successional stages should be more successfully colonized by wind-dispersed species (Janzen 1988). The higher proportion of species dispersing during the dry season at early successional stages suggests that more wind-dispersed species are successfully colonizing these stages. Once some vegetation is established within areas of early succession, these areas may be used by animals as resting or feeding areas, bringing seeds of endozoochorous plants. In intermediate stages, the proportion of species dispersed during the dry season is lower than that during the early stages. Further analysis on differences in dispersal syndromes among successional stages will shed light on how seed dispersal affects TDF ecological succession.

7.5 Conclusions

Despite there being differences in phenological patterns among countries, we found differences among successional stages within countries, and some general patterns across countries. The intensity of the leaf phenophase differed between stages. In Costa Rica leafing intensity was lower along the dry season in early successional stages, whereas in Mexico it was higher during the dry season. In Brazil, leafing intensity tended to be higher at the start of the rainy season in early successional stages. We propose that these differences may be explained by different functional responses of the plant communities that follow differences in precipitation regimes among countries. Flowering phenological patterns consistently differed among successional stages, being more intense in early and intermediate stages than in late stages. This higher intensity is partly explained by a consistently longer flowering duration of individuals at earlier stages and partly by plant species traits (flowering frequency, regularity, and duration) at different successional stages. Nevertheless, different traits contribute to explain the higher intensity for the different countries. Finally, following flowering intensity patterns, fruiting intensity tended to be higher for early and intermediate stages, but only in Mexico and Costa Rica. A higher proportion of species fruited during the dry season across countries, and the highest peaks were in this season for early stages, suggesting a higher proportion of wind-dispersed species at these sites. An analysis of the phenological responses and abundance of common species and of species exclusive to the different successional stages should unravel the extent to which differences in phenological patterns are associated with species replacement along successional gradients and with species plasticity. Further work is needed regarding the physiological processes and environmental cues responsible for the observed differences in phenological patterns along succession.

Appendix

TABLE 7.1

Results of the Analysis of Variance Comparing the Number of Individuals and Species Falling on the Phenology Transects on Three TDFs in a Latitudinal Gradient: Chamela, Mexico; Santa Rosa, Costa Rica; and Mata Seca, Brazil.

	Response Variable	*F*	Model d.f., Error d.f.	*p* Value
Mexico	Number of species	23.79	2, 6	0.0014
	Number of individuals	56.64	2, 6	0.0001
Costa Rica	Number of individuals	0.57	2, 6	0.5914
	Number of species	6.40	2, 6	0.0326
Brazil	Number of individuals	21.64	2, 6	0.0018
	Number of species	2.90	2, 6	0.1316

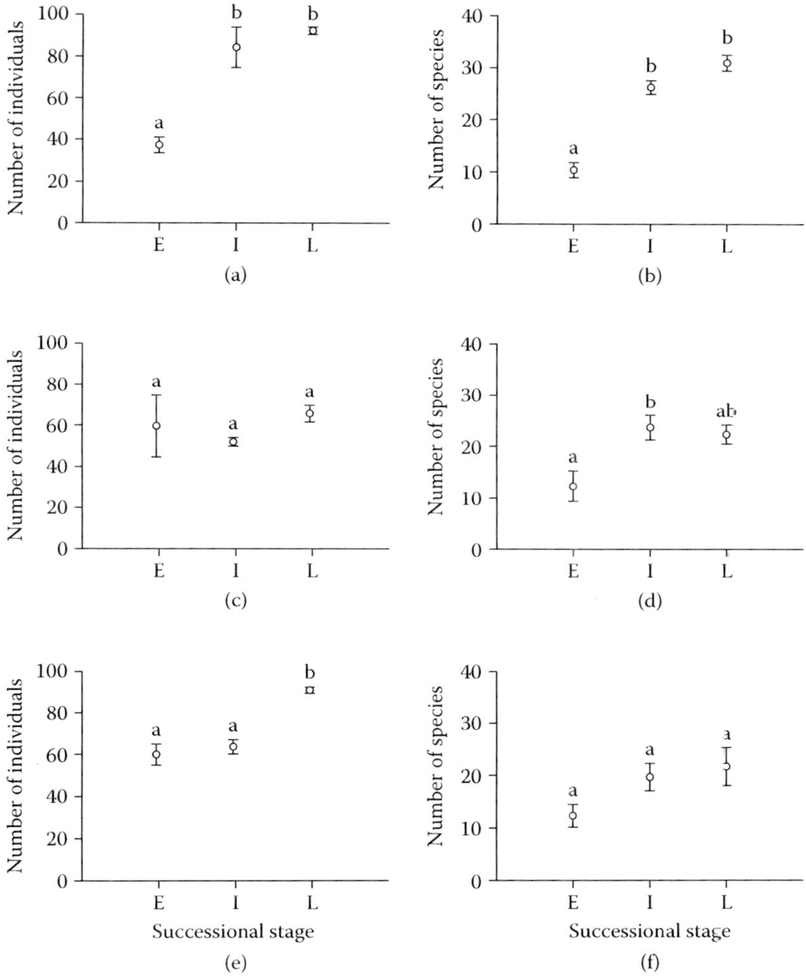

FIGURE 7.11
Mean (±SE) number of individuals (a, c, e) and species (b, d, f) at the phenology transects of different successional stages for Chamela, Mexico (a, b), Santa Rosa, Costa Rica (c, d), and Mata Seca, Brazil (e, f): different letters above the bars denote significant differences among successional stages.

FIGURE 7.12
(See color insert.) Photographs of the TDF of Chamela, Mexico, taken from the same position on 6 consecutive days at the start of the rainy season in 2009. Leaf expansion of the closest trees happening within a few days can be appreciated; in a week's time, the forest turned green.

8

Plant Functional Trait Variation in Tropical Dry Forests: A Review and Synthesis

Catherine M. Hulshof, Angelina Martínez-Yrízar, Alberto Burquez, Brad Boyle, and Brian J. Enquist

CONTENTS

8.1 Introduction

Plant functional traits are an integrative measure of plant fitness in different environments and strongly impact ecosystem processes such as primary productivity, decomposition, and nutrient cycling (Diaz et al. 2004). Across environmental gradients, shifts in functional traits at the whole-plant level are consistent with a trade-off between rapid biomass production and efficient nutrient conservation (Reich et al. 1997). Species in resource-rich environments tend to be fast growing and have rapid rates of resource acquisition and tissue turnover, whereas species in resource-poor environments tend to have a resource conservation strategy with slow rates of growth and tissue turnover and high longevity (Westoby et al. 2002). The diversity and spacing (or dispersion) of plant functional traits along environmental gradients also provides insight into community assembly and species coexistence mechanisms (Weiher and Keddy 1995). Despite the explosion of trait-based studies in ecology, most studies emphasize trait patterns in temperate, Mediterranean, or tropical moist forest ecosystems. Surprisingly, there are relatively few studies addressing patterns of functional trait diversity in tropical dry forests (TDFs) or other extensive water-limited ecosystems (Chaturvedi et al. 2011). Thus, our understanding of the underlying drivers of plant trait variation and community assembly in TDFs is unclear.

TDFs are characterized by pronounced seasonality in precipitation, with several months of prolonged drought, 80% of annual precipitation occurring during a four- to six-month rainy season, and high interannual rainfall variability (Mooney et al. 1995; Pennington et al. 2009; Maass and Burgos 2011). The marked temporal availability of soil water gives rise to adaptive traits associated with the avoidance, resistance, or tolerance of water stress (Olivares and Medina 1992). Unsurprisingly, adaptations to strong seasonal drought and limited water supply, high solar irradiance, and high evaporative demand largely determine the physiological activity, growth, reproduction, and survival of dry forest plant species (Borchert 1994b). A review of previous functional trait studies within TDFs and an overview of important traits in dry forests have recently been described (e.g., Chaturvedi et al. 2011). In this chapter, we highlight three important functional traits related to water stress. Specifically, leaf phenology (i.e., deciduous, semideciduous, evergreen, etc.), wood-specific gravity (i.e., wood density), and specific leaf area (SLA) are strongly linked to soil water availability and whole-plant water status in TDF trees.

8.1.1 Variation in Deciduousness

One striking aspect of TDFs is the range of species variation in deciduousness (Sobrado 1991), for example, dry tropical forests usually consist of both evergreen and deciduous species. The degree of deciduousness and patterns of leaf renewal and senescence in dry forest trees are thought to be strongly linked to the extent and intensity of seasonal drought (Borchert 1994) as well as to spatial heterogeneity in soil water availability (Janzen 1986a; Powers et al. 2009a; but see also Aide 1992). A rich area of research is whether plants with different leaf phenologies (i.e., degrees of deciduousness or leaf habit) also possess different suites of functional traits (Eamus 1999; Powers and Tiffin 2010). For example, in TDFs, broadleaf deciduous trees were reported to have higher nitrogen (N) content, SLA, and rates of photosynthesis compared with broadleaf evergreen trees (Sobrado 1991; Prior et al. 2003; Ishida et al. 2006; Kushwaha et al. 2011; Pringle et al. 2011). In contrast, recent studies found that interspecific variation in leaf phenology is insufficient to predict leaf function and physiology within TDFs (Brodribb and Holbrook 2004; Powers and Tiffin 2010) and highlight the diversity of functional strategies within phenological categories (see Giraldo and Holbrook 2011). In addition, leaf deciduousness can vary within species depending on local microhabitat conditions (Elliott et al. 2006) such as differences in temperature, soil water availability, and rooting depth (Singh and Kushwaha 2006). Since previous studies often categorize a species' deciduousness based on literature reviews or local expert knowledge, it is difficult to determine the actual variability of deciduousness within species. An individual approach in which both trait data and phenological observations are recorded from the same individuals over time would further clarify the interaction between leaf phenology and functional strategies (e.g., Kushwaha et al. 2011).

8.1.2 Wood Density

An example of how plant functional traits can elucidate key ecological and physiological strategies is the linkage between plant distribution, leaf phenology, and wood density. Wood density is an important factor in determining carbon stocks (Chave et al. 2009) and resource turnover and is thus a critical functional trait for linking species-level physiology to ecosystem-level function. Wood density is strongly related to soil water availability (Borchert 1994), with a larger proportion of denser woods in drier soils (Borchert 1994), and is known to respond to gradients of soil fertility and climate (Swenson and Enquist 2007). For example, in TDFs of Costa Rica and western Mexico, dry upland forest sites primarily comprise deciduous hardwood trees and water-storing softwood trees (hardwoods such as *Lysiloma microphyllum*, *Tabebuia* spp., and *Guaiacum coulteri*; softwoods like *Cochlospermum vitifolium*, *Bursera* spp., and several species of columnar cacti), whereas evergreen and semideciduous, hard- and softwood trees (hardwoods such as *Enterolobium cyclocarpum*, *Brosimum alicastrum*, and *Hymenaea courbaril*; softwoods like *Ceiba pentandra*, *Bursera instabilis*, and *Cedrela* spp.) are confined to moist lowland forests (Janzen 1986; Borchert 1994). Thus, tree water status, wood density, and phenology depend on functional adaptations to drought in addition to local environmental variables (Borchert 1994). In the context of TDFs, wood density is particularly informative because it represents a fundamental trade-off between growth, mechanical support, hydraulic conductivity, and water storage capacity (Niklas 1995). Denser wood is thought to convey stronger mechanical stability (Niklas 1995), higher resistance to xylem embolism and cavitation (Hacke et al. 2001a), slower stem radial growth, longer life span, and lower rates of mortality (Enquist et al. 1999). As a result, high-density wood may allow heavy-wooded plant species in TDFs to continue growing when light-wooded species, particularly those that lack water storage tissues, are forced to shut down water transport during drought (Chave et al. 2009). Indeed, wood density emerges from vascular properties such as diameter and frequency of xylem conduits; wood density increases with decreasing conduit diameter and frequency and increasing fiber density (Russo et al. 2010). Some studies suggest that diameter growth rate may be more strongly related to xylem traits (i.e., lumen area and vessel density) than wood density (Russo et al. 2010), and xylem conductivity is known to critically determine plant water balance (Gleason et al. 2012).

Differences in plant functional traits and trade-offs between different trait axes are thought to cause functional divergence and niche separation of tree species along gradients of light and water availability (Poorter 2005;

Sterck et al. 2011). A better understanding of spatial and temporal variation in functional traits including wood density and wood hydraulic properties will further clarify the role of trade-offs in plant physiological functioning and species distributions in TDFs.

8.1.3 Specific Leaf Area

SLA, defined as the light-capturing surface area per unit of dry biomass, reflects a trade-off between resource capture and conservation (Poorter 2009) and has been shown to correlate with net photosynthetic capacity, leaf longevity, relative growth rate, wood density, and competitive ability (Reich et al. 1997). SLA is thought to vary with leaf phenology, with deciduous species having higher SLA and evergreen species having lower SLA values, representing a trade-off between short leaf life span with high resource acquisition and long leaf life span with slow resource acquisition (Reich et al. 1997). As already mentioned, although the correlation between SLA and leaf phenology has been shown for a collection of evergreen and deciduous tropical species (e.g., Sobrado 1991, Ishida et al. 2006, Pringle et al. 2011) the generality of this relationship is still unclear (see Williams-Linera 2000, Brodribb and Holbrook 2004, and Powers and Tiffin 2010). SLA has also been shown to shift across large-scale latitudinal gradients (Reich et al. 1997). Specifically, SLA was found to be higher in TDFs compared with wetter tropical forests (Gotsch et al. 2010). In addition, leaf traits such as leaf mass per area (i.e., the inverse of SLA) were found to have lower variation in dry forests in response to light compared to wetter forests. This finding suggests that light-related variation is not as important for dry forest plant species as for wet forests and the low soil water availability likely constrains the magnitude of leaf trait variation (Markesteijn et al. 2011). However, a comparison of SLA values between contrasting ecosystems across the world, including tropical dry and rain forests, showed that dry forest ecosystems have some of the highest variation in SLA (Villar and Merino 2001). Differences in SLA are likely important for the coexistence of species in TDFs, because higher variation in SLA within a community might allow different species to partition resources across the highly variable environment of TDFs in both time (i.e., rainfall seasonality) and space (i.e., variation in topography and soil water content). The magnitude and patterns of trait variation in TDFs compared with other ecosystems is an interesting area of research for understanding the assembly of dry forest communities as well as for understanding species distributions within and between TDFs and across dry–wet forest transitions.

8.1.4 Trait Variation, Nutrient Cycling, and Ecosystem Functioning

The functional composition and phenology of TDFs has direct consequences for ecosystem functioning, particularly for seasonal variation in primary productivity and biogeochemistry (Giraldo and Holbrook 2011; Jaramillo et al. 2011), and energy, water, and carbon cycles (Sanches et al. 2008; Maass and Burgos 2011). Although studies on net primary productivity and biogeochemistry in TDFs are still very scarce (Jaramillo et al. 2011), ecosystem functioning can be related to several key functional traits (Murphy and Lugo 1986b). For example, it has been reported that TDF species generally have higher leaf N content, SLA, and light-saturated photosynthetic rates and lower leaf life span and N:phosphorous ratio compared to desert, tropical evergreen, temperate evergreen, and tundra biomes (Chaturvedi et al. 2011). These characteristics have direct consequences for the timing and nutrient composition of leaf fall (Martínez-Yrízar et al. 1999; Xuluc-Tolosa et al. 2003), rates of decomposition (Harmon et al. 1995; Sundarapandian and Swamy 1999; Powers et al. 2009), as well as carbon and nutrient cycling (Jaramillo et al. 2011). Specifically, leaf N concentration or lignin:N ratio are among the best predictors of decomposition rates (Cornwell et al. 2008), yet due to dry conditions decomposition is drastically reduced during the dry season compared to wetter forests (Powers et al. 2009). Further, water availability is known to regulate nutrient dynamics in litter and soil in TDFs (Jaramillo et al. 2011; Anaya et al. 2012).

Perhaps because water plays such a dominant role in the regulation of structure and dynamics of TDFs, very little attention has been given to the interaction between soil nutrients and plant functional traits within TDFs (Murphy and Lugo 1986; Jaramillo et al. 2011). Leguminous species are particularly dominant in TDFs (Gentry 1995; Pennington et al. 2009) and are critical in determining the levels and rates of N fixation within communities. However, detailed studies relating the contribution of fixed N to ecosystems is surprisingly rare, although, given the seasonality, water availability regimes may affect nutrient acquisition more strongly in species without root symbionts than those with N-fixing bacteria or mycorrhizae. Among the few studies relating the amount of fixed N in tropical forests, the majority have taken place in humid regions except for a few recent studies in TDFs (Freitas et al. 2010). There is thus a large knowledge gap related to the effects of leguminous, N-fixing plant species on ecosystem N dynamics.

8.2 Trait Variation and Environmental Gradients

8.2.1 Environmental Gradients within Tropical Dry Forests

The degree of deciduousness and other plant adaptations to seasonal drought varies both within TDFs as a result of topography, forest structure, and land-use history (Janzen 1986; Powers et al. 2009) and between TDFs due to differences in annual rainfall, temperature, and solar radiation regimes. In addition, on a landscape scale geological substrates may decouple the expected relationship between rainfall and deciduousness (Bohlman 2010). TDFs are characterized by high environmental complexity (Mooney et al. 1995) and, as a result, the inherent variability of dry forest ecosystems allows only coarse generalizations regarding the ecological characteristics and classifications of dry forests (Murphy and Lugo 1986; Burquez and Martínez-Yrízar 2010). Topography within seasonal dry forests is a major driver of insolation and hydrologic processes (Martínez-Yrízar et al. 2000) and results in patchy availability of soil nutrients (Roy and Singh 1994) and water (Oliveira-Filho et al. 1998; Daws et al. 2002; Segura et al. 2003). For example, gravity-driven runoff causes a larger moisture accumulation in lower topographic positions (Daws et al. 2002; Markesteijn et al. 2010). Temporal variation is also another obvious source of variation in water availability; during the wet season most of the water is found in the top soil layers, whereas during the dry season more water is available in deep soil layers (Markesteijn et al. 2010).

Within TDFs, variation in forest structure, species diversity, and distributions reflect differences in soil water availability (Borchert 1994; Oliveira-Filho et al. 1998, Balvanera et al. 2011; Enquist and Enquist 2011). Due to increased resource partitioning within TDFs as a result of the spatiotemporal heterogeneity of soil water availability, functional shifts in water use and acquisition should be apparent across gradients of soil moisture (Poorter 2005; Sterck et al. 2011). Although this has been shown for wood density, phenology, and other key plant traits for TDF species in northwestern Costa Rica (Borchert 1994, Gotsch et al. 2010) and Bolivia (Markesteijn et al. 2010; Sterck et al. 2011), this remains an interesting area of research for understanding the maintenance of species diversity and functional diversity within TDFs. SLA, for example, plays an important role in the assembly of plant communities (Weiher and Keddy 1995). Within other forests, SLA has been shown to play a large role in the assembly of species across gradients of light and water availability (Poorter 2009). Although temporal and spatial variation in water availability drive habitat associations and niche partitioning within tropical forests (see Segura et al. 2003 and Balvanera et al. 2010), whether variation in plant functional traits such as SLA reflect shifts in plant distributions both within and between TDFs is unknown; yet it would provide a possible way to distinguish TDFs.

Another primary source of environmental variation within TDFs is land-use and anthropogenic disturbances. TDFs are widely cited as the most fragmented and endangered tropical forest type (e.g., Janzen 1988a; Gentry 1995). Variation in land-use history, time since last disturbance, frequency of disturbance, and size of disturbance, among other key factors (Chazdon et al. 2007), strongly influence the functional variation and assembly of resulting plant communities in TDFs (Lebrija-Trejos et al. 2010). Understanding how plant- and ecosystem-level functioning change over time could enhance the ability to predict and manage successional forests to recover key ecosystem services or functions long before they recover, if any, floristic similarity to previous conditions (Alvarez-Yépiz et al. 2008). On a fundamental level, secondary succession is community assembly in action (Lebrija-Trejos et al. 2011). Studies relating plant functional diversity along successional trajectories within TDFs argue that abiotic processes restrict initial colonization stages while competitive interactions constrain later successional stages (Lebrija-Trejos et al. 2011). The majority of successional studies to date, however, use chronosequences or a series of sites of different ages (e.g., Kennard et al. 2002; Lebrija-Trejos et al. 2011). These studies are based on the assumption that environmental conditions, site history, and seed availability are similar across sites and over time (Chazdon et al. 2007), which rarely reflects true site conditions. Thus, long-term studies are needed to monitor the functional properties and assembly of specific sites over time in TDFs (e.g., Chazdon et al. 2007; Enquist and Enquist 2011). In summary, more research is needed to quantify the spatiotemporal variation in functional composition and ecosystem processes in TDFs over large spatial and temporal scales (Maass and Burgos 2011; Anaya et al. 2012).

8.2.2 Geographical Gradients: A Trait-Based Approach for Comparing TDFs

A primary research goal within TDF studies is to identify the current extent and the degree of fragmentation of TDFs (Sánchez-Azofeifa et al. 2005c). However, as mentioned, the inherent variability of dry forest ecosystems allows only coarse generalizations regarding the ecological characteristics and classifications of dry forests (Murphy and Lugo 1986; Burquez and Martínez-Yrízar 2010). Attempts to characterize differences among TDFs by the degree of deciduousness and to explain them by variations in the amount and seasonality of annual rainfall have been challenging (Murphy and Lugo 1986; Burquez and Martínez-Yrízar 2010). A functional trait–based approach offers insight into the functional differences between and within TDFs across large geographical gradients. At local scales, previous studies have been successful in linking coordinated changes in plant functional traits to seasonality and changes in mean annual precipitation over space (e.g., Poorter 2009; Gotsch et al. 2010; Sterck et al. 2011) and time (e.g., Rentería and Jaramillo 2011). To test whether functional diversity can indeed describe functional differences between and within TDFs across large geographical gradients, here we compare functional diversity of one key leaf trait (SLA) between neotropical forests and climatic factors. First, we compare the magnitude of functional diversity in TDFs with other tropical forest ecosystems and, second, we relate functional diversity to climatic variables across broad spatial scales.

There are two contrasting hypotheses as to how climate and seasonality should influence functional diversity in TDFs. On the one hand, increased seasonality and decreased precipitation are strong environmental filters for drought-adaptive traits (Poorter 2005) and should thus limit trait variation at the community level. Specifically, the extended periods of drought that are characteristic of dry forests (usually from 6 to 8 months) can be thought of as a selective filter, allowing only those species with traits that confer drought avoidance, resistance, or tolerance to coexist in these water-limited ecosystems. As a result, we expect functional diversity to decrease in drier, more seasonal environments. On the other hand, if seasonality results in an increase in temporal and spatial variation in soil moisture then we might expect an increase in functional diversity in more seasonal forests. Indeed, TDFs are characterized by highly heterogeneous habitats in both space and time. The differing abilities of plant species to tolerate drought leads to niche differentiation, for example, within forests species sort out along slope gradients of water availability (Balvanera et al. 2010). Thus, spatiotemporal heterogeneity within TDFs can allow for increased functional diversity despite a shorter growing season compared to wetter tropical forests.

To test these hypotheses, we paired plant inventory and trait collections across multiple tropical forest sites throughout the Americas. We obtained inventory and trait data from a total of 26 plots ranging from tropical dry to wet forests throughout Mexico and Costa Rica. We used the online database SALVIAS (www.salvias.net) to download plot and trait data; we also used plot and trait data from the studies of Gómez-Sapiens (2001) and Hulshof et al. (2013). Based on the latitude and longitude coordinates of each plot, we extracted mean annual temperature, temperature seasonality (coefficient of variation of mean monthly temperatures), total annual precipitation, and precipitation seasonality (coefficient of variation of monthly precipitation totals) in addition to 15 other bioclimatic variables at a 2.5 minute resolution from WorldClim (Hijmans et al. 2005).

To quantify functional diversity, we calculated the functional dispersion of SLA (in square centimeters per gram) for each plot using the *FD* package in *R* (http://cran.r-project.org/web/packages/FD/). Functional dispersion is the mean distance in multidimensional trait space between individual species, weighted by relative abundances, and the centroid of all species within a community (Laliberté and Legendre 2010). Among the many metrics of functional diversity, we chose dispersion because it describes the distribution of individuals in trait space, can be used for single and multiple traits, is not strongly influenced by outliers, is independent of species richness, and incorporates species' abundances (Laliberté and Legendre 2010). We calculated functional dispersion for SLA because it represents a fundamental trade-off axis between resource capture and conservation (Poorter 2009). SLA was also widely sampled for the majority of species in each plot in our database, eliminating the need to extrapolate species mean trait values (i.e., assigning a species mean trait value collected in one site to individuals of that species found in another site), which underestimates the total amount of variation between species and sites. In general, one to five leaves were collected from a minimum of three individuals of each species across all plots.

To understand how functional traits vary with climate, we performed a forward stepwise multiple regression model of functional dispersion of SLA against all 19 bioclimatic parameters provided by WorldClim (see Hijmans et al. 2005 and the website www.worldclim.org/bioclim). A standard least-squares regression was then performed between functional dispersion of

SLA and the parameters that minimized the Akaike's information criterion (AIC) score in the stepwise model. All statistical analyses were conducted in *R* (R Development Core Team 2011).

Of the 19 bioclimatic parameters provided by WorldClim, the parameters that minimized the AIC score in the stepwise model included precipitation seasonality, precipitation of the wettest month, and precipitation of the warmest quarter (Table 8.1). The standard least-squares regression of functional dispersion against precipitation seasonality had the lowest AIC score (Table 8.2). Specifically, functional dispersion was positively correlated with precipitation seasonality (Figure 8.1). Interestingly, the plots located in the northern distribution of TDFs at 28°N (San Javier and La Colorada, Sonora, Mexico) (Gómez-Sapiens 2001) were among the sites with the highest precipitation

TABLE 8.1

Summary Statistics of the Stepwise Model Showing the Parameters Used in the Standardized Least-Squares Regression of SLA Functional Dispersion and the Bioclimatic Variables

Bioclimatic Variable	Parameter	Estimate	*P* Value	SS	*F* Value
	Intercept	–0.655	1.000	0	0
Annual mean temperature	BIO_1	0	0.51	0.027	0.43
Mean diurnal range	BIO_2	0	0.92	0.001	0.01
Isothermality	BIO_3	0	0.89	0.001	0.02
Temperature seasonality	BIO_4	0	0.70	0.009	0.15
Maximum temperature of warmest month	BIO_5	0	0.59	0.018	0.29
Minimum temperature of coldest month	BIO_6	0	0.53	0.025	0.41
Temperature annual range	BIO_7	0	0.93	0.000	0.007
Mean temperature of wettest quarter	BIO_8	0	0.49	0.030	0.50
Mean temperature of driest quarter	BIO_9	0	0.67	0.011	0.18
Mean temperature of warmest quarter	BIO_10	0	0.51	0.027	0.44
Mean temperature of coldest quarter	BIO_11	0	0.52	0.026	0.42
Annual precipitation	BIO_12	0	0.95	0.000	0.004
Precipitation of wettest month	BIO_13	0.002	0.003	0.575	9.52
Precipitation of driest month	BIO_14	0	0.76	0.006	0.097
Precipitation seasonality	BIO_15	0.011	0.001	0.792	13.10
Precipitation of wettest quarter	BIO_16	0	0.62	0.015	0.25
Precipitation of driest quarter	BIO_17	0	0.99	0.00	0.00
Precipitation of warmest quarter	BIO_18	0.000	0.065	0.216	3.58
Precipitation of coldest quarter	BIO_19	0	0.76	0.006	0.098

Note: SS = sum of squares.

seasonality and the highest levels of functional diversity (Figure 8.1). In addition, the dry forest sites with similar functional dispersion as wetter forests tend to be characterized by evergreen and semideciduous species, typical of the *bosque húmedo* (humid forest) or wetter microhabitats commonly found within the TDF life zone. Finally, species richness was not correlated with precipitation seasonality (Figure 8.2), which highlights the challenges of using a species-centric approach for distinguishing TDFs based on patterns of species richness and composition. Together, these findings support the hypothesis that functional diversity increases with increasing rainfall seasonality due to plant adaptations to seasonal drought. Further analyses that include other functional traits (e.g., wood density, hydraulic traits, and seed mass) will help to determine whether TDFs can be distinguished based on the amount and seasonality of annual rainfall. Using a single functional trait, SLA, we show here that it may be possible to use a trait-based approach to distinguish and classify TDFs based on climatic factors, primarily, precipitation seasonality.

TABLE 8.2

Summary Statistics for Each Standardized Linear Regression Model of SLA Functional Dispersion against the Bioclimatic Variables That Minimized the Model AIC Score

Bioclimatic Variables	AIC	r^2	*P* Value	*F* Value
BIO15	20.00	0.39	0.0007	15.10
BIO15 + BIO13 + BIO18	24.32	0.55	0.0004	9.12
BIO15 + BIO13	24.67	0.53	0.0002	12.72

Note: BIO13 = precipitation of wettest month; BIO15 = precipitation seasonality; BIO18 = precipitation of warmest quarter.

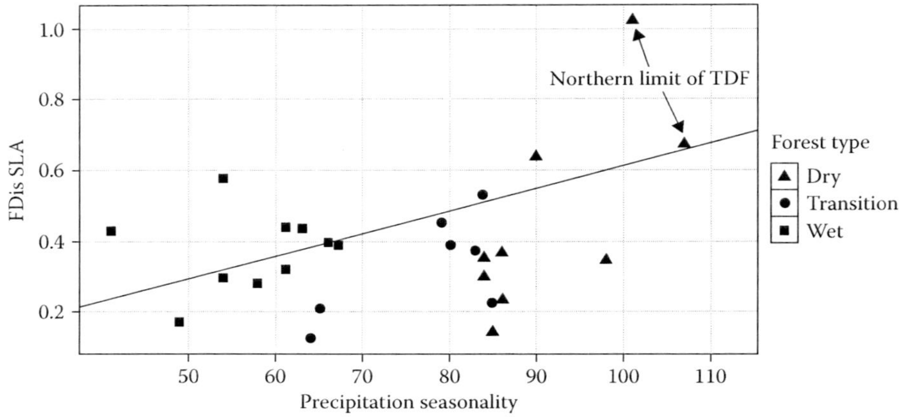

FIGURE 8.1
Functional dispersion (FDis) of specific leaf area (SLA) (in square centimeters per gram) for each plot as a function of precipitation seasonality (coefficient of variation of monthly precipitation) ($r^2 = 0.39$, $P = 0.0007$, $F_{1,24} = 15.1$ for 1 and 24 degrees of freedom).

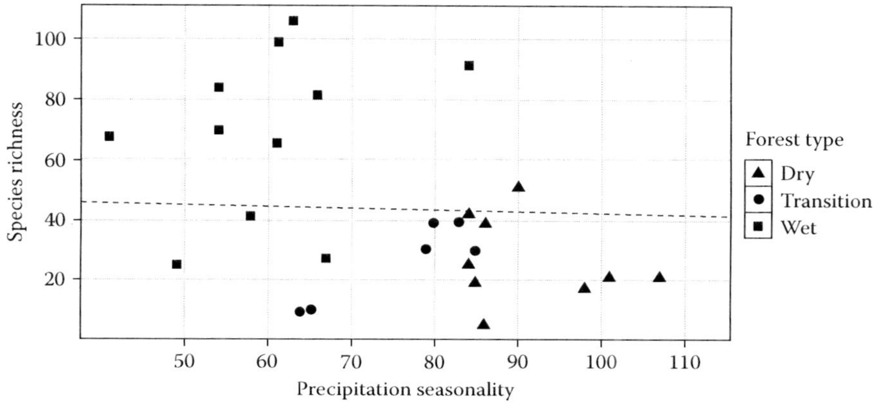

FIGURE 8.2
Species richness within each plot as a function of precipitation seasonality ($r^2 = 0.0013$, $P = 0.86$, $F_{1,24} = 0.032$).

8.3 Conclusions

In summary, TDFs are characterized by strong spatiotemporal variation in environmental factors, which directly impacts plant- and ecosystem-level function. We briefly outline how three key plant functional traits vary in response to environmental variability including local topographical and land-use gradients and broad-scale geographical gradients within and between TDFs. Trait-based ecological research within TDFs will advance our understanding of how these fragmented ecosystems interact with and differ from other tropical ecosystems. We identify key research priorities where a trait-based approach can be easily implemented to quantify adaptive plant strategies and ecosystem functioning within TDFs.

Many regions that encompass TDFs are predicted to experience more frequent/extended periods of drought, warming temperatures, and decreased mean precipitation (Meir and Pennington 2011). Thus, a major research priority is to understand the effect of climatic changes on dry forest plant communities. Although the effects of climatic change on the biodiversity, structure, and functioning of TDFs is discussed in detail elsewhere in the literature (Meir and Pennington 2011), a trait-based approach can provide additional insight into plant responses to changing environmental conditions (Enquist and Enquist 2011). Phenological shifts in response to changing climates have been widely documented in temperate regions; yet the complex interactions between plant phenology (both leaf and reproduction phenologies), pollination, dispersal, herbivory, and changing climate are relatively underexplored in TDFs (Sánchez-Azofeifa et al. 2005). Although TDFs are considered to be drought resistant (Meir and Pennington 2011), the decoupling of pollinators, dispersers, herbivores, and plant reproduction can have large effects on the plant population dynamics of many species (Janzen 1967; Bawa and Dayanandan 1998; Elzinga et al. 2007). This is particularly true for TDFs as dry-season plant reproduction coincides with increased pollinator and disperser activity (Janzen 1967).

The importance of connectivity between dry and wet forests is another rich area of research. Research relating dry forest ecology and conservation often refers to TDFs as distinct ecosystems even though the exchange of organisms and genetic material between dry and wetter ecosystems is an important source of genetic diversity (Bawa and Dayanandan 1998). In addition, nearby rain forests and mesic microhabitats within TDFs are essential sources of refugia for dry forest species, permitting migration out of the dry forest during the dry season (Janzen 1987; Hanson 2011). Research linking functional diversity within and between species across tropical landscapes that include TDFs will broaden our understanding of the diverse interactions taking place between tropical dry and wet forest ecosystems as well as the biotic and abiotic drivers of plant trait variation and community assembly.

9

Nutrient Cycling in Tropical Dry Forests

Maria Gabriela Gei and Jennifer S. Powers

CONTENTS

9.1 Introduction

Nutrient elements such as nitrogen (N) and phosphorus (P) enter terrestrial ecosystems such as forests through a variety of abiotic and biotic input sources, are transformed and cycled within ecosystems, and ultimately are eliminated from ecosystems through a variety of abiotic and biotic mechanisms (Figure 9.1) (Attiwill and Adams 1993). As a consequence of human activities, the inputs of N are increasing both globally (Gruber and Galloway 2008) and regionally within the tropics (Hietz et al. 2011). Altered nutrient inputs have the potential to change both species composition (Bobbink et al. 2010) and ecosystem processes (Hall and Matson 1999). Moreover, the global climate is changing as a result of anthropogenic greenhouse gas emissions (IPCC 2007a), and understanding coupled carbon (C) and nutrient cycles and their drivers is critically important for predicting climate–biosphere feedbacks (Hungate et al. 2003; Gruber and Galloway 2008; Townsend et al. 2011). Thus, investigating nutrient cycling in tropical forests has moved beyond a

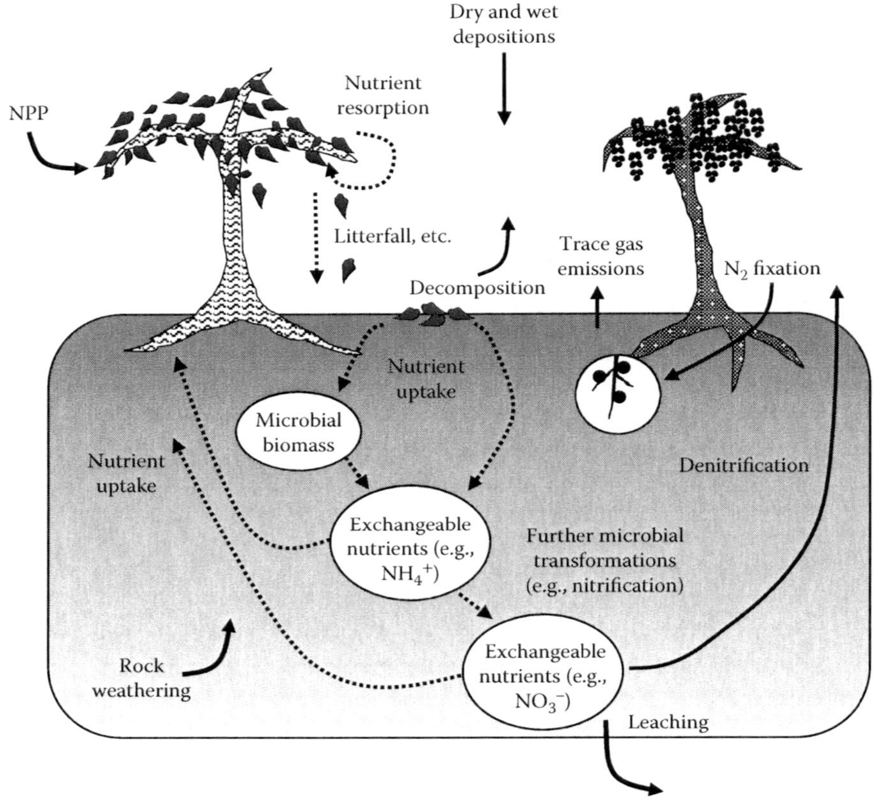

FIGURE 9.1
Biotic and abiotic processes through which nutrient elements (i.e., nitrogen and phosphorus) enter, cycle within, and are eliminated from terrestrial ecosystems. NPP = net primary production; N = nitrogen.

question of academic curiosity to becoming essential for predicting earth system dynamics.

Our purpose in this chapter is not to exhaustively review all nutrient cycling processes or studies in tropical dry forests (TDFs). Rather, we discuss recent advances in our understanding of inputs, outputs, and nutrient transformations in TDFs; highlight the differences between wet and dry forests that are the consequences of strong seasonal drought and/or low precipitation; review studies that document nutrient limitation of ecosystem processes in TDFs; and conclude by suggesting promising avenues for future work. We emphasize the importance of N and P, as they are two elements that frequently limit plant growth and differ in the degree to which their dynamics are controlled by biological (N) versus geological processes (P).

9.2 Inputs, Outputs, and Transformations

Studies of nutrient inputs, outputs, and transformations in TDFs can be loosely divided into two different themes. Many past studies have focused on documenting seasonal nutrient dynamics and intra-annual variability, for example, how at the onset of each rainy season large pulses of nutrients are released from the mineralization of microbial cells (Singh et al. 1989; Raghubanshi 1992; Campo et al. 2001). A second major focus of research has been on understanding how the extensive and intensive transformation of TDFs into managed lands such as pastures and cropped fields, as well as subsequent secondary forest succession, affects element cycling processes (Valdespino et al. 2009). Both of these types of studies contribute to our understanding of nutrient cycling processes and how they are shaped by climate and human activities.

9.2.1 Inputs

N is notable among the nutrients that are essential for plant growth in that the dominant pathway for entry into ecosystems is biological and not geological (Figure 9.1). The plant family Fabaceae (i.e., legumes) is one of the most abundant and species-rich families in Neotropical TDFs (Pennington et al. 2009). Because many legume species have the capacity to form symbiotic associations with N_2-fixing bacteria (Corby 1988), the prominence of this family suggests that TDFs have the potential for high N_2 fixation rates. Nevertheless, surprisingly few studies have attempted to quantify N_2 fixation in TDFs.

Freitas et al. (2010) used the natural abundance isotope method to estimate the percentages of N derived from N_2 fixation for different legume species and then extrapolated these estimates to the ecosystem scale using data on forest composition and foliar N concentrations in two sites in Brazil that had an annual rainfall of 700–768 mm with 4–7 month dry seasons. Their data suggested that between 28% and 88% of the N content of legume species with the capacity to fix N_2 was derived from fixation, and presumably the rest came from mineral N in the soil. At the ecosystem scale, annual quantities of N inputs from symbiotic biological N_2 fixation (BNF) depended on the percentages of N_2-fixing plants present in the forests and ranged from ~2.5 to 11.2 kg $N \cdot ha^{-1}$. They considered these fluxes of BNF to be low and speculated that BNF rates would be perhaps an order of magnitude higher in early successional vegetation with greater legume abundance. Similarly, in planted stands of the N_2-fixing legume *Acacia koa* in a region of Hawaii that receives 1800 mm of annual rainfall, BNF estimated using the acetylene reduction assay declined from 23 kg $N \cdot ha^{-1} \cdot year^{-1}$ in 6-year-old regenerating forests to 1.5 kg $N \cdot ha^{-1} \cdot year^{-1}$ in 20-year-old forests as soil N accumulated (Pearson and Vitousek 2001). These studies together clearly demonstrate the potential of BNF to add significant quantities of N to ecosystems, but they also underscore that much more work is needed on quantifying the patterns and controls on N_2 fixation to make robust generalizations across TDFs. In fact, despite the abundance and diversity of Fabaceae in TDFs, the limited research to date indicates that the contribution of this group of species to N inputs by N_2 fixation is modest.

Additional studies that simultaneously quantify other important N inputs are urgently needed to have a complete N budget of TDFs. Of particular concern is the complete lack of studies on atmospheric deposition in TDFs. N inputs through dry and wet depositions are likely to increase in tropical ecosystems and have subsequent impacts on N cycling and C storage (Bobbink et al. 2010; Chen et al. 2010). The long dry season of TDFs and their vulnerability to fire-induced losses of N emphasize that estimates of wet and dry deposition of N in these ecosystems should be a research priority.

In contrast to N, the dominant sources of P in ecosystems include dust deposition on shorter timescales and rock weathering on longer timescales (rock weathering inputs occur at rates that are typically so slow that they are assumed to be negligible on the typical annual to decadal timescale of most ecosystem studies). Theoretical models have suggested that P from atmospheric dust deposition and fog precipitation helps to maintain dry forest vegetation and productivity and that the strong feedback between forest vegetation and its ability to trap P from these sources induce alternative stable states of dry forests, such as savannas or grasslands (DeLonge et al. 2008). A detailed study of P inputs in dry and wet deposition and outputs via stream water over a six-year period at the Chamela Biological Station in Mexico showed that P inputs exceeded outputs every year and ranged from a net gain of 0.011–0.166 $kg \cdot ha^{-1}$ over the years (Campo et al. 2001). The magnitude of internal P cycling (through litterfall and throughfall) was high relative to inputs, and the authors concluded that P may not be limiting at this site.

9.2.2 Transformations

The N cycle also exhibits strong seasonality in TDFs. Anaya et al. (2007) found that N dynamics are not only influenced by rainfall seasonality but also controlled by labile C availability in Chamela, Mexico. In the conceptual model they developed, at the onset of rains soluble organic matter from leaf litter becomes available for microbial activity. Labile C then peaks, promoting N immobilization followed by N mineralization (Anaya et al. 2007). In contrast, during the dry season, microbial activity is limited by the lack of soluble organic matter.

It is expected that in mature lowland tropical forests, the patterns of N cycling exhibit "leaky" properties that suggest relative N abundance, whereas the P cycle shows more conservative processes (Vitousek 1984). Because of the potential for N fixation (see Section 9.2.1) and the evidence for considerable N losses from these ecosystems (see Section 9.2.3), we consider that the N cycle of TDFs is consistent with the expected pattern of an "open" cycle. Nevertheless, studies of nutrient cycling in TDFs need to always be considered in light of the nuance that rainfall seasonality imposes to the patterns of cycling.

By contrast, P cycles relatively tightly in TDFs. For example, TDFs in Mexico display fast mean residence times in soil P and P inputs exceed outputs (Campo et al. 2001; Valdespino et al. 2009). Additionally, Read and Lawrence (2003) examined P nutrient use efficiency (NUE) in several TDFs along a precipitation gradient, and they found high P NUE values that decreased with precipitation, with the highest P NUE in the driest site. These results suggest that TDFs might exhibit a P cycle that can be efficient or conservative in sites with low precipitation.

The internal cycling of P in TDFs appears to be closely linked to both successional stage of forests and seasonal patterns in rainfall. In a study of a TDF in Morelos, Mexico, Valdespino et al. (2009) found that the P concentrations in litterfall increased from 0.63 mg $P \cdot g^{-1}$ in an early successional forest to 0.90 mg $P \cdot g^{-1}$ in a late successional forest and 0.92 mg $P \cdot g^{-1}$ in a primary forest. This same study highlighted the seasonal differences in P cycling: total P in litter was higher during the dry season than during the wet season across all forest ages. Also, in intermediate to late successional stages of the forest, bicarbonate-extractable soil P was more available during the dry season. The results of the study by Valdespino et al. (2009) match what Campo et al. (2001) found in Chamela in that they found similar seasonal patterns in P concentrations in litterfall, as well as similar mean residence times (1.2 years for P in both forests, 1.4 years for organic matter in Morelos, and 1.3 years for organic matter in Chamela). Altogether these data suggest that in these two forests P availability follows the typical dry forest short-term pulsed cycle: during the dry season plant uptake and nutrient losses are minimal so that nutrients accumulate in the litter mass, and at the beginning of the rainy season decomposition starts and these nutrients become available for plant uptake as they are released from the leaf litter.

9.2.3 Outputs

The few studies of nutrient outputs from TDFs have focused on quantifying gaseous emissions from these ecosystems and to a lesser extent inorganic N leaching. In this section, we discuss the most important conclusions.

In the tropics, soils with high N availability often represent the ideal environment to support nitrifying bacteria that convert ammonium to nitrate (Matson et al. 2002). Therefore, it is likely that in TDFs high rates of nitrification generate significant emissions of nitric oxide (NO) and the greenhouse gas nitrous oxide (N_2O). Erickson et al. (2002) measured N oxide emissions in a TDF in Guánica, Puerto Rico, and found that emissions of NO were greater than those of N_2O with maximum values of 23.0 ng $N \cdot cm^{-2} \cdot h^{-1}$ for NO and 3.25 ng $N \cdot cm^{-2} \cdot h^{-1}$ for N_2O. In one of their sites, they measured the largest mean soil–atmosphere NO flux measured for a tropical forest

(11.9 kg N · ha^{-1} · year^{-1}), and it occurred in sites with low leaf-litter C:N ratios, high rates of N mineralization, and high soil nitrate (NO_3^-) to ammonium (NH_4^+) ratios. These results confirm that N availability influences N oxide emissions; however, moisture played an important role since these fluxes increased with rainfall. Another important conclusion was that postagricultural successional forests dominated by legumes were the main source of N oxide emissions. In Amazonian forests, Brazil, N_2O emissions increased with forest age (Davidson et al. 2007). These examples show how N oxide emissions accompany the gradual accumulation of N and the shift from a more conservative to a more active N cycle during succession. The emergence of a dominant N_2-fixing species, like what happened in Guánica, could accelerate N cycling rates and therefore generate rapid increases in N oxide emissions.

Because of the abundance of deciduous species in TDFs, it would be reasonable to expect that N leaching from the leaf litter layer exhibits a pronounced seasonality in synchrony with rainfall. Recently, Yamashita et al. (2010) quantified inorganic N leaching from a TDF in northeastern Thailand. They found annual fluxes of total inorganic N of 6.88×10^{-7} kg N · ha^{-1} · year^{-1} from the leaf litter layer and 2.08×10^{-7} kg N · ha^{-1} · year^{-1} from the 40 cm soil layer. The fluxes coming from the litter and the upper soil layer (0–5 cm) peaked during the early wet season. This seasonal pattern suggested that the high levels of NO_3^- and NH_4^+ leaching during the early wet season were favored by the accumulation of litter during the dry season, followed by the decomposition of the litter layer with the start of the rainy season, leading to a high diffusion of dissolved organic carbon (DOC), a substrate for N mineralization. Later into the wet season, they observed a decrease in N leaching that was accompanied by an increase in N mineralization possibly regulated by plant uptake.

Losses of P from TDFs also tend to follow a seasonal pattern with high P leaching potential during the early wet season but plant retention of nutrients during the end of the rainy period and the dry season. Campo et al. (1998) performed a laboratory experiment to determine the fate and retention of added P in the TDF of Chamela. They collected intact cores from the upper soil profile and surface litter at the peak of the dry season and during the rainy period, and they added two different volumes of simulated rain with labeled P (^{32}P) and a control treatment with no water addition. They observed higher P concentrations in all litter and soil fractions of samples collected during the dry season than on samples collected during the rainy season. After adding water, the rainy season cores retained more ^{32}P (99.9%–94%) than the dry season cores (98.9%–80%). Interestingly, the simulation of rainfall rapidly stimulated the release of P mineralized on cores collected during the dry season, whereas P immobilization increased on the rainy season samples. These results show that there are less efficient P retention mechanisms during the dry season and beginning of the rainy season and that they can be explained by the presence of inactive, dead, and slowly responding microbes.

9.3 Patterns along Precipitation Gradients and Consequences of Seasonal Drought

TDFs are different from rain forests in that they have lower annual precipitation levels, and in many TDFs the rain that does fall has a pronounced seasonal distribution (Murphy and Lugo 1986a), although there are examples of aseasonal TDFs that lack a dry season (e.g., Hawaii [Sandquist and Cordell 2007]). Furthermore, many models suggest decreased rainfall in many tropical areas (Neelin et al. 2006) and a shift from wetter to drier forests (Malhi et al. 2009). Understanding the sensitivity of nutrient cycling to rainfall is important for predicting how tropical forests and the ecosystem services they provide may change in the future. How does rainfall amount and seasonality affect nutrient cycling and availability within dry forests?

9.3.1 Insights from Precipitation Gradients

Precipitation gradients are useful for determining the influence of changing patterns of rainfall on C and nutrient cycling. Quantifying nutrient cycling at different sites that vary in annual rainfall is one way to evaluate the overall effects of rainfall amount (Table 9.1). There are a number of caveats that accompany observational studies along precipitation gradients. It is often difficult to locate sites that vary in precipitation but not other ecosystem state factors such as parent material, weathering time, and species composition. Nevertheless, these studies are useful for identifying dominant patterns and also for suggesting mechanisms through which precipitation affects ecosystem processes and nutrient cycling among sites. Literature reviews have shown clear patterns for C uptake and storage in biomass along precipitation gradients. Net primary productivity in tropical forests increases with rainfall until ~2450 mm mean annual precipitation (MAP) beyond which it decreases (Schuur 2003), whereas tree biomass stocks increase linearly with rainfall over a range of 500–2000 mm MAP and presumably reach an asymptote at higher MAP values (Becknell et al. 2012).

Despite clear relationships between rainfall and C cycling, there are contrasting predictions for how nutrient cycles should vary among sites as a function of MAP. On one hand, lower total rainfall quantities and soil moisture may directly constrain microbial processes and result in lower overall rates of nutrient cycling in dry forests compared to wetter forests, except in very wet forests where anoxia can limit biological activity (Schuur 2001). For example, a study of litter decomposition in 23 tropical forests showed that decomposition rates increased with MAP over a gradient of 760–5800 mm (Powers et al. 2009b). On the other hand, higher precipitation quantities may indirectly affect nutrient availability through increased inputs (e.g., rock weathering) and/or outputs (e.g., leaching) in areas with higher rainfall, and the balance between these processes may result in depleted nutrient stocks in wetter forests (Austin and Vitousek 1998).

TABLE 9.1

Observational Studies of Nutrient Cycling along Precipitation Gradients in Tropical Forests

Location	Range of Annual Rainfall (mm)	Response Variables	Findings	Reference
Hawaii	500–5500	Total and extractable soil nutrients; foliar nutrients and stable C and N isotopes	Extractable soil nutrients and foliar nutrients were greater at drier sites, suggesting that leaching outputs at wetter sites deplete nutrients; no trend for N mineralization	Austin and Vitousek (1998)
Montane Forest, Hawaii	1000–2000	Basal area, foliar nutrients, N mineralization, and P fractionation	Lowest foliar nutrients at low-rainfall site; highest N mineralization rate at low-rainfall site; P fractions did not vary monotonically across the rainfall gradient	Idol et al. (2007)
Panama	1800–3500	Total and extractable soil nutrients; foliar and litter nutrients	Foliar nutrients and extractable soil P decreased with increasing MAP; extractable N increased with MAP; no trend for N mineralization; species and soil fertility effects can overwhelm MAP effects	Santiago et al. (2005)
Amazon/Brazil	1900–3400	Foliar, litter, and total and available soil nutrients and stable isotopes	N was mineralized in the drier site and immobilized in all wetter sites; nutrient availability and foliar nutrients were higher in the drier site	Nardoto et al. (2008)

Although many studies have investigated soil and foliar nutrients and other aspects of nutrient cycling along wetter precipitation gradients (Schuur and Matson 2001; Alvarez-Clare and Mack 2011), only a few studies have included sites with MAP <2000 mm (Table 9.1). The studies that have investigated nutrient cycling along rainfall gradients in both dry and wet forests show that foliar nutrients and extractable soil nutrients (PO_4, potassium [K], calcium, and magnesium [Mg]) tend to decrease with increasing MAP (Santiago et al. 2005; Nardoto et al. 2008), but this can vary by nutrient (Austin and Vitousek 1998). There are no consistent directional trends in N mineralization rates across precipitation gradients (Table 9.1), but this might be due to the fact that daily mineralization rates, which are typically measured, do not reflect yearly cumulative totals or that N mineralization is controlled by factors other than MAP. Collectively, it is difficult to draw broad generalizations from the studies in Table 9.1 because they are limited in geographic scope and include few dry forests. What these studies do suggest is that even though litter decomposition rates and rock weathering may be higher in wetter forests, over the course of ecosystem development the higher leaching rates that accompany higher rainfall result in lower nutrient stocks and availability as well as lower foliar nutrient concentrations in wetter forests. However, the most salient conclusion from Table 9.1 is the need for additional comparative studies of nutrient cycling that include more sites from dry tropical forests and time-integrated measurements of nutrient availability.

9.3.2 Nutrient Pulsing and Intra-Annual Variation

The cyclic pulses of water due to rainfall seasonality have important implications for how nutrients cycle within seasonally dry ecosystems. As episodes of drought and extreme events of precipitation are increasing in frequency, it is critical to understand the consequences of wetting and drying cycles for ecosystem processes.

During the dry season, plant uptake is reduced and the main source of available nutrients is resorbed nutrients. At this time, litterfall peaks and coincides with the lowest levels in both soil P and soil N concentrations, such as in a TDF in Yucatán, Mexico (Read and Lawrence 2003). When rewetting occurs, soil water potential rapidly changes and results in the lysis of soil microbial biomass. As diffusion increases, dead microbial biomass with low C:N ratios along with accumulated organic matter become an available substrate for C and N mineralization. For example, a TDF in Belize had high levels of C:N ratios in soil organic matter throughout the year that dropped at the beginning of the rainy season (Lambert et al. 1980). Microbes, in turn, immobilize some of the released nutrients by increasing their biomass; the nutrients that are in excess of microbial demand become available to plants (Lodge et al. 1994; Austin et al. 2004). At this point nutrient availability increases, which explains why TDFs usually have the highest N and P levels in the soil during this period (Lambert et al. 1980). However, there is also a great potential for nutrient losses via N leaching or denitrification, or through leaching of available P pools (Yamashita et al. 2010). Synchrony between nutrient pulses and plant uptake through fine root proliferation can prevent these losses.

It has been shown that aspects of the wet–dry cycle, such as microbial survival, can be affected by characteristics of the cycle itself: temperatures reached during the dry period, duration of the dry period, and the time between wet and dry periods (Austin et al. 2004). Several studies have addressed seasonal litter nutrient dynamics in TDFs, which also reveal how closely coupled these fluxes are to precipitation patterns. Campo et al. (1998) found in Chamela that during the dry season, reduced leaching and plant uptake promote the accumulation of soluble forms of P in litter and topsoil. When the rainy season starts, these nutrients are taken up to support biomass growth until the soil P pool is reduced. The authors emphasized how critical the length and severity of the dry season are as antecedent conditions that affect nutrient cycling during wet episodes.

Their conclusion stems from results that suggest that litter and soil moisture contents prior to a rainfall event may determine the pattern of microbially mediated nutrient fluxes. Other studies (some also based at Chamela) have found that N and P resorption efficiencies respond to water availability but not to soil nutrient availability (Campo et al. 1998; Rentería et al. 2005; Yamashita et al. 2010; Rentería and Jaramillo 2011). Finally, Read and Lawrence (2003) found differences in retranslocation throughout successional stages of a forest in Yucatán: the reduction in litter nutrients during peak litterfall tended to be higher in older forests. Altogether these results suggest TDFs are ecosystems that are particularly vulnerable to changes in temperature and especially precipitation.

9.4 Evidence for Nutrient Limitation

No discussion of nutrient cycling would be complete without considering the question of which nutrients most limit ecosystem processes such as net primary production (NPP) and decomposition. Much of our understanding of this question derives from the conceptual model advanced by Walkers and Syers, who hypothesized that ecosystem processes would be limited by atmospherically-derived elements, such as N during early stages of primary succession, and by rock-derived elements, such as P or base cations on highly weathered parent materials during late stages of ecosystem development (Walker and Syers 1976). At a coarse geographic scale, soils in tropical latitudes are usually more weathered than those in higher latitudes, and it is commonly hypothesized that NPP is limited by N in temperate zone soils and by P in tropical soils. A recent meta-analysis found evidence for a causal link between soil P availability and NPP, supporting the generalization that tropical forests are generally limited by P and not N (Cleveland et al. 2011). However, this analysis was restricted to moist and wet tropical forests with rainfall >1300 mm MAP. Given the lower overall precipitation levels and presumably lower weathering and leaching rates in dry tropical forests, it is unclear whether TDFs should show patterns similar to wetter forests or processes such as NPP and decomposition should be limited by the same nutrients.

There are both direct and indirect methods for assessing nutrient limitation. The gold standard for demonstrating nutrient limitation studies is the large-scale fertilization experiment, where individual or multiple nutrients are applied in an inorganic or organic form to large plots and ecosystem responses are monitored relative to unfertilized control plots. Other direct methods include fertilizing mesocosms in the laboratory or the use of fertilized root ingrowth cores in the field (Raich et al. 1994). Indirect methods for inferring nutrient limitation are based on observations such as the magnitude of nutrient resorption prior to senescence, indices of NUE calculated from nutrient return in litterfall, or ratios of foliar nutrients (Vitousek 1984; Vitousek and Sanford 1986; Townsend et al. 2007). What evidence is there to suggest whether TDFs experience nutrient limitation?

Most of the direct tests of nutrient limitation in dry forests have been conducted within the context of secondary succession (Table 9.2), which is not surprising given how much of TDFs are recovering from previous land use. The few plot-scale fertilization experiments in regenerating dry forests on limestone in Yucatán (Campo and Vázquez-Yanes 2004) and in Amazonia, Brazil (Davidson et al. 2004), have shown that forest recovery rates increase with nutrient addition (Table 9.2). Moreover, these studies have shown that added N as well as P can affect ecosystem processes, perhaps because the repeated fires that often accompany land uses such as grazing deplete N stocks (Davidson et al. 2007). Thus, patterns of nutrient limitation are more complex than the simple dogma of P limitation of tropical forests, and nutrient cycles may interact with each other. For example, addition of N increased P concentrations in litterfall in a young forest in Mexico (Campo et al. 2007). Nutrient addition studies in the laboratory have also shown ecological responses to a variety of nutrients. One study that evaluated how both macro- and micronutrients affect leaf litter decomposition found stimulatory effects of added P and zinc (Zn), whereas N and Mg decreased decomposition rates (Powers and Salute 2011).

Plants often respond to nutrient limitation by increasing the efficiency with which a limiting resource is utilized (Chapin 1989). According to this, plants retranslocate more of the nutrient that is the most limiting to plant growth. Resorbed nutrients represent an important source of nutrient supply especially during leaf initiation and expansion, enabling the trees to take full advantage of the abundance of water and nutrients in the rainy season (Lal et al. 2001a, b). In a TDF in India, the efficiency of N and P resorption in senescing leaves did not differ. However, the same study reported species differences in nutrient resorption, which could explain the lack of a broad-scale trend (Lal et al. 2001b). Another common, indirect way to infer the type of nutrient limitation is to use the N:P ratio in plant tissues (Koerselman and Meuleman 1996), where N:P ratios >16 suggest that plant growth is P limited, N:P ratios <14 that the plant community is N limited, and N:P ratios between 14 and 16 that there is colimitation by N and P. We summarized foliar N:P ratios measured from TDFs in Costa Rica, Mexico, and Brazil (Table 9.3).

TABLE 9.2

Direct Evidence of Nutrient Limitation in TDFs

Site Location and Description	Mean Annual Rainfall and Temperature	Fertilization Treatment	Response Variable	Findings	Reference
Young (6 years) secondary forest, Pará, Amazonia, Brazil	1800 mm	N (100 $kg \cdot ha^{-1}$), P (50 $kg \cdot ha^{-1}$), NP, and control plots (size: $20 \times 20\ m^2$)	Vegetation biomass (including grasses), microbial C and N, trace gas emission, litter decomposition	N and NP but not P addition increased tree biomass accumulation; repeated fires may have decreased N availability	Davidson et al. (2004)
Young (10 years) and older (60 years) secondary forest, Yucatán Peninsula, Mexico	760–946 mm, 25.8°C	N (220 $kg \cdot ha^{-1}$), P (75 $kg \cdot ha^{-1}$), NP, and control plots (size: $12 \times 12\ m^2$)	Tree seedling dynamics	Recruitment and survival under fertilization differed among species and interacted with light availability	Ceccon et al. (2003)
Young (10 years) and older (60 years) secondary forest, Yucatán Peninsula, Mexico	760–946 mm, 25.8°C	N (220 $kg \cdot ha^{-1}$), P (75 $kg \cdot ha^{-1}$), NP, and control plots (size: $12 \times 12\ m^2$)	Extractable, microbial, total N and P, and potential N mineralization	Responses depended on forest age; 15%–30% of added N and P were incorporated into microbial biomass at the young site but not at the old site	Solís and Campo (2004)

Young (10 years) and older (60 years) secondary forest, Yucatán Peninsula, Mexico	760–946 mm, 25.8°C	N (220 kg · ha^{-1}), P (75 kg · ha^{-1}), NP, and control plots (size: 12 × 12 m^2)	Annual tree diameter growth, litterfall, litter standing crop	Litterfall increased in NP treatment only in both forests after 3 years; in general, NPP in young sites was colimited by N and P, and older sites were just limited by P	Campo and Vázquez-Yanes (2004)
Young (10 years) and older (60 years) secondary forest, Yucatán Peninsula, Mexico	760–946 mm, 25.8°C	N (220 kg · ha^{-1}), P (75 kg · ha^{-1}), NP, and control plots (size: 12 × 12 m^2)	N and P fluxes in litterfall	Increased N and P fluxes in litterfall and decreased mean resident times of nutrients on the forest floor	Campo et al. (2007)
Two litter types (*Gliricidia sepium* and *Quercus oleoides*) collected from Area de Conservacíon Guanacaste, Costa Rica	1580 mm, 25°C	Short-term lab incubation of leaf litter receiving single nutrient addition at low and high concentrations (N, P, Mg, K, nickel [Ni], Zn)	Leaf litter mass loss and CO_2 mineralization rates	Added N and Mg tended to slow decomposition, whereas P and Zn increased it; Mg and Ni had no effects	Powers and Salute (2011)

TABLE 9.3

Indirect Evidence of Nutrient Limitation in TDFs

Location	N:P, Average	Range of Variation	Reference
Santa Rosa, Costa Rica	22.5	12.6–43.5	Powers and Tiffin (2010)
Chamela	18.2	8.6–30.2	Rentería and Jaramillo (2011)
Amazonia	30.1	24.8–33.8	Nardoto et al. (2008)
Hawaii	25.3	21.2–27.7	Idol et al. (2007)

Most of the N:P ratios from these TDFs are higher than 16, suggesting a trend toward P limitation in biomass production, but other species show ratios consistent with N limitation or colimitation by N and P. Analogous to studies in tropical wet forests (Townsend et al. 2007), the variation in N:P ratio in each TDF indicates intrinsic species differences in nutrient cycling that can potentially make nutrient limitation more heterogeneous at a local scale. In fact, the extent to which these ratios remain consistent across different scales, from individuals to ecosystems, has been questioned (Rentería and Jaramillo 2011). Other authors have also examined whether leaf nutrient concentrations reflect soil fertility. For example, in Chamela P concentration in leaves was well correlated with soil P availability (Rentería et al. 2005). Here, foliar N:P ratios also indicate a plant community more limited by P. Last, comparisons of nutrient ratios and resorption efficiency raise the intriguing possibility that the nature of nutrient limitation in a TDF may depend on water as a key factor. In the TDF of Chamela, P pools were measured at the onset of the rainy season and they were large enough to support the annual biomass production (Campo et al. 1998). Rentería and Jaramillo measured N and P resorption in 21 woody species of the same forest during a three-year period and found not only that leaf P concentrations decreased with water availability but also that higher P resorption occurred during dry years compared to a wet year. These results show that P dynamics are tightly linked to rainfall availability.

9.5 Knowledge Gaps and Future Directions

Much of our knowledge of nutrient cycling in TDFs comes from studies at either the wet end of the precipitation spectrum (i.e., sites receiving >1800 mm MAP) or the very dry end. Fertilization experiments in Mexico and Brazil have yielded fundamental insights into the nature of nutrient limitation and how it may change during secondary succession, but the small number of sites covered by the studies makes it difficult to extrapolate these results across TDFs. We also argue that in spite of making considerable efforts in quantifying nutrient pools and fluxes in a number of TDFs, our ability of understanding nutrient dynamics in this biome is severely limited by the lack of sufficient information on nutrient inputs, particularly N fixation and deposition. TDFs are regions vulnerable to increasing N inputs, but we cannot forecast possible effects of N deposition before having a decent understanding of the patterns and mechanisms of N cycling in unmodified forests as a benchmark. Because N inputs are increasing at a global scale, N outputs such as leaching and N oxide emissions should receive further attention. Also, additional efforts should be made to determine the major controls on nutrient dynamics within and among sites, with particular attention to the role of water availability. Such data are essential for improving the representation of TDFs in ecosystem models. To achieve this, we need to continue building a body of data on nutrient pools and transformations from an integrative approach that links individual species to ecosystems.

10

Edge Influence on Canopy Openness and Understory Microclimate in Two Neotropical Dry Forest Fragments

Carlos Portillo-Quintero, Arturo Sánchez-Azofeifa, and Mário Marcos do Espírito-Santo

CONTENTS

10.1 Introduction

One of the most widespread anthropogenic changes to ecosystem integrity is the fragmentation and degradation of continuous vegetation through deforestation (Aizen and Feisinger 1994). Deforestation affects biological diversity through direct destruction of habitats and through habitat fragmentation (Skole and Tucker 1993). In addition, one of the major changes brought about by habitat fragmentation to tropical forests is an increase in the proportion of forest edge exposed to other habitats (Kapos et al. 1997; Laurance and Curran 2008). Abrupt exposure to different environmental conditions in open areas indirectly alters the microclimate, consequently increasing tree mortality rates and decreasing plant species recruitment (Laurance et al. 1998; Asquith and Mejia-Chang 2005). Direct changes resulting from forest edge exposure include alterations in air temperature, soil moisture, relative humidity, and the amount of light penetrating the forest understory (Kapos et al. 1997; Pohlman et al. 2007).

Another effect of fragmentation is the exposure of the forest fragment perimeter to windthrow. Winds striking an abrupt forest edge can exert strong lateral-shear forces on exposed trees and create considerable downwind turbulence for at least 2–10 times the height of the forest edge (Laurance and Curran 2008). These conditions make treefall gaps very frequent near the edges of a forest fragment (Kapos et al. 1997; Laurance and Curran 2008). In general, tree damage leads to reduced canopy cover and greater abundance of snags and logs at edges (Harper et al. 2005). Typical vegetation responses to these changes in edge environment result in an increased presence of exotic and disturbance-adapted species, increased sapling and tree densities, increased shrub cover, and higher species richness (Gelhausen et al. 2000; Laurance et al. 2002; Harper et al. 2005). Animal communities are also affected by these changes. Near forest edges, disturbance-adapted butterfly and beetle species increase, the species composition of leaf-litter invertebrates changes, small mammal populations develop alterations in their behavioral and biological characteristics, and the number and structure of bird communities alter (Laurance et al. 2002; Laurance 2004; Manu et al. 2007; Fuentes-Montemayor 2009). Ecological processes such as seed dispersal and predation, nest predation, brood parasitism, and herbivory are also affected by distance from the edge of the forest fragment (Murcia 1995). For most abiotic and biotic edge effects in tropical rain forests, Laurance et al. (2002) and Harper et al. (2005) estimated that some ecological processes can be affected up to 300–400 m from the forest edge. These authors also report that edge effects have shown higher magnitude and penetration distance in recently created edges (<7 years old) and tend to be reduced and become stabilized with time in older tropical rain forest edges (≥7–12 years).

Edge influence on forest structure and composition has been studied in tropical rain forest, temperate, and boreal ecosystems. In the tropics, long-term systematic research on edge effects has been conducted mostly in fragmented rain forests, especially within the framework of the Biological Dynamics of Forest Fragments Project in the Brazilian Amazon, a project launched in 1979 by the World Wildlife Fund and Brazil's National Institute for Research in Amazonia (Bierregard et al. 1992). Tropical dry forest (TDF) ecosystems, though highly productive and biodiversity rich, have not been assessed for their response to the ecological dynamics following edge creation. Further, TDFs have been transformed and occupied by urban and agricultural land use at significantly higher rates than tropical rain forests (Murphy and Lugo 1986a) and the majority of TDF extents occur within lowland human-dominated landscapes (Portillo-Quintero and Sánchez-Azofeifa 2010). However, the impact of edge effects on TDFs largely remains an open question.

A perspective on the subject has been proposed by Harper et al. (2005), suggesting that forests that are subject to frequent natural disturbances may exhibit lower magnitude or distance of edge influence. Here, edge creation may have relatively little impact on forest composition and structure. Such resilience may be characteristic of the TDF ecosystem, which is characterized by ecological processes adapted to seasonal fluctuations of water availability. Also, the TDF ecosystem has been historically subjected to pressures from human disturbance and is considered to be capable of recovering more quickly after disturbance (Murphy and Lugo 1986a; Segura et al. 2003). Quesada et al. (2009) point out that, in fact, evidence suggests that TDFs are more susceptible to human disturbance because growth rate and regeneration in this ecosystem are slow, reproduction is highly seasonal, and most plants are mainly outcrossed and dependent on animal pollination. Quesada et al. (2009) also point out the importance of light dynamics (gap dynamics) in the regeneration of TDFs after disturbance and suggest more efforts to understand the fundamental ecological processes of this ecosystem.

As a contribution to the understanding of biological dynamics in TDF fragments, this study shows results from eight forest edge-to-interior transects surveyed in two TDF fragments (old maintained edges) located in Venezuela and Brazil. The specific objective of this study is to make a systematic evaluation of forest edge influence on the amount of visible light penetrating the canopy and edge influence on understory microclimatic conditions.

10.2 Methods

10.2.1 Study Sites

We selected two TDF fragments within which to conduct edge effect surveys. The first is located within Parque Estadual da Mata Seca in Brazil and the second within the Hato Piñero (HP) private wildlife refuge in northern Venezuela (Figure 10.1a and c). Both fragments represent the remnants of mature TDFs within agricultural and cattle ranching land-use matrices (Alves 2008; Bertsch and Barreto 2008). A full description of the characteristics of each site is provided in Chapter 1.

10.2.2 Sampling Design

We conducted this study using a sampling scheme adapted from the studies of Runkle (1992), Kapos (1997), Laurance (1997), Gerwing (2002), Kalácska et al. (2005c), Souza et al. (2005), and Pohlman et al. (2007). At four locations for each site, we established 500 m transects extending from the edge of the fragment into the forest interior in four locations at each site. Each transect allowed us to perform surveys of the following variables: total canopy gap

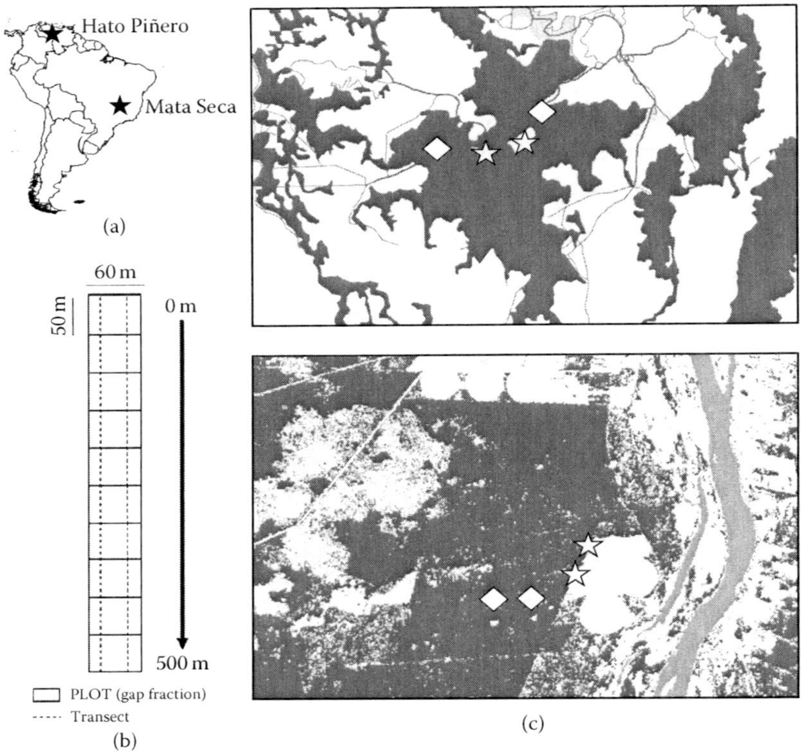

FIGURE 10.1
(a) Relative location of study sites in South America, (b) design of the parallel transects and 50 m × 60 m plots established at each forest edge at each site to study edge influence, and (c) location of transects at each site: surveys were performed at each transect at several intercepts. Two contiguous parallel transects were established to include more sampling effort for each transect intercept. In the figure, open pastures (stars) and linear openings (diamonds) are shown.

fraction; fraction of intercepted photosynthetically active radiation (FiPAR); canopy openness; plant area index (PAI); and microclimatic variables such as temperature, vapor pressure deficit, and photosynthetically active radiation (PAR) in the forest understory. Two contiguous parallel transects were established at each location to include more samples for each transect intercept. Total forest area surveyed at each location covered 0.03 km^2. For each site, two transects were located in edges exposed to linear openings (LOs) (unpaved roads ~10 m wide) and two transects were located in edges facing open pastures (OPs) (large forest clearings with little to no forest vegetation regrowth) to include exposure to different conditions at the agricultural matrix. Due to a limited amount of comparable edges for mature TDFs, transect selection did not consider effects of the edge aspect and exposure to prevailing wind and solar radiation conditions. Figure 10.1b shows the sampling design for each transect.

Marked differences between sites for climatic and hydrological conditions caused forest canopy and understory tree deciduousness to be much higher in Mata Seca (MS) than in HP during the dry season. Therefore, measurements at each site were made during the wet season (between May and June 2007 for HP and between December 2007 and January 2008 for the MS site), when fully expanded canopy leaves and edge effects on light environment and forest structural properties were comparable. Data collection of each variable was undertaken once per transect, as quickly as possible, in days with clear or relatively clear weather (overcast skies but not rainy) for a total of 16 days of data collection per site. An explanation of the variables measured is given in Sections 10.2.2.1 through 10.2.2.4.

10.2.2.1 Gap Fraction

A gap refers to an area within the forest where canopy (leaf height of tallest stems) height is noticeably diminished (Runkle 1992). Gaps in a forest are generally created by the death of one or more canopy trees. In forest edges, increased tree mortality and damage as a consequence of windthrow result in more frequent treefall gaps (Laurance et al. 2002). To measure gap size, 60 m (width) × 50 m (length) plots were surveyed along transects extending from the forest's edge to 500 m into the forest (for a total of 10 plots, see Figure 10.1b). The dimensions (width and length) of all forest canopy gaps—areas where maximum canopy height averaged 3–5 m (following the study by Kapos et al. 1997)—were measured within each plot. Gap size was calculated following the Runkle (1992) formula, which assumes elliptical dimensions for every treefall gap, and the percentage of gap area per total plot area (gap fraction) was found using the following formula:

$$\text{Gap fraction} = \left\{ \frac{\left[\frac{(\pi \times \text{gap length} \times \text{gap width})}{4} \right]}{\text{plot area}} \right\} \tag{10.1}$$

10.2.2.2 Fraction of Intercepted Photosynthetically Active Radiation

FiPAR has been used to assess forest canopy light absorption dynamics and canopy structure (Gamon et al. 2005; Olofsson et al. 2007; Serbin et al. 2009). FiPAR is measured by contrasting below-canopy downwelling PAR readings with above-canopy downwelling PAR readings. FIPAR is calculated using the following formula:

$$\text{FiPAR} = \frac{(\text{Id A} - \text{Id B})}{\text{Id A}} \tag{10.2}$$

where Id A represents downwelling PAR radiation above the canopy and Id B represents downwelling PAR radiation below the canopy. To measure FiPAR, transects were surveyed at 13 line intercepts located at 0, 10, 25, 50, 100, 150, 200, 250, 300, 350, 400, 450, and 500 m from the edge of the forest. At each line intercept, we used a Li-190 Line Quantum Sensor (LI-COR Biosciences, Lincoln, Nebraska) to measure Id B at the intercept and at 7.5 m from the intercept (left and right). Data collection was made at breast height. Id A was monitored on a large clearing outside the forest using a PAR sensor (S-LIA-M003) (Onset Computer Corporation, Bourne, Massachusetts) connected to a HOBO data-logging weather station (Onset Computer Corporation). The line quantum sensor averages photosynthetic photon flux density over its 1 m length (Licor Inc. 2010), which excluded the possibility of spatial autocorrelation between measurements.

10.2.2.3 Canopy Openness and Plant Area Index

We estimated canopy openness in 13 transect intercepts located at 0, 10, 25, 50, 100, 150, 200, 250, 300, 350, 400, 450, and 500 m from the edge of the forest in each transect. At each line intercept, we took hemispherical photographs of the forest canopy (using a Nikon CoolPix 995 camera) at 7.5 m from the intercept (left and right). Photographs were taken at a uniform camera height of 1.5 m. According to Chong et al. (2008), this is the optimal camera position to record wood and foliage for as much as 11 m around the sampling location in mature TDFs. Therefore, the distance maintained between measurements also excludes the possibility of spatial autocorrelation between measurements. The photographs were processed with the Canopy Gap Light Analyzer v. 2.0 (Frazer et al. 1999) to extract canopy structure information from the hemispherical photographs. From each hemispherical photograph, we quantified the percentage of canopy openness (percentage of open sky seen from beneath the forest canopy) and PAI, PAI 4 Ring (Kalácska et al. 2005c), which is the effective leafy and woody area index integrated over the zenith angles 0°–60°.

10.2.2.4 Understory Microclimate

At forest edges, elevated light penetration and windthrow affect environmental conditions at the understory (Kapos et al. 1997; Laurance et al. 2002; Pohlman et al. 2007). Changes in vapor pressure deficit (VPD), light availability, and temperature at the understory can affect sapling and tree densities, favoring disturbance-adapted plant species and changing species richness and composition (Gelhausen et al. 2000; Laurance et al. 2002; Pohlman et al. 2009). In this study, edge gradients of temperature, VPD, and PAR were investigated following the methods of Pohlman et al. (2006) and Turton and Freiburguer (1997). Temperature, relative humidity, and PAR values were measured in nine transect intercepts located at 0, 4, 8, 12, 16, 20, 30, 50, and 100 m. Measurements were taken at four heights from the ground: 30 cm, 1 m, 2 m, and 3 m. A HOBO PAR smart sensor (Onset Computer Corporation) and a HOBO temperature/relative humidity smart sensor (Onset Computer Corporation) were mounted on an extendible pole and connected to a HOBO data-logging weather station. The monitoring system was allowed to record microclimatic measurements for a full 2 minutes at each height. At each measurement height, the PAR sensor was leveled to a horizontal fixed position. Data collection was undertaken as quickly as possible, in clear or relatively clear weather (cloudy or overcast but not rainy), between 8:00 a.m. and 10:00 a.m. when the light environment at the understory is relatively stable (Chazdon and Fetcher 1984).

10.2.2.5 Statistical Analyses

We examined differences in canopy gap fraction, FiPAR, canopy openness, PAI, and understory microclimate as a function of distance from the forest edge using parametric one-way analyses of variance (ANOVAs), and nonparametric Kruskal–Wallis one-way ANOVAs. Most canopy structural changes due to tree damage and mortality (as determined by gap fraction, FiPAR, canopy openness, and PAI) occur within the first 100 m from the edge, and forest interior conditions are reached after 300–400 m from the edge (Laurance et al. 2002; Harper et al. 2005). Following Turton et al. (1997), we grouped values in categories of "distance from edge" for the statistical analyses. Comparison was made at three different categorical levels: 0–100, 150–300, and 350–500 m from the forest edge.

Regarding microclimatic variables, previous studies have shown that most edge-related changes occur within 40 m from the edge (Laurance et al. 2002; Harper et al. 2005; Pohlman et al. 2007) and conditions at 100 m are similar to those of the undisturbed forest interior (Turton et al. 1997; Pohlman et al. 2007). Therefore, statistical differences in temperature, VPD, and PAR were analyzed as a function of distance from the forest edge in three different categories: 0–10, 10–30, and 50–100 m. VPD was calculated from temperature and relative humidity following the methodology by Allen et al. (1998b).

10.3 Results and Discussion

Overall, results show that TDF fragments display both physical and structural responses to edge exposure. All observed variables were affected significantly corresponding to distance from forest edge in at least one of the treatments and sites sampled. A description of results for each variable is given in Sections 10.3.1 through 10.3.5.

10.3.1 Gap Fraction

Gap fraction was noticeably affected by distance from edge in all dry forest transects studied. Abrupt changes in canopy height due to treefall gaps were more frequent in the first 300 m from edge. Generally, gaps reached dimensions of 10–15 m in width or length created by the collapse of one single tree. Other gaps were formed by the collapse of two or more trees, creating a larger gap. These larger gaps were ~30–40 m in width or length. Figure 10.2 shows that the percentage of area in forest gaps increases with distance and reaches a peak at 150–300 m from the edge (mid distance). This pattern has been previously identified for tropical rain forests (Laurance et al. 2002) and has been associated with the edge-closure phase where vegetation regrowth partially seals the edge, but factors such as increased windthrow produce a proliferation of treefall gaps. In MS, gap fraction values in both treatments (OPs and LOs) varied significantly with distance from the forest edge ($F_{2,7} = 6.57_{OP}$; $F_{2,7} = 21.13_{LO}$, all $P < 0.05$), showing a similar pattern of mid-distance increase in canopy gap area. In HP, OP gap fraction values did not vary significantly with distance from edge (all $P > 0.05$). Here, gaps showed little increase at mid distances, and values remained relatively uniform along the 500 m transects. Although LOs in HP showed an increase in accumulated gap fraction at mid distances, those at the HP site did not show any statistically significant difference with distance from the forest edge (see Figure 10.2 and Table 10.1).

10.3.2 Fraction of Intercepted Photosynthetically Active Radiation

In contrast to gap fraction, in which actual forest structural and physical changes at the canopy level were measured, FiPAR measurements reflect the amount of photosynthetically active light intercepted by canopy and subcanopy leaves. FiPAR is thus an indicator of forest canopy and subcanopy openness, which is affected by not only the frequency and size of treefall gaps but also the exposure to lateral light penetration at the forest edge. For the OP treatment at the MS site, FiPAR values varied significantly with distance from forest edge ($H = 34.37$, $P < 0.01$) with the amount of intercepted light increasing toward the forest interior. At the MS site, in LOs FiPAR did not differ significantly with distance since light penetration remained low as far as 300 m into the forest interior. The effect of distance in this treatment, however, closely approached statistical significance ($H = 4.9$, $P = 0.08$), supporting the strong trend of FiPAR values increasing as forest interior conditions become predominant (see Figure 10.2). At the HP site, FiPAR values in both treatments were affected by distance to edge (all $P < 0.05$). FiPAR showed sensitivity to both lateral light penetration from edge exposure and changes in downwelling PAR reaching the forest understory as a result of canopy gaps. This is evident in Figure 10.2, where abrupt decreases occur in FiPAR values at the forest edge (0–10 m) and at

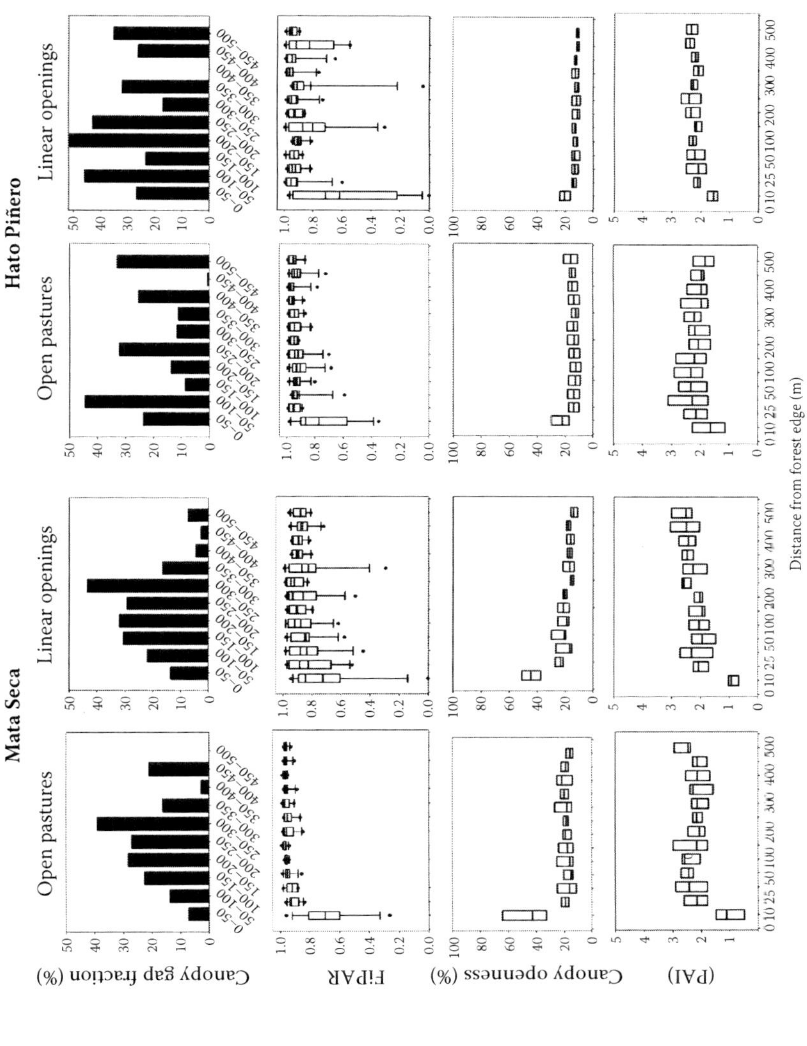

FIGURE 10.2
Variation in forest structure parameters (percentage of canopy gap fraction, fraction of intercepted photosynthetically active radiation [FiPAR], percentage of canopy openness, and plant area index [PAI]) with distance from the forest edge.

TABLE 10.1

One-Way ANOVAs (F-ratio and Kruskal–Wallis tests) Showing the Effects of Distance from the Edge on Gap Fraction, FiPAR, Canopy Openness, and PAI for Each Site and Edge Type

	Site/treatment		p Value
Gap fraction	MS-OP	$F_{2,7} = 6.572$	<0.05
	MS-LOs	$F_{2,7} = 21.131$	<0.05
	HP-OP	$F_{2,7} = 0.764$	>0.05
	HP-LOs	$F_{2,7} = 0.844$	>0.05
FiPAR	MS-OP	$H = 34.37$	<0.001
	MS-LOs	$H = 4.906$	>0.05
	HP-OP	$H = 11.351$	<0.05
	HP-LOs	$H = 6.894$	<0.05
Canopy openness	MS-OP	$H = 0.125$	>0.05
	MS-LOs	$H = 27.291$	<0.001
	HP-OP	$H = 0.458$	>0.05
	HP-LOs	$H = 10.131$	<0.05
PAI	MS-OP	$H = 0.305$	<0.05
	MS-LOs	$H = 22.018$	<0.001
	HP-OP	$H = 1.181$	>0.05
	HP-LOs	$F_{2,101} = 4.059$	<0.05

mid distances in all treatments. The decline of values at mid distances was closely related to the occurrence of large canopy gaps as registered by gap fraction measurements.

10.3.3 Canopy Openness and Plant Area Index

Differences in canopy openness as a result of edge influence (measured from hemispherical photographs of the forest canopy) showed dissimilar responses between treatments. In OP edges at the MS and HP sites, the range of percentage of canopy openness was very similar throughout the 500 m transect with only an abrupt increase in the first 10 m as a result of exposure to lateral light penetration from the edge. In OP edges, analysis showed no significant difference in canopy openness values corresponding to distance from the edge (all $P > 0.05$). In LOs at both sites, however, canopy openness varied significantly with distance from edge (all $P < 0.01$), showing an abrupt increase in the first 10 m and slightly decreasing toward the forest interior's low openness conditions (canopy openness of 10%–15%). Similar to canopy openness, the PAI, as calculated from hemispherical photographs, was responsive to increased lateral light exposure at the very edge of the forest (0–10 m) in all treatments and showed no significant difference with distance from edge in OP edges. In LOs, calculations reflected a slight continuous increase in PAI values toward the forest interior conditions (~2.5 PAI).

In general, hemispherical photographs showed canopy openness and PAI measurements to be sensitive to lateral light penetration, but they showed no obvious response to canopy disturbance from treefall gaps at mid distances from the edge. The results obtained at these mid distances might have been affected by a higher density of understory vegetation. When hemispherical photographs are taken at ~1.5 m from the ground, final canopy openness and PAI estimates usually include the woody and foliage component from the understory for as much as 11 m around the sampling location (Chong et al. 2008). Abundant foliage from the understory can mask a large proportion of a hemispherical photograph, thereby yielding low values of canopy openness in areas where the canopy is actually open and affected by treefall gaps. Even though this study did not include a survey of understory vegetation biomass, an increased abundance of understory vegetation at mid distances from the edge was indeed observed. The increased diversity and abundance of saplings, herbs, and shrubs is, in fact, one of the most generalized responses to edge creation in tropical forests (Kapos et al. 1997; Laurance et al. 2002; Harper et al. 2005). Therefore, contrary to FiPAR measurements, canopy openness and PAI values might have been affected by a higher abundance of understory biomass at mid distances.

10.3.4 Understory Microclimate

In the TDF sites studied, understory microclimatic variables were affected significantly by distance from the forest edge (Figure 10.3 and Table 10.2). One primary observation is that temperature, VPD, and PAR ranges were different across sites ($T \geq 22$ and ≤ 31, VPD ≥ 0.6 and ≤ 2, and PAR ≥ 25 and ≤ 1400 at the MS site; and $T \geq 25$ and ≤ 28, VPD ≥ 0.2 and ≤ 1.4, and PAR ≥ 25 and ≤ 400 at the HP site). These dissimilarities might be related to differences in leaf area, species composition, and forest structural properties driven by climatic and hydrological conditions inherent to each site. Nonetheless, variables showed a similar response to edge conditions. In all sites and treatments, temperature, VPD, and PAR showed decreasing values at 4–16 m from edge and then a swift increase toward 100 m conditions. Also, when results were compared between sites response to edge influence showed similar directionalities. This is shown in Table 10.2, where median values increase significantly with distance from edge (all $P < 0.01$). The results show that for all transects in MS and HP, temperature, VPD, and PAR increase as a function of distance from the forest edge and clearly demonstrate that the distribution of values of the three variables changes toward the forest interior following the same direction.

Figures 10.3 also shows that light conditions at the MS site were highly variable even at 100 m from the forest edge. In contrast to the low and stable light environment found at the HP site, in the MS site we found values of PAR at the 100 m intercept that did not resemble tropical forest interior conditions (Chazdon et al. 1984; Chaves and Avalos 2008). Such values can be the result

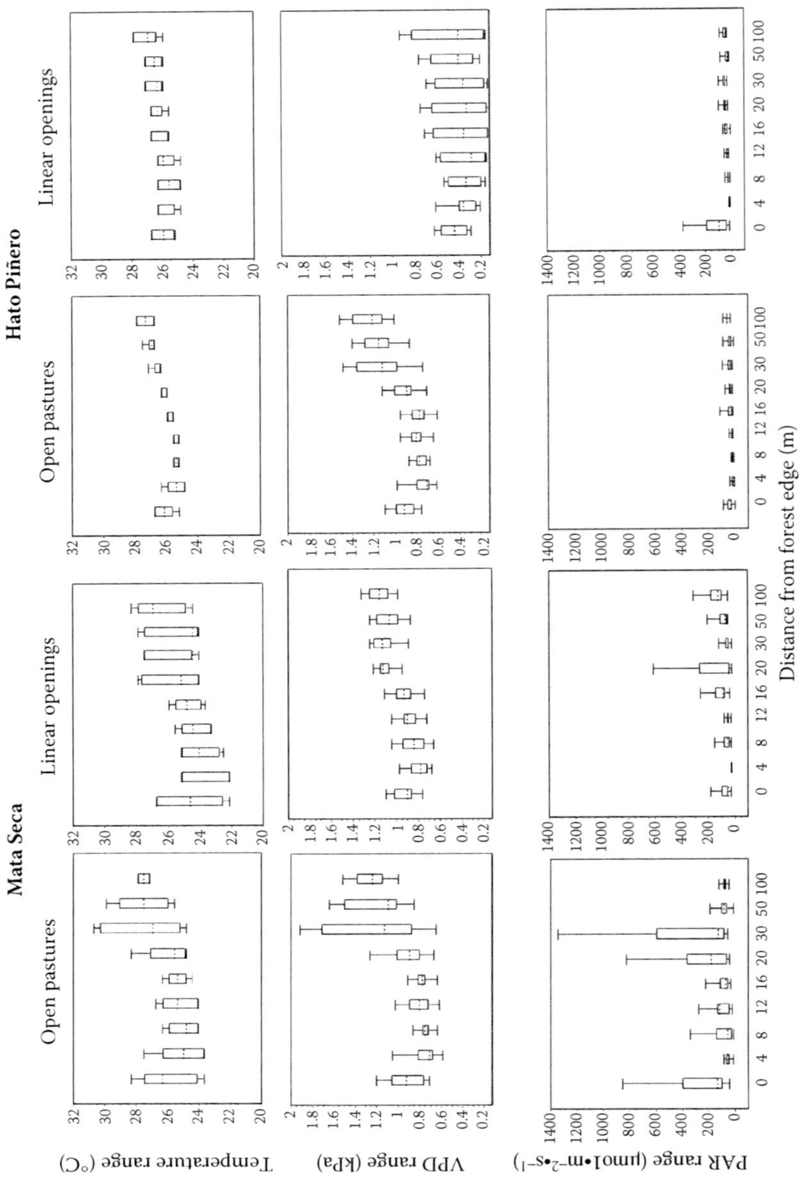

FIGURE 10.3
Variation in understory microclimatic parameters (temperature, vapor pressure deficit [VPD], and photosynthetically active radiation [PAR]) with distance from the forest edge.

TABLE 10.2

One-Way ANOVAs (Kruskal–Wallis Tests) Showing the Effects of Distance from the Edge on Understory Microclimatic Parameters (Temperature, VPD, and PAR)

		ANOVA	
Distance from edge	**Median**	*H*	***p* Value**
HP site			
PAR (μmol·m^{-2}·s^{-1})			
0–10 m	21.2	2425.733	<0.001
12–30 m	36.2		
50–100 m	46.2		
VPD (kPa)			
0–10 m	0.62	1064.95	<0.001
12–30 m	0.69		
50–100 m	0.90		
Temperature (°C)			
0–10 m	25.56	5817.269	<0.001
12–30 m	25.95		
50–100 m	26.925		
MS site			
PAR (μmol·m^{-2}·s^{-1})			
0–10 m	58.7	961.059	<0.001
12–30 m	76.2		
50–100 m	91.2		
VPD (kPa)			
0–10 m	0.79	5947.63	<0.001
12–30 m	0.91		
50–100 m	1.13		
Temperature (°C)			
0–10 m	24.79	2890.868	<0.001
12–30 m	25.17		
50–100 m	27.12		

of particular characteristics in dry forest structure and biological dynamics (Murphy and Lugo 1986a; Quesada et al. 2009) or the result of persisting disturbance from edge influence. Further research on understory microclimate should evaluate the use of longer transects (~500 m) to corroborate if light conditions (and other variables) stabilize toward the interior of the forest fragment.

10.3.5 Overall Analysis

Based on our observations and the data collected, some generalities related to edge effects can be drawn across transects and sites. In the first 10 m from the forest edge, lateral light penetration through the canopy and subcanopy layers was high. Generally, logs and branches from damaged and fallen trees in the forest floor were visibly more frequent in the first 0–25 m. Vegetation abundance at the forest subcanopy and understory layers was also noticeably higher here than in the forest interior (>400 m from edge). After 100 m, canopy gaps start to appear more often and are larger in size, usually created by fallen tall trees (15–20 m height) that lie on the forest floor, in some cases having brought down adjacent trees or branches as well. The occurrence of larger treefall gaps at the 150–300 m range was seen in all treatments and all sites resembling the edge response of humid forests to increased windthrow (Laurance and Curran 2008). In these gaps, lianas and tree saplings were present and generally reached a height of 3–5 m. After the 300 m intercept, forest understory vegetation was sparser and evenly distributed, whereas canopy and subcanopy layers were more continuous and less interrupted by treefall gaps, which at these distances were less frequent and smaller in size.

The edge effect profile observed was similar in all sites. Measured canopy gap fraction clearly showed alterations of canopy and subcanopy structures as a result of edge influence. FiPAR measurements were able to capture these variations, whereas canopy openness and PAI measurements from hemispherical photographs only captured changes occurring at the very edge of the forest. The effect of edge influence on temperature, PAR, and VPD was stronger at the MS site and seemed to penetrate further than 100 m. At the HP site, values were relatively stable throughout the transect showing only subtle decreases in PAR, VPD, and temperature at the 4–16 m intercept. A decrease in VPD, PAR, and temperature at these distances was also registered for the MS site. Overall, microclimatic conditions were similarly affected by distance from edge at all sites. A similar confounded response of these variables has already been reported in edges of tropical humid forests, especially related to a gradient of subcanopy vegetation density (Jose et al. 1996; Didham and Lawton 1999). In the case of the MS and HP sites, edge exposure favors the abundance of rapid-growth trees, shrubs, and lianas in a margin extending up to 20 m from the edge and creates dense vegetation conditions with slightly lower temperature, VPD, and PAR values in the subcanopy and understory layers. As the edge effect diminishes its impact, more stable temperature, VPD, and PAR conditions seem to be reached, usually after 30 m from the edge. This behavior is, however, uncertain and should be studied further using longer transects (>100 m from the edge).

In general, our results indicate that the edges studied in this chapter might be situated in the postclosure phase according to Harper et al. (2005). In this phase, secondary responses often result in the development of a sidewall of dense vegetation that fills open spaces at the edge (Laurance et al. 2002;

Harper et al. 2005). Tree damage and canopy gap proliferation are still high due to persistent elevated windthrow. Gap fraction and FiPAR data show that edge influence at these sites extends at least 300 m into these dry forest fragments. The edges studied in these forest fragments are older (>25 years since creation). According to Harper et al. (2005), edge effects in older humid forest edges should be considered as attenuated and affecting less than 50–100 m into the forest. However, the dry forest edges studied here showed similar responses to young or newly created humid forest edges. A plausible explanation is that biophysical and ecological characteristics of TDFs play an important role in determining resilience to edge effects. A TDF will have lower basal area and stature than a humid forest (Pennington et al. 2006b), which can make it structurally more vulnerable to lateral light penetration, lateral winds, and gap formation following edge creation. Severe drought conditions and seasonality might also be important factors perpetuating the strong edge influence. Further research on the response of understory microclimatic conditions, as well as changes in biotic properties (e.g., species composition, tree mortality, phenology patterns, and nest predation), is important to understand the degree of disturbance affecting TDF remnants after deforestation events. These studies should consider the evaluation of several TDF sites with similar seasonal, hydrological, and climatic conditions (e.g., similar forest phenology patterns). Measuring variables in transects >500 m into the forest interior for canopy openness and >100 m for understory microclimatic conditions will also shed more light on the magnitude and distance of edge effects in TDF fragments.

10.4 Conclusions

More than half of TDFs have been transformed globally (Miles et al. 2006). In the Americas, only 44% is left (Portillo-Quintero and Sánchez-Azofeifa 2010). Most TDFs survive in lowland human-dominated landscapes with limited legal protection in comparison with rain forests and are regularly threatened by crop cultivation, cattle ranching, and tourism expansion (Portillo-Quintero and Sánchez-Azofeifa 2010). Deforestation for land development and road construction has converted dry forest areas that were once continuous and relatively uninterrupted into fragmented landscapes of small and large forest remnants embedded in agricultural and road matrices. One of the major changes brought about by habitat fragmentation to tropical forests is an increase in the proportions of edge exposed to other habitats (Kapos et al. 1997; Laurance and Curran 2008). However, most research on edge evolution through time (structural and biological dynamics during edge creation, closure, and sealing processes) has evaluated humid forests. In this chapter, we presented results from edge-to-interior surveys in two TDF fragments where we assessed the edge influence on important indicators of canopy structure, integrity, and functionality (gap fraction, FiPAR, canopy openness, and PAI) and the edge influence on understory microclimatic variables that are critical in determining ecological processes such as plant growth, decomposition, and nutrient cycling (Turton et al. 1997; Laurance et al. 1998; Sizer and Tanner 1999).

Measurements of canopy gap fractions across transects clearly indicate a physical and structural impact of edge exposure that extends up to 300 m from the forest edge into the forest interior in both TDF fragments studied. Treefall gaps proliferate in the margin 150–300 m from the forest edge. FiPAR results showed that light penetration to the understory increases at the edge of the forest and around created treefall gaps. This increment in light availability in the understory might favor an observed (but not documented) increase in abundance of understory vegetation at the edge and around treefall gaps. Temperature, VPD, and PAR were also affected by edge conditions in the first 100 m from edge. Although edges selected in this study were created >25 years ago, penetration of primary processes (tree damage) following edge creation can be compared to the response of young or newly created humid forest edges. Tree biomass, seasonality, plant growth rate, and reproduction dynamics might play an important role in determining restoration (Quesada et al. 2009) and edge evolution in TDF fragments. We suggest that further research on biological dynamics arising from edge creation is needed to better understand the resilience and regeneration capacity of TDFs in response to fragmentation.

11

Fruit-Eating Bats and Birds of Three Seasonal Tropical Dry Forests in the Americas

Jafet Nassar, Kathryn E. Stoner, Luis Ávila-Cabadilla, Mário Marcos do Espírito-Santo, Carla I. Aranguren, José A. González-Carcacía, Juan Manuel Lobato-García, Lemuel Olívio-Leite, Mariana Álvarez-Añorve, Hugo N. de Matos Brandão, Luiz Alberto Dolabela-Falcão, and Jon Paul Rodríguez

CONTENTS

11.1 Introduction

Many of the introductory contents in this book have emphasized that seasonally tropical dry forests (STDFs) are among the most threatened ecosystems in the tropics due to anthropogenic stresses, which have caused dramatic impacts on the natural landscapes and transformed once-continuous forests into a complex mosaic of vegetation units, with a variable degree of human intervention. In response to this complexity, multidisciplinary studies, which cover all ecological aspects of this poorly understood ecosystem, are needed to formulate local, regional, and global policies and strategies that are aimed at guaranteeing forest regeneration while seeking their own sustainable use (Sánchez-Azofeifa et al. 2005c).

The identification and evaluation of the pollination and seed dispersal networks, including the characterization of the species groups that perform these functions, are among the key aspects for understanding the dynamics of the regeneration of TDFs. A major proportion of the vertebrate biomass in tropical forests is supported by fruits (Fleming et al. 1987; Levey 1988), and the seeds of these fruits are effectively dispersed by many of these vertebrates (Howe and Smallwood 1982; Wunderle 1997). Between 50% and 90% of the shrub and tree species in the tropics are dispersed by animals (Howe and Smallwood 1982; Janzen 1983). The importance of animal-mediated seed dispersal is accentuated in the case of fragmented landscapes, where birds and small terrestrial and flying mammals play critical roles in forest regeneration (Fleming and Heithaus 1981; Guevara et al. 1986; Janzen 1988; Galindo-González et al. 2000; Griscom et al. 2007).

Birds and bats are considered among the best seed vectors in terms of both the amount of seeds dispersed and their dispersal distance, due to the great variety of fruits consumed, the long retention time of ingested seeds, and high vagility (Howe 1990; Medellín and Gaona 1999; Galindo-González et al. 2000; Ortiz-Pulido et al. 2000). Birds have been described as the most important seed dispersers of many angiosperms because of their high species diversity (Fleming and Kress 2011). They represent the largest number of frugivorous species in the neotropics (Ferreira Fadini and De Marco Jr. 2004), and a large proportion of tropical plants rely on them for their propagation (Terborgh 1986; Murray 1988; Howe 1990). Birds frequently move seeds from forests to disturbed habitats and isolated forest remnants (Howe and Smallwood 1982; Guevara and Laborde 1993; Bianconi et al. 2007), including isolated trees in pastures (Guevara and Laborde 1993; Galindo-González et al. 2000). Neotropical fruit-feeding bats almost entirely belong to the family *Phyllostomidae* (Fleming et al. 1987). They are also considered important seed dispersal agents (Geiselman et al. 2007), especially for small-seeded pioneer plants in early successional stages (Humphrey and Bonaccorso 1979; Charles-Dominique 1986; Fleming 1988; Medellín and Gaona 1999; Muscarella and Fleming 2007).

Studies that compare the seed dispersal attributes of birds and bats support the hypothesis that both these groups have complementary functions as seed vectors (Charles-Dominique 1986; Muscarella and Fleming 2007; Jacomassa and Pizo 2010; Mello et al. 2011). Bats defecate seeds inside forests, in forest gaps, and in open spaces during flight; whereas birds defecate most seeds under trees while they are perched. Bats can also remove comparatively more fruits than do birds for a given foraging effort unit, but birds more than bats tend to favor the germination performance of seeds. So, in order to understand and enhance the regeneration potential in a tropical forest under a particular regime of disturbance, it is important to describe the characteristics of the ensembles of fruit-eating birds and bats present in it. This is particularly crucial in the case of heterogeneous landscapes that are composed of a mosaic of successional stages and agricultural lands, where dispersion of plant propagules between vegetation units requires the intervention of highly mobile seed vectors.

The majority of studies that are conducted to characterize the species composition and structure of bat and bird assemblages in TDFs have not focused on fruit-feeding guilds in particular, but rather on the entire assemblages of these vertebrates (Moreno and Halffter 2001; Stoner 2005; Sánchez et al. 2007; Avila-Cabadilla et al. 2009; Losada-Prado and Molina-Martínez 2011; Pech-Canche et al. 2011), and only a few have addressed the problem of seed dispersal (Ortiz-Pulido et al. 2000; Griscom et al. 2007). This is despite the fact that animal-mediated seed dispersal is substantially high (35%–70%) in dry forests (Willson et al. 1989). The available evidence indicates that both bats and birds effectively disperse seeds between different vegetation units within fragmented TDFs, and that forest gaps and open areas receive a substantial amount of seeds that are transported from primary and secondary growth forests. However, TDFs vary considerably in terms of plant species composition, structure, and physiognomy throughout the Americas (Bullock et al. 1995; Mooney et al. 1995; Vieira and Scariot 2006), and we do not understand well how this variation translates into changes in the ensembles of fruit-feeding bats and birds. Poulin et al. (1994) detected marked differences in bird species abundances in three adjacent dry habitats in northeastern Venezuela that exhibited different levels of vegetation complexity and resource availability. Stoner (2005) showed that dry forests with contrasting precipitation regimes (Palo Verde, Costa Rica vs. Chamela, Mexico) differed considerably in their bat faunas, and, more precisely, in their attributes of fruit-feeding ensembles. Increases in species richness and abundance of both frugivorous bats and frugivorous birds were positively associated with increases in vegetation complexity and seasonal resource availability. These responses to vegetation complexity and resource availability have been well documented for ensembles of tropical fruit-eating bats and birds across a broad range of ecosystems (Karr et al. 1982; Stoner 2005; Sánchez et al. 2007; Mello 2009; Klingbeil and Willig 2010; Pech-Canche et al. 2011; Avila-Cabadilla et al. 2012).

The present study documents the species composition and structure of the ensembles of fruit-eating bats and birds that are present in different successional stages of three TDFs across a latitudinal gradient in the neotropics, comprising study sites in Mexico, Venezuela, and Brazil. To the best of our knowledge, no equivalent comparisons of the ensembles of vertebrate frugivores were conducted earlier across a latitudinal gradient of dry forests. We aim at responding to the following questions: What attributes characterize the patterns of species richness and the abundance of these ensembles across regions, successional stages, and seasons? How structured are these ensembles in terms of the relative importance of fruits in the animals' diets? Can we identify species or groups of species that could be considered of particular importance for the regeneration of TDFs?

11.2 Materials and Methods

11.2.1 Study Sites

Study sites are located in seasonal dry forests within three protected areas: Chamela–Cuixmala Biosphere Reserve (CCBR), Mexico; Unidad Productiva Socialista Agropecuaria–Piñero (UPSAP), Venezuela; and Mata Seca State Park (MSSP), Brazil. Their geographic location, climate, and vegetation characterization are described in Chapter 1.

11.2.2 Sampling Plots

A series of plots were established at each study site following the protocols of the *Tropi-Dry Manual of Methods* (Nassar et al. 2008). At each study location, we selected 12 sampling sites that represented four successional stages (three sampling sites per treatment): three pastures (open areas used as pastures before the beginning of the project, protected from fire with fire breaks and with fences to keep cattle outside); three early stage sites (areas that were logged and used for agricultural and/or cattle-raising purposes and which were then protected against human intervention and fire, allowing vegetation recovery for the last 3–5 years); three intermediate stage sites (forests that were logged and used for agricultural and/or cattle raising and which were then protected against human intervention and fire, allowing vegetation recovery for the last 8–15 years); and three late stage sites (old forests, with more than 50 years after the last known human intervention).

Each sampling plot in Mexico consisted of a 90 m × 120 m area inside a matrix of vegetation of a similar kind with all margins of at least 50-m width and surrounded by fire breaks and fences to keep cattle outside. For

Venezuela and Brazil, the plots were 30 m × 60 m. A description of the vegetation characteristics is presented in Table 1.1, In Chapter 1. In Mexico, the plots that had been established in secondary vegetation were located at 1–5 km from mature forests in CCBR. In Venezuela, intermediate successional plots were at a variable distance from late successional plots (500 m–4 km), and early plots were separated from them by 15 km. The plots that had been established in the pastures were set in open areas that were distributed among the other plot sites. In the study sites in Brazil, all plots were established within a 2.5-km radius, across an area of forest fragments and abandoned pastures under similar topographic and microclimatic characteristics. Plots from the same successional stage were located at 0.2–1.0 km from each other.

Vegetation attributes corresponding to the sampling plots in each region are described in Chapter 1.

11.2.3 Bat and Bird Sampling

Samplings of bats and birds took place during four surveys per year for 2 years, scheduled according to the occurrence of the dry and rainy season at each study site, trying to cover the beginning and advanced stages of both these seasons during each year (Nassar et al. 2008). Surveys consisted of a minimum of 12 days of capturing bats and birds in each of the 12 plots established per region. Sampling could extend for more days if a particular plot had to be repeated because of bad weather. The sequence in which the different plots were sampled within a study site was randomly chosen, having as the only condition that two plots of the same successional stage could not be sampled in subsequent days. Sampling was avoided on rainy days and moonlit nights to minimize the effects of wet mist nets and lunar phobia (Morrison 1978), respectively.

For animal captures, we used mist nets placed at ground level. Although this technique is biased toward sampling birds and bats that are flying low and is inappropriate for many tropical moist/rain forests (Bernard et al. 2001), the comparatively lower canopy height observed in these dry forests allows us to assume that most small- and medium-sized frugivore birds and bats that are foraging in the studied forests frequently fly at ground level. Phyllostomid bats, which include the majority of fruit-feeding species in the neotropics, tend to be well sampled by using mist netting at ground level (Kingston 2009). Large birds flying above the nets were not captured, and their occurrences were considered anecdotic records. In all cases, with the exception of bats in Mexico and birds in Brazil, we used ten 12-m long × 2.6-m high mist nets (38-mm mesh; Avinet Inc., Dryden NY) that were placed along the edges of the successional plots. For bats in Mexico, we used two 6-m long × 2.6 m-high nets, two 9-m long × 2.6-m high nets, and one 12-m long × 2.6-m high mist net. For birds in Brazil, we used fifteen 12-m long × 2.6-m high mist nets on each sampling day. After being opened, the nets were monitored every 30 minutes. During each survey period, we conducted five hours mist netting sessions on each successional plot to capture bats (7:00 p.m. to 12:00 a.m.) and birds (6:00 a.m. to 11:00 a.m.). These time intervals corresponded to the peak foraging times of most phyllostomid bats (Aguiar and Marinho-Filho 2004) and most diurnal birds (Sutherland et al. 2004). In the case of bats, the sampling design described was used in Venezuela and Brazil, from July 2007 to April 2009 and from April 2007 to August 2009 (one extra session in Brazil), respectively. Bats sampled in Mexico were collected under a variation of the general protocol (Avila-Cabadilla et al. 2009). In Mexico, sites were sampled approximately every 46 days from June 2004 to August 2006 (16 sampling periods), with the exception of pasture sites, which were sampled less than the other sites due to low capture rates. In the case of birds, the general sampling protocol was used across the three regions with one variant in Brazil, where mist netting sessions lasted six hours (6:00 a.m. to 12:00 p.m.) instead of five hours. In Mexico, birds were sampled from March 2007 to December 2008; in Venezuela, from July 2007 to April 2009; and in Brazil, from October 2007 to June 2009. Total sampling effort at each site was determined by multiplying the total area of mist nets used per hour by the total number of hours sampled during the entire study: 77,532 $h \cdot m^2$ for bats in Mexico; 149,760 $h \cdot m^2$ for bats and birds in Venezuela, birds in Mexico, and bats in Brazil; and 269,568 $h \cdot m^2$ for birds in Brazil.

Captured animals were temporally placed in cotton bags to obtain fecal samples. We identified them using dichotomous keys, taxonomic descriptions, and field guides (Linares 1986, 1998; Ridgely and Tudor 1989, 1994; del Hoyo et al. 1992; Medellín et al. 1997; Sick 1997; Timm and Laval 1998; Gregorin and Taddei 2002; Hilty 2003; dos Reis et al. 2007; Restall et al. 2007). For each animal that was processed, we recorded weight, gender, age class, reproductive condition, forearm length (bats), and molt condition (birds). Bat and bird species were assigned to general guild categories based on dietary attributes reported in the literature cited earlier, documents describing the species diet (Appendix A), online databases (www.iucnredlist.org, http://animaldiversity.ummz. umich.edu, http://www.museodelasaves.org/, http://www.inbio.ac.cr/es/default.html), and the analysis of fecal and stomach contents (unpublished data). Bats were assigned to one of the following general guilds: frugivores, insectivores, omnivores, hematophagous, and carnivores; birds were assigned to one of the following categories: frugivores, insectivores, carnivores/insectivores, nectarivores/insectivores, granivores, granivores/insectivores, and omnivores. Following a classification previously used for frugivorous birds and bats based on their levels of dependence on fruits (Kissling et al. 2009; Mello et al. 2011), we assigned species to three categories: primary frugivores (strong dependence on fruits), secondary frugivores (fruits represent a secondary food item in the diet), and occasional frugivores (fruits are rarely included in the diet).

11.2.4 Sampling Completeness of Fruit-Feeding Bats and Birds

We assessed the completeness of surveys of fruit-feeding bats and birds by calculating the percentage of the total estimated species richness that effectively was covered by sampling on each plot. Total species richness was estimated by computing the second-order Jackknife index (Jack 2; EstimateS v.7.5.0) (Colwell 2005), because this method corrects best for underestimation (Zahl 1977; Heltshe and Forrester 1983). The Jackknife procedure is a technique that is used for reducing the bias of estimates (Colwell and Coddington 1994), and designed for data analyses with a small sample size (<100 individuals per site). Ninety percent of completeness was considered a satisfactory level of sampling efficiency (Moreno and Halffter 2001). The species richness estimates were computed from a species abundance matrix with 999 randomizations for the animals captured at each study site.

11.2.5 Statistical Analyses

Rank–abundance (dominance–diversity) graphs were built following the methodology described by Feinsinger (2001). These graphs are a useful tool for visualizing some attributes of a community of organisms, such as species richness (number of points), evenness (slope), number of rare species (tail of the curve), and relative abundance of each species (order of the species in the graph), and have been proposed as an alternative way of comparing communities (Feinsinger 2001), particularly in the case of bat assemblages (Kingston 2009). These graphs are constructed by plotting the rank of each species on the x-axis from most to least abundant and $\log_{10} p_i$ for each one of the species on the y-axis, where p_i is the proportion of individuals of a given species to the total number of individuals captured. These qualitative representations of species assemblages were used to compare the ensembles of fruit-feeding bats and birds at three levels: regions (pooling all samples within each region), seasons (pooling all samples within each season in each region), and successional stages (pooling all samples within each successional stage in each region).

To determine whether the season or successional stage affected the characteristics of the species ensembles of fruit-feeding bats and birds examined in each region, we conducted nonparametric multifactorial analyses of variance with the program PerMANOVA (Anderson 2001; McArdle and Anderson 2001). PerMANOVA uses permutations to generate a probability distribution, and it does not assume a particular distribution of the data to operate. Our categorical independent variables were season (dry and rainy) and successional stage ones (pasture, early, intermediate, and late). Three response variables were examined: (1) the Jackknife 2 estimator of total species richness per sampling plot; (2) the number of species captured per sampling day (night in the case of bats); and (3) the number of individuals captured per day. We performed 9999 permutations for the analyses conducted. Only in the case of Mexico, pastures were not included as a treatment of the successional stage factor when examining variables (2) and (3). The research team responsible for captures in Mexico did not conduct mist netting sessions during the dry season due to low or no capture of bats in those plots during that period of the year.

11.3 Results

11.3.1 Species Richness and Abundance of Animals Captured

Birds had higher species richness and abundance than bats across regions (Tables 11.1 and 11.2). This difference was more accentuated in the case of Mexico and Brazil for species richness, where up to four times more species of birds than bats were trapped. In the case of the number of animals sampled, the greatest differences were found in Mexico and Venezuela, where between two and three times more birds than bats were captured. The highest number of species was found in Venezuela (bats: $N = 47$, birds: $N = 117$), followed by Brazil (bats: $N = 25$, birds: $N = 93$) and Mexico (bats: $N = 22$, birds: $N = 83$). Mexico and Venezuela topped the lists of bird (Mexico: $N = 2755$, Venezuela: $N = 2424$) and bat (Mexico: $N = 947$, Venezuela: $N = 970$) individuals captured, respectively; whereas Brazil ranked lowest in both groups (birds: $N = 1011$, bats: $N = 787$).

Based on the general feeding guilds assigned, we determined that 48.9%–63.6% of the bat species and 45.2%–67.5% of the bird species include fruits in their diets (frugivores and omnivores) (Table 11.1). The remaining taxa were classified as insectivores (28.0%–44.7%), carnivores (0%–4.3%), and hematophagous (2.1%–8.0%) in the case of bats, and as insectivores (16.9%–36.5%), carnivores/insectivores (0%–6%), nectarivores/insectivores (5.1%–9.6%), granivores (0%–2.6%), and granivores/insectivores (4.8%–7.6%) in the case of birds. In terms of the number of fruit-feeding bats and birds represented in the samples of the three regions, we found that between 53.1% and 82% of all bats and between 33.4% and 77.7% of all birds captured include fruits in their diets (frugivores and omnivores) (Table 11.2).

11.3.2 Sampling Completeness of Fruit-Feeding Bats and Birds

The sampling effort was sufficient to characterize the ensembles of fruit-feeding bats across regions, because we obtained sampling completeness levels above 90% in the early, intermediate, and late successional plots (Table 11.3). Only in the pastures, completeness was slightly below that threshold value. In the case of the fruit-feeding birds, all the Mexican sites had sampling completeness levels at or above 90% of completeness, whereas sampling plots in Venezuela and Brazil were between 1% and 5% points below that threshold value. Overall, the level of representativeness achieved in this study was within adequate limits to characterize the ensemble attributes of fruit-eating bats and birds.

11.3.3 Ensemble Attributes of Fruit-Feeding Bats and Birds

Substantially more (up to threefold) species of fruit-eating birds were recorded in this study than fruit-eating bats for the three regions (Table 11.1), and in terms of relative abundance, fruit-eating birds outnumbered bats in Mexico, but were similar for Venezuela and Brazil (Table 11.2).

TABLE 11.1

Species Richness of Bats and Birds Captured with Mist Nets at the Study Sites of Mexico, Venezuela, and Brazil According to General Guild Categories

	Mexico	Venezuela	Brazil
Bats			
Number of species	22	47	25
Insectivores	7 (31.8%)	21 (44.7%)	7 (28.0%)
Carnivores	0 (0%)	2 (4.3%)	1 (4.0%)
Hematophagous	1 (4.6%)	1 (2.1%)	2 (8.0%)
Frugivores	7 (31.8%)	10 (21.3%)	4 (16.0%)
Omnivores	7 (31.8%)	13 (27.6%)	11 (44.0%)
Number of species of fruit consumers	14	23	15
Primary frugivores	8 (57.2%)	12 (52.2%)	5 (33.3%)
Secondary frugivores	5 (35.7%)	7 (30.4%)	7 (46.7%)
Occasional frugivores	1 (7.1%)	4 (17.4%)	3 (20.0%)
Birds			
Number of species	83	117	93
Insectivores	14 (16.9%)	32 (27.4%)	34 (36.5%)
Carnivores/insectivores	0 (0.0%)	7 (6.0%)	4 (4.3%)
Nectarivores/insectivores	8 (9.6%)	6 (5.1%)	6 (6.5%)
Granivores	1 (1.2%)	3 (2.6%)	0 (0.0%)
Granivores/insectivores	4 (4.8%)	9 (7.6%)	7 (7.5%)
Frugivores	4 (4.8%)	12 (10.3%)	5 (5.4%)
Omnivores	52 (62.7%)	48 (41.0%)	37 (39.8%)
Number of species of fruit consumers	56	60	42
Primary frugivores	4 (7.1%)	10 (16.7%)	3 (7.1%)
Secondary frugivores	29 (51.8%)	33 (55.0%)	25 (59.5%)
Occasional frugivores	23 (41.1%)	17 (28.3%)	14 (33.4%)

Note: In the case of fruit-eating species, a subdivision of the number of taxa identified based on the relative importance of fruits in their diets is also included.

Species richness of fruit-eating bats differed between the dry forests examined in the three regions (Figures 11.1 and 11.2; Table 11.1). Venezuela presented the highest total number of species ($N = 23$), followed by Brazil ($N = 15$) and Mexico ($N = 14$). Average Jack 2 estimates of species richness per plot were also considerably higher for the Venezuelan plots (Figure 11.1). Primary frugivore species were more numerous (52.2%–57.2%) than secondary frugivore ones (30.4%–35.7%) in Venezuela and Mexico, respectively (Table 11.1). In Brazil, secondary frugivore bats were the most important category (46.7%), followed by primary (33.3%) and occasional (20.0%) fruit-eating species.

TABLE 11.2

Number of Bats and Birds Captured with Mist Nets at the Study Sites of Mexico, Venezuela, and Brazil According to General Guild Categories.

	Mexico	Venezuela	Brazil
Bats			
Number of individuals	947	970	787
Insectivores	324 (34.2%)	125 (12.8%)	123 (15.6%)
Carnivores/insectivores	0 (0%)	2 (0.2%)	5 (0.6%)
Hematophagous	120 (12.7%)	48 (5.0%)	188 (23.9%)
Frugivores	327 (34.5%)	611 (63.0%)	301 (38.3%)
Omnivores	176 (18.6%)	184 (19.0%)	170 (21.6%)
Number of fruit consumers	503	795	471
Primary frugivores	328 (65.2%)	615 (77.3%)	304 (64.5%)
Secondary frugivores	174 (34.6%)	123 (15.5%)	161 (34.2%)
Occasional frugivores	1 (0.2%)	57 (7.2%)	6 (1.3%)
Birds			
Number of individuals	2755	2424	1011
Insectivores	319 (11.6%)	617 (25.5%)	345 (34.1%)
Carnivores/insectivores	0 (0.0%)	32 (1.3%)	8 (0.8%)
Nectarivores/insectivores	158 (5.7%)	189 (7.8%)	106 (10.5%)
Granivores	83 (3.0%)	48 (2.0%)	0 (0.0%)
Granivores/insectivores	54 (2.0%)	649 (26.7%)	215 (21.2%)
Frugivores	23 (0.8%)	243 (10.0%)	43 (4.3%)
Omnivores	2118 (76.9%)	646 (26.7%)	294 (29.1%)
Number of fruit consumers	2141	889	337
Primary frugivores	23 (1.1%)	154 (17.8%)	4 (1.2%)
Secondary frugivores	1313 (61.3%)	437 (48.7%)	188 (55.8%)
Occasional frugivores	805 (37.6%)	298 (33.5%)	145 (43.0%)

Note: In the case of fruit-eating species, a subdivision of the number of individuals identified based on the relative importance of fruits in their diets is also included.

The total number of fruit-eating bats captured differed between regions (Table 11.2). Primary frugivore bats numerically dominated the three regions (64.5%–77.3%), followed by secondary frugivores (15.5%–34.6%) and occasional frugivores (0.2%–7.2%) (Table 11.2).

Evenness of species was considerably higher in the Venezuelan ensemble than in Mexico and Brazil (Figure 11.2). The dominance of a particular species of bats was more marked in these two countries. More rare species were found in Mexico than in Brazil and Venezuela. Appendix B shows the list of bat species captured in the three study locations that include fruits in their diets according to the literature. The fruit-feeding bats more frequently captured were *Phyllostomus elongatus, Platyrrhinus vittatus, Carollia brevicauda, Uroderma magnirostrum,* and *Glossophaga longirostris* in Venezuela; *Artibeus*

TABLE 11.3

Sampling Completeness (%) of Fruit-Eating Taxa Based on the Estimation of Total Species Richness in Each Plot Using the Second-Order Jackknife Index (Jack 2)

Animal Group	Country	Successional Stage	Mean	Range
Bats	Mexico	Pasture	87.7	87.0–88.5
		Early	92.1	89.3–94.2
		Intermediate	91.7	88.9–94.5
		Late	90.3	87.1–94.0
	Venezuela	Pasture	89.4	87.1–92.0
		Early	92.8	92.1–93.9
		Intermediate	90.0	88.5–91.2
		Late	90.6	88.2–92.8
	Brazil	Pasture	88.8	86.1–92.4
		Early	90.0	87.0–92.0
		Intermediate	90.3	88.6–92.2
		Late	92.3	91.3–93.8
Birds	Mexico	Pasture	92.1	91.0–93.8
		Early	92.3	90.7–93.7
		Intermediate	91.4	90.0–93.0
		Late	89.5	88.5–90.2
	Venezuela	Pasture	85.2	80.6–90.9
		Early	88.8	86.8–92.4
		Intermediate	88.9	88.4–89.5
		Late	88.4	87.8–88.9
	Brazil	Pasture	86.7	86.7–90.9
		Early	88.0	88.0–90.6
		Intermediate	87.2	87.2–89.3
		Late	87.5	86.2–87.5

Note: Three plots were sampled per successional stage.

jamaicensis, *Artibeus intermedius*, *Glossophaga soricina*, and *Leptonycteris yerbabuenae* in Mexico; and *Artibeus planirostris*, *Carollia* spp., *Phyllostomus discolor*, and *G. soricina* in Brazil. Occasional frugivores were infrequently captured. One exception was *Noctilio albiventris*, an insect-feeding species that was often captured in Venezuela.

Species richness of fruit-eating birds differed among regions (Figures 11.1 and 11.2; Table 11.1). Overall, Venezuela (N = 60) and Mexico (N = 56) had more species than Brazil (N = 42), despite the fact that a considerably higher sampling effort was invested in Brazil. At plot level, Mexico presented comparatively higher average values of the Jack 2 estimate of species richness than Venezuela and Brazil (Figure 11.1). The proportions of the different fruit-feeding categories evaluated were quite consistent across regions (Table 11.1). Secondary frugivore species always dominated (51.8%–59.5%), followed by occasional fruit eaters (28.3%–41.1%) and primary frugivore taxa (7.1%–16.7%).

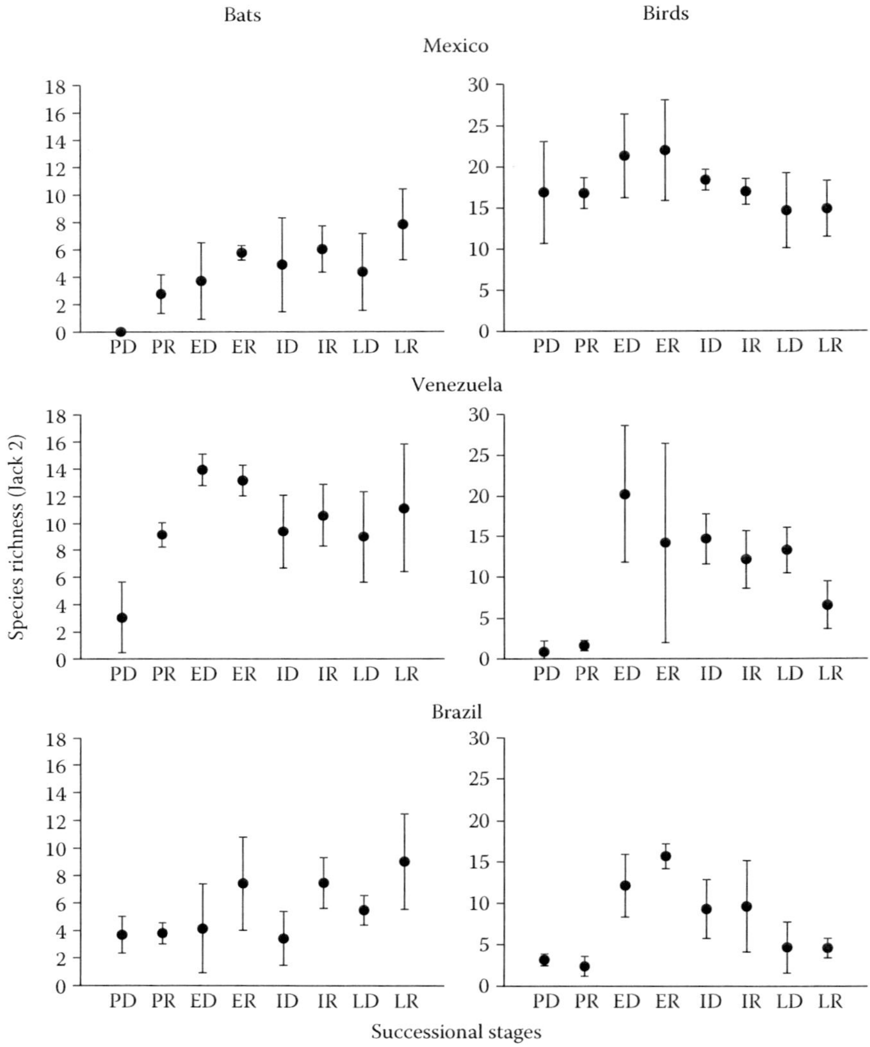

FIGURE 11.1
Average species richness (mean ± 1 SD) estimated per successional plot and season with the second-order jackknife estimator (Jack 2). Notation of each successional plot during the dry and rainy season comprises two letters, with the first referred to the *successional stage* (P: pasture, E: early, I: intermediate, L: late) and the second referred to as the *season* (D: dry, R: rainy).

The total number of fruit-eating birds captured differed substantially among regions (Table 11.2). The number of birds captured in Mexico was twice and five times the amount captured in Venezuela and Brazil, respectively; this was despite the fact that total sampling effort in Mexico was equivalent to that in Venezuela and lower than that in Brazil. Secondary

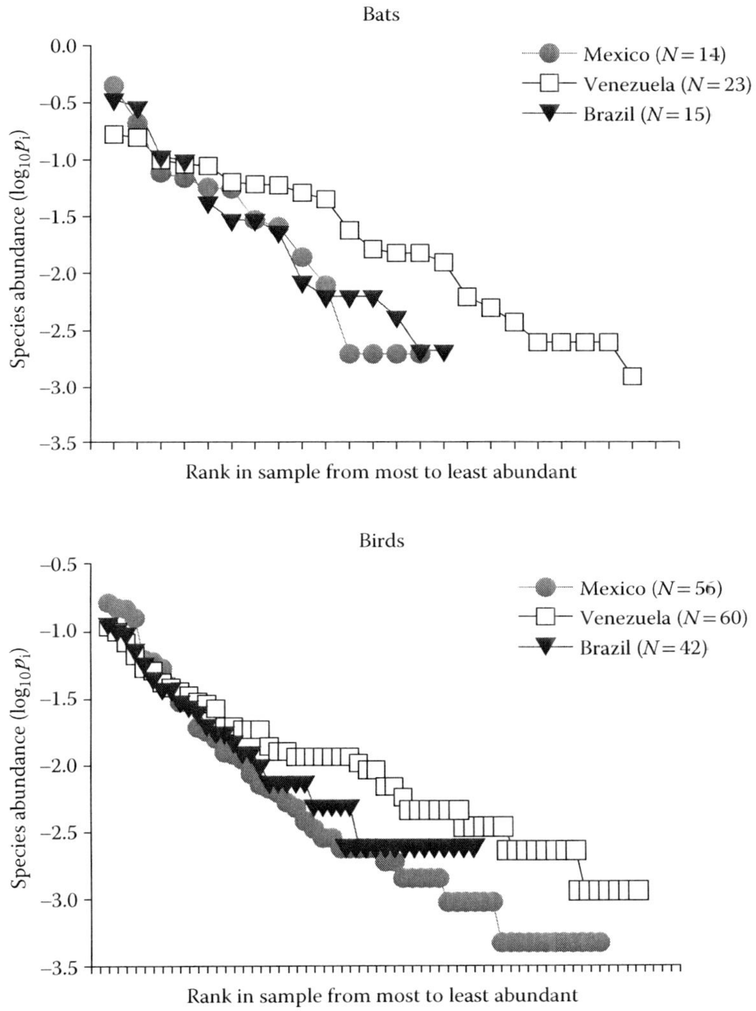

FIGURE 11.2
Rank–abundance (dominance–diversity) graphs of species of fruit-eating bats and birds pooled across all sampling plots within each region: CCBR (Mexico), UPSAP (Venezuela), MSSP (Brazil). p_i: relative abundance of each species.

frugivore birds dominated across regions (48.7%–61.3%), followed by occasional frugivores (33.5%–43.0%) and primary frugivores (1.1%–17.8%) in the last place (Table 11.2).

Species evenness of the Venezuelan taxa was comparatively higher than for bird species of Brazil and Mexico (Figure 11.2). The rare species tails in the rank–abundance curves were longer for Brazil and Mexico. Appendix C shows the list of birds captured in the three study locations that include fruits in their diets according to the literature. The fruit-feeding birds most frequently captured were *Chiroxiphia lanceolata*, *Myiarchus tyrannulus*, *Coereba flaveola*, *Columbina talpacoti*, and *Tolmomyias flaviventris* in Venezuela; *Vireo flavoviridis*, *Passerina leclancherii*, *Cyanocompsa parellina*, *Myiarchus nuttingi*, and *Columbina passerina* in Mexico; and *Turdus amaurochalinus*, *M. tyrannulus*, *Myiodynastes maculatus*, and *Myiopagis viridicata* in Brazil.

11.3.4 Effect of Seasons

Fruit-feeding bats increased in species richness, abundance, or both of these parameters during the rainy season (Figure 11.3; Table 11.4). The PerMANOVA comparisons supported this trend for the Jack 2 estimate of species richness per plot in the case of Mexico and Brazil (Appendix D), and for the number of species and individuals captured per night in the case of Mexico and Venezuela (Appendices E and F). Brazil did not present significant differences between seasons for the last two variables.

Most species of fruit-feeding bats captured in Mexico and Venezuela during the two seasons were primary frugivores; whereas in Brazil, secondary frugivores were more numerous. In terms of the abundance of individuals, the most

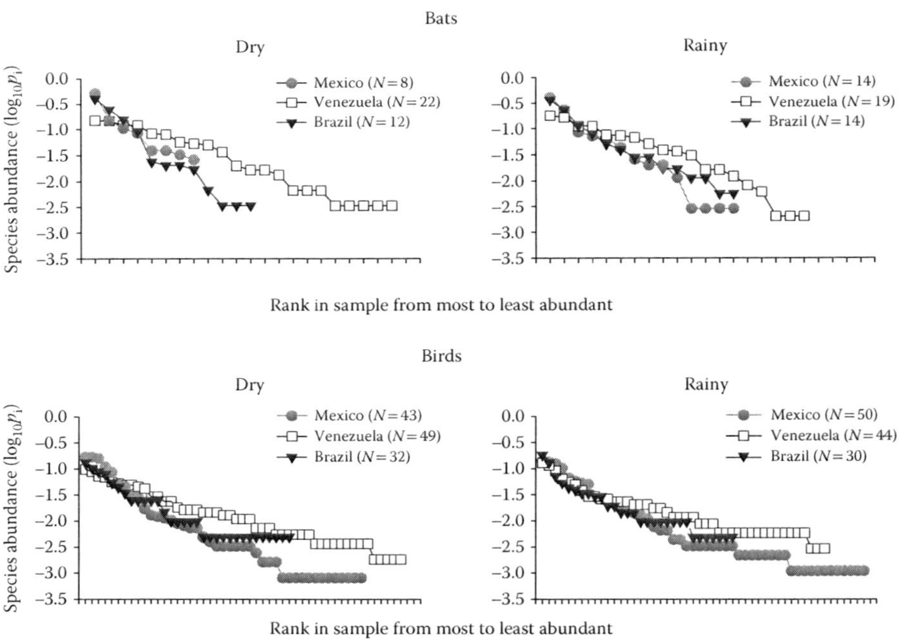

FIGURE 11.3
Rank–abundance (dominance–diversity) graphs of species of fruit-eating bats and birds pooled across all sampling plots within each season at each region: CCBR (Mexico), UPSAP (Venezuela), MSSP (Brazil). p_i: relative abundance of each species.

TABLE 11.4

Number of Species (*S*) and Number of Bats and Birds (*N*) Captured with Mist Nets in Mexico, Venezuela, and Brazil as a Function of Season (Dry and Rainy)

		Mexico	Venezuela	Brazil
Bats				
Dry season	Total species	$S = 8$ $N = 153$	$S = 22$ $N = 301$	$S = 10$ $N = 144$
	PF	$S = 5$ (62.5%) $N = 121$ (79.1%)	$S = 12$ (54.5%) $N = 237$ (78.7%)	$S = 3$ (30.0%) $N = 72$ (50.0%)
	SF	$S = 3$ (37.5%) $N = 32$ (20.9%)	$S = 6$ (27.3%) $N = 51$ (16.9%)	$S = 7$ (70.0%) $N = 72$ (50.0%)
	OF	$S = 0$ (0.0%) $N = 0$ (0.0%)	$S = 4$ (18.2%) $N = 13$ (4.3%)	$S = 0$ (0.0%) $N = 0$ (0.0%)
Rainy season	Total species	$S = 14$ $N = 350$	$S = 19$ $N = 494$	$S = 14$ $N = 327$
	PF	$S = 8$ (57.1%) $N = 207$ (59.1%)	$S = 12$ (63.2%) $N = 378$ (76.5%)	$S = 4$ (28.6%) $N = 232$ (71.0%)
	SF	$S = 5$ (35.7%) $N = 142$ (40.6%)	$S = 5$ (26.3%) $N = 72$ (14.6%)	$S = 7$ (50.0%) $N = 89$ (27.2%)
	OF	$S = 1$ (7.1%) $N = 1$ (0.2%)	$S = 2$ (10.5%) $N = 44$ (8.9%)	$S = 3$ (21.4%) $N = 6$ (1.8%)
Birds				
Dry season	Total species	$S = 43$ $N = 1228$	$S = 49$ $N = 547$	$S = 32$ $N = 161$
	PF	$S = 4$ (9.3%) $N = 7$ (0.6%)	$S = 9$ (18.4%) $N = 82$ (15.0%)	$S = 3$ (9.4%) $N = 3$ (1.9%)
	SF	$S = 22$ (51.2%) $N = 763$ (62.1%)	$S = 29$ (59.2%) $N = 285$ (52.1%)	$S = 18$ (56.3%) $N = 98$ (60.9%)
	OF	$S = 17$ (39.5%) $N = 458$ (37.3%)	$S = 11$ (22.4%) $N = 180$ (32.9%)	$S = 11$ (34.3%) $N = 60$ (37.2%)
Rainy season	Total species	$S = 50$ $N = 913$	$S = 44$ $N = 342$	$S = 30$ $N = 176$
	PF	$S = 3$ (6.0%) $N = 16$ (1.8%)	$S = 6$ (13.6%) $N = 72$ (21.1%)	$S = 1$ (3.3%) $N = 1$ (0.5%)
	SF	$S = 28$ (56.0%) $N = 550$ (60.2%)	$S = 21$ (47.7%) $N = 152$ (44.4%)	$S = 17$ (56.7%) $N = 89$ (50.6%)
	OF	$S = 19$ (38.0%) $N = 347$ (38.0%)	$S = 17$ (38.6%) $N = 118$ (34.5%)	$S = 12$ (40.0%) $N = 86$ (48.9%)

Note: Frugivore categories: primary frugivores (PF), secondary frugivores (SF), and occasional frugivores (OF). Percentages of total species and individuals are included within parentheses.

numerous bats during the dry and rainy season in Mexico and Venezuela were primary frugivores. The same pattern occurred for Brazil during the rainy season, but during the dry season, primary and secondary frugivores were equally numerous (Table 11.4). During both seasons, evenness of species was comparatively higher in the Venezuelan ensembles than in Mexico and Brazil

(Figure 11.3). The dominance of a particular species of bats was more marked in Mexico and Brazil. With regard to rare species, the number increased for Venezuela and Brazil during the dry season and for Mexico during the rainy season. Overall, the taxonomic composition of the most abundant species of fruit-feeding bats did not change substantially between seasons, but several species with intermediate to low abundances either changed their relative importance or were not captured depending on the season (Appendix G). The most outstanding seasonal changes were associated to *L. yerbabuenae* (dry: $N = 5$, rainy: $N = 30$) in Mexico, *Uroderma bilobatum* (dry: $N = 37$, rainy: $N = 15$) and *N. albiventris* (dry: $N = 5$, rainy: $N = 36$) in Venezuela, and *Phyllostomus discolor* (dry: $N = 45$, rainy: $N = 6$) in Brazil.

There seems to be no particular trend for species richness and abundance of fruit-feeding birds in relation to seasons in the three regions (Figure 11.3; Table 11.4). Of all PerMANOVA comparisons conducted, only in Venezuela we detected a significant increase in the number of species captured per day during the dry season (Appendix E). With regard to the relative importance of the different categories of fruit-feeding birds during the two seasons, we observed a consistent pattern across regions and seasons, with dominance of secondary frugivores, followed by occasional fruit eaters and primary frugivores as the less rich group (Table 11.4). The rank–abundance curves did not show noticeable differences between seasons (Figure 11.3); however, it is important to note that species evenness was comparatively higher in Venezuela during the dry and rainy season, and also in Mexico, the number of rare species of birds increased during the rainy season. Overall, the taxonomic composition of the most abundant species of fruit-feeding birds did not change substantially between seasons (Appendix H). A few outstanding seasonal changes in the abundance of a particular species included *M. nuttingi* (dry: $N = 107$, rainy: $N = 21$) and *Passerina ciris* (dry: $N = 10$, rainy: $N = 49$) in Mexico; *M. tyrannulus* (dry: $N = 38$, rainy: $N = 8$), *C. passerina* (dry: $N = 29$, rainy: $N = 2$), and *Elaenia parvirostris* (dry: $N = 27$, rainy: $N = 2$) in Venezuela; and *Pachyramphus polychopterus* (dry: $N = 0$, rainy: $N = 15$) in Brazil.

11.3.5 Effect of Successional Stages

Depending on the region and the animal group examined, we observed several cases of pronounced variation in species richness among plots within particular successional stages (Figure 11.1). In Mexico, marked inter-plot variation was observed in all successional stages, mainly during the dry season, for both bats and birds, and in the early stage during the two seasons in the case of birds. In Venezuela, inter-plot variation in species richness was accentuated in the intermediate and late stages during both seasons for bats, and in the early stage during both seasons for birds. Finally, in Brazil, the most marked variation in species richness among plots occurred in the early and late successional stages for bats and in the intermediate stage for birds. These marked levels of variation among plots within stages limited our capacity to detect differences in the variables examined as a response to increments in vegetation complexity in dry forests. Despite this, several effects were detected.

Species richness and abundance of fruit-eating bats in pastures were comparably lower than in the other stages (Figures 11.1 and 11.4; Table 11.5). The PerMANOVA comparisons supported this trend for the Jack 2 estimate of species richness in the case of Mexico and Venezuela (Appendix D), and for the number of species and individuals captured per day in the case of Venezuela. Although late successional plots in Venezuela had on average ($S = 10$) a higher Jack 2 estimate of species richness than pastures ($S = 6$), no significant differences were detected between these two stages, because there was a high variation in the estimates of number of species between sampling plots in the late stage (Figure 11.1). Additional differences among successional stages were also evidenced in the case of Venezuela and Brazil. Early plots in Venezuela presented a significantly higher number of species of fruit-eating bats than did intermediate and late successional plots (Appendix E). In Brazil, the late successional plots had a comparatively higher number of species and individuals captured per night than did plots in the other stages (Appendices E and F).

Most fruit-feeding bats captured in Mexico and Venezuela in the four successional stages were primary or secondary frugivores, in this order of importance (Table 11.5). On the contrary, in Brazil, secondary frugivores were proportionally more important in number of species than bats in the other two categories across all successional stages. In terms of abundance of individuals, in the three regions, the most abundant bats across successional stages were primary frugivores, with only one exception, the early stage in Mexico, where secondary frugivores were numerically superior. The other particularity observed was that occasional frugivores (*N. albiventris*) were relatively abundant (30%) in the pastures of Venezuela, a condition not observed in the other countries. In general, evenness of species within the ensembles was relatively low across stages in the three regions, but we can mention, in particular, the late stage in Mexico and the early stage in Venezuela, where ensembles presented a comparatively more even distribution of several species (Figure 11.4). The dominance of a particular species of bats was more marked in Mexico and Brazil. In Mexico, *A. jamaicensis* and *G. soricina* numerically dominated across all successional stages. *L. yerbabuenae* and *G. commissarisi* were relatively abundant in the early stage, became scarce in more advanced stages, and disappeared in the pastures. Six species were found only in the late stage, and the remaining species were present in three or four stages (Appendix B). In Venezuela, the dominant species varied in part according to the successional stage: *N. albiventris, P. elongatus,* and *U. magnirostrum* dominated the pastures; *P. elongatus, U. bilobatum,* and *U. magnirostrum* in the early stage; *P. vittatus, P. elongatus,* and *U. magnirostrum* in the intermediate stage; and *C. brevicauda* and

TABLE 11.5

Number of Species (*S*) and Number of Bats (*N*) Captured with Mist Nets in Mexico, Venezuela, and Brazil as a Function of Successional Stage (Pastures, Early, Intermediate, and Late)

		Mexico	Venezuela	Brazil
Pastures	Total species	$S = 4$ $N = 12$	$S = 13$ $N = 89$	$S = 9$ $N = 68$
	PF	$S = 3$ (75.0%) $N = 9$ (75.0%)	$S = 8$ (61.5%) $N = 51$ (57.3%)	$S = 4$ (44.4%) $N = 48$ (70.6%)
	SF	$S = 1$ (25.0%) $N = 3$ (25.0%)	$S = 3$ (23.1%) $N = 11$ (12.4%)	$S = 5$ (55.6%) $N = 20$ (29.4%)
	OF	$S = 0$ (0%) $N = 0$ (%)	$S = 2$ (15.4%) $N = 27$ (30.3%)	$S = 0$ (0.0%) $N = 0$ (0.0%)
Early	Total species	$S = 8$ $N = 159$	$S = 19$ $N = 281$	$S = 12$ $N = 101$
	PF	$S = 5$ (62.5%) $N = 50$ (31.4%)	$S = 10$ (52.6%) $N = 215$ (76.5%)	$S = 3$ (25.0%) $N = 68$ (67.3%)
	SF	$S = 3$ (37.5%) $N = 109$ (68.6%)	$S = 6$ (31.6%) $N = 52$ (18.5%)	$S = 7$ (58.3%) $N = 30$ (29.7%)
	OF	$S = 0$ (0%) $N = 0$ (0%)	$S = 3$ (15.8%) $N = 14$ (5.0%)	$S = 2$ (16.7%) $N = 3$ (3.0%)
Intermediate	Total species	$S = 8$ $N = 115$	$S = 18$ $N = 211$	$S = 11$ $N = 78$
	PF	$S = 5$ (62.5%) $N = 90$ (78.3%)	$S = 9$ (50.0%) $N = 178$ (84.4%)	$S = 3$ (27.3%) $N = 44$ (56.4%)
	SF	$S = 3$ (37.5%) $N = 25$ (21.7%)	$S = 5$ (27.8%) $N = 24$ (11.4%)	$S = 7$ (63.6%) $N = 33$ (42.3%)
	OF	$S = 0$ (0%) $N = 0$ (0%)	$S = 4$ (22.2%) $N = 9$ (4.2%)	$S = 1$ (9.1%) $N = 1$ (1.3%)
Late	Total species	$S = 14$ $N = 217$	$S = 18$ $N = 213$	$S = 12$ $N = 224$
	PF	$S = 8$ (57.1%) $N = 179$ (82.5%)	$S = 10$ (55.5%) $N = 170$ (79.8%)	$S = 4$ (33.3%) $N = 144$ (64.3%)
	SF	$S = 5$ (35.6%) $N = 37$ (17.1%)	$S = 5$ (27.8%) $N = 36$ (16.9%)	$S = 7$ (58.3%) $N = 78$ (34.8%)
	OF	$S = 1$ (7.1%) $N = 1$ (0.5%)	$S = 3$ (16.7%) $N = 7$ (3.3%)	$S = 1$ (8.3%) $N = 2$ (0.9%)

Note: Frugivore categories: primary frugivores (PF), secondary frugivores (FF), and occasional frugivores (OF). Percentages of total species and individuals are included within parentheses.

P. vittatus in the late stage. The great majority of species were captured in more than one stage (Appendix B). In Brazil, two species dominated across all stages: *A. planirostris* and *Carollia* spp. The majority of species were present in more than one stage (Appendix B).

In relation to fruit-eating birds, the first and most outstanding trend observed was that early successional plots presented a comparatively higher number of species and more individuals than did the other successional

TABLE 11.6

Species Richness (*S*) and Number (*N*) of Birds Captured with Mist Nets in Mexico, Venezuela, and Brazil According to Successional Stage (Pastures, Early, Intermediate, and Late)

		Mexico	Venezuela	Brazil
Pastures	Total species	$S = 33$ $N = 520$	$S = 5$ $N = 7$	$S = 5$ $N = 35$
	PF	$S = 0$ (0.0%) $N = 0$ (0.0%)	$S = 0$ (0.0%) $N = 0$ (0.0%)	$S = 0$ (0.0%) $N = 0$ (0.0%)
	SF	$S = 16$ (48.5%) $N = 323$ (62.1%)	$S = 3$ (60.0%) $N = 5$ (71.4%)	$S = 2$ (40.0%) $N = 9$ (25.7%)
	OF	$S = 17$ (51.5%) $N = 197$ (37.9%)	$S = 2$ (40.0%) $N = 2$ (28.6%)	$S = 3$ (60.0%) $N = 26$ (74.3%)
Early	Total species	$S = 35$ $N = 795$	$S = 43$ $N = 464$	$S = 34$ $N = 159$
	PF	$S = 2$ (5.7%) $N = 5$ (0.6%)	$S = 7$ (16.3%) $N = 104$ (22.4%)	$S = 2$ (5.9%) $N = 3$ (1.9%)
	SF	$S = 19$ (54.3%) $N = 481$ (60.5%)	$S = 24$ (55.8%) $N = 227$ (48.9%)	$S = 22$ (64.7%) $N = 125$ (78.6%)
	OF	$S = 14$ (40.0%) $N = 309$ (38.9%)	$S = 12$ (27.9%) $N = 133$ (28.6%)	$S = 10$ (29.4%) $N = 31$ (19.5%)
Intermediate	Total species	$S = 34$ $N = 457$	$S = 31$ $N = 274$	$S = 21$ $N = 106$
	PF	$S = 3$ (8.8%) $N = 9$ (2.0%)	$S = 5$ (16.1%) $N = 22$ (8.0%)	$S = 1$ (4.8%) $N = 1$ (0.9%)
	SF	$S = 19$ (55.9%) $N = 279$ (61.0%)	$S = 17$ (54.8%) $N = 137$ (50.0%)	$S = 11$ (52.4%) $N = 43$ (40.6%)
	OF	$S = 12$ (35.3%) $N = 169$ (37.0%)	$S = 9$ (29.0%) $N = 115$ (42.0%)	$S = 9$ (42.8%) $N = 62$ (58.5%)
Late	Total species	$S = 34$ $N = 369$	$S = 24$ $N = 144$	$S = 14$ $N = 37$
	PF	$S = 3$ (8.8%) $N = 9$ (2.4%)	$S = 2$ (8.3%) $N = 28$ (19.4%)	$S = 0$ (0.0%) $N = 0$ (0.0%)
	SF	$S = 18$ (52.9%) $N = 230$ (62.3%)	$S = 15$ (62.5%) $N = 68$ (47.2%)	$S = 6$ (42.9%) $N = 10$ (27.0%)
	OF	$S = 13$ (38.2%) $N = 130$ (35.3%)	$S = 7$ (29.2%) $N = 48$ (33.4%)	$S = 8$ (57.1%) $N = 27$ (73.0%)

Note: Frugivore categories: Primary frugivores (PF), secondary frugivores (SF), and occasional frugivores (OF). Percentages of total species and individuals are included within parentheses.

stages (Appendices D through F; Figures 11.1 and 11.4; Table 11.6). We observed a reduction in the total number of birds captured as vegetation complexity increased, from early to late successional stage (Table 11.6). Besides this, the pastures in Venezuela and Brazil presented comparatively lower values of the Jack 2 estimate of species richness, number of species, and number of individuals captured per day. Contrary to these regions, Mexico presented a comparatively large number of species and individuals in the pastures.

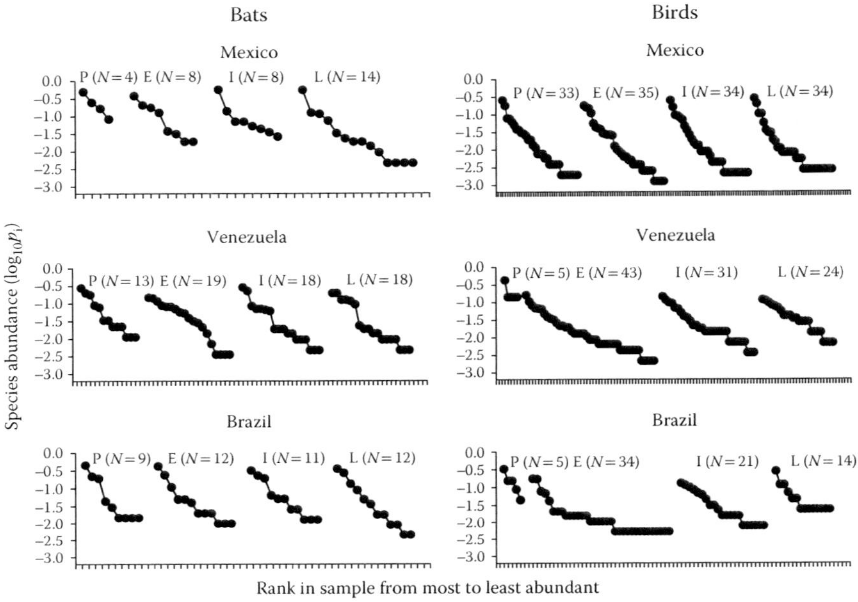

FIGURE 11.4
Rank–abundance (dominance–diversity) graphs of species of fruit-eating bats and birds pooled across sampling plots within each successional stage at each region: CCBR (Mexico), UPSAP (Venezuela), MSSP (Brazil). P: pasture, E: early, I: intermediate, L: late. p_i: relative abundance of each species.

For the three regions, the majority of species of fruit-eating birds captured across successional stages were classified as secondary frugivores (Table 11.6). Only three exceptions were observed: In the pastures of Mexico and Brazil, there were more species of occasional than secondary frugivores; a similar situation was also observed in the late successional stage in Brazil. For the three regions, primary frugivores were the minority across all successional stages. In terms of the abundance of individuals, dominance of secondary or occasional frugivores occurred across regions and successional stages. In pastures, secondary frugivores dominated in Mexico and Venezuela, and occasional frugivores did so in Brazil. In the early successional stage, secondary frugivores dominated across regions. In the intermediate stage, occasional frugivores dominated in Venezuela and Brazil, whereas secondary frugivores dominated in Mexico. In the late stage, secondary frugivores dominated in Mexico and Venezuela, whereas occasional frugivores dominated in Brazil. Species evenness within ensembles varied among regions (Figure 11.4). Venezuela had a more even distribution of species in early, intermediate, and late stages than Mexico and Brazil, and Brazil had a more even distribution than Mexico. Mexico presented the most marked uneven distribution of species, with few dominant ones per stage. Comparatively more rare species were detected in Brazil and Mexico than in Venezuela. In Mexico, *C. parellina* and *V. flavoviridis*, both of which are secondary frugivores, dominated from the early to the late successional stage. Along with them, we found *C. passerina* and *P. leclancherii* in the early stage, which also dominated the pastures. Eighteen species (32%) were captured in only one successional stage, and the majority corresponded to the late stage and the pastures; the remaining species were found in multiple stages (Appendix B). In Venezuela, the identity of the dominant species changed according to the stage. No species dominated the pastures; *C. lanceolata* and *C. flaveola* dominated the early stage; *C. talpacoti* and *C. flaveola* did so in the intermediate stage; and *T. flaviventris*, *Pipra filicauda*, *M. tyrannulus*, and *C. lanceolata* were the most abundant taxa in the late stage. Thirty-four (57%) species were captured in a single successional stage, with the majority of them being in the early one (Appendix B). In Brazil, *M. viridicata* was dominant across stages. *T. flaviventris* was rare in the early stage and abundant at more complex sites; on the other hand, *P. polychopterus*, *M. maculatus*, and *M. tyrannulus* were common in early and intermediate stages and less common in the late stage. It is important to mention that several species which were captured at a high frequency were occasional frugivores: *M. viridicata* (early, intermediate, and late), *T. flaviventris* (intermediate), and *C. talpacoti* (pasture). Twenty-two species (52.4%) were captured in low numbers in a single successional stage, in most cases within early plots (Appendix B). Species captured in the three regions included *T. melancholicus*, which was mostly found in pastures; *Phaeomyias murina* and *Myiozetetes similis*, both of which were in the early plots.

11.4 Discussion

The present study represents the most complete comparative analysis of ensembles of fruit-eating bats and birds (small and medium size) that has been conducted in TDFs in the Americas until present. In addition to this, this investigation illustrates how we can contribute to characterizing the availability of potential seed dispersal agents across ecosystems of the same type, by following comparable sampling procedures and by taking into consideration the effects of vegetation complexity and seasonal regimes in tropical regions. Based on our results, TDFs in the tropical Americas host an important number of species of flying vertebrates potentially capable of contributing to their own maintenance and regeneration. Species richness, abundance, and the structure of fruit-feeding classes in these ensembles respond to the geographic location of forest, seasons, and level of complexity of the vegetation units that integrate the landscape.

11.4.1 Species Richness and Abundance of Fruit-Eating Bats and Birds in TDFs

The TDFs examined host a pool of fruit-eating bats and birds comprising more than 60 and 80 taxa, respectively. The animals captured were not only species that were traditionally classified as frugivores, but also species that feed on fruits facultatively or on rare occasions. In three regions, fruit-eating birds exceeded fruit-eating bats in terms of the number of families and species, a pattern that was concordant with the general trend proposed for the tropics, where passerine birds dominate the frugivore faunas (Fleming and Kress 2011).

The number of species of fruit-eating bats captured at the three sites varied between 14 and 23, estimates that were within the range reported for other dry forests in the region (Yucatán, Mexico: $N = 9$ [Pech-Canche et al. 2011; Chicamocha and Patía]; Colombia: $N = 10–11$ [Sánchez et al. 2007]; Central Veracruz, Mexico: $N = 10–14$ [Moreno and Halffter 2001]; Palo Verde, Costa Rica: $N = 21$ [Stoner 2005]; Córdoba, Colombia: $N = 14$ [Vela-Vargas and Pérez-Torres 2012]). Available reports for neotropical wet forests indicate similar or slightly superior values (14–30) of species of fruit-feeding bats per location (Goncalves da Silva et al. 2008; Rex et al. 2008; Klingbeil and Willig 2010). In the case of fruit-eating birds, we captured between 46 and 60 species per site, not including large frugivorous birds (e.g., Cracidae) that were seen or which have been reported locally (Noguera et al. 2002; Ascanio and García 2005). These levels of species richness are equal to or above the few estimates available for dry and thorny forests in the neotropics (Araya Peninsula, Venezuela: $N = 16$ [Poulin et al. 1994]; Los Santos, Panama: $N = 34$ [Griscom et al. 2007]; Veracruz, Mexico: $N = 54$ [Ortiz-Pulido et al. 2000]), and are within the estimates ($N = 29–88$) obtained in different variants of rain forests in the Americas (Levey 1988; Cardoso da Silva et al. 1996; Ferreira-Fadini and De Marco Jr. 2004; Peters et al. 2010).

The dry forest in UPSAP, Venezuela, presented the highest species richness of fruit-feeding bats and birds in the study. The species numbers recorded should increase slightly considering the fact that between 8% and 15% of the taxa theoretically present in the plots were not captured. Of the three regions examined, Venezuela has the lowest latitudinal position. Several authors have indicated that bat and bird assemblages increase in species richness with decreasing latitude (Rahbek and Graves 2001; Stevens et al. 2006). The total sampling effort in Mexico was comparatively smaller than the sampling effort in Brazil for both bats and birds; however, similar species richness of fruit-eating bats was observed in both countries, and Mexico surpassed Brazil in terms of species richness of fruit-eating birds. In terms of the total number of birds captured, Mexico ranked first among the three regions.

Differences in species richness and abundance in assemblages of bats and birds between localities in the tropics have been directly or indirectly linked to differences in precipitation regimes, plant species diversity, structural complexity of the vegetation, and landscape heterogeneity (Stoner 2005; Sánchez et al. 2007; Rex et al. 2008; Pech-Canche et al. 2011). In the three sites examined, habitat heterogeneity was present as a mosaic of vegetation units that include pastures, secondary growth areas of different ages, and fragments of mature forest. With regard to rainfall patterns, all study sites share the marked seasonality characteristic of TDFs, but UPSAP in Venezuela has twice the annual precipitation of the study sites in Mexico and Brazil, which translates into higher water availability to the plant communities established at that site. In terms of the plant community attributes of the three forests, with the exception of the early successional stage, Mexico presented the highest values of plant densities, number of woody species, and structural complexity, followed by Venezuela, and, finally, Brazil. Besides this, the successional plots established in Mexico were much larger than those in Venezuela and Brazil, and covered a broader habitat extension that could increase the chances of capturing a larger variety and abundance of bats and birds. All these differences among sites are concordant with the patterns of species richness and abundance described for the ensembles examined.

Bats and birds differed in terms of the relative importance of the three fruit-feeding classes represented. In the case of bats, primary frugivores were slightly more important than secondary frugivores in Mexico and Venezuela; only in Brazil, species in the latter category were more numerous. However, in terms of abundance, primary frugivores dominated the three sites. Other studies conducted in dry forests have also shown this trend, with some exceptions. In five deciduous and semi-deciduous forests examined in the Yucatán Peninsula, Mexico, primary frugivorous bats dominated the number of species, but in terms of abundance, they shared the first positions with secondary frugivores (Pech-Canche et al. 2011). In two dry forests of Colombia, primary frugivorous bats dominated the most humid forest, whereas a combination of primary and secondary frugivores dominated the driest one (Sánchez et al. 2007). The most abundant species recorded by us in each forest, although taxonomically different between sites, belonged to two important functional groups identified in neotropical forests: understory and canopy frugivores (Baker et al. 2003; Rex et al. 2008). In the case of understory frugivores, the most abundant species included *Carollia brevicauda*, *Uroderma magnirostrum*, *Glossophaga longirostris*, and *Glossophaga soricina*; and in the case of canopy frugivores, the most abundant species included *Phyllostomus elongatus*, *Phyllostomus discolor*, *Platyrrhinus vittatus*, *Artibeus jamaicensis*, *Artibeus intermedius*, and *Artibeus planirostris*. Their presence in abundance guarantees seed dispersal of chiropterochoric shrubs and trees in these forests, and their roles as seed vectors should be complemented by a number of species belonging to the three fruit-feeding categories, especially species of secondary and occasional frugivores, which are represented by several taxa in Venezuela and Brazil. The case of *Noctilio albiventris* is worth mentioning, because this occasional frugivore was frequently captured in early plots and pastures in Venezuela. This species, an insect-feeding bat that feeds on fruits seasonally (Aranguren et al. 2011), contributes to seed dispersal (*Maclura tinctoria*, *Ficus* sp., and *Piper* sp.) in pastures with more intensity than several fruit-eating phyllostomids.

In the case of birds, secondary frugivores dominated in terms of both number of species and abundance across sites, followed by occasional frugivores, and in last place, primary frugivores. This pattern suggests that fruit-feeding birds associated with dry forests need to be able to rely on a broad spectrum of food items besides fruits. The same hierarchy of importance of fruit-feeding classes was found by Ortiz-Pulido et al. (2000) in a study aimed at characterizing frugivory and seed dispersal in the bird community associated with a fragmented dry forest in Veracruz, Mexico, and by Poulin et al. (1994) in a spiny scrub-dry forest complex in the Araya Peninsula, Venezuela. The dominance of birds that feed on fruits facultatively or rarely has also been evidenced for several tropical wet forests (Levey 1988; Peters et al. 2010).

Fruit specialist birds (*Chiroxiphia lanceolata* and *Pipra filicauda*) were captured in abundance only in Venezuela. Due to their qualities as specialized fruit-feeding species (Ferreira Fadini and de Marco Jr. 2004), they should be considered key seed dispersal agents in this type of forest; however, they represent only a small fraction of the available pool of seed vectors in this ecosystem. As in the case of bats, the most abundant species of fruit-eating birds captured in the three forests included taxa that forage in the understory (*Coereba flaveola*, *C. lanceolata*, *Camptostoma obsoletum*, *Cyanocompsa parellina*, *Passerina versicolor*, *Passerina ciris*, and *Leptotila verreauxi*) and the canopy frugivores (*Icterus pustulatus*, *M. maculates*, *C. lanceolata*, *Icteria virens*, *Myiarchus tyrannulus*, and *Vireo flavoviridis*). Opportinistic fruit-feeding taxa such as *Arremonops rufivirgatus*, *Empidonax difficilis*, *Myiarchus nuttingi*, *Columbina passerina*, *Columbina inca*, *Tolmomyias flaviventris*, and *Columbina talpacoli* were relatively abundant wherever found. Several of these species feed mainly on insects, and are able to forage inside dense forests, in canopies, and at forest edges, and they can eventually eat berries and disperse their seeds in forest clearings and open areas surrounding forest patches. Doves move through secondary growth vegetation, forest clearings, and pastures surrounding forest. Although well known by their granivorous habits, pigeons and doves do not destroy all the seeds they swallow (Corlett and Primack 2011); therefore, in relatively high abundances, they could also contribute as occasional seed dispersers in open habitats with some vegetation to perch.

11.4.2 Effect of Seasons

Ensembles of tropical fruit-eating bats and birds tend to be relatively dynamic and variable in time across a broad range of ecosystems, but particularly in forests in which resources tend to be more variable seasonally (Karr et al. 1982; Stoner 2005; Sánchez et al. 2007; Mello 2009; Klingbeil and Willig 2010;

Pech-Canche et al. 2011). Species richness and abundance of fruit-eating taxa tend to increase when fruit and insect availability are higher, and this condition is usually associated with the rainy season, especially in the case of dry forests (Bullock 1995; Klingbeil and Willig 2010). In this study, differences in ensemble attributes between seasons were consistent with the trend mentioned earlier only in the case of bats. The estimated number of bat species per plot (Mexico and Brazil) and the number of species and individuals captured per night (Mexico and Venezuela) were comparatively higher during the rainy season. Only in the case of Venezuela, more species of bats were captured during the dry season, but these additional species (*Tonatia saurophila*, *Micronycteris hirsuta*, and *Mimon bennettii*) were found in low numbers and feed mostly on insects. Other studies of bat assemblages in TDFs have reported the same pattern of seasonal variations (Sánchez et al. 2007 in Colombia and Pech-Canche et al. 2011 in Mexico).

Disappearance of some species of bats during part of the year and a reduction in their abundances at some sites are attributed to movements among locations, searching for resources, or escaping low temperatures (Giannini 1999; Mello et al. 2009). In this study, several primary frugivores were not captured during the dry season for two consecutive years in Mexico (*Carollia* sp. and *Chiroderma salvini*) and Brazil (*Artibeus obscurus* and *Chiroderma villosum*); however, the more relevant inter-seasonal differences were the significant reductions in the proportion of several species of bats in the three regions during the dry season (*G. soricina* and *G. commissarisi* in Mexico, *P. hastatus* in Venezuela, and *Carollia* sp. and *Phylloderma stenops* in Brazil). On the other hand, a few species of bats (*Platyrrhinus helleri*, *Uroderma bilobatum*, and *Phyllostomus discolor*) increased their abundance during the dry season, suggesting that some available resources attracted them to the dry forest during that part of the year. *P. helleri* and *U. bilobatum* feed heavily on *Cecropia* fruits (Palmeirim et al. 1989; Linares and Moreno-Mosquera 2010), which are available year round (Klingbeil and Willig 2010), and *P. discolor* feeds primarily on nectar and pollen (Kwiecinski 2006), resources that can be found in abundance during the dry season, when many species bloom in TDFs (Borchert 1996).

In the case of birds, our results do not support the seasonal pattern reported for many tropical ecosystems, including dry forests. No significant differences in the variables examined were observed between seasons, with the exception of Venezuela, for which the number of species captured per day was higher during the dry season. Contrary to the marked seasonal changes in abundance reported for fruit specialists such as the manakins (Levey 1988), in Venezuela, representatives of this group of birds remained equally abundant between seasons (*C. lanceolata*, dry: $N = 49$, rainy: $N = 44$; *P. filicauda*, dry: $N = 9$, rainy: $N = 8$). Proportionally few cases of sizable changes in abundance were detected between seasons, and in most of them, species abundance increased during the dry season (e.g., *Myiarchus nuttingi* in Mexico and *M. tyrannulus* in Venezuela and Brazil).

Differences observed between responses of fruit-feeding bats and birds to the seasonal regime in the dry forests examined are connected with the relative importance of the fruit-feeding classes represented in each of these groups. As mentioned earlier, the ensembles of fruit-eating bats at the three sites were dominated by primary frugivores, either in terms of number of taxa, abundance of individuals, or both. High dependence on fruit can constrain the permanence of these bats in dry forests during periods of fruit scarcity, particularly in the case of species specialized in fruits produced during the rainy season. According to Stoner (2005), in Chamela, Mexico, the highest peaks of chiropterochoric and chiropterophilic resources occur during the rainy season. Despite this, several primary frugivorous bats were able to remain in the study sites at lower abundances during the dry season. This was possible because other plant resources also used by fruit-eating bats, such as nectar and pollen, are available during the dry season (Stoner 2005). In addition, in the three regions, humid areas (riparian and semi-deciduous forests) were relatively close to the study sites, where more fruits and also insects could be available during the dry months of the year. In the case of birds, the ensembles of fruit-eating species were dominated by secondary frugivores, which have a broad and flexible diet and can ingest a wide range of plant and animal resources (mostly insects), allowing them to remain in the dry forest year round. In the case of the three dry habitats surveyed in northeastern Venezuela, Poulin et al. (1994) showed how, with the exception of several insectivore species exclusively dependent on arthropods, the great majority of species of birds captured included both animal and plant tissues in their diets. In UPSAP, Aranguren (2011) found that insects were the dominant food item in the fruit-eating birds captured during most of the year. Another factor that could contribute to the pattern observed for birds is related with changes in their foraging behavior driven by food scarcity. Rangel-Salazar et al. (2009) suggested that food scarcity during the dry season could promote in birds a higher investment in time dedicated to foraging, spending more time flying, and, therefore, increasing their chances of being captured. Besides this, because of the loss of foliar biomass at the canopy level during the dry season, resident birds could be forced to forage at lower heights, again increasing their chances of being captured.

11.4.3 Effect of Successional Stages

Significant variations in ensemble attributes were observed among successional stages for both fruit-feeding bats and birds across regions. With one exception, pastures across regions had the lowest values of species richness and abundance for both groups of vertebrates. Other studies that compared bat and bird assemblages along gradients of vegetation complexity and anthropogenic disturbance have also associated pastures with the lowest species richness and abundance of these vertebrates (O'Dea and Whittaker 2007; Avila-Cabadilla et al. 2009). These studies link their observations to the extreme simplification of the vegetation structure and reduced resource availability characteristic of that successional stage; however, numerous species of bats and birds are capable of foraging in pastures, especially in the presence of solitary trees and shrubs (Guevara et al. 1986; Guevara and Laborde 1993; Cardoso da Silva et al. 1996; Ortiz-Pulido et al. 2000; Griscom et al. 2007). In our study, several species of bats that are associated with secondary growth, forest gaps, and disturbed habitats were often captured in pastures (Venezuela: *U. magnirostrum*, *N. albiventris*; Brazil: *Carollia* sp.), but also species usually associated with the canopy and old-growth forests were frequently trapped in pastures (Venezuela: *P. elongatus*; Mexico: *A. jamacensis*; Brazil: *A. planirostris*, *P. discolor*). Their presence could be explained either because bats were feeding or resting on isolated trees (Bianconi et al. 2007) or because they were captured when moving from one forest fragment to another.

In the case of fruit-eating birds, only a few species were captured in low numbers in the pastures of Venezuela and Brazil; however, in Mexico, the figures obtained were totally contrasting and unexpectedly high. In the pastures of Chamela, 16 and 17 species of secondary and occasional fruit eaters, respectively, were captured; and the total number of individuals trapped was above the values found for intermediate and late successional stages. The pastures of that study site had shrubs and dispersed isolated trees that were never removed when the land was cleared. Besides this, those pastures were relatively close to other vegetation types, and could be frequently used by birds as pathways between forest fragments. The most abundant species in these pastures, *C. passerina*, *Passerina leclancherii*, *Passerina versicolor*, *Cyanocompsa parellina*, and *V. flavoviridis*, were also present in older successional stages. Other studies have demonstrated that a substantial number of species of birds move freely and frequently among vegetation units in heterogeneous landscapes in the tropics (Ortiz-Pulido et al. 2000; Griscom et al. 2007; Losada-Prado and Molina-Martínez 2011). It has also been demonstrated that abandoned pastures with presence of woody elements attract substantially more birds and bats than clean areas dominated by herbaceous vegetation (Cardoso da Silva et al. 1996; Medina et al. 2007).

Mature forests are at the other extreme of the successional gradient. High species richness of bats and birds has been associated with high plant diversity and habitat complexity in tropical ecosystems (Loisele and Blake 1994; Medellín et al. 2000). The old-growth forest fragments considered in this study possessed the highest plant diversity and structural complexity in relation to the other successional stages; however, in the majority of cases, this complexity did not translate into higher species richness and abundance of fruit-feeding bats and birds. Only in the case of Mexico, the two parameters reached maximum estimates for bats in the late successional stage. In the case of Brazil, the bat ensemble reached maximum abundance in that stage.

The early and intermediate successional plots of the forests examined in the three regions had levels of species richness and abundance of bats and birds equal to or superior to levels found in the late successional plots. The observed pattern did not correspond with our expectations of a progressive increase in species richness and abundance of fruit-eating taxa as the structural complexity of the vegetation increased. Absence of differences among early, intermediate, and late stages was reported by Castro-Luna et al. (2007) when comparing bat assemblages in a tropical rain forest in Agua Blanca State Park, Tabasco, Mexico. Several studies have found that the bird assemblages associated with the most advanced successional stages in tropical forests are not always the ones showing higher complexity in community parameters. In some instances, the highest complexity has been reported in intermediate successional stages (Johns 1991; Blake and Loiselle 2001), where a combination of plant species of all successional stages can converge. A high percentage of species of fruit-feeding bats and birds in our study was found to be exclusively associated with the early successional stages in Venezuela and Brazil. In the case of Venezuela, we can explain this pattern by the unusually high vegetation complexity recorded in those plots and their proximity to permanent water bodies. However, early plots in Brazil and Mexico were comparatively less complex in terms of vegetation, but species richness and abundance of bats and birds were equal to or higher than intermediate and late successional plots. Many species of bats and birds in fragmented tropical forests could be attracted to early and intermediate growth vegetation patches for particular resources (insects, seeds, and fruits) present in them. As an example, at the Venezuelan study site, species dispersed by bats and birds in early plots included *Protium heptaphilum* (birds), *Annona purpurea* and *A. jahnii* (bats), *Hecastotemon completus* (birds), *Cecropia peltata* (bats), *Curatella americana* (birds), *Inga interrupta* (bats), and *Ficus maxima* (bats). Several studies have evidenced high use of forest gaps by birds in tropical and subtropical ecosystems, and have correlated this pattern to greater availability of food resources (Blake and Hoppes 1986; Levey 1988). More recently, Champlin et al. (2009) have suggested that, besides resources abundance, bird use of early successional habitats could be stimulated by within-habitat structural characteristics. This suggests that the architecture of early successional tropical forests could be particularly attractive to birds in TDFs. Finally, birds could frequently use semi-open areas as pathways between older forest fragments or between different forest types.

11.4.4 Implications for the Regeneration of TDFs

Despite the high importance of anemochoric and autochoric plants (30%–63%) in TDFs in comparison with other forest types (Bullock 1995; Gentry 1995; Vieira and Scariot 2006), many fleshy-fruited species that rely on animal-mediated seed dispersal are also part of these forests across the Americas (Gentry 1995; Quesada et al. 2009).

Many TDFs consist of a mosaic of patches of mature forest, secondary vegetation, and open areas, where lack of seed arrival to the latter represents a major limitation for forest regeneration. Fruit-eating birds and bats should be the most important seed dispersal vectors of zoochoric plants in TDFs because they can easily move abundant seeds from trees and shrubs to pastures and early successional habitats (Humphrey and Bonaccorso 1979; Howe and Smallwood 1982; Charles-Dominique 1986; Fleming 1988; Guevara and Laborde 1993; Medellín and Gaona 1999; Muscarella and Fleming 2007), and because there is an important absence of specialized frugivores (i.e., specialized frugivores mammals) in comparison with wet and moist forests (Ceballos 1995). However, only rarely, seed dispersal by bats and birds has been empirically evidenced in dry forests (Ortiz-Pulido et al. 2000; Griscom et al. 2007; Avila-Cabadilla et al. 2012). The current study does not include reports of the seeds found in the fecal contents of the bats and birds captured at the study sites or of seed rains associated with them, but it offers a relatively complete description of the ensembles of fruit-eating bats and birds that inhabit TDFs across a latitudinal gradient in the tropical Americas, and from this information, we can formulate inferences regarding the status of the seed dispersal service.

Because of their simpler floristic composition, lower number of seral stages, and faster recovery to pre-disturbed heights, TDFs are considered more resilient than many tropical rain forests (Vieira and Scariot 2006; but see Lebrija-Trejos 2009). The characteristics of the ensembles of fruit-eating bats and birds present in the three forests examined contribute to supporting this attribute of TDFs, and suggest that zoochoric species with small- and medium-sized seeds count on a wide range of potential seed vectors.

The three study sites have a good representation of bats and birds that feed on fruits regularly, facultatively, or occasionally. All these sites include bats that are associated with canopy and understory plants; what changes is the set of bat species present in each region. In all cases, fruit specialists comprised an important component of the ensembles of fruit-feeding bats, and this guarantees the service of seed dispersal to the chiropterochoric flora present in those forests when fleshy fruits are available. Besides this, there are a number of secondary and occasional frugivore bats, several of which are in high abundance, that can also contribute to the dispersal of the seeds of those plants. More importantly, 63% and 83% of all fruit-eating bats reported in this study ($N = 35$) visited the pastures and early plots, respectively, with the concomitant possibility of transporting and dispersing seeds into them. In the case of birds, primary frugivores were the minority in the three forests examined; however, lack of fruit specialists was significantly compensated by a relatively large number of species of secondary and occasional frugivores that can contribute to an assurance of the seed dispersal of the ornithochoric flora. Besides this, the high capability of many of these species to move through different seral stages guarantees seed deposition in early successional habitats and pastures with isolated trees. Nearly 30% and 72% of all fruit-eating bird taxa recorded in this study ($N = 137$) visited pastures and early plots, respectively. Altogether, if we acknowledge the long-standing hypothesis, which proposes that bats and birds have complementary roles as seed dispersal agents (Palmeirim et al. 1989; Medellín and Gaona 1999; Muscarella and Fleming 2007; Jacomassa and Pizo 2010), we have solid grounds to conclude that the current combinations of fruit-feeding bats and birds present in the three TDFs examined, in addition to large frugivorous birds (e.g., *Crax* and *Ortalis*), should fulfill the dispersal needs of zoochoric plants in these ecosystems. It is important to mention, however, that several species of plants that are being dispersed toward the early stages by this group of dispersers come from riparian vegetation (based on the seeds collected from feces in the three regions) (Avila-Cabadilla et al. 2012). This is the case, for example, of several species belonging to the genera *Ficus* and *Piper*. Due to their ecological requirements, these species would probably not germinate and grow in early successional stages, where there is a low level of water availability and high levels of radiation (Vieira and Scariot 2006; Alvarez-Añorve et al. 2012). In general, more studies are needed in order to quantify the real contribution of frugivorous bats and birds to the regeneration and successional process in dry forests.

Considering actions that could help maintain and catalyze this fundamental service provided by bats and birds to TDFs, we can think in terms of three basic measures. First, we should conduct efforts to help preserve the levels of species richness documented in this study for both groups of vertebrates. One way to do so is by enforcing the protection of large- and medium-sized fragments of mature forest that are broadly distributed within the TDF matrix. These protected fragments, besides functioning as refuges, nesting areas, and sources of food to many fruit-eating bats and birds in the locality, represent the sources of seeds that will be used to regenerate disturbed areas. Secondly, we need to emphasize the protection of primary frugivores, with particular attention to birds, a group that resulted in being underrepresented in Mexico and Brazil. Investments on the conservation of focal species of frugivores can be translated into the preservation of other species with similar living requirements. These species should be chosen for their morphological attributes, ecological roles, and ecosystem services they provide in their areas of distribution. One group of birds that could be used with this purpose is the manakins. In the dry forest at UPSAP, *Chiroxiphia lanceolata* fed on fruits of 25 species of plants, and was recognized as the most important frugivorous bird (Aranguren 2011). Another outstanding group comprises the colorful trogons. In Chamela, Mexico, *Trogon citreolus*, a species captured during this study, eats the fruits of 25 species of plants in 18 families (Noguera et al. 2002). Finally, the fundamental role of bats and birds as forest regenerators can be amplified and accelerated by promoting the formation of the so-called "nuclei of regeneration" in abandoned pastures (Guevara et al. 1986; Galindo-González et al. 2000). Isolated trees in pastures can function as centers of regeneration, because they concentrate the seed rain produced by bats and birds that feed and perch on them. In our study, the difference in terms of species richness and abundance of fruit-feeding birds between pastures with isolated shrubs and trees (Mexico) and those without them (Venezuela and Brazil) was remarkable. We can help speed the process of secondary succession in fragmented TDFs through planned planting of native species of fast-growing fruiting trees that are used by both bats and birds in abandoned pastures.

Appendix A

References consulted to assign general guild categories to bats and birds captured in Mexico, Venezuela, and Brazil:

Acosta S.L. and A.F. Aguanta. 2006. New contribution on the diet of the frugivorous bats *Aartibeus lituratus* and *A. jamaicensis*. *Kempffiana* 2: 127–133.

Alvarez J., M.R. Willig, J. Knox Jones Jr. and Wm. D. Webster. 1991. *Glossophaga soricina*. *Mammalian Species* 379: 1–7.

Alvarez T. and N. Sánchez-Casas. 1997. Notas sobre la alimentación de *Musonycteris* y *Choeroniscus* (Mammalia: Phyllostomidae) en México. *Revista Mexicana de Mastozoología* 2: 113–115.

Angulo S.R., J.A. Ríos and M. Díaz. 2008. *Sphaeronycteris toxophyllum* (Chiroptera: Phyllostomidae). *Mammalian Species* 814: 1–6.

Aranguren, C.I., J.A. González-Carcacía, H. Martínez and J.M. Nassar. 2011. *Noctilio albiventris* (Noctilionidae), a potential seed disperser in disturbed tropical dry forest habitats. *Acta Chiropterologica* 13: 189–194.

Ascorra, C.F., S. Solari and D.E. Wilson. 1996. Diversidad y ecología de los quirópteros en Pakitza. In *Manu, the Biodiversity of Southeastern Peru*, eds. D.E. Wilson and A. Sandoval, 585–604. Lima: Ed. Horizonte, Smithsonian Institute.

Caballero-Martínez L.A., I.V. Rivas Manzano and L.I. Aguilera Gómez. 2009. Hábitos alimentarios de *Anoura geoffroyi* (Chiroptera: Phyllostomidae) en Ixtapan del Oro, Estado de México. *Acta Zoológica Mexicana* (n.s.) 25: 161–175.

Cramer M. Jr., M.R. Willig and C. Jones. 2001. *Trachops cirrhosus*. *Mammalian Species* 656: 1–6.

Curson, J., D. Quinn and D. Beadle. 1994. *Wrablers of the Americas*. Boston, MA: Houghton Mifflin Company. 262 p.

Ferrell, C.S. and D.E. Wilson. 1991. *Platyrrhinus helleri*. *Mammalian Species* 373: 1–5.

Fleming, T.H. 1988. *The short-tailed fruit bat: a study in plant-animal interactions*. Chicago, IL: University of Chicago Press. 365 p.

Gannon, M.R, M.R. Willig and J. Knox Jones Jr. 1989. *Sturnira lilium*. *Mammalian Species* 333: 1–5.

Goncalves da Silva, A., O. Gaona and R.A. Medellín. 2008. Diet and trophic structure in a community of fruit-eating bats in Lacandon forest, Mexico. *Journal of Mammalogy* 89: 43–49.

Goodwin, G.G. and A.M. Greenhal. 1961. A review of the bats of Trinidad and Tobago. *Bulletin of the American Museum of Natural History* 122: 187–302.

Haynes M.A. and E.L. Thomas Jr. 2004. *Artibeus obscurus*. *Mammalian Species* 752: 1–5.

Howell S. E. and S. Webb. 2005. *A guide to the birds of mexico and northern Central America*. California: Oxford Univesity Press. 1010 p.

Isler, M. and P. Isler. 1999. *The Tanagers*. Washington, DC: Smithsonian Institution Press. 406 p.

Kwiecinski, G.G. 2006. *Phyllostomus discolor*. *Mammalian Species* 801: 1–11.

Medellín, R. and O. Gaona. 1999. Seed Dispersal by Bats and Birds in Forest and Disturbed Habitats of Chiapas, Mexico. *Biotropica* 31: 478–485.

(*Continued*)

Molinari, J., A. de Ascencao and E.K.V. Kalko. 2000. *Bat Research News* 41: 130.

Moojen, J., J.C.M. Carvalho and H.S. Lopes. 1941. Observações sobre o conteúdo gástrico das aves brasileiras. *Memórias do Instituto Oswaldo Cruz* 36: 405–444.

National Geographic. 2005. *Field Guide to the Birds of North America* (4th ed.). Washington, DC: National Geographic. 504 p.

Ortega J. and Arita H.T. 1997. *Mimon bennettii. Mammalian Species* 549: 1–4.

Ortega J. and I. Castro-Arellano. 2001. *Artibeus jamaicensis. Mammalian Species* 662: 1–9.

Pyle, P., S.N.G. Howell, R.P. Yunick and D.F. DeSante. 1987. *Identification guide to North American passerines*. Bolinas: Slate Creek Press. 278 p.

Remsen, J.V. Jr. and S.K. Robinson. 1990. A classification scheme for foraging behavior of birds in terrestrial habitats. *Studies in Avian Biology* 13: 144–160.

Restall, R. 1975. *Finches and other Seed-Eating Birds*. London: Faber & Faber. 333 p.

Robinson, S.K. and R.T. Holmes. 1982. Foraging behavior of forest birds: the relationships among search tatics, diet, and habitat struture. *Ecology* 63: 1918–1931.

Santos, M., L.F. Aguirre, L.B. Vázquez and J. Ortega. 2003. *Phyllostomus hastatus. Mammalian Species* 722: 1–6.

Schubart, O., A.C. Aguirre and H. Sick. 1965. Contribuição para o conhecimento da alimentação das aves brasileiras. *Arquivos de Zoologia de São Paulo* 12: 95–249.

Scultori C., D. Dias and A.L. Peracchi. 2009. Notes on geographic distribution. Mammalia, Chiroptera, Phyllostomidae, *Artibeus cinereus*: First record in the state of Paraná, Southern Brazil *Check List* 5: 325–329.

Sick, H. 1997. *Ornitologia brasileira*. Rio de Janeiro: Editora Nova Fronteira. 862 p.

Webster Wm. D. and J. K. Jones Jr. 1993. *Glossophaga commissarisi. Mammalian Species* 446: 1–4.

Webster Wm. D., Ch. O. Handley Jr., and P.J. Soriano. 1998. *Glossophaga longirostris. Mammalian Species* 576: 1–5.

York, H.A. 2008. Observations of frugivory in *Phylloderma stenops* (Chiroptera: Phyllostomidae). *Caribbean Journal of Science* 44: 257–260.

Appendix B

List of species of bats captured with mist nets in Mexico, Venezuela, and Brazil that include fruits in their diets according to the literature consulted. The number of individuals captured for each species is also indicated. Letters within parentheses indicate the successional stages where species were captured (p: pasture, e: early, i: intermediate, l: late):

Family	Species	Mexico	Venezuela	Brazil
Primary Frugivores				
Phyllostomidae	*Artibeus intermedius*	39 (p,e,i,l)	—	—
Phyllostomidae	*Artibeus jamaicensis*	224 (p,e,i,l)	48 (e,i,l)	—
Phyllostomidae	*Artibeus obscurus*	—	—	3 (p,l)
Phyllostomidae	*Artibeus phaeotis*	28 (e,i,l)	—	—

Family	Species	Mexico	Venezuela	Brazil
Phyllostomidae	*Artibeus planirostris*	—	—	160 (p,e,i,l)
Phyllostomidae	*Artibeus watsoni*	15 (p,e,i,l)	—	—
Phyllostomidae	*Carollia brevicauda*	—	80 (p,e,i,l)	—
Phyllostomidae	*Carollia perspicillata*	—	49 (e,i,l)	—
Phyllostomidae	*Carollia spp.*	7 (l)	—	137 (p,e,i,l)
Phyllostomidae	*Centurio senex*	1 (l)	—	—
Phyllostomidae	*Chiroderma salvini*	1 (l)	2 (l)	—
Phyllostomidae	*Chiroderma villosum*	—	2 (e,l)	3 (e,i,l)
Phyllostomidae	*Phyllostomus elongatus*	—	133 (p,e,i,l)	—
Phyllostomidae	*Platyrrhinus helleri*	—	12 (p,e)	—
Phyllostomidae	*Platyrrhinus vittatus*	—	125 (p,e,i,l)	—
Phyllostomidae	*Sphaeronycteris toxophyllum*	—	5 (p,i)	—
Phyllostomidae	*Sturnira lilium*	13 (e,i,l)	36 (p,e,i,l)	1 (p)
Phyllostomidae	*Uroderma bilobatum*	—	52 (p,e,i,l)	—
Phyllostomidae	*Uroderma magnirostrum*	—	71 (p,e,i,l)	—
Secondary Frugivores				
Phyllostomidae	*Artibeus lituratus*	4 (l)	3 (p,e)	11 (e,i,l)
Phyllostomidae	*Choeroniscus godmani*	1 (l)	—	—
Phyllostomidae	*Glossophaga commissarisi*	29 (e,i,l)	—	—
Phyllostomidae	*Glossophaga longirostris*	—	73 (p,e,i,l)	—
Phyllostomidae	*Glossophaga soricina*	105 (p,e,i,l)	12 (e,i,l)	47 (p,e,i,l)
Phyllostomidae	*Leptonycteris yerbabuenae*	35 (e,i,l)	—	—
Phyllostomidae	*Lonchophylla mordax*	—	—	14 (p,e,i,l)
Phyllostomidae	*Phylloderma stenops*	—	1(i)	20 (p,e,i,l)
Phyllostomidae	*Phyllostomus discolor*	—	—	51 (p,e,i,l)
Phyllostomidae	*Phyllostomus hastatus*	—	19 (e,i,l)	14 (p,e,i,l)
Phyllostomidae	*Tonatia saurophila*	—	2 (e,l)	4 (e,i,l)
Occasional Frugivores				
Phyllostomidae	*Lophostoma brasiliense*	—	—	1 (i)
Phyllostomidae	*Micronycteris hirsuta*	—	4 (e,l)	—
Phyllostomidae	*Micronycteris megalotis*	—	10 (p,e,i,l)	—
Phyllostomidae	*Mimon bennettii*	—	2 (e,i)	3 (e,l)
Phyllostomidae	*Musonycteris harrisoni*	1 (l)	—	—
Noctilionidae	*Noctilio albiventris*	—	41 (p,e,i,l)	2 (e)

Appendix C

List of species of birds captured with mist nets in Mexico, Venezuela, and Brazil that include fruits in their diets according to the literature consulted. The number of individuals captured for each species is also indicated. Letters in parentheses indicate the successional stages where species were captured (p: pasture, e: early, i: intermediate, l: late):

Family	Species	Mexico	Venezuela	Brazil
Primary Frugivores				
Turdidae	*Catharus ustulatus*	4 (e,l)	11 (e,i)	—
Pipridae	*Chiroxiphia lanceolata*	—	93 (e,i,l)	—
Thraupidae	*Euphonia trinitatis*	—	2 (e)	—
Pipridae	*Pipra filicauda*	—	17 (i,l)	—
Cardinalidae	*Saltator maximus*	—	1 (e)	—
Cardinalidae	*Saltator orenocensis*	—	8 (i)	—
Cardinalidae	*Saltator similis*	—	—	1 (e)
Thraupidae	*Tangara cayana*	—	9 (e)	
Thraupidae	*Thraupis episcopus*	—	10 (e)	—
Thraupidae	*Thraupis sayaca*	—	—	2 (e)
Trogonidae	*Trogon citreolus*	5 (i,l)	—	—
Trogonidae	*Trogon mexicanus*	1 (i)	—	—
Turdidae	*Turdus assimilis*	3 (e,l)	—	—
Turdidae	*Turdus leucomelas*	—	1 (e)	1 (i)
Turdidae	*Turdus nudigenis*	—	2 (i)	—
Turdidae	*Turdus rufopalliatus*	14 (e,i,l)	—	—
Secondary Frugivores				
Psitacidae	*Aratinga cactorum*	—	—	3 (e)
Tyrannidae	*Attila spadiceus*	2 (i,l)	—	—
Icteridae	*Cacicus cela*	—	26 (i)	—
Icteridae	*Cacicus melanicterus*	5 (i,l)	—	—
Tyrannidae	*Camptostoma obsoletum*	—	44 (e,i,l)	10 (p,e,i,l)
Tyrannidae	*Capsiempis flaveola*	—	5 (e)	—
Picidae	*Celeus flavescens*	—	—	3 (e,i)
Thraupidae	*Coereba flaveola*	—	85 (e,i,l)	1 (i)
Picidae	*Colaptes melanochloros*	—	—	1 (e)
Tynamidae	*Crypturellus noctivagus*	—	—	1 (i)
Corvidae	*Cyanocorax cyanopogon*	—	—	7 (e,i,l)
Corvidae	*Cyanocorax sanblasianus*	2 (i,l)	—	—
Cardinalidae	*Cyanocompsa parellina*	304 (p,e,i,l)	—	—
Parulidae	*Dendroica petechia*	1 (l)	3 (e)	—
Parulidae	*Dendroica striata*	—	2 (i)	—
Picidae	*Dryocopus lineatus*	—	1 (i)	—

Family	Species	Mexico	Venezuela	Brazil
Tyrannidae	*Elaenia cristata*	—	10 (e,i,l)	—
Tyrannidae	*Elaenia flavogaster*	—	36 (e,i,l)	—
Tyrannidae	*Elaenia parvirostris*	—	29 (e,i,l)	—
Thraupidae	*Eucometis penicillata*	—	10 (l)	—
Psitacidae	*Forpus passerinus*	—	2 (e)	—
Columbidae	*Geotrygon montana*	1 (l)	—	—
Icteridae	*Gnorimopsar chopi*	—	—	1 (e)
Icteridae	*Icteria virens*	37 (p,e,i)	—	—
Icteridae	*Icterus bullocki*	1 (p)	—	—
Icteridae	*Icterus cucullatus*	13 (p,e)	—	—
Icteridae	*Icterus icterus*	—	—	1 (e)
Icteridae	*Icterus pustulatus*	62 (p,e,i,l)	—	—
Icteridae	*Icterus spurius*	2 (p)	—	—
Tyrannidae	*Inezia tenuirostris*	—	23 (e,i,l)	—
Tyrannidae	*Megarynchus pitangua*	—	—	2 (e)
Picidae	*Melanerpes chrysogenys*	8 (e,i,l)	—	—
Mimidae	*Melanotis caerulescens*	3 (l)	—	—
Mimidae	*Mimus gilvus*	—	4 (p,e)	—
Mimidae	*Mimus polyglottos*	3 (e)	—	—
Tyrannidae	*Myiarchus ferox*	—	17 (e,i)	—
Tyrannidae	*Myiarchus swainsoni*	—	10 (e,l)	3 (e)
Tyrannidae	*Myiarchus tyrannulus*	5 (p,e,i)	46 (p,e,i,l)	29 (e,i,l)
Tyrannidae	*Myiodynastes maculatus*	—	4 (i)	38 (e,i,l)
Tyrannidae	*Myiopagis caniceps*	—	—	1 (e)
Tyrannidae	*Myiopagis gaiimardii*	—	3 (e,l)	—
Tyrannidae	*Myiophobus fasciatus*	—	2 (e)	—
Tyrannidae	*Myiozetetes cayanensis*	—	3 (e)	—
Tyrannidae	*Myiozetetes similis*	—	1 (e)	1 (e)
Thraupidae	*Nemosia pileata*	—	6 (e)	—
Cotingidae	*Pachyramphus aglaiae*	6 (e,i,l)	—	—
Cotingidae	*Pachyramphus polychopterus*	—	3 (e)	15 (e,i,l)
Thraupidae	*Paroaria gularis*	—	4 (e)	—
Thraupidae	*Paroaria dominicana*	—	—	3 (e)
Cardinalidae	*Passerina caerulea*	5 (p,e)	—	—
Cardinalidae	*Passerina ciris*	59 (p,e,i,l)	—	—
Cardinalidae	*Passerina leclancherii*	311 (p,e,i,l)	—	—
Cardinalidae	*Passerina versicolor*	81 (p,e,i,l)	—	—
Tyrannidae	*Phaeomyias murina*	—	3 (e)	4 (e,i)
Cardinalidae	*Pheucticus chrysopeplus*	1 (i)	—	—
Cardinalidae	*Pheucticus melanocephalus*	6 (p,e,i)	—	—
Tyrannidae	*Pitangus sulphuratus*	—	6 (i,l)	11 (p,e)
Tyrannidae	*Poecilotriccus sylvia*	—	10 (e,i,l)	—
Icteridae	*Psarocolius decumanus*	—	4 (e)	—
Furnariidae	*Pseudoseisura cristata*	—	—	2 (e)

(Continued)

Family	Species	Mexico	Venezuela	Brazil
Thraupidae	*Rhodinocichla rosea*	3 (l)	—	—
Cardinalidae	*Saltator coerulescens*	7 (e,i)	—	—
Parulidae	*Setophaga ruticilla*	—	16 (i,l)	—
Thraupidae	*Thlypopsis sordida*	—	—	1 (e)
Turdidae	*Turdus albicollis*	—	—	3 (e)
Turdidae	*Turdus amaurochalinus*	—	—	45 (e,i,l)
Turdidae	*Turdus fumigatus*	—	8 (i,l)	—
Tityridae	*Tityra inquisitor*	—	—	1 (e)
Tyrannidae	*Tyrannus savana*	—	1 (p)	—
Vireonidae	*Vireo olivaceus*	—	10 (e,i,l)	—
Vireonidae	*Vireo atricapilla*	2 (p,i)	—	—
Vireonidae	*Vireo bellii*	26 (p,e,i,l)	—	—
Vireonidae	*Vireo flavoviridis*	339 (p,e,i,l)	—	—
Vireonidae	*Vireo gilvus*	10 (p,e,i,l)	—	—
Vireonidae	*Vireo hypochryseus*	5 (e,l)	—	—
Occasional Frugivores				
Emberizidae	*Arremonops rufivirgatus*	110 (p,e,i,l)	—	—
Parulidae	*Basileuterus lachrymosus*	1 (l)	—	—
Tyrannidae	*Camptostoma imberbe*	18 (p,e,l)	—	—
Tyrannidae	*Casiornis fuscus*	—	—	12 (i,l)
Tyrannidae	*Cnemotriccus fuscatus*	—	—	5 (i,l)
Columbidae	*Columbina inca*	40 (p,e,i,l)	—	—
Columbidae	*Columbina passerina*	260 (p,e,i,l)	31 (e)	—
Columbidae	*Columbina picui*	—	—	8 (p,e)
Columbidae	*Columbina talpacoti*	33 (p,e,i)	71 (p,e,i)	18 (p,e)
Tyrannidae	*Contopus cinereus*	—	4 (i)	—
Cuculidae	*Crotophaga ani*	—	10 (e)	—
Cuculidae	*Crotophaga major*	—	2 (l)	—
Cuculidae	*Crotophaga sulcirostris*	—	2 (e)	—
Cardinalidae	*Cyanocompsa brissonii*	—	—	1 (e)
Tyrannidae	*Deltarhynchus flammulatus*	15 (p,e,i,l)	—	—
Parulidae	*Dendroica petechia*	1 (l)	3 (e)	—
Tyrannidae	*Empidonax difficilis*	123 (p,e,i,l)	—	—
Tyrannidae	*Empidonax minimus*	11 (p,e,i)	—	—
Parulidae	*Geothlypis poliocephala*	3 (p)	—	—
Tyrannidae	*Hemitriccus margaritaceiventer*	—	—	5 (i,l)
Dendrocolaptidae	*Lepidocolaptes souleyetii*	—	25 (e,i,l)	—
Tyrannidae	*Leptopogon amaurocephalus*	—	11 (e,l)	2 (e,l)
Columbidae	*Leptotila verreauxi*	25 (p,e,i,l)	33 (e,i)	6 (e,i,l)
Falconidae	*Milvago chimachima*	—	1 (i)	—
Momotidae	*Momotus mexicanus*	1 (l)	—	—
Tyrannidae	*Myiarchus cinerascens*	4 (p,e,l)	—	—
Tyrannidae	*Myiarchus nuttingi*	128 (p,e,i,l)	—	—
Tyrannidae	*Myiarchus tuberculifer*	23 (p,e,i,l)	—	—

Family	Species	Mexico	Venezuela	Brazil
Tyrannidae	*Myiodynastes luteiventris*	1 (p)	—	—
Tyrannidae	*Myiopagis viridicata*	1 (i)	16 (e,i,l)	40 (e,i,l)
Emberizidae	*Oryzoborus angolensis*	—	4 (i)	—
Tyrannidae	*Pachyramphus validus*	—	—	2 (e)
Tyrannidae	*Pachyramphus viridis*	—	—	1 (e)
Tyrannidae	*Phelpsia inornata*	—	2 (e,l)	—
Picidae	*Picoides scalaris*	2 (e)	—	—
Emberizidae	*Sporophila minuta*	2 (p,e)	—	—
Emberizidae	*Sporophila torqueola*	1 (p)	—	—
Tyrannidae	*Todirostrum cinereum*	—	16 (e)	1 (i)
Tyrannidae	*Tolmomyias flaviventris*	—	57 (e,i,l)	23 (e,i,l)
Tyrannidae	*Tolmomyias sulphurescens*	—	12 (e,i,l)	15 (e,i,l)
Tyrannidae	*Tyrannus melancholicus*	1 (p)	1 (p)	7 (p,e)
Columbidae	*Zenaida asiatica*	1 (i)	—	—

Appendix D

Results of PerMANOVA tests comparing variations of the Jackknife 2 estimate of total number of species of fruit-eating bats and birds captured per sampling plot among four successional stages (pasture "p," early "e," intermediate "i," "late "l") and two seasons (dry "d," rainy "r"). Only significant (*$p \leq 0.05$) main effects and interactions are shown, including results of pairwise tests between groups showing significant differences, where t is a multivariate analogue to the univariate Student's t-test, and p is the p-value associated to this statistic, obtained using permutations, not tables. Averages corresponding to each of the groups compared are indicated within parentheses following the acronym for each group:

Animal/ Region		PerMANOVA Terms				
Bats						
Mexico		*df*	*SS*	*MS*	*F*	*p*
	Season	1	31.29	31.29	6.67	0.021
	Successional stage	3	75.54	25.18	5.36	0.011
	Pairwise significant tests		Groups		*t*	*p*
	Season		R > D		2.58	0.027
	Successional stage		E > P		3.67	0.007
			I > P		3.47	0.012
			L > P		4.06	0.004

(Continued)

Animal/ Region	PerMANOVA Terms					
	Group averages D (3.1), R (5.6), P (1.4), E (4.7), I (5.5), L (6.4), DP (0), DE (3.7), DI (4.9), DL (4.4), RP (2.8), RE (5.8), RI (6.0), RL (7.8)					
Venezuela		*df*	*SS*	*MS*	*F*	*p*
	Successional stage	3	167.41	55.80	7.90	0.004
	Pairwise significant tests		Groups		*t*	*p*
	Successional stage		E > I		3.17	0.013
			E > P		8.10	0.001
			I > P		3.00	0.020
	Group averages D (8.8), R (10.9), P (6.0), E (13.5), I (9.9), L (10.0), DP (3.0), DE (13.9), DI (9.3), DL (8.9), RP (9.1), RE (13.1), RI (10.5), RL (11.1)					
Brazil		*df*	*SS*	*MS*	*F*	*p*
	Season	1	45.21	45.21	8.07	0.017
	Pairwise significant tests		Groups		*t*	*p*
	Season		R > D		2.84	0.013
	Group averages D (4.2), R (6.9), P (3.8), E (5.8), I (5.5), L (7.3), DP (3.7), DE (4.1), DI (3.4), DL (5.5), RP (3.8), RE (7.4), RI (7.5), RL (9.0)					
Birds						
Mexico		*df*	*SS*	*MS*	*F*	*p*
	No significant effects					
	Group averages D (17.8), R (17.6), P (16.8), E (21.6), I (17.6), L (14.7), DP (16.9), DE (21.3), DI (18.3), DL (14.6), RP (16.7), RE (21.9), RI (16.9), RL (14.8)					
Venezuela		*df*	*SS*	*MS*	*F*	*p*
	Successional stage	3	836.87	278.96	8.54	0.001
	Pairwise significant tests		Groups		*t*	*p*
	Successional stage		E (17.1) > P (1.2)		3.71	0.008
			I (13.3) > P (1.2)		8.53	0.001
			L (9.9) > P (1.2)		7.00	0.001
	Group averages D (12.2), R (8.5), P (1.2), E (17.1), I (13.3), L (9.8), DP (0.8), DE (20.1), DI (14.6), DL (13.2), RP (1.5), RE (14.1), RI (12.1), RL (6.5)					
Brazil		*df*	*SS*	*MS*	*F*	*p*
	Successional stage	3	459.12	153.04	16.74	0.001
	Pairwise significant tests		Groups		*t*	*p*
	Successional stage		E > L		6.14	0.002
			E > P		9.01	0.003
			I > P		3.47	0.013
	Group averages D (7.3), R (8.0), P (2.7), E (13.9), I (9.5), L (4.6), DP (3.1), DE (12.1), DI (9.3), DL (4.6), RP (2.3), RE (15.7), RI (9.6), RL (4.5)					

Appendix E

Results of PerMANOVA tests comparing variations of number of species of fruit-eating bats and birds captured per day among four successional stages (pasture "p," early "e," intermediate "i," late "l") and two seasons (dry "d," rainy "r"). Only significant (*$p \leq 0.05$) main effects and interactions are shown, including results of pairwise tests between groups showing significant differences. Averages corresponding to each of the groups compared are indicated within parentheses following the acronym for each group:

Animal/ Region	**PerMANOVA Terms**					
Bats						
Mexico		*df*	*SS*	*MS*	*F*	*p*
	Season	1	11.58	11.58	5.45	0.019
	Pairwise significant tests		Groups		*t*	*p*
	Season		R > D		2.33	0.013
	Group averages					
	D (1.0), R (1.6), P (-), E (1.2), I (1.1), L (1.7), DP (-), DE (0.7), DI (1.0), DL (1.3), RP (-), RE (1.7), RI (1.1), RL (2.0)					
Venezuela		*df*	*SS*	*MS*	*F*	*p*
	Season	1	30.38	30.38	6.73	0.010
	Successional stage	3	162.08	54.03	11.97	0.001
	Pairwise significant tests		Groups		*t*	*p*
	Season			R > D	2.59	0.011
	Successional stage			E > I	2.58	0.013
				E > L	2.68	0.009
				E > P	5.80	0.001
				I > P	4.19	0.001
				L > P	3.04	0.006
	Group averages					
	D (3.4), R (4.5), P (2.1), E (5.8), I (4.0), L (3.8), DP (1.0), DE (6.0), DI (3.3), DL (3.2), RP (3.2), RE (5.5), RI (4.8), RL (4.4)					
Brazil		*df*	*SS*	*MS*	*F*	*p*
	Successional stage	3	50.28	16.76	8.40	0.001
	Pairwise significant tests		Groups		*t*	*p*
	Successional stage		L > E		2.79	0.008
			L > I		3.51	0.001
			L > P		4.85	0.001
	Group averages					
	D (1.4), R (1.9), P (0.9), E (1.7), I (1.4), L (2.8), DP (1.3), DE (1.0), DI (0.9), DL (2.5), RP (0.7), RE (2.2), RI (1.8), RL (3.0)					

(*Continued*)

Animal/ Region	PerMANOVA Terms					
Birds						
Mexico		*df*	*SS*	*MS*	*F*	*p*
	Successional stage	3	134.28	44.76	5.47	0.002
	Pairwise significant tests		Groups		*t*	*p*
	Successional stage		E > I		2.45	0.010
			E > L		3.83	0.001
			E > P		2.35	0.024
	Group averages					
	D (7.3), R (7.2), P (7.0), E (9.1), I (7.1), L (6.0), DP (7.3), DE (8.6), DI (7.1), DL (6.2), RP (6.4), RE (9.9), RI (7.0), RL (5.6)					
Venezuela		*df*	*SS*	*MS*	*F*	*p*
	Season	1	64.21	64.21	7.42	0.010
	Successional stage	3	325.58	108.53	12.55	0.001
	Pairwise significant tests		Groups		*t*	*p*
	Season		55 pt		2.72	0.005
	Successional stage		E > L		2.53	0.012
			E > P		5.04	0.001
			I > P		6.35	0.001
			L > P		6.55	0.001
	Group averages					
	D (4.1), R (2.6), P (0.4), E (5.1), I (4.3), L (3.0), DP (0.3), DE (7.3), DI (4.8), DL (3.8), RP (0.4), RE (3.5), RI (3.7), RL (2.0)					
Brazil		*df*	*SS*	*MS*	*F*	*p*
	Successional stage	3	204.38	68.13	32.00	0.001
	Pairwise significant tests		Groups		*t*	*p*
	Successional stage		E > L		7.18	0.001
			E > P		8.97	0.001
			I > L		4.48	0.002
			I > P		6.10	0.001
			L > P		2.37	0.027
	Group averages					
	D (2.2), R (2.5), P (0.6), E (4.3), I (3.3), L (1.3), DP (0.8), DE (3.5), DI (3.3), DL (1.2), RP (0.5), RE (5.0), RI (3.3), RL (1.4)					

Appendix F

Results of PerMANOVA tests comparing variations of number of fruit-eating bats and birds captured per day among four successional stages (pasture "p," early "e," intermediate "i," late "l") and two seasons (dry "d," rainy "r"). Only significant (*$p \leq 0.05$) main effects and interactions are shown, including results of pairwise tests between groups showing significant differences. Averages corresponding to each of the groups compared are indicated within parentheses following the acronym for each group:

Animal/ Region	PerMANOVA Terms					
Bats						
Mexico		*df*	*SS*	*MS*	*F*	*p*
	Season	1	138.52	138.52	4.15	0.039
	Pairwise significant tests		Groups		*t*	*p*
	Season		R > D		2.04	0.042
	Group averages					
	D (2.1), R (4.4), P (-), E (3.1), I (2.4), L (4.9), DP (-), DE (1.0), DI (1.8), DL (4.4), RP (-), RE (5.1), RI (3.0), RL (5.2)					
Venezuela		*df*	*SS*	*MS*	*F*	*p*
	Season	1	388.01	388.01	7.94	0.007
	Successional stage	3	794.20	264.73	5.41	0.001
	Pairwise significant tests		Groups		*t*	*p*
	Season		R > D		2.82	0.008
	Successional stage		E > P		5.04	0.001
			I > P		3.12	0.001
			L > P		2.57	0.013
	Group averages					
	D (6.3), R (10.3), P (3.7), E (11.7), I (8.8), L (8.9), DP (1.5), DE (12.3), DI (6.3), DL (5.0), RP (5.9), RE (11.0), RI (11.3), RL (12.9)					
Brazil		*df*	*SS*	*MS*	*F*	*p*
	Successional stage	3	408.45	136.15	5.87	0.004
	Pairwise significant tests		Groups		*t*	*p*
	Successional stage		L > E		2.58	0.011
			L > I		3.10	0.004
			L > P		2.90	0.006
	Group averages					
	D (3.0), R (4.3), P (2.3), E (3.4), I (2.6), L (7.0), DP (3.1), DE (1.3), DI (1.3), DL (6.3), RP (1.7), RE (4.7), RI (3.4), RL (7.6)					

(*Continued*)

Animal/ Region	**PerMANOVA Terms**					
Birds						
Mexico		*df*	*SS*	*MS*	*F*	*p*
	Successional stage	3	4053.7	1351.2	4.53	0.005
	Pairwise significant tests		Groups		*t*	*p*
	Successional stage		E > I		2.47	0.013
			E > L		3.24	0.003
	Group averages					
	D (24.4), R (20.0), P (22.4), E (33.1), I (19.1), L (16.2), DP (24.5), DE (33.1), DI (20.1), DL (19.8), RP (18.6), RE (33.2), RI (17.4), RL (10.6)					
Venezuela		*df*	*SS*	*MS*	*F*	*p*
	Successional stage	3	3601.30	1200.40	7.27	0.001
	Pairwise significant tests		Groups		*t*	*p*
	Successional stage		E > L		2.50	0.009
			E > P		3.78	0.001
			I > P		4.07	0.001
			L > P		4.13	0.001
	Group averages					
	D (11.4), R (7.1), P (0.4), E (15.9), I (11.9), L (6.3), DP (0.3), DE (22.4), DI (14.2), DL (8.7), RP (0.6), RE (11.4), RI (9.5), RL (3.6)					
Brazil		*df*	*SS*	*MS*	*F*	*p*
	Successional stage	3	446.61	148.87	20.99	0.001
	Pairwise significant tests		Groups		*t*	*p*
	Successional stage		E > I		2.53	0.014
			E > L		6.93	0.001
			E > P		5.76	0.001
			I > L		4.68	0.001
			I > P		3.69	0.002
	Group averages					
	D (3.4), R (3.7), P (1.5), E (6.6), I (4.4), L (1.5), DP (1.8), DE (6.1), DI (4.3), DL (1.3), RP (1.2), RE (7.2), RI (4.5), RL (1.8)					

Appendix G

Number of individuals of each species of fruit-feeding bats captured at each study site during the dry and rainy season:

Species	Fruit-Feeding Class	Dry	Rainy
Mexico			
Artibeus intermedius	Fruit dominant	16	23
Artibeus jamaicensis	Fruit dominant	80	144
Artibeus phaeotis	Fruit dominant	13	15
Artibeus watsoni	Fruit dominant	6	9
Carollia sp	Fruit dominant	0	7
Centurio senex	Fruit dominant	0	1
Chiroderma salvini	Fruit dominant	0	1
Sturnira lilium	Fruit dominant	6	7
Artibeus lituratus	Fruit partial	0	4
Choeroniscus godmani	Fruit partial	0	1
Glossophaga commissarisi	Fruit partial	4	25
Glossophaga soricina	Fruit partial	23	82
Leptonycteris yerbabuenae	Fruit partial	5	30
Venezuela			
Artibeus jamaicensis	Fruit dominant	15	33
Carollia brevicauda	Fruit dominant	25	55
Carollia perspicillata	Fruit dominant	24	25
Chiroderma salvini	Fruit dominant	1	1
Chiroderma villosum	Fruit dominant	1	1
Phyllostomus elongatus	Fruit dominant	45	88
Platyrrhinus helleri	Fruit dominant	11	1
Platyrrhinus vittatus	Fruit dominant	44	81
Sphaeronycteris toxophyllum	Fruit dominant	1	4
Sturnira lilium	Fruit dominant	17	19
Uroderma bilobatum	Fruit dominant	37	15
Uroderma magnirostrum	Fruit dominant	16	55
Artibeus lituratus	Fruit partial	0	3
Glossophaga longirostris	Fruit partial	36	37
Glossophaga soricina	Fruit partial	6	6
Phylloderma stenops	Fruit partial	1	0
Phyllostomus hastatus	Fruit partial	1	18
Tonatia saurophila	Fruit partial	2	0
Trachops cirrhosus	Fruit partial	5	8
Micronycteris hirsuta	Fruit rare	4	0
Micronycteris megalotis	Fruit rare	2	8
Mimon bennettii	Fruit rare	2	0
Noctilio albiventris	Fruit rare	5	36

(Continued)

Species	Fruit-Feeding Class	Dry	Rainy
Brazil			
Artibeus obscurus	Fruit dominant	0	3
Artibeus planirostris	Fruit dominant	44	116
Carollia spp.	Fruit dominant	27	110
Chiroderma villosum	Fruit dominant	0	3
Sturnira lilium	Fruit dominant	1	0
Artibeus lituratus	Fruit partial	3	8
Glossophaga soricina	Fruit partial	11	36
Lonchophylla mordax	Fruit partial	3	11
Phylloderma stenops	Fruit partial	3	17
Phyllostomus discolor	Fruit partial	45	6
Phyllostomus hastatus	Fruit partial	6	8
Tonatia saurophila	Fruit partial	1	3
Lophostoma brasiliense	Fruit rare	0	1
Mimon bennettii	Fruit rare	0	3
Noctilio albiventris	Fruit rare	0	2

Appendix H

Number of individuals of each species of fruit-feeding birds captured at each study site during the dry and rainy season:

Species	Fruit-Feeding Class	Dry	Rainy
Mexico			
Trogon citreolus	Frugivore dominant	4	1
Trogon mexicanus	Frugivore dominant	1	0
Turdus assimilis	Frugivore dominant	1	2
Turdus rufopalliatus	Frugivore dominant	1	13
Attila spadiceus	Frugivore partial	0	2
Cacicus melanicterus	Frugivore partial	2	3
Catharus ustulatus	Frugivore partial	2	2
Cyanocompsa parellina	Frugivore partial	210	94
Cyanocorax sanblasianus	Frugivore partial	0	2
Geotrygon montana	Frugivore partial	0	1
Icteria virens	Frugivore partial	13	24
Icterus bullocki	Frugivore partial	0	1
Icterus cucullatus	Frugivore partial	1	12
Icterus pustulatus	Frugivore partial	40	22
Icterus spurius	Frugivore partial	0	2
Melanerpes chrysogenys	Frugivore partial	5	3
Melanotis caerulescens	Frugivore partial	1	2

Species	Fruit-Feeding Class	Dry	Rainy
Mimus polyglottos	Frugivore partial	0	3
Myiarchus tyrannulus	Frugivore partial	4	1
Pachyramphus aglaiae	Frugivore partial	3	3
Passerina caerulea	Frugivore partial	4	1
Passerina ciris	Frugivore partial	10	49
Passerina leclancherii	Frugivore partial	196	115
Passerina versicolor	Frugivore partial	35	46
Pheucticus chrysopeplus	Frugivore partial	1	0
Pheucticus melanocephalus	Frugivore partial	4	2
Rhodinocichla rosea	Frugivore partial	0	3
Saltator coerulescens	Frugivore partial	4	3
Vireo atricapilla	Frugivore partial	1	1
Vireo bellii	Frugivore partial	11	15
Vireo flavoviridis	Frugivore partial	210	129
Vireo gilvus	Frugivore partial	4	6
Vireo hypochryseus	Frugivore partial	2	3
Arremonops rufivirgatus	Frugivore rare	57	53
Basileuterus lachrymosus	Frugivore rare	0	1
Camptostoma imberbe	Frugivore rare	14	4
Columbina inca	Frugivore rare	21	19
Columbina passerina	Frugivore rare	136	124
Columbina talpacoti	Frugivore rare	6	27
Deltarhynchus flammulatus	Frugivore rare	9	6
Dendroica petechia	Frugivore rare	0	1
Empidonax difficilis	Frugivore rare	62	61
Empidonax minimus	Frugivore rare	9	2
Geothlypis poliocephala	Frugivore rare	0	3
Leptotila verreauxi	Frugivore rare	15	10
Momotus mexicanus	Frugivore rare	0	1
Myiarchus cinerascens	Frugivore rare	0	4
Myiarchus nuttingi	Frugivore rare	107	21
Myiarchus tuberculifer	Frugivore rare	16	7
Myiopagis viridicata	Frugivore rare	1	0
Myiodynastes luteiventris	Frugivore rare	0	1
Picoides scalaris	Frugivore rare	1	1
Sporophila minuta	Frugivore rare	1	1
Sporophila torqueola	Frugivore rare	1	0
Tyrannus melancholicus	Frugivore rare	1	0
Zenaida asiatica	Frugivore rare	1	0
Venezuela			
Catharus ustulatus	Fruit dominant	2	9
Chiroxiphia lanceolata	Fruit dominant	49	44

(Continued)

Species	Fruit-Feeding Class	Dry	Rainy
Euphonia trinitatis	Fruit dominant	0	2
Pipra filicauda	Fruit dominant	9	8
Saltator maximus	Fruit dominant	1	0
Saltator orenocensis	Fruit dominant	8	0
Tangara cayana	Fruit dominant	7	2
Thraupis episcopus	Fruit dominant	3	7
Turdus leucomelas	Fruit dominant	1	0
Turdus nudigenis	Fruit dominant	2	0
Cacicus cela	Fruit partial	4	22
Camptostoma obsoletum	Fruit partial	25	19
Capsiempis flaveola	Fruit partial	2	3
Coereba flaveola	Fruit partial	54	31
Dendroica petechia	Fruit partial	3	0
Dendroica striata	Fruit partial	2	0
Dryocopus lineatus	Fruit partial	1	0
Elaenia cristata	Fruit partial	2	8
Elaenia flavogaster	Fruit partial	23	13
Elaenia parvirostris	Fruit partial	27	2
Eucometis penicillata	Fruit partial	10	0
Forpus passerinus	Fruit partial	0	2
Inezia tenuirostris	Fruit partial	13	10
Mimus gilvus	Fruit partial	2	2
Myiarchus ferox	Fruit partial	13	4
Myiarchus swainsoni	Fruit partial	8	2
Myiarchus tyrannulus	Fruit partial	38	8
Myiophobus fasciatus	Fruit partial	2	0
Myiodinastes maculatus	Fruit partial	0	4
Myiopagis gaiimardii	Fruit partial	0	3
Myiozetetes cayanensis	Fruit partial	3	0
Myiozetetes similis	Fruit partial	0	1
Nemosia pileata	Fruit partial	6	0
Pachyramphus polychopterus	Fruit partial	1	2
Paroaria gularis	Fruit partial	2	2
Phaeomyias murina	Fruit partial	3	0
Pitangus sulphuratus	Fruit partial	6	0
Poecilotriccus sylvia	Fruit partial	7	3
Psarocolius decumanus	Fruit partial	4	0
Setophaga ruticilla	Fruit partial	16	0
Turdus fumigatus	Fruit partial	3	5
Tyrannus savana	Fruit partial	1	0
Vireo olivaceus	Fruit partial	4	6
Columbina passerina	Fruit rare	29	2
Columbina talpacoti	Fruit rare	31	40
Contopus cinereus	Fruit rare	0	4

Species	Fruit-Feeding Class	Dry	Rainy
Crotophaga ani	Fruit rare	3	7
Crotophaga major	Fruit rare	0	2
Crotophaga sulcirostris	Fruit rare	0	2
Lepidocolaptes souleyetii	Fruit rare	16	9
Leptopogon amaurocephalus	Fruit rare	9	2
Leptotila verreauxi	Fruit rare	27	6
Milvago chimachima	Fruit rare	0	1
Myiopagis viridicata	Fruit rare	9	7
Oryzoborus angolensis	Fruit rare	2	2
Phelpsia inornata	Fruit rare	0	2
Todirostrum cinereum	Fruit rare	6	10
Tolmomyias flaviventris	Fruit rare	40	17
Tolmomyias sulphurescens	Fruit rare	8	4
Tyrannus melancholicus	Fruit rare	0	1
Brazil			
Saltator similis	Fruit dominant	1	0
Thraupis sayaca	Fruit dominant	1	1
Turdus leucomelas	Fruit dominant	1	0
Aratinga cactorum	Fruit partial	1	2
Camptostoma obsoletum	Fruit partial	7	3
Celeus flavescens	Fruit partial	2	1
Coereba flaveola	Fruit partial	1	0
Colaptes melanochloros	Fruit partial	1	0
Crypturellus noctivagus	Fruit partial	0	1
Cyanocorax cyanopogon	Fruit partial	5	2
Gnorimopsar chopi	Fruit partial	1	0
Icterus icterus	Fruit partial	1	0
Megarynchus pitangua	Fruit partial	2	0
Myiarchus swainsoni	Fruit partial	1	2
Myiarchus tyrannulus	Fruit partial	21	8
Myiodynastes maculatus	Fruit partial	27	11
Myiopagis caniceps	Fruit partial	0	1
Myiozetetes similis	Fruit partial	1	0
Pachyramphus polychopterus	Fruit partial	0	15
Paroaria dominicana	Fruit partial	1	2
Phaeomyias murina	Fruit partial	1	3
Pitangus sulphuratus	Fruit partial	5	6
Pseudoseisura cristata	Fruit partial	2	0
Thlypopsis sordida	Fruit partial	0	1
Tityra inquisitor	Fruit partial	0	1
Turdus albicollis	Fruit partial	0	3
Turdus amaurochalinus	Fruit partial	18	27

(Continued)

Species	Fruit-Feeding Class	Dry	Rainy
Casiornis fuscus	Fruit rare	5	7
Cnemotriccus fuscatus	Fruit rare	3	2
Columbina picui	Fruit rare	2	6
Columbina talpacoti	Fruit rare	9	9
Cyanocompsa brissonii	Fruit rare	1	0
Hemitriccus margaritaceiventer	Fruit rare	5	0
Leptopogon amaurocephalus	Fruit rare	0	2
Leptotila verreauxi	Fruit rare	2	4
Myiopagis viridicata	Fruit rare	0	40
Pachyramphus validus	Fruit rare	0	2
Pachyramphus viridis	Fruit rare	0	1
Todirostrum cinereum	Fruit rare	1	0
Tolmomyias flaviventris	Fruit rare	16	7
Tolmomyias sulphurescens	Fruit rare	11	4
Tyrannus melancholicus	Fruit rare	5	2

12

Spatiotemporal Dynamics of Insects in a Brazilian Tropical Dry Forest

Frederico de Siqueira Neves, Jhonathan O. Silva, Tatianne Marques, João Gabriel Mota-Souza, Bruno Madeira, Mário Marcos do Espírito-Santo, and Geraldo Wilson Fernandes

CONTENTS

12.1 Introduction

Understanding the distribution patterns of organism diversity as well as the mechanisms that determine them is the main objective of ecological studies (Fernandes and Price 1991; Lawton 1999; Price 2002b). The structure of a community can be determined by several factors that act in different temporal and spatial scales (Ricklefs and Schluter 1993; Godfray and Lawton 2001), such as abiotic conditions and the availability and quality of resources (Strong et al. 1984; Bell et al. 1991; Ricklefs and Schluter 1993). Although the analysis of different scales is essential to understand diversity patterns, this approach is rarely used for tropical insect communities (Lewinsohn et al. 2005). The majority of studies with these organisms were performed for short periods of time or only in a local scale (Lewinsohn et al. 2005), especially in tropical regions where tropical dry forests (TDFs) are largely neglected in terms of scientific investigation (Sánchez-Azofeifa et al. 2005c; Quesada et al. 2009).

A timescale rarely used in studies with insects in tropical regions is that of ecological succession, since a long time is required to characterize the community changes associated with this process. An alternative to evaluate the effects of successional processes on insects without resorting to long-term sampling approaches is to use chronosequences. In this case, time is replaced by space and analyzed by comparing areas with different successional ages but similar abiotic characteristics (e.g., soil type and topography) and land-use history (Kalácska et al. 2004, 2005c; Arroyo-Mora et al. 2005a; Madeira et al. 2009; Quesada et al. 2009; Lebrija-Trejos et al. 2010a; Dupuy et al. 2012b).

As secondary succession progresses, gradual changes occur in the abiotic conditions, vegetation structure, and composition (Guariguata and Ostertag 2001; Kalácska et al. 2004; Ruiz et al. 2005; Madeira et al. 2009) through a directional and continuous process of colonization and extinction of species populations (Quesada et al. 2009; Morin 2011). Usually, there is an increase in environmental complexity, associated with a decrease in light availability at the ground level and an increase in soil nutrients. These factors affect the first trophic level (e.g., plant quality) and cause cascading effects to higher trophic levels, influencing the performance of herbivorous insects and their predators (Brown and Ewel 1987; Neves et al. 2010a; Silva et al. 2012), in addition to detritivorous insects (Neves et al. 2010b).

One of the main structural changes observed along a successional gradient is the vertical stratification of the forest with a well-defined distinction between canopy and understory (Basset et al. 2003). Tropical rain forests have a wide variety of habitats and resources distributed along the vertical strata, from the ground to the canopy (Longino and Nadkarni 1990; Schaefer et al. 2002; Basset et al. 2003). Usually, the distribution of insects responds to this stratification (Basset et al. 2003; Novotny et al. 2003; Grimbacher and Stork 2007), with a decrease in insect diversity from the canopy to the understory (DeVries and Walla 2001; Basset et al. 2003; Grimbacher and Stork 2007). However, such patterns are not known for TDFs, which are structurally simpler environments compared to rain forests (Murphy and Lugo 1986a).

Possible differences in spatial distribution patterns of the insect community in dry compared to wet forests may be exacerbated by the marked seasonal changes in TDF vegetation (Janzen 1981; Filip et al. 1995; Boege 2004; Silva et al. 2012). In these ecosystems, up to 95% of leaves are lost during the dry season (Leigh and Windsor 1982; Pezzini et al. 2008), strongly affecting the quality and quantity of the resources available for insects. Also, variations in foliage quality to herbivorous insects may occur in smaller temporal scales during the wet season (Janzen 1981; Barone 2000; Boege 2004; Oliveira et al. 2012; Silva et al. 2012). Although some studies have investigated such variations in leaf traits in TDFs (e.g., the studies by Poorter et al. [2004] and Lebrija-Trejos et al. [2010]), their consequences to the insect communities associated with these environments are scarcely addressed.

Insects represent the most diverse group of the Animalia kingdom and are fundamentally involved in several ecological processes in terrestrial and aquatic ecosystems (Ødegaard 2000; Novotny et al. 2002; Lewinsohn and Prado 2005). Insects can be divided into many functional groups, including phytophagous insects, predators, parasites, aquatic filter feeders, detritivores, and saprophagous insects (Gullan and Cranston 2000). These organisms are essential in trophic chains and take part in various ecosystem services such as nutrient cycling, plant pollination, and seed dispersal, as well as positive and negative associations with other organisms, including those from parasitism to mutualism (Gullan and Cranston 2000). Thus, studies involving insects are very important to understand ecosystem functioning and can provide valuable information for practical applications such as restoration programs and conservation strategies.

Although we have experienced an increase in studies on the ecology of TDFs in recent years, the same has not been observed for the important group of insects (Neves et al. 2008; Santos et al. 2011a). Therefore, this study aimed at detecting patterns and understanding the mechanisms that affect the insect community at different spatial and temporal scales. We studied the insects' spatial distribution patterns in a TDF, both horizontally (using a chronosequence design) and vertically (comparing vertical strata in a forest at an advanced stage of succession). The chronosequence approach also allowed the development of inferences about changes that may occur along the timescale of ecological succession. Finally, we investigated the seasonal variations in the insect community associated with the aforementioned environments. These comparisons were conducted for the community of insects belonging to different trophic guilds: chewing and sucking herbivorous insects, frugivorous butterflies, ants, and dung beetles.

12.2 Study Area

This chapter presents a compilation of results from the studies conducted between 2006 and 2009 in Mata Seca State Park (MSSP), Minas Gerais, Brazil. For a better description of the study area, see Chapters 1 and 5. All samples were taken from five plots of 20 m × 50 m (0.1 ha) in early, intermediate, and late successional stages (see the details of this chronosequence in the study by Madeira et al. [2009]). Samples of the different trophic guilds occurred during the rainy and dry seasons, using specific methodologies for each guild of insects.

12.3 Free-Feeding Insect Herbivores (Sap Sucking and Leaf Chewing)

To verify the effects of secondary succession on the fauna of herbivorous insects, Neves (2009) sampled 293 trees distributed in five plots per successional stage, 100 trees in the early stage, 100 trees in the intermediate stage, and 93 trees in the late stage, during the rainy season (2007–2008). Access to the canopy was achieved by free rope-climbing techniques for trees higher than 7 m and by using a ladder in trees less than 7 m in height. The insects were sampled in the upper canopy for all sampled trees by the branch beating method, followed by capture using an entomological umbrella (Neves et al. 2010c; Costa et al. 2011).

To confirm whether there is a vertical and seasonal stratification in the community of herbivorous insects, Neves (2009) randomly selected 185 trees (93 reaching the canopy and 92 in the understory) at the end of the rainy season (March 2007; monthly precipitation = 13.7 mm) and 140 trees (70 reaching the canopy and 70 in the understory) at the beginning of the rainy season (December 2007; monthly precipitation = 81.2 mm). All the trees evaluated for vertical stratification were inside the five late successional plots.

Neves (2009) sampled 955 free-feeding herbivorous insects, distributed in 25 families, from which 501 were sap suckers and 454 were chewers. Among the sap-sucking insects Tingidae and Cicadellidae were the families with higher morphospecies richness and abundance, whereas Chrysomelidae and Curculionidae were the most representative chewing insects. This study verified a change in the family composition of free-feeding canopy herbivores for the early stage compared to advanced successional stages (intermediate and late) (Figure 12.1). The four dominant insect families listed earlier

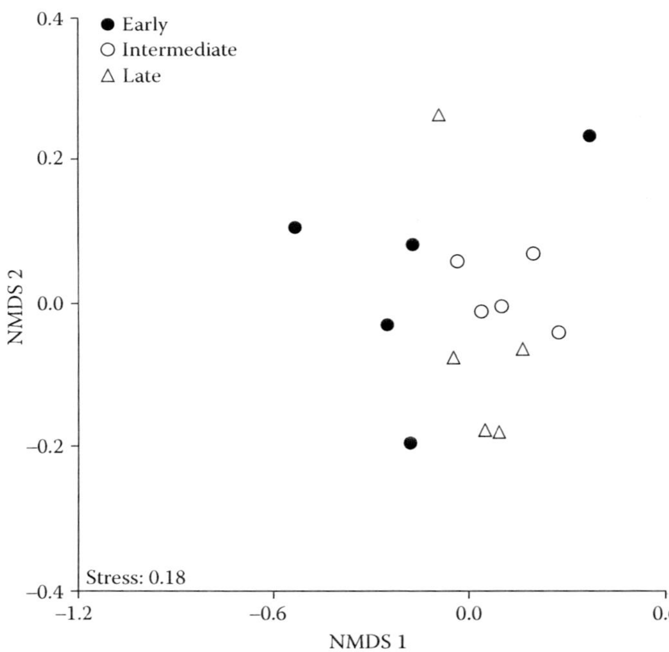

FIGURE 12.1
Ordination of 15 plots from three successional stages (early, intermediate, and late) through a nonmetric multidimensional scaling (NMDS) based in their family composition of free-feeding herbivores ($p < 0.05$).

contributed more than 57% of the observed differences between early stage and intermediate and late stages.

These authors also observed an increase in the richness of sap-sucking and chewing insects along the successional gradient (Figure 12.2). The change observed in the insect community is probably determined by alterations in floristic composition and forest structure, such as host plant height and canopy complexity (see the studies by Novotny et al. [2007] and Neves et al. [2013]). In the studied chronosequence, forest structural complexity (as indicated by the Holdridge complexity index) increased approximately 92 times from the early to the late stage. Also, the floristic composition differs greatly between the early stage and intermediate and late stages, which are similar to each other (Madeira et al. 2009) (Chapter 18). It is likely that these differences are due to more suitable microclimatic conditions and a greater variety of niches and available resources in more advanced successional stages.

Neves (2009) also found differences between vertical forest strata. In this study, a total of 2119 free-feeding insect herbivores were sampled in two forest strata. Among the sap-sucking insects, the family Cicadellidae exhibited the greatest richness of adult insect morphospecies. For chewing herbivores, Chrysomelidae and Curculionidae were the families with the highest richness and abundance of morphospecies. The pattern of vertical stratification varied between guilds and throughout the rainy season. For sap-sucking insects, the canopy showed a higher richness at the beginning of the rainy season, but at the end of the rainy season the richness became higher in the understory (Figure 12.3a). For chewing insects, the richness was higher in the canopy, despite the sampling period (Figure 12.3b).

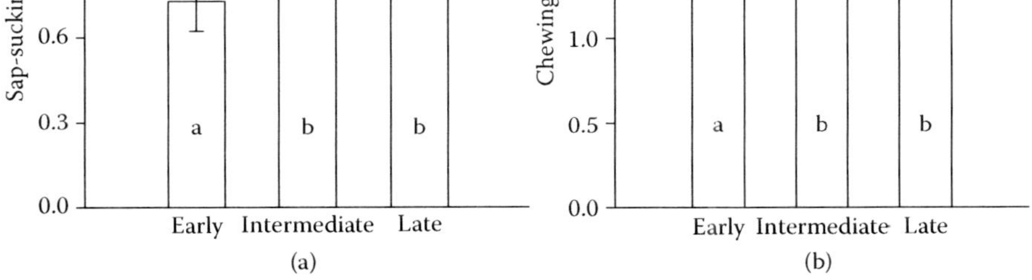

FIGURE 12.2
Free-feeding herbivores richness of sap sucking (a) and chewing (b) along a successional gradient at a canopy tree in a tropical dry forest in Brazil. Error bars indicate one standard error. Different letters above the bars represent statistically different means ($p < 0.05$).

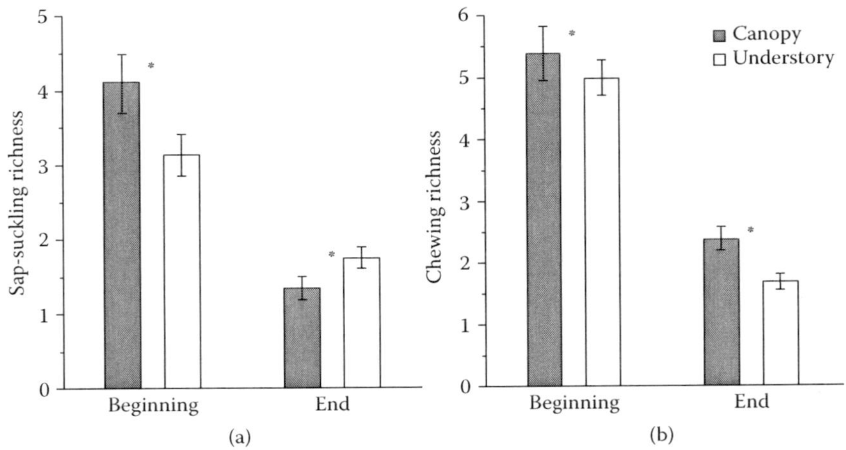

FIGURE 12.3
Free-feeding herbivores richness of sap sucking (a) and chewing (b) within two forest strata (canopy and understory) at the beginning and the end of the rainy season in a tropical dry forest in Brazil. Error bars indicate one standard error. In the figure, "*" indicates $p < 0.05$.

Other studies in tropical forests have found a higher richness of herbivorous insects in adult trees that reached the canopy compared to juvenile understory trees (Basset et al. 2001; Campos et al. 2006; Neves et al. 2013). When encountering a large host tree, a herbivorous insect is supplied with a great amount of resources, thus reducing energy costs of locomotion involved in foraging. In addition, bigger and more complex trees reduce the exposure of herbivorous insects to their natural enemies (Boege 2005; Neves et al. 2013). For chewing herbivores, the forest canopy provides a higher availability of food resources (e.g., density of new leaves) over the entire rainy season, since this habitat possesses higher primary productivity and plant growth rates compared to the understory (Basset et al. 2003). Moreover, the dispersal of these insects between neighboring trees can be facilitated by interconnected crowns in the canopy. Sap-sucking insects probably found higher resource availability at the beginning of the rainy season in the canopy. However, a possible reduction in the hydraulic pressure in the vessels that reach the canopy probably decreased the quantity and quality of available sap in higher plant parts (Neves 2009), resulting in a decrease in sap-sucking insects in the canopy compared to the understory.

In general, a reduction in the richness and abundance of free-feeding herbivorous insects was observed along the rainy season in the studied TDF (Neves 2009; Silva et al. 2012), confirming the pattern described by other studies in dry forests (Janzen 1981; Dirzo and Domínguez 1995; Boege 2005). The high density of Lepidoptera larva and adult chewing beetles at the beginning of the rainy season can cause a peak of herbivory by chewing insects in this period (Janzen 1981). This pattern is usually attributed to the synchronization between herbivore attack and host plant production of new, highly nutritious tissues (Janzen and Waterman 1984; Filip et al. 1995; Boege 2005) and a possible temporal escape from natural enemies (Dirzo and Domínguez 1995). There is evidence that the diversity of herbivorous insects in the TDF studied here is controlled by both bottom-up factors (foliage quality) and top-down pressures (Oliveira et al. 2012), such as predation by spiders (Silva et al. 2012), birds, and bats and parasitoid attacks.

12.4 Fruit-Feeding Butterflies (Lepidoptera: Nymphalidae)

The diversity of fruit-feeding butterflies was compared along the successional gradient, vertical strata, and seasons at the MSSP by Madeira (2008). This study consisted of two one-week sampling periods: the first was in February 2006, during the wet season, and the second was in September 2006, during the dry season. Twenty plots were sampled (six at early, six at intermediate, and eight at late successional stages, including the plots sampled for herbivorous insects). In each early forest plot, we placed two Van Someren–Rydon traps (DeVries 1987), which were positioned on the vegetation at breast height. In the intermediate- and late-stage plots, we also placed two traps in the understory at 1.0–1.5 m from the forest floor. In addition to understory traps, we also placed two traps in the canopy, at a minimum distance of 10 m above the ground, in each of the intermediate- and late-stage plots. Understory traps could be serviced directly from the ground, whereas canopy traps were suspended from thin ropes run over the branches of an emergent tree, such that all traps could be raised and lowered from the ground. Traps were baited with fresh bananas mashed and mixed with fermented sugarcane juice. On the day before the beginning of each sampling period bait was put on plastic dishes and placed in the traps, which was replaced by fresh bait at each subsequent day of sampling.

In total, 35 butterfly species (5915 individuals) were sampled: 16 species were encountered in early successional plots, 24 in intermediate plots, and 30 in late-stage plots (Madeira 2008). All 16 species sampled in the early stage were also found on intermediate and late successional stages, whereas the intermediate and late stages shared 19 species. The greatest similarity was observed between the early and intermediate stages. The intermediate stage presented 5 exclusive species, and 11 species were sampled only on late-stage plots.

The species composition of butterflies differed between the three stages of succession (Figure 12.4); the average butterfly species richness per plot was significantly higher in late successional stage plots and did not vary between early and intermediate successional stages (Figure 12.5a); and abundance did not differ between the stages of succession (Figure 12.5b). The change in species composition and higher butterfly richness in late successional habitats is a consequence of their greater plant species richness (see Madeira et al. 2009) and

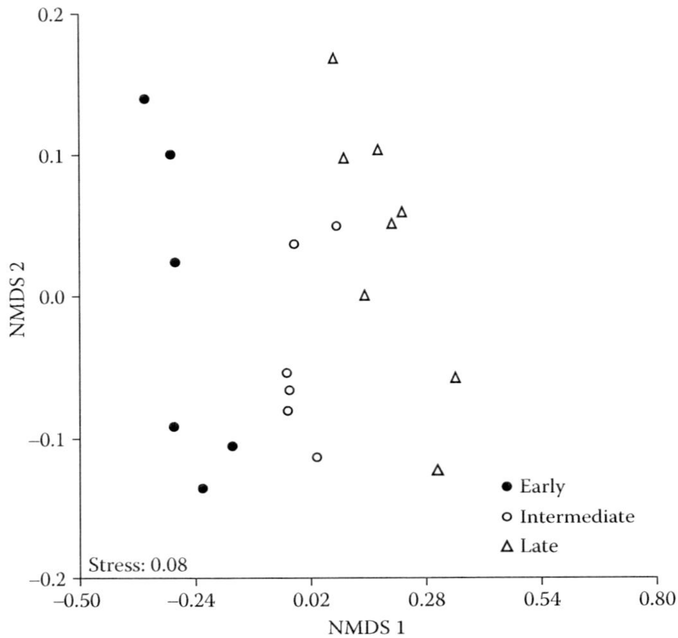

FIGURE 12.4
Ordination of 20 plots from three successional stages (early, intermediate, and late) through a nonmetric multidimensional scaling (NMDS) based in their species composition of fruit-feeding butterflies ($p < 0.001$).

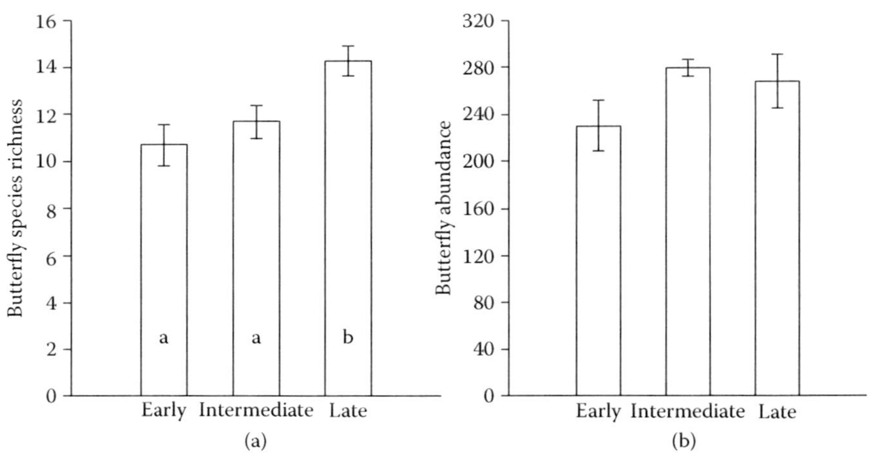

FIGURE 12.5
Fruit-feeding butterfly species richness (a) and abundance (b) at plots of three successional stage fragments in a tropical dry forest in Brazil. Error bars indicate one standard error. Different letters above the bars represent statistically different means ($p < 0.05$).

variety of microhabitats. In early regeneration stages many species of caterpillars may be absent due to the exclusion of a particular plant species (Janzen 1984), resulting in a lower richness of adult butterflies. This result indicates that fragments with lower disturbance intensity supported higher butterfly diversity, highlighting the importance of this group as a bioindicator of habitat quality (Uehara-Prado et al. 2007; Madeira 2008).

Madeira (2008) verified that butterfly diversity also varied between forest strata, but the pattern was opposite to that observed for herbivorous insects. In total 30 species (5119 individuals) were collected in understory traps, whereas only 20 species (796 individuals) were sampled in canopy traps at the MSSP. From these, 5 species were found only in the canopy (*Hamadryas arete, Historis odius, Libytheana carinenta, Prepona omphale*, and *Smyrna blomfildia*), whereas the remaining 15 species were sampled in both understory and canopy traps. Thus, a large fraction of the community can be considered to be strata "generalists," occurring in the forest floor as well as in the canopy. On the other hand, only a small portion of the species can be considered to be canopy "specialists." Butterfly diversity also varied significantly between strata in time, mainly related to changes in the understory. In this stratum, species richness decreased from wet to dry seasons, and the abundance decreased during the same period (Figure 12.6). In the canopy, butterfly richness and abundance did not differ between seasons. Similarly, DeVries et al. (1997) found higher butterfly diversity in the understory in an Ecuadorian rain forest and DeVries (1988) found higher fruit-feeding nymphalid abundance in the understory during the dry season in a Costa Rican rain forest. The considerable decrease in species richness from the rainy to the dry season verified in the understory may

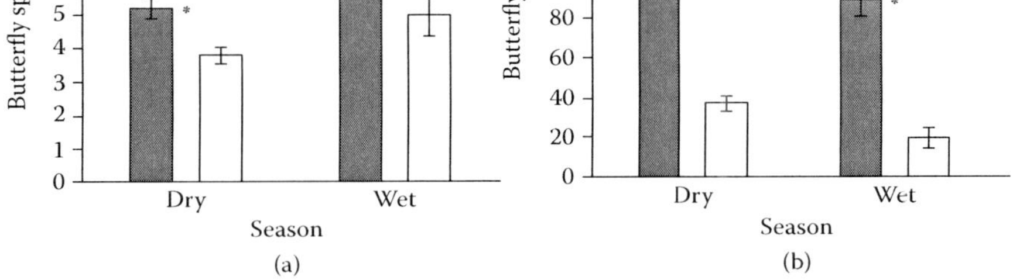

FIGURE 12.6
Fruit-feeding butterfly species richness (a) and abundance (b) per plot between seasons (dry × wet) and vertical strata (understory × canopy). Error bars indicate one standard error. In the figure, "*" indicates $p < 0.05$.

account for the massive leaf fall that occurred in this stratum. For instance, some nymphalid female species are rare in the canopy and are known to oviposit on host plants occurring near the ground (DeVries 1987). Moreover, temperature and humidity are milder in the understory and rotting fruits fall to the ground, positively affecting butterfly abundance in that stratum.

In general, butterfly richness was higher in the rainy season, but abundance was higher in the dry season. Madeira (2008) sampled 32 species (2207 individuals) in total during the rainy season and only 9 species (3708 individuals) during the dry season (Figure 12.6). Whereas 6 species were sampled in both rainy and dry seasons, a total of 26 species were sampled only in the rainy season and only 3 species were exclusively found during the dry season (an unidentified Satyrinae, *Siderone* sp., and *Fountainea ryphea*). These results indicate that the community of fruit-feeding butterflies in the MSSP undergoes great changes in species composition and richness between rainy and dry seasons. In fact, only a small fraction of the community withstands the harsh conditions of the dry season. The unexpected higher abundance in the dry season may be explained by the occurrence of a single species, *Hamadryas februa*. This species alone corresponded to more than 50% of all individuals sampled in the study, and its abundance was much higher in the dry season (2313 individuals) than in the rainy season (666 individuals). In fact, when we removed this species from the analysis, we could not detect any seasonal difference in butterfly abundance. Neotropical butterflies of the genus *Hamadryas* typically spend much time perching on trees and are believed to be cryptically patterned and colored with respect to tree trunks and branches that they use as perching sites (Monge-Nájera et al. 1998). Thus, it is likely that *H. februa* avoids predators (e.g., lizards and birds) during the dry season since the tree bark ("background") color tended to match their wing coloration.

12.5 Ants (Hymenoptera: Formicidae)

Ants were sampled during two different studies in the MSSP. The first study sampled only ants foraging on tree trunks (1.3 m; "arboreal" ants) in February and September 2007 (Neves et al. 2010a). The second study sampled ants foraging on three different microhabitats in September 2008 and February 2009 (Marques 2011a). The sampling period for both studies was the end of the dry season (September) and the mid-rainy season (February). Ants were sampled using pitfall traps (Ribas et al. 2003; Schmidt and Solar 2010) in the same five plots of each successional stage (early, intermediate, and late). Neves et al. (2010a) used only arboreal pitfalls baited with honey and sardines to collect the ants. Marques (2011) used pitfalls without baits to sample the ant community of arboreal, epigeic, and hypogeic microhabitats. The arboreal ants forage in trees, epigeic ants forage on the ground or litterfall, and hypogeic ants forage inside belowground tunnels.

Neves et al. (2010a) sampled 43 species of arboreal ants, and Marques (2011) collected 95 ant species in the three microhabitats (tree, epigeic, and hypogeic) together in the same plots. Both studies detected the influence of different factors on ant richness along the successional gradient, such as plant species richness, habitat conditions, and ant interspecific interactions. Neves et al. (2010a) verified a change in species composition along the secondary succession, but species richness was not significantly different between stages. However, Marques (2011) also observed a change in species composition (Figure 12.7) but detected an increase in total ant species richness in advanced successional stages (Figure 12.8a). When each microhabitat was analyzed separately, the authors found no effects of succession on the richness of arboreal and epigeic ants (Figure 12.8b). Therefore, the total increase in ant species is due to the colonization of new species of hypogeic ants, linked to changes in soil characteristics and a higher plant diversity in intermediate and late stages.

The arboreal and epigeic ant communities are probably regulated by habitat conditions and interspecific competition. Marques (2011) recorded the occurrence of aggressive arboreal ant species that dominate their resources (such as *Crematogaster obscurata* and *Cephalotes atratus*) only in advanced stages of succession. Such dominance may be caused by the presence of specific resources for structuring large nests, such as trees with larger basal areas. Furthermore, it is likely that the presence of dominant species inhibits the establishment of subdominant species and prevents the increase in the number of species in these habitats.

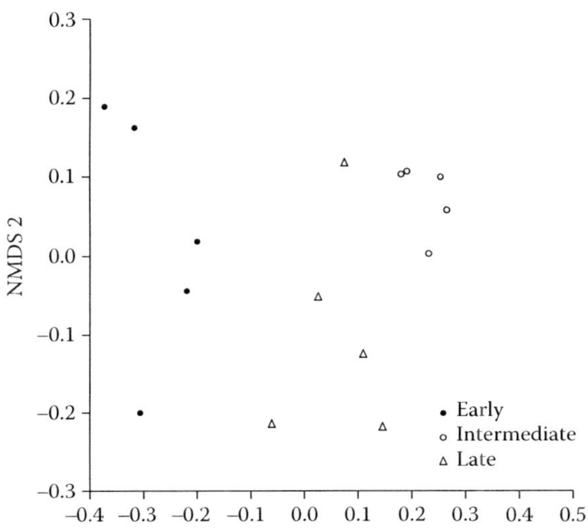

FIGURE 12.7
Ordination of 15 plots from three successional stages (early, intermediate, and late) through an NMDS based in their species composition of ants ($p < 0.001$).

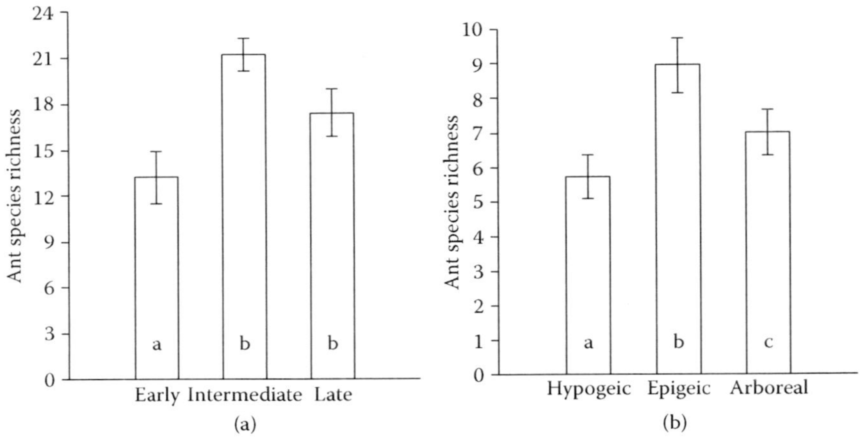

FIGURE 12.8
Ant richness at plots of three successional stages (a) and microhabitats (b) in a tropical dry forest in Brazil. Error bars indicate one standard error. Different letters represent statistically different means ($p < 0.05$).

As observed for herbivorous insects and butterflies, there was also a vertical stratification of ant communities in the studied TDF, although the comparison was not between the canopy and the understory. Such differences between strata were only observed in the dry season, with greater species richness in the epigeic microhabitat (40 species) followed by arboreal and hypogeic microhabitats (36 species each). This pattern is likely caused by the displacement of arboreal ants to the epigeic stratum as a result of the lack of resources in the canopy during the dry season.

Seasonal variations in ant community structure were detected in the two studies conducted at the MSSP. Neves et al. (2010a) found similar species richness of arboreal ants between the dry and the rainy seasons but significant changes in species composition. On the other hand, Marques (2011) detected a higher number of ant species (considering the three microhabitats) during the dry season (25.1 ± 1.71 species per plot) than in the wet season (17.8 ± 1.06). It is probable that due to resource scarcity during the dry season the ant species increased their foraging area and reduced aggressive behaviors to defend territories, increasing pitfall captures in this period.

The same pattern of seasonal variation described by Marques (2011) was found in other studies conducted in different TDFs. Litter ant communities and epigeic ants in the Argentinean Chaco (Delsinne et al. 2008) and communities of arboreal and ground ants in a Mexican TDF (Gove et al. 2005) also showed a higher species richness in the dry season. However, there is still a paucity of studies concerning hypogeic and canopy habitats, precluding a deeper comprehension of the mechanisms underlying seasonal and successional influences on tropical ant communities.

12.6 Dung Beetles (Coleoptera: Scarabaeidae)

Dung beetles were also sampled in two different studies in the MSSP, conducted by Neves et al. (2010b) and Evangelista et al. (in prep.). In the first study, beetles were sampled in two periods in 2007: February (wet season) and September (dry season). Soil pitfall traps baited with human feces or decomposing chicken liver were used. In the study by Evangelista et al. (in prep.), samples were collected using pitfalls baited only with human feces, along the rainy season of 2009–2010 in three periods: January, February, and April (beginning, middle, and end of the season, respectively). In both studies, samples were collected in the same plots in which ants were sampled (see Section 12.5). The stratification effects were not examined for dung beetles, since they forage predominantly up to 2 m from the ground level (Davis 1999).

Neves et al. (2010b) sampled 2752 dung beetles belonging to 38 species and 14 genera. Clear changes were observed in the composition, richness, and abundance of dung beetle species along the successional gradient. In the early stage 41 individuals of 13 species (1 exclusive species) were sampled, whereas 2010 individuals belonging to 32 species (10 exclusive species) were captured in the intermediate stage. Finally, in the late stage 701 individuals and 27 species were collected (5 exclusive species). There was a change in beetle community composition between early and advanced stages of regeneration. Evangelista et al. (in prep.) sampled a total of 3075 dung beetles belonging to 41 species and 15 genera. There was a change in composition (Figure 12.9a) and an increase in species richness along the successional gradient (Figure 12.10a).

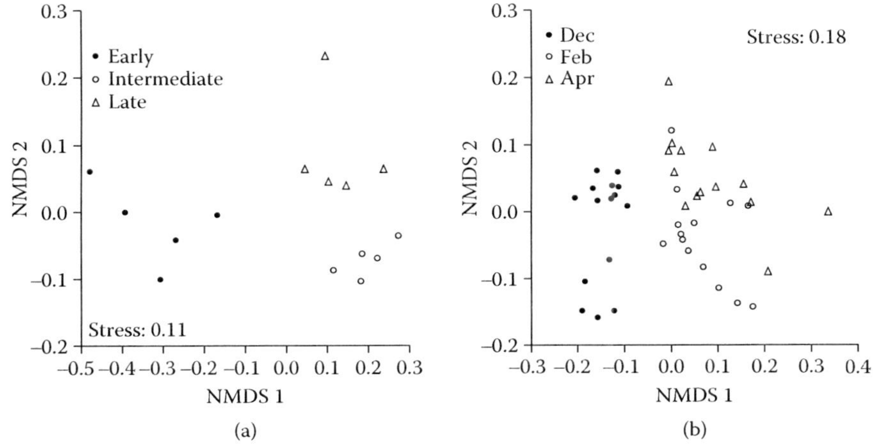

FIGURE 12.9
Ordination of 15 plots from three successional stages (a) and sampled periods (b) through an NMDS based in their species composition of dung beetle ($p < 0.001$).

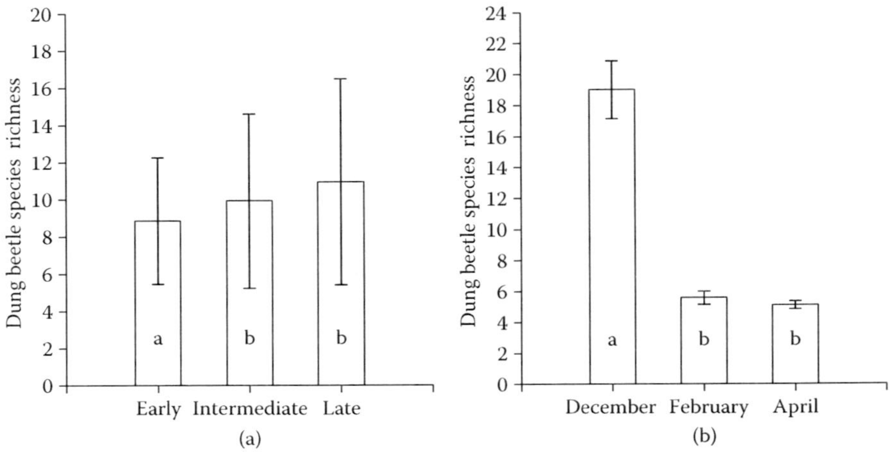

FIGURE 12.10
Dung beetle richness at plots of three successional stages (a) and sampled periods (b) in a tropical dry forest in Brazil. Error bars indicate one standard error. Different letters represent statistically different means ($p < 0.05$).

The community of dung beetles sampled in the MSSP showed a strong seasonal variation (Neves et al. 2010b). In the wet season 2748 individuals belonging to 38 species were sampled, whereas only 4 individuals of 1 species (*Uroxys* sp.) were collected in the dry season. A similar pattern was observed in a study with dung beetles in the caatinga biome (Liberal et al. 2011) in northeastern Brazil, an ecosystem very similar to the TDF studied here (Särkinen et al. 2011; Werneck et al. 2011). With leaf fall during the prolonged dry season, the top layer of the soil becomes dry and compact, causing rapid desiccation of feces and carcasses. Furthermore, due to the scarcity of food resources, the vertebrates that mainly produce the feces consumed by dung beetles probably migrate to more humid habitats such as riparian forests. Finally, the low diversity of dung beetles during the dry season is likely related to their typical life cycle in TDFs (Andresen 2005). Most dung beetle species are annual (Gill 1991) and accumulates resources along the rainy season for nest formation and consequent oviposition (Camberfort 1991). Adults usually die at the end of this period and larvae develop during the dry season, emerging as adults at the onset of the following rainy season.

Evangelista et al. (in prep.) also detected temporal alterations in the community of dung beetles at a finer scale: species composition changed along the rainy season (Figure 12.9b), and species richness decreased along this period (Figure 12.10b). Similar results were observed in a TDF in Mexico (Andresen 2005). The richness and the abundance observed early in the rainy season are consistent with the biological cycle described for dung beetles, for which the emergence of adults is synchronized with the first rains in seasonal environments (Kohlmann 1991).

12.7 Conclusions

Our results show that TDF insect communities change in multiple spatial and temporal scales. For all the studied groups, there was a general increase in species richness and significant alterations in species composition along the successional gradient. The mechanisms that determine these patterns have not been fully clarified and certainly depend on specific resource (food and refuges from natural enemies) requirements and abiotic conditions of the habitat occupied by each group of insects. However, the increasing structural complexity of vegetation from early to late stages seems to be the ultimate factor influencing niche availability for the entomofauna in tropical forests.

At the smallest spatial scale, we observed the occurrence of vertical stratification for all groups that use different compartments of the forest. It is possible that at this scale the responses of each group are more idiosyncratic, without a clear pattern of species richness, which is the case for different guilds of insects. In each forest strata, species richness varied between seasons and within the same season depending on the life history of the most abundant species and the phenological differences between canopy and understory plants. In the case of ants, the differences between forest strata (arboreal, epigeic, and hypogeic) are even more distinct, with each group being affected by contrasting environmental factors.

In the smallest timescale, it was possible to detect remarkable seasonal differences in richness and species composition for all the studied groups. Such temporal variations occurred not only between the dry and the rainy seasons, when environmental conditions change drastically, but also in the course of the rainy season. Therefore, it is possible to conclude that insects are well-adapted to the strong seasonality of TDFs, responding to changes in resource availability and environmental conditions at small temporal scales.

All the taxa studied in this chapter can be used as bioindicators of disturbance and regeneration status in areas of TDFs, since they respond to the effects of secondary succession for at least one of the biotic components analyzed (richness, abundance, and/or composition). However, if we consider the cost–benefit relationship in terms of sampling effort, available collections, and the time spent in identifying species or sorting morphospecies, dung beetles would be the best terrestrial bioindicators of habitat quality and for monitoring the recovery of degraded areas, as shown in other studies for TDFs (Neves et al. 2008) and tropical rain forests (Gardner et al. 2008).

Considering the successional patterns observed in the present study, we conclude that in about 20–25 years of TDF regeneration significant restoration of the insect community occurs, confirming the importance of secondary forests for the development of strategies for managing and conserving biodiversity. Nevertheless, there are still striking differences between intermediate and late successional stages, and further studies based on functional groups are required to evaluate the capacity and time necessary for the recovery of integrity of ecosystem functions of TDFs.

13

Arbuscular Mycorrhizal Fungi: Essential Belowground Organisms in Tropical Dry Environments

Ricardo Louro-Berbara, Bruno T. Goto, Camila P. Nobre, Fernanda Covacevich, and Henrique M.A.C. Fonseca

CONTENTS

13.1 Introduction

Many natural systems are being affected by climate change, a process that has been sped up by human beings in recent years (Thomas et al. 2004). Current scientific evidence indicates that global warming is progressing due to an increase in anthropogenic activities; that is, man-made greenhouse gas concentrations, which are mostly due to industrial development and population growth (Mearns et al. 1999). People in South America are heavily dependent on the continent's natural resources (rangelands, plants and animals, fisheries, etc.) as well as on agricultural production (crops, forestry production, and livestock) (Solbrig 2005). Agriculture is the major transforming agent of natural landscapes, and it spreads over an extensive area of the Brazilian and Argentinean territories and remains an important part of the economy in both countries. In Brazil, the expansion of soybean farming is concentrated in Goiás and Mato Grosso, displacing the rich native biota of the savannas and the Amazonian forest (Solbrig 2005). Between May 2000 and August 2006, Brazil lost nearly 150,000 km^2 of Amazonian rainforest (Malhi et al. 2008) that have been devastated by the actions of poor subsistence cultivators, land clearing for pastureland, commercial exploitation of forest resources, and government subsidized agriculture and colonization programs. In Argentina, the original vegetation structure was also displaced by agriculture (Solbrig 2005). In comparison to the rest of the agroecological areas, grain crops had faster expansion in the Rolling Pampas, replacing the pristine natural lands. Since 1990, soybean production has increased by 66% worldwide (from 108 million tons in 1990 to 179 million tons in 2002). The Americas, Brazil, and Argentina account for most of the increase (88%). The major change in the agricultural production of Argentina in the last 20 years has been the dominance of soybean as the principal crop of the Pampa (Solbrig 2005). According to the National Directorate of Forests, Argentina is experiencing the most intense deforestation in its history due to the replacement of its forests with soybean plantations, and Córdoba is the province with the most devastating environmental damage. The high deforestation rate in the country resulted in numerous areas that suffer water shortages due to climate change caused by the felling of the forests (Castrillo Marin 2009).

Studies on how climate is changing, and how these changes differ along latitudes, are much more advanced than studies on how global change will affect plant community structure and how it relates to soil symbionts along different ecosystems. The main reason for this is basically that changes occurring aboveground are easier to measure than those occurring belowground. However, there is now emerging consensus that aboveground and belowground compartments are intimately linked. Thus, any changes occurring aboveground (in the atmosphere) will undoubtedly affect plant, animal, and belowground microbial biodiversity (van der Putten et al. 2009).

Arbuscular mycorrhizal fungi (AMF) are belowground plant symbionts; they are a powerful link in the chain of transfers by which carbon (C) moves from the atmosphere to the plant and finally sinks into the soil (Staddon et al. 2003). Mycorrhizal fungi are able to develop a symbiotic association with most (more than 90%) of the vascular terrestrial plants (Fitter et al. 2004), including most agricultural crops, and usually, their symbiosis with the host plant is not host specific. Moreover, AMF are a diverse taxon, both systematically and functionally, which display abundant ecological differentiation and specialization in both their biotic and abiotic environments, including types that are highly specialized to unusual ecological niches (Fitter et al. 2004). Mycorrhizal fungi form a uniformly distributed mycelium in soil, and hyphal proliferation occurs in response to several types of organic material deposition, near potential host plants, improving physical soil quality. The contribution of AMF to phosphorus nutrition of plants is widely accepted and documented (Covacevich and Echeverría 2009). It is accepted that the symbiosis links the biotic and geochemical components of the ecosystem, enhancing the uptake of not only nutrients with low mobility, like phosphorus, but also ammonium, zinc, and others (Smith and Read 2008). The sustainability of agricultural production is linked to the beneficial effects of nutrition on mycorrhizal plants, particularly with regard to the phosphorus uptake, which is a nonrenewable natural resource. Several plant species respond positively to inoculation with AMF; among them are coffee, soybean, corn, sweet potato, cassava, and sugarcane, in addition to several Argentinean and Brazilian fruit and forest trees. However, the benefits granted by the fungus go beyond increases in mineral nutrition for individual plants.

In addition, AMF provide other benefits to their host plants besides enhancing mineral nutrition as well as stabilizing soil structure; increasing tolerance to water stress, soil salinity and drought, soil compaction, root pathogens and heavy metals, or other toxic substances present in the soil (Covacevich and Echeverría 2009). The high amount of hyphae produced by AMF is correlated with significant increases in the aggregate stability in soils (Jastrow et al. 1998; Treseder and Turner 2007). It modifies the structure of the soil's ability to mobilize nutrients, water content, the penetration of roots, and soil erosion potential. The AMF mycelium also interconnects the root system of neighboring plants of the same species or different species. In this sense, mycorrhizal networks can create indefinitely large numbers of fungal linkages that connect many plants in a community (Giovannetti et al. 2006). This suggests that AMF formation is an important element in the definition of plant succession in ecosystems. Consequently, AMF are increasingly of interest for agriculture, agroforestry, restoration of degraded lands, and conservation of natural ecosystems.

In this chapter, we discuss the mycorrhizal associations in the wider context that goes beyond their impact on plant mineral nutrition, as important as they are, apparently, more complex aspects are still to be revealed. In addition, we preset yet unpublished results that are aimed at evaluating the mycorrhizal potential in soils from different environmental sites from tropical dry forest (TDF) ecosystems. Mycorrhiza–soil interactions can result in the modification of soil structural properties (Rillig and Mummey 2006) as well as the enhanced availability of nutrients. So, we evaluate these interactions and consider their implications for agriculture and ecology in semiarid Caatinga.

13.2 Changing Environment and Mycorrhiza

Soil conditions can affect the activity of microorganisms as well as the activity and interactions among rhizospheric microorganisms can influence soil conditions and, hence, plant growth and microorganisms' activities (Lisboa et al. 2011). Arbuscular mycorrhizal fungi are recognized as the most important and influential soil microbes, because they significantly affect both plants and soil microorganism's growth and activity. Thus, the region around the mycorrhizal roots is called *mycorrhizosphere* (Johansson et al. 2004). The mycorrhizosphere soil is larger than the rhizosphere soil and in such regions, most of the actively absorbing rootlets are connected with the surrounding soil.

Environmental changes such as fertility also alter the community structure of mycorrhizal fungi (Covacevich et al. 2007; Covacevich and Echeverría 2009).

Furthermore, several authors have reported a negative relationship between plant diversity and AM fungal colonization (reviewed by Kernaghan 2005). Changing environments can induce the predominance of intermediate-sized spores, as *Glomus* species, which, in general, modulate the production of spores directly to C availability, showing direct investment in reproduction as expected for r-strategists (Ijdo et al. 2010), as well as reduced abundance and richness species. There is significant evidence pointing to the importance of soil conditions and plant diversity in the control of mycorrhizal fungal communities (Kernaghan 2005).

Several studies indicate that mycorrhizal growth response of AMF-inoculated plants depends on genetic and functional compatibility between plant species and strain of fungus participants of the symbiosis, as well as on prevailing environmental conditions, such as soil type, pH, and nutrient availability, especially P (Covacevich and Echeverría 2009). Besides these variables, in natural conditions where more than one species of fungus colonizes roots of host plants simultaneously, the benefits of mycorrhizae depend on the community of fungi and competition. Mycorrhizal fungi cause impacts ranging from their relationships with plants (mainly for uptake of nutrients), with plant communities (and acting on their diversity and abundance), and, finally, with processes related to ecosystem stability by participating actively and significantly in the dynamics of C and soil aggregation. Thus, perceived not only from the perspective of the plant, but also from the soil in its multiple relationships, mycorrhizal formation is now recognized as a fundamental and integral component in the construction and stability of ecosystems around the world (Rilling 2004).

13.3 Relationship between Mycorrhizal Infection Capacity and Environmental Degradation: Tropical Dry Forests in Brazil

Observations on the patterns of plant succession in semiarid regions reported that AMF play an important role in the composition and stability of plant communities (Rilling 2004). Thus, changes in vegetation as a result of environmental degradation may affect the infectivity of mycorrhizal fungi native ecosystems. However, there are still no conclusive results on the relationship between infectivity and the spatial heterogeneity of environmental degradation products. Mycorrhizal fungi have a widespread presence in all environments, and within tropics. However, little is known about the possible relationship between the infectivity of fungi that are native to the TDFs and those associated with different states of degradation of ecosystems. In a study (Covacevich and Berbara, unpublished data) aimed at evaluating the infectivity of indigenous AMF of environmental sites with different states of degradation, soil samples were collected from the State Park Dry Forest, in the north of Minas Gerais, Brazil (43° 97′02″S-14° 64′ 09″ and 44° 00′ 05″ S-14° 53′ 08″ W). The climate is classified as tropical wet Aw with an annual rainfall of 660 mm and a well marked dry season in winter (rainfall less than 60 mm/month, temperature 24°–26° C). The rains are irregular and concentrated in summer (December–March). Soil was collected in February in areas with different stages of succession as detailed: initial (area with 8 years in spontaneous regeneration process), intermediate (area with 17 years without human activity), late (area without any human activity), and pasture (area with 5 years without human activity, covered by grasses with *Panicum maximum* as the main cover crop). For each area of study, plots of 1000 m^2 were delimited. Soil samples (about 1500 g each) were randomly collected with tree replications/area, and each individual sample was composed by nine subsamples of each area. Field soil was sieved (1 cm); large root fragments were cut (1 cm) and returned to the soil. The soil was air dried in darkness (10°C). Before filling the pots, the soil of each area was mixed with washed–autoclaved (twice) river sand (2:1, soil–sand). Experimental units (80) comprised white plastic pots containing 0.8 kg of soil for the first collected plants (45 days after planting DAP), and 1 kg of soil for the late collected plants (90 DAP). The experiment was set up with six replications: four degradation areas (initial, intermediate, final, and pasture) and two harvested ones (45 DAP, 90 DAP). Plants were grown in a glasshouse under natural light conditions, were daily watered with distilled water, and the substrate was maintained at water-holding capacity (65% w/s). Hoagland Solutions were supplied at 30 and 60 DAP. Shoot dry matter at 45 DAP was higher in intermediate and pasture soils (Figure 13.1). At 90 DAP, plants growing in intermediate soils

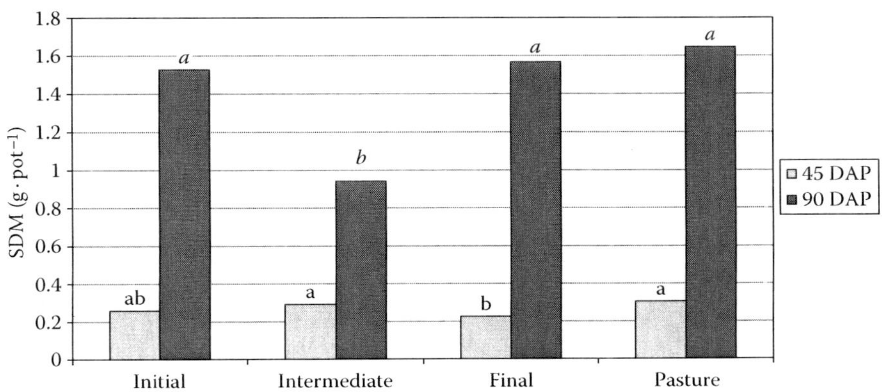

FIGURE 13.1
Shoot dry matter (SDM) production of *Brachiaria* at 45 and 90 days after planting (DAP) grown in soils with different stages of succession. For each harvest date (45 DAP normal font, 90 DAP italics), columns with same letters show no significant differences in SDM between soils (LSD < 0.05).

showed lowest SDM production. This may be the result of the fact that intermediate had available soil P (0.30 mg P kg^{-1}) and Ca (1.96 cmol kg^{-1}) contents that were two or three times lower than in the other areas as well as for presenting acid pH (5.06), which was about 1.5 units lower than in other soils.

For root mycorrhizal colonization and percentage of arbuscules (Figures 13.2 and 13.3), the trend was toward increases in formation of the symbiosis for pastures followed by the initial and final states. Lower formation of symbiosis was detected for plants grown in intermediate soil. This could be probably

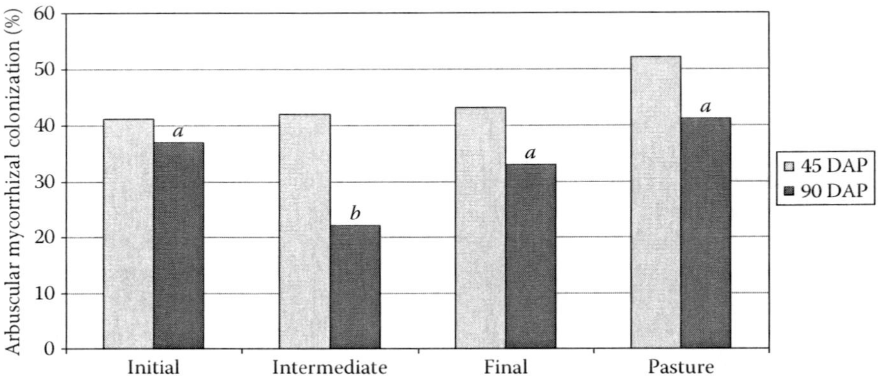

FIGURE 13.2
Arbuscular mycorrhizal colonization of *Brachiaria* at 45 and 90 DAP grown in soils with different stages of succession. Columns with same letters show no significant differences in SDM between soils; at 45 DAP, the data were nonsignificantly different (LSD < 0.05).

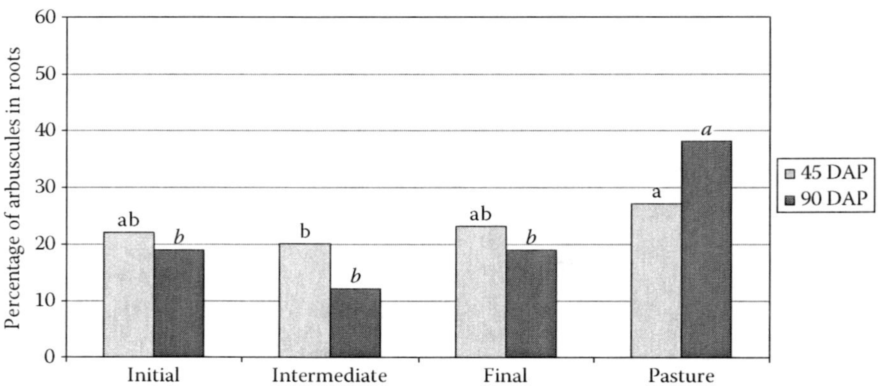

FIGURE 13.3
Arbuscular mycorrhizal colonization of *Brachiaria* at 45 and 90 DAP grown in soils with different stages of succession. For each harvest date (45 DAP normal font, 90 DAP italics), columns with same letters show no significant differences in SDM between soils (LSD < 0.05).

associated with the pH that was more acidic than the other sites. Covacevich et al. (2006) mentioned that soil acidity interferes with the colonization of roots as well as in the absorption of nutrients and, therefore, adversely affects plant growth.

Belowground diversity of AMF is a major factor that contributes to the maintenance of plant biodiversity as well as to ecosystem functioning. Bever et al. (2001) found for the same area of this study that a higher number of mycorrhizal spores was found for the soil from pastures, followed by the initial, intermediate, and late successional stage plots. According to Bever et al. (2001), this would be a sign of a more stable ground state for intermediate and final sites compared with the other two states. Our results agree with those of Bever et al. (2001) in relation to the pastures that also showed higher mycorrhizal colonization (and certainly more active AMF). Carrenho et al. (2001) developed a study of restoration areas along the river Moji-Guaçu, Brazil, from a mixture of pioneer plant species, secondary (early and late) and climax. The study found that the rhizosphere of pioneer species (*Croton urucurana*) had the lowest number of spores, and this increased progressively until the climax plant (*Genipa americana*) was attained. They concluded that the number of AMF spores increases with the stage of succession; whereas the indices of diversity, richness, and evenness of Glomalean species tend to decrease in climax-established communities.

Most ecosystems are subject to periodic natural disturbances. In these cases, ecosystems tend to recover naturally due to environmental adaptation to changes. This adaptation is largely mediated by soil microorganisms whose activity is controlled by the presence of AMF. However, unexpected disruptions may occur in natural sites due to human interventions. They can also accelerate natural processes of disturbance due to climate change, where human activity has an important responsibility.

Certain main factors control the transformation of soil organic matter and nutritional status. In this sense, the monitoring of microbial populations, and in particular mycorrhizal fungi, as well as the physicochemical properties of soil can serve as useful tools in determining that the states of disturbance have different ecological areas. Studies with AMF suggest that the establishment and maintenance of plant diversity in communities relies on the presence of a wide diversity of AMF (Viglizzo et al. 2003). Further, plant productivity is in general greater when a large array of AMF is associated with plants. Presumably, each of the fungi has a degree of specificity with hosts and the environment, resulting in a complex interaction by which the fungal population that contributes to plant growth and development changes over time and space. However, the mechanisms that regulate and maintain biodiversity, AMF, and plant species composition, are not well understood. Studies need to be aimed at identifying the mechanisms involved to ensure successful management for conservation and restoration of diverse natural ecosystems.

13.4 Tropical Dry Forests—Caatinga

The systematic taxonomy of mycorrhizal fungi underwent major changes after the Glomeromycota had been proposed in 2001 (Oehl et al. 2011). Currently, the new phylum presents three classes (*Archaeosporomycetes*, *Glomeromycetes*, and *Paraglomeromycetes*), five orders (*Archaeosporales*, *Diversisporales*, *Glomerales*, *Gigasporales*, and *Paraglomerales*), 15 families, and 31 genera (see Table 13.1), and many of these taxa, especially those

TABLE 13.1

Arbuscular Mycorrhizal Species in the Semiarid Caatinga, Brazil

Gênero	Genus	Total Species	Percentage
Acaulospora	16	18	88.88
Albahypha	0	1	0
Ambispora	3	10	30
Archaeospora	1	3	33.33
Cetraspora	2	7	28.57
Dentiscutata	4	9	44.44
Diversispora	1	14	7.14
Entrophospora	1	1	100
Funneliformis	4	10	40
Fuscutata	1	5	20
Geosiphon	0	1	0
Gigaspora	6	8	75
Glomus	21	79	26.58
Intraornatospora	1	1	100
Intraspora	0	1	0
Kuklospora	0	2	0
Orbispora	1	2	50
Otospora	0	1	0
Pacispora	1	7	14.28
Paradentiscutata	1	2	50
Paraglomus	3	6	50
Quatunica	1	1	100
Racocetra	7	12	58.33
Redeckera	0	1	0
Sacculospora	0	1	0
Scutellospora	3	7	42.85
Septoglomus	1	4	25
Simiglomus	0	1	0
Tricispora	0	1	0
Viscospora	0	1	0

at a higher level (genera to class), have been recently described (Goto et al. 2010a; Oehl et al. 2011). Only in the last 10 years, 68 (6.8 species per year) new species were described, which is equivalent to 27.4% of all known species, significantly expanding the knowledge of the richness of this special fungal group. The majority of the described species in Glomeromycota was from temperate countries (61%) in which diversity is admittedly lower than in tropical areas with 38% of known species. This is basically the distribution of specialists who until recently were only in temperate areas (Americas, United Kingdom, Poland, Canada, Germany, and Switzerland).

According to Blackwell (2011), molecular data support the existence of approximately 5.1 million fungal species. This number is related to the exorbitant amount of sequences in genbank with no corresponding taxa described as yet (Heijden et al. 2008). Of this total, only 100,000 species of fungi are recognized (Kirk et al. 2008), which is a very small fraction of the fungal diversity. This scenario is further aggravated in tropical areas, many still unexplored and even in an advanced process of degradation, where arid and semiarid areas suffer from desertification processes (Mello et al. 2012). Bever et al. (2001) demonstrated that in a hectare that had been abandoned for more than 60 years, and used as a pasture, one out of three of all species were unknown, showing the richness of this group of organisms in temperate areas. Of the 248 known species of Glomeromycota (Table 13.1), most form a symbiotic association (endomycorrhiza) with plant roots, except *Geosiphon piriformis*, which forms a unique association with cyanobacteria of the genus *Nostoc* (Schüssler 2002). Despite this apparent limited wealth in Glomeromycota when compared with other fungal groups, recent estimates support 32,000–78,000 species.

Of the total, 168 currently recognized species are from temperate areas; 107, from tropical areas; and 2, from polar climate areas. In the case of dry forests or regions of semiarid or arid climate, where we can see this type of plant physiognomy, there are limited data, especially those which describe new species (Tchabi et al. 2009; Goto et al. 2010b). Within this scenario, the Caatinga currently owns 36% of all species, with six new species being described from material collected in the areas of dry forest (Goto et al. 2012a). This number is likely to increase shortly with three new species, that is, two new *Acaulospora* and a new *Septoglomus* being published (Goto et al. unpublished data). Dry forest areas may be the cradle of new taxa as the new family (*Intraornatosporaceae*), two new genera (*Intraornatospora* and *Paradentiscutata*), and a new species *P. bahiana* (Goto et al. 2012b). This scenario becomes even more promising when cultivated areas serve as potential for the discovery of new species. During inventory of the diversity of AMF in areas planted with sisal (*Agave sisalana* Perr.) in the Caatinga of Bahia, three new species were found. This shows the potential Glomeromycota diversity in this biome.

The Brazilian semiarid region presents an area of 844,453 km^2, covering 11% of the country with various land-cover types such as Savana, shrubby Caatinga, Caatinga trees, and high land swamps (Sampaio 1995). The name "Caatinga" is of Indian origin (Tupi-Guarani) and means "White Forest," all reference to the appearance of the vegetables in the dry season when plants lose their leaves and get lighter coloration. The Caatinga is a unique biome in the world, and it is limited exclusively to the Brazilian north through rainforests, to the east by the Atlantic Forest, and to the south by Cerrado, forming large areas that are also very rich in biodiversity.

For many years, this biome has been neglected by both public policy and research institutions that saw Caatinga as poor in biodiversity, as 8 of the 12 months of the year are considered severe due to drought (Sampaio 1995). Of the 844,453 km^2, 518,317 km^2 are native forest vegetation and non-forest (equivalent to 62.77% of Caatinga) and 299,616 km^2 of disturbed areas (36.28%), with only 2% of these areas of environmental reserves protected by law, which generally remain exploited or impacted by the resident population. This bleak scenario has changed in the last 20 years, mainly due to public policies that encourage research that has launched edicts series in support of knowledge of biological diversity in the Brazilian semiarid region, such as Millennium Institute of Semiarid (INSEAR) and two editions of Research Program on Biodiversity in Semiarid (PPBio) that have significantly expanded knowledge about the organisms in this biome (Maia et al. 2010).

In the Caatinga, the 31 currently recognized genera, *Intraspora*, *Geosiphon*, *Otospora*, *Tricispora*, *Sacculospora*, *Albahypha*, *Simiglomus*, *Viscospora*, and *Redeckera* have not yet been reported (29%). Actually, 71% of the genera currently accepted occur in areas of Caatinga. In this current scenario, using new taxonomic criteria, areas of dry forest (Caatinga) show wide diversity (Table 13.1). According to Goto et al. (2012a, 2012b), tropical areas may be centers of endemism for some groups of FMA, mainly in order Gigasporales. This has been recently supported with many new species, and genera are described (Goto et al. 2010a, 2012b; Mello et al. 2012).

In total, the Caatinga biome presents 90 species that have been recognized during the last 20 years, equivalent to 36.2% of global diversity. In Brazil, de Souza et al. (2010) present a Brazilian checklist with 119 AM species reported, representing 48% of global diversity found in Brazilian soils. The Caatinga accounts for 75.6% of all Brazilian species. The state of the art in Brazilian semiarid forests has been advanced in the last 10 years, but many new species and new records will be found in this biome in the future.

13.5 Conclusions and Perspectives

Disturbance of soil, plant diversity, and soil conditions are the main contributors to determining fungal diversity and biomass at any one location. Different agricultural practices adversely affect AM functioning in the field (soil tillage, chemical fertilization, biocides, monocropping, and nonmycorrhizal plants). The loss of biodiversity in soils represents a poorly understood field of research that requires more attention. This is especially urgent in TDF regions that tend to be areas associated with high agricultural production. The present reduction in biodiversity on Earth and its potential threat to ecosystem stability and sustainability can only be reversed or stopped if whole ecosystems, including ecosystem components other than plants, are protected and conserved (Heijden et al. 2008). The current production and environmental trajectories indicate that many farming systems still resemble intensive models. In response, we need a new productive and environmental view that replaces the traditional one. This explains why in previous decades, some Brazilian and Argentinean farmers have become increasingly interested in managing their soils more ecologically for minor risks of erosion. Therefore, the selection of appropriate agricultural practices is a must for the enhanced efficiency of AM symbiosis in the field when using biological fertilization.

However, our understanding of the basic biology of the AMF is still limited, and to report the general conclusions about the effects of climate change and human practices on mycorrhizal symbiosis is still arduous. The complexity of the interactions (nutrients, moisture, temperature, plants, fungi, etc.) involved requires interdisciplinary work and coordination to design and implement the necessary measures that are used to reduce the negative effects of climate change and anthropogenic practices on environmental sustainability.

14

An Evaluation of Rainfall Interception in Secondary Tropical Dry Forests

César D. Jiménez-Rodríguez and Julio Calvo-Alvarado

CONTENTS

14.1 Introduction

Evaporation rates can be modified by morphological and structural characteristics of plant species. These effects alter the water amount traveling toward mineral soil (Klaassen et al. 1998). Miralles et al. (2011) remark that the main limiting factors of evaporation from terrestrial surface are energy availability and precipitation volume. Rainfall interception plays an important role in partitioning the rain into water available for plants and water to be evaporated directly from plant surfaces. This process has been widely evaluated under tropical conditions (Tobon-Marin et al. 2000; Arcova et al. 2003; Ferreira et al. 2005; Balieiro et al. 2007; Cuartas et al. 2007; Calvo-Alvarado et al. 2009b, 2012a; Moura et al. 2009; Siles et al. 2010). These studies show the broad quantity of parameters influencing rainfall interception: forest structure, floristic composition, and local and regional climatic characteristics.

Rainfall characteristics (volume and intensity) and monthly distribution of rainfall affect water availability in tropical forest ecosystems. Additionally, the occurrence of a defined dry season is an important ecological driver in tropical dry forests (TDFs), controlling the vegetative dynamics and phenology of plant species (Murphy and Lugo 1986a; Calvo-Alvarado et al. 2012a). Deciduous tree species are adapted to the dry season in TDF reducing water consumption, while the leaf unfolding periods at the end of the dry seasons are influenced by the remaining soil water (Yoshifuji et al. 2006). Rain seasonality alters leaf area dynamics (Maass et al. 1995; Reich 1995), modifying transpiration and evaporation rates within forest ecosystems (Cavelier and Vargas 2002).

Herwitz (1985) emphasizes the importance of floristic and demographic data of plant species in tropical rain forests to accurately estimate rainfall interception. Morphological plant characteristics such as branch positions, leaf types, pubescence presence, and bark tissue influence the water storage capacity at the canopy layer (Herwitz 1985; Llorens and Gallart 2000; Garcia-Estringana et al. 2010). Moreover, as TDF tree species are susceptible to water availability (Balvanera et al. 2011), the throughfall and rainfall interception spatial distributions depend on the presence or absence of plant species. Some plant species such as babassu palm (*Orbignya phalerata*) have the capacity to redistribute the rainfall within their canopy (Germer et al. 2006) on rain forests, expecting a similar effect by dry ecosystem palm species.

Rainfall interception quantification according to plant composition and morphology is essential for hydrological studies. According to Jetten (1996), water flux modeling in forest ecosystems is affected by the canopy saturation storage capacity during small rainfall events, underestimating the total rainfall interception. Hence, the evaporation of intercepted rainfall by the canopy depends on the canopy storage capacity, evaporation rates from plant surfaces, and storage emptying frequency (Gash 1979). Lloyd et al. (1988) modeled these processes under tropical conditions and did not find significant differences between model estimates and measured interception losses. Thus, it is important to emphasize the need for reliable water flux estimations, in which the effects of forest cover type and land-use change are the major factors modifying the net water fluxes within any ecosystem (Bruijnzeel 2004).

Classical models such as that by Rutter (Rutter et al. 1975) require a wide input of data, whereas approaches such as that of Gash (1979) simplify the information to be used in a wide range of environments. Some stochastic models representing the "process from single leaf to a branch segment, and then up to the individual tree level" (Xiao et al. 2000, p. 29, 173) describe the interception process step by step in forest structures. Muzylo et al. (2009) summarized more than 100 physically-based rainfall interception models used worldwide and remarked on the need for more modeling of deciduous forests.

Nowadays, rainfall interception prediction based on satellite information provides realistic values when it is done at landscape level (Nieschulze et al. 2009). Remote sensing in complex tropical regions such as TDFs facilitates the identification of forest types, forest secondary stages, and main characteristics of forest composition and structure (Kalácska et al. 2004b, 2007a; Arroyo-Mora et al. 2005b; Castillo-Núñez et al. 2011; Castillo et al. 2012), enabling a better estimation of land-use change effects on local and regional watershed and forest hydrology.

The northwest Pacific coast of Costa Rica encloses characteristic TDF patches from the Central American Pacific Coast, which are considered to be endangered forest ecosystems (Janzen 1988a). The deforestation processes in Costa Rica between 1940 and 1991 affected the TDFs, reducing the forest cover from 67% to 19% (Sader and Joyce 1988; Sánchez-Azofeifa 2000). The drop in meat prices exerted a strong pressure on the economy after the 1970s, diminishing livestock activities in Costa Rica and increasing land abandonment in Guanacaste, Costa Rica (Sánchez-Azofeifa 2000). This situation provided the required socioeconomic conditions for secondary forest regrowth, reinforced by implemented policies such as the payment of environmental services and the creation of conservation areas (Calvo-Alvarado et al. 2009a).

After the 1980s, the economic development of Guanacaste increased the use of and demand for water, especially in the Tempisque River Basin. Activities such as pisciculture, agro-industry, crop irrigation, and tourism are the main water consumers in this watershed (Guzmán and Calvo-Alvarado 2012). Therefore, an assessment analysis of the impact of land use and forest cover restoration is highly required for this region.

This study quantifies the rainfall interception rates along three successional stages of TDF in the Santa Rosa National Park, Guanacaste. Further, a trial was carried out by Calvo-Alvarado et al. (2012b) in a tropical wet forest near the park. The aim of both studies was to evaluate and calibrate the measurement protocols for forest rainfall interception studies under tropical conditions.

14.2 Methodology

14.2.1 Experimental Design

The study was performed during the rainy season from August to November 2009 at the Santa Rosa National Park (Chapter 1, Figure 1.2). All measurements were carried out on a daily basis from 5:00 to 8:00 AM, following the successional stage order: early, intermediate, and late. The instrumentation established in each plot considered the same variables: rainfall (millimeters per day), throughfall (millimeters per day), and stemflow (millimeters per day).

Early and intermediate rainfalls were collected with a 11.5 cm diameter rain gauge at 1.5 m above the ground. For collecting early-stage rainfall a rain gauge was placed within a plot, whereas intermediate-stage rainfall was collected in an open area 25 m from the plot edge. Due to the absence of a nearby open area, rainfall for the late-stage plot was measured using a funnel of 15.3 cm diameter placed at 1.5 m above the forest canopy. The collected water from the funnel was conducted by a hose to a 5 L container placed at ground level for daily measurement (Figure 14.1).

FIGURE 14.1
Instrumentation employed in the rainfall interception trial conducted at the Santa Rosa National Park: (a) throughfall rain gauge, (b) plastic collector ring and digital meter for stemflow measurements, (c) rain gauge employed for rainfall measurements in early and intermediate stages, and (d) rain gauge employed for rainfall measurements in the late stage.

Underneath the canopy of each plot, 40 rain gauges were placed to measure throughfall. The pluviometers were randomly located within each plot after five rain events. Each plot contains 400 possible locations to place the throughfall collectors. Stemflow measurements were performed in four trees per plot. The selected trees corresponded to the 20, 40, 60, and 80 percentiles of the plot tree diametric distribution. A plastic collar ring was adhered to tree stems and sealed with transparent silicon to prevent water losses. The total water volume collected from each tree was measured with a digital meter (Figure 14.1).

The water density was considered to be 1 $g \cdot mL^{-1}$ in all volume measurements. Rainfall, throughfall, and stemflow were translated from water volume (in milliliters) into water depth (in millimeters) through the collector area of each funnel. Horizontal crown area projection was used to translate the stemflow volume into water depth. The horizontal projection was computed on the basis of six measured radii taken from the stem to the crown edge.

14.3 Data Analysis

Effective rainfall (E_{RF}) (millimeters per day) was computed by adding stemflow and throughfall, whereas the intercepted rainfall percentage was obtained with Equation 14.1. A linear regression was performed between rainfall and throughfall to evaluate the linear dependency of E_{RF} measurements. Analysis of covariance (ANCOVA) was applied ($\alpha = 0.05$) to evaluate the differences between daily interception percentages and successional stages. A Fisher least significant difference (LSD) test was used to determine homogeneous groups when statistical differences were found. All statistical analyses were performed with the software STATISTICA 6.0 (StatSoft Inc. 2003). Rainfall interception (in millimeters per day) percentage was computed as follows:

$$I_i = \frac{1}{n}\sum_{i=1}^{n}\left(\frac{\left(\mathrm{RF}_i - \left(\mathrm{TF}_i + \mathrm{SF}_i\right)\right)}{\mathrm{RF}_i}\right) * 100 \tag{14.1}$$

where I_i is the rainfall interception percentage, RF_i is the rainfall (in millimeters per day), TF_i is the throughfall (in millimeters per day), SF_i is the stemflow (millimeters per day), and n is the total number of registered events.

Due to the lack of instrumentation to measure evapotranspiration on site, we used the Hargreaves equation (Hargreaves et al. 1985) to compute actual evapotranspiration rates (in millimeters per day). The selected equation was modified by Droogers and Allen (2002) and improved by Farmer et al. (2011):

$$\mathrm{ET} = 0.0019 \times (0.408 \times \mathrm{RA}) \times (T + 21.0584) \times (\Delta T - 0.0874 \times P)^{0.6278} \tag{14.2}$$

where ET is the actual evapotranspiration (in millimeters per day), RA is the radiation (megajoules per square meter per day), T is the mean daily temperature (in degree Celsius), ΔT is the daily temperature difference between minimum and maximum temperatures (in degree Celsius), and P is the daily rainfall (in millimeters per day).

The estimation of throughfall entails an error for each individual event. To evaluate the precision of a sampling design, this error was computed for each registered event per successional stage. The analysis was based on the determination of the minimum number of rain gauges required to guarantee errors of 5%, 10%, 15%, and 20%. The analysis used the equations employed by Ortiz and Carrera (2002) to determine the sampling error for each rainfall event (Equation 14.3) and the minimum number of samples required (Equation 14.4).

$$E_{(1-\alpha)} = \mathrm{SE} \times t_{\frac{\alpha}{2}, n-1} \tag{14.3}$$

$$n = \frac{t^2 CV^2}{E\%^2} \tag{14.4}$$

where $E_{(1-\alpha)}$ is the sampling error, SE is the standard error, $t_{\frac{\alpha}{2},n-1}$ is the t value from the Student distribution, n is the total number of samples, CV is the coefficient of variation (in percentage), and $E\%$ is the sampling error percentage with respect to the average value.

14.4 Results

14.4.1 Successional Stage Characteristics

Tree densities (D) registered according to early, intermediate, and late stages are 720, 1990, and 1860 $n \cdot ha^{-1}$ (n is the number of trees), respectively. An increase in diameter implies a reduction in tree density in all the plots. All the trees in each forest stage have diameters below 0.20 m (early), 0.35 m (intermediate), and 0.45 (late). The intermediate stage registered a high abundance of trees between the diameters 0.05 and 0.10 m, whereas the early stage has the lowest tree abundance in all the diametric categories. Table 14.1 shows the increment in structural variables when stage maturity increases. This pattern is depicted for basal area (G), tree densities (D), canopy cover (CC), vegetal area index (VAI), and Holdridge complexity index (HCI).

Table 14.2 describes the important value index (IVI) for each successional stage. *Rhedera trinervis* is the most important species for the early and intermediate stages, in contrast to the late stage, which is dominated by *Sebastiana emerus*. Some species are present in specific stages: *Cochlospermum vitifolium*,

TABLE 14.1

Descriptive Variables for Forest Structure and Floristic Composition of Three Successional Stages of TDFs, Santa Rosa National Park

	Successional Stage		
Variables	**Early**	**Intermediate**	**Late**
Tree density ($n \cdot ha^{-1}$)	720	1990	1860
Basal area ($m^2 \cdot ha^{-1}$)	6.3	23.2	30.6
Species density ($n \cdot ha^{-1}$)	13	17	45
HCI (–)	0.2	15.4	106.58
CC (%)	28.9[a]	88.7[b]	92.0[c]
VAI ($m^2 \cdot m^{-2}$)	1.5[a]	4.3[b]	6.3[c]

Note: Significant differences are shown by different letters in the same row ($p = 0.05$).

TABLE 14.2

IVI Distribution per Species in Three Successional Stages of TDFs, Santa Rosa National Park

Early		Intermediate		Late	
Species	**IVI**	**Species**	**IVI**	**Species**	**IVI**
R. trinervis	120.3	*R. trinervis*	114.9	*Sebastiana emerus*	61.4
C. vitifolium	76.1	*S. mexicanum*	56.0	*Exostema mexicanum*	28.1
B. crassifolia	17.3	*Gliricidia sepium*	17.5	*L. divaricatum*	21.0
A. collinsii	17.0	*Dalbergia retusa*	16.6	*L. candida*	14.1
S. mexicanum	15.3	*B. simarouba*	16.3	*M. sapota*	13.7
C. cujete	14.2	*G. americana*	13.5	*Ardisia pitieri*	13.2
Swietenia microphylla	7.4	*Machaerium biovulatum*	10.3	*S. mexicanum*	11.7
Cordia guanacastensis	7.1	*Lonchocarpus minimiflorus*	9.9	*Eugenia oerstediana*	11.0
M. biovulatum	6.2	*D. salicifolia*	7.7	*Casearia sylvestris*	10.4
Pisonia aculeata	5.2	*Tabebuia ochracea*	7.4	*Astronium graveolens*	8.0
Subtotal 10 species	286.4	Subtotal 10 species	269.9	Subtotal 10 species	192.5
Other species (3)	13.7	Other species (7)	30.1	Other species (35)	107.5

Acacia collinsii, and *Crescentia cujete* in the early stage; *Genipa americana*, *Dyospirus salicifolia* y, and *Ateleia herbert-smithii* in the intermediate stage; and *Sebastiana emerus*, *Exostema mexicanum*, and *Lysiloma divaricatum* in the late stage. *Semialarium mexicanum* is present in all plots and is seen as a dynamic species through all the stages.

14.4.2 Rainfall Interception

Total registered precipitation in 2009 was 1489.4 $mm \cdot year^{-1}$, with a mean annual temperature of 26.8°C ± 1.2°C, in the Santa Rosa National Park meteorological station (Figure 14.2). Rainfall collected during the 59 sampling days was 589.2 mm. This value differs considerably from the collected rainfall at each plot. Plot rainfall differs from the rainfall registered in the meteorological station during the sampling period. The early, intermediate, and late plots collected 44.5%, 53.8%, and 23.5% less water than the meteorological station record, respectively. The reference evapotranspiration (ET) in 2009 was 1024.74 $mm \cdot year^{-1}$. During the sampling period, the evapotranspiration was 157.9, 158.8, and 153.1 mm in early, intermediate, and late stages, respectively (Table 14.3). Figure 14.3 shows the percentage of intercepted rainfall

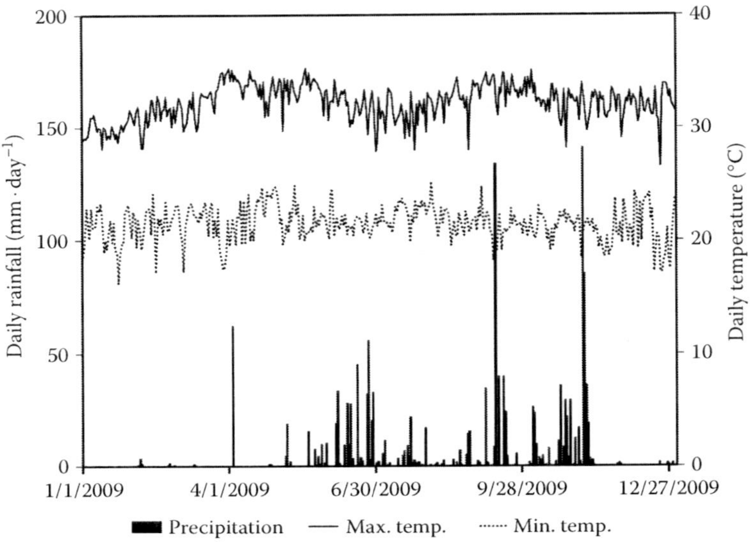

FIGURE 14.2
Daily rainfall and temperature (minimum and maximum) values registered at the Santa Rosa National Park meteorological station (10°50′23″; W 085°37′03″): the period recorded was 2009.

TABLE 14.3

Daily Average Values of Effective and Intercepted Rainfall and Reference Evapotranspiration Computed for Three Successional Stages of TDFs, Santa Rosa National Park

	Rainfall			
Successional Stage	**Intercepted ($mm \cdot day^{-1}$)**	**Effective ($mm \cdot day^{-1}$)**	**Total (mm)**	**Daily ($mm \cdot day^{-1}$)**
Early	0.6[a]	8.5[b]	157.8	2.7
Intermediate	1.7[a]	6.3[a]	158.8	2.7
Late	7.5[b]	6.6[a]	153.1	2.6

Note: Significant differences are shown by different letters in the same column ($p = 0.05$).

and evapotranspiration with respect to daily rainfall amount; both variables decrease as the rainfall amount increases.

In general, forest interception capacity does not imply 100% of rainfall interception in small events (Figure 14.3). Only early and intermediate stages registered events with 100% rainfall interception (one and three events, respectively). Total average rainfall interception differs statistically among stages ($F = 20.7328$; $p = 0.0000$; $\alpha = 0.05$). The late stage intercepted

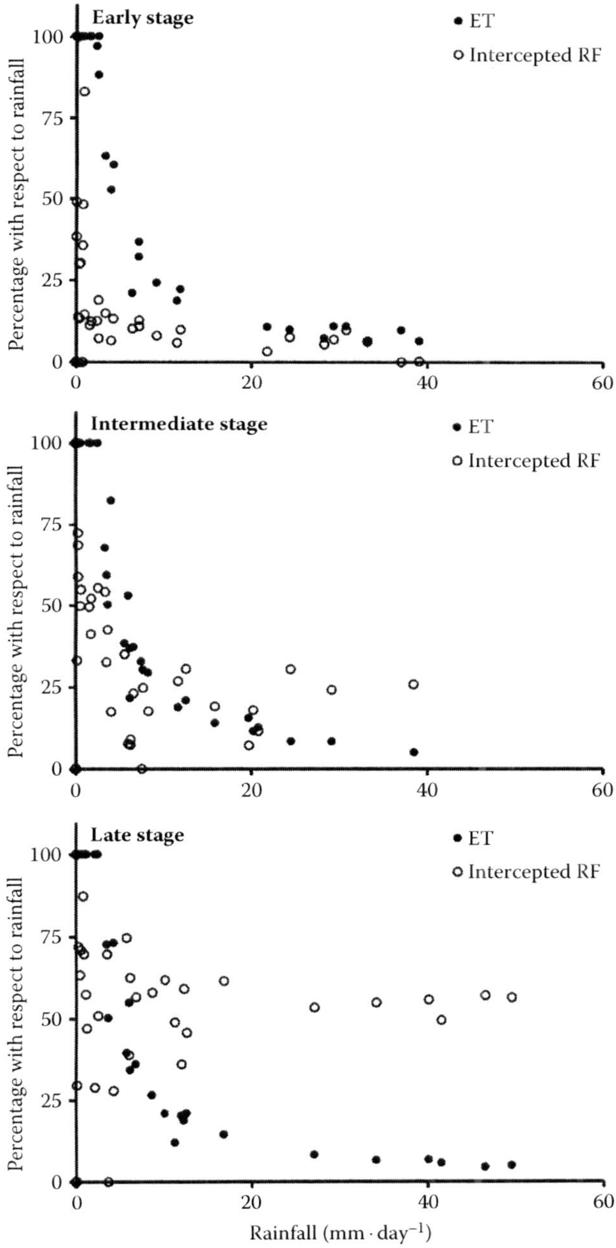

FIGURE 14.3
Rainfall interception (empty circles) and evapotranspiration (filled circles) percentages distribution with respect to rainfall size: data obtained from three successional stages of tropical dry forests (TDFs), Santa Rosa National Park. ET = evapotranspiration; RF = rainfall.

$7.5 \text{ mm} \cdot \text{day}^{-1}$, which is a much higher value than those of early and intermediate stages (0.6 and $1.7 \text{ mm} \cdot \text{day}^{-1}$ respectively), whereas daily evapotranspiration among stages does not differ statistically ($F = 0.071$; $p = 0.9314$; $\alpha = 0.05$) (Table 14.3).

Computed evapotranspiration at each plot does not represent the actual vapor flux to the atmosphere, but it can be considered as the minimal atmospheric vapor requirement for the sites. An adjusted Hargreaves equation (Farmer et al. 2011) was applied with daily rainfall collected at each plot. Daily rainfall interception rates of 0.6 and $1.7 \text{ mm} \cdot \text{day}^{-1}$ in early and intermediate stages, respectively, are below the evapotranspiration computed for each plot ($2.7 \text{ mm} \cdot \text{day}^{-1}$ in both cases). Interception rates in the late stage ($7.5 \text{ mm} \cdot \text{day}^{-1}$) exceed the computed evapotranspiration for this plot ($2.6 \text{ mm} \cdot \text{day}^{-1}$).

Effective rainfall underneath the forest canopy shows significant linear tendencies in all stages (Figure 14.4). This linear tendency is dominated by throughfall, whereas stemflow provides only 1.7%, 2.9%, and 0.43% of registered rainfall in early, intermediate, and late stages, respectively. These fluxes do not modify the total registered interception but produce certain variations in the linear coefficients at each plot (Figure 14.4).

Tobon-Marin et al. (2000), Oliveira Júnior and Dias (2005), and Moura et al. (2009) have used linear regressions, using the equation slope value as a global E_{RF} parameter for the evaluated plot. Computed slopes for early, intermediate, and late stages are 96.2%, 78.2%, and 46.0% of rainfall, respectively (Figure 14.4).

Daily interception values are affected by rainfall amount ($F = 7.9630$; $p = 0.0057$; $\alpha = 0.05$) and differ significantly among successional stages ($F = 26.1035$; $p = 0.0000$; $\alpha = 0.05$). Daily average percentages of rainfall interception are 18.0%, 38.2%, and 54.9% for early, intermediate, and late stages, respectively. Table 14.4 shows the total water fluxes per stage. It is important to note that the highest stemflow value is in the intermediate stage, whereas the late stage has a lower stemflow value.

This study shows that rainfall interception variability is directly related to successional forest structure. The early stage possesses a low water-holding capacity as a result of low CC (28.97%), where $1.54 \text{ m}^2 \cdot \text{m}^{-2}$ of VAI are covered by leaves of mainly *R. trinervis*, *C. vitifolium* y, and *Byrsonima crassifolia*. These leaf types have a low water retention capacity by low surface pubescence. Moreover, stemflow values in this stage are affected by bark type and small crown diameters (≈3.91 m). Dominant tree species such as *R. trinervis*, *B. crassifolia*, and *S. mexicanum* can hold more water in their exfoliating bark than *A. collinsii* or *C. vitifolium*.

In intermediate-stage species with a smooth bark such as *Bursera simarouba* and *G. americana*, water flow is facilitated stemflow, as well as middle crown diameter (≈5.21 m) branches. Even if the canopy layer is dominated by *R. trinervis*, the presence of mildly pubescent species in the canopy layer obstructs water flow toward mineral soil.

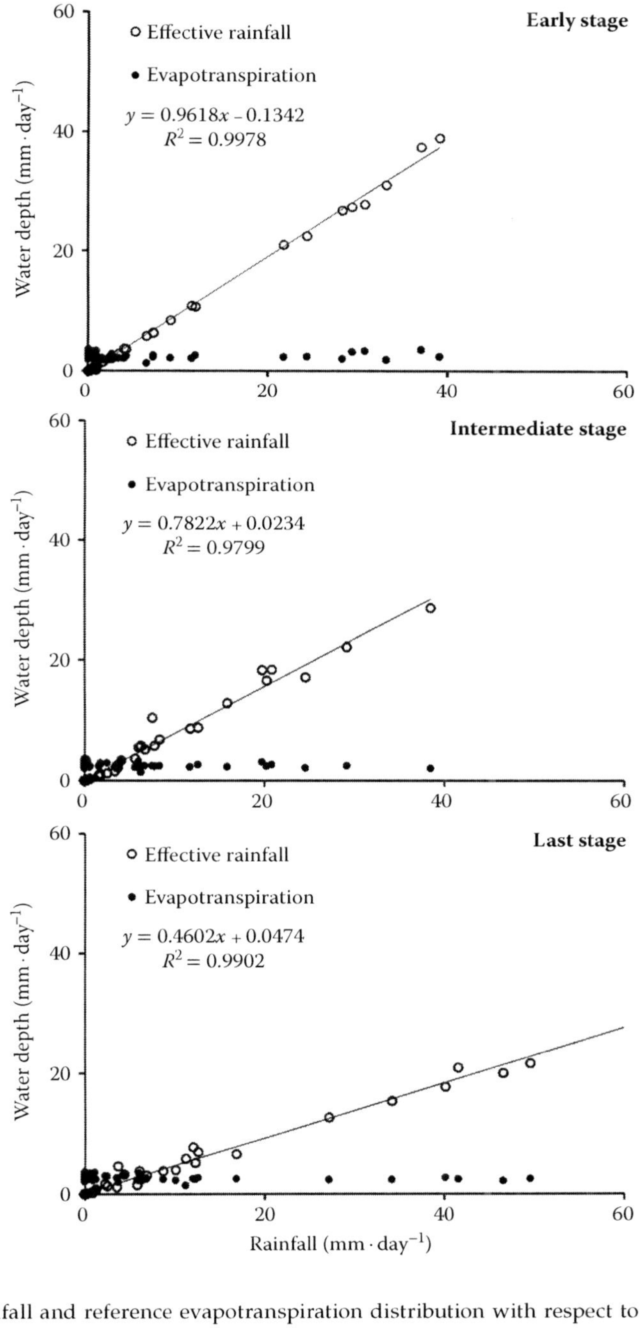

FIGURE 14.4
Effective rainfall and reference evapotranspiration distribution with respect to rainfall size: data obtained from three successional stages of TDFs, Santa Rosa National Park.

TABLE 14.4
Total Water Values of Rainfall, Throughfall, Stemflow, and Intercepted Rainfall Measured in Three Successional Stages of TDFs, Santa Rosa National Park

Successional Stage	Rainfall (mm)	Throughfall (mm)	Stemflow (mm)	Intercepted Rainfall (mm)
Early	327.2	301.2	5.6	20.4
Intermediate	272.1	206.2	8.0	27.9
Late	450.9	208.3	2.0	240.7

High rainfall interception rates in the late stage are related to high canopy leaf density, where VAI reaches the maximum value, 6.3 $m^2 \cdot m^{-2}$, explaining the high interception percentage at this stage. In addition, the presence of exfoliating bark species such as *Luehea candida, L. divaricatum* y, and *Manilkara sapota* influences stemflow values (the lowest among the three successional stages).

The mean average errors in throughfall estimation were 13.9%, 17.1%, and 16.3% for early, intermediate, and late stages, respectively, where the error decreases as the rainfall increases (Figure 14.5). Error reduction is possible by placing more rain gauges below the forest canopy, as shown in Table 14.4. The total number of rain gauges required to reduce error depends on the forest structure; the more complex the forest structure, the greater the number of rain gauges required. Table 14.5 shows the total rain gauge numbers required to reduce the sampling error below 5%, 10%, 15%, and 20%.

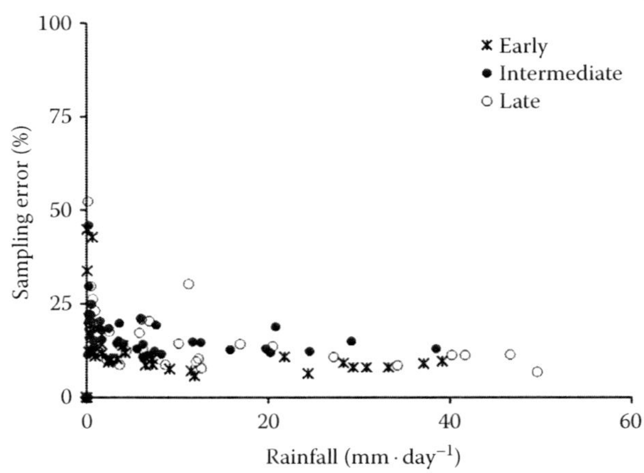

FIGURE 14.5
Sampling errors of throughfall estimation in three successional stages of TDFs, Santa Rosa National Park.

TABLE 14.5

Rain Gauges Required in Throughfall Estimation for Three Successional Stages of TDFs, Santa Rosa National Park

	Early				Intermediate				Late			
Rainfall	**5%**	**10%**	**15%**	**20%**	**5%**	**10%**	**15%**	**20%**	**5%**	**10%**	**15%**	**20%**
0–5	633	158	70	40	711	178	79	44	721	180	80	45
5–10	126	32	14	8	336	84	37	21	441	110	49	28
10–15	67	17	7	4	333	83	37	21	446	112	50	28
20–25	123	31	14	8	254	63	28	16	304	76	34	19
25–30	114	28	13	7	342	85	38	21	240	60	27	15
30–35	101	25	11	6	350	88	39	22	189	47	21	12
35–40	134	34	15	8	264	66	29	17	118	29	13	7
40–45									199	50	22	12
45–50									136	34	15	8
Average	185	46	21	12	370	93	41	23	311	78	35	19

Note: Amounts required to obtain sampling errors of 5%, 10%, 15%, and 20%.

14.5 Discussion

Evaporation from tropical forests represents 53% of rainfall at the global scale (Miralles et al. 2011), and a proportion of this amount is retained by canopies as rainfall interception. Water losses from canopies depend on weather and forest composition and structure (Carlyle-Moses and Gash 2011); as a consequence of this complexity, various authors underline the practical difficulties related to rainfall interception estimation, including Calder et al. (1986); Crockford and Richardson (2000); Marin et al. (2000); Tobon-Marin et al. (2000); Arcova et al. (2003); Ferreira et al. (2005); Fleischbein et al. (2005); Germer et al. (2006); Cuartas et al. (2007); Calvo-Alvarado et al. (2009b); and Holwerda et al. (2010). The studies by these authors provide some suggestions and broad methodological scopes to be employed in tropical regions.

Rainfall interception data for tropical forests contemplate a wide range of ecosystems, but TDFs are characterized by a lack of information, particularly when successional stages are considered. Only a few deciduous forest ecosystems have been evaluated to determine rainfall interception rates. Mediterranean deciduous forests convey 67.1%–71.5% of rainfall toward mineral soil (Šraj et al. 2008), whereas semideciduous forests in Venezuelan tropics intercept up to 30% of rainfall (Córcega Pita and Silva Escobar 2011). Both studies conducted under seasonal rainfall conditions showed that rainfall interception can be higher than 30% of total rainfall. These values represent the annual rainfall interception in some seasonal forests, but as CC varies along time a better quantification during the entire growing season is desired. Seasonal changes in canopy storage of water have been reported in temperate seasonal forests (Link et al. 2004), mixed deciduous woodlands (Herbst et al. 2008), and single beech forests (Staelens et al. 2008), depicting the strong correlation between forest phenology and canopy storage capacity.

Total volumes of small rain events are almost fully intercepted by the canopy layer in all stages; the aforementioned pattern is a direct effect of foliage seasonality (André et al. 2011), whereas the rainfall duration is important in leafless periods in deciduous forests (Mużyło et al. 2011). Additionally, VAI differs as intercepted water does in the three stages analyzed.

Tanaka et al. (2008) reported the daily evapotranspiration values of deciduous ecosystems in Thailand obtained by Attarod et al. (2006) and Toda et al. (2002). Attarod et al. (2006) found for a teak forest estimates with maximum rates of 4.1 $mm \cdot day^{-1}$ during the growing season, whereas in the dry season vapor fluxes reach 2.5 $mm \cdot day^{-1}$. The dipterocarp deciduous forest investigated by Toda et al. (2002) transfers to the atmosphere 0.6–3.3 $mm \cdot day^{-1}$. Young successional stages in the Santa Rosa National Park evaporate less water from the canopy surfaces than the calculated evapotranspiration (early: 0.56 $mm \cdot day^{-1}$; intermediate: 1.7 $mm \cdot day^{-1}$), but in the late stage they are able to intercept up to 7.5 $mm \cdot day^{-1}$, exceeding the computed evapotranspiration of 2.6 $mm \cdot day^{-1}$.

Larger dominant tree species increase vegetal surfaces to retain more water for evaporation (Figure 14.6). These results support the statement of rainfall interception augment with the age increment of secondary tropical wet forests proposed by Calvo-Alvarado et al. (2012b). This statement explain that effective rainfall underneath the canopy decreases as forest structure and species composition increase in the Guanacaste National Park. Both this study and the study by Calvo-Alvarado et al. (2012b) were carried out in Área de Conservación Guanacaste (Guanacaste Conservation Area) and

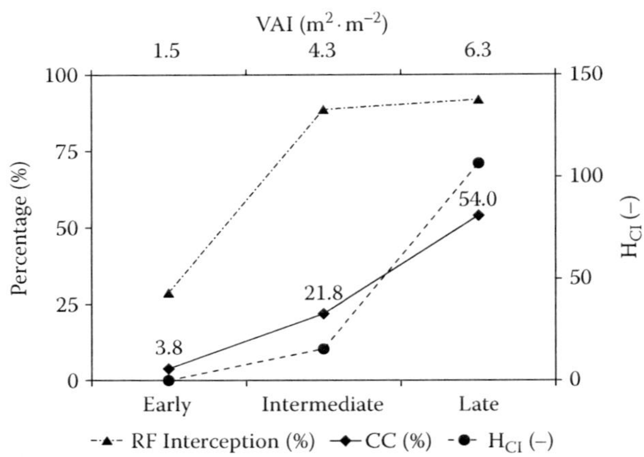

FIGURE 14.6
Rainfall interception percentages and their relationship with forest structure and floristic composition in three successional stages of TDFs, Santa Rosa National Park.

depict the clear influence of forest stage characteristics on rainfall interception rates. This pattern of forest age effect on rainfall interception has also been reported for Douglas fir forests in the Gifford Pinchot National Forest in southern Washington (Pypker et al. 2005); Monterey pine forests in southern Chile (Huber and Iroumé 2001); and lower montane cloud forests in Veracruz, Mexico (Holwerda et al. 2010).

The increment in rainfall interception as successional maturity increases is a direct consequence of forest structure and species composition. Free water movement from the canopy toward mineral soil is obstructed by leaves, branches, understory vegetation, and canopy species. Water movement as stemflow in tropical rain forest ecosystems has been reported with values below 5% of gross rainfall (Tobon-Marin et al. 2000; Cavelier and Vargas 2002; Arcova et al. 2003; Calvo-Alvarado et al. 2009b, 2012c; Friesen et al. 2012), whereas in a TDF in Mexico the stemflow unlikely exceeded 10% of gross rainfall (Kellman and Roulet 1990).

Forest structural variables such as VAI, leaf area index (LAI), and HCI describe forest structure and species composition in a practical numerical way that correlates very well with remote sensing techniques, as demonstrated by Running et al. (1986), Kalacska et al. (2004b, 2005c, 2007a), Arroyo-Mora et al. (2005c), and Chambers et al. (2007). Canopy interception in watershed hydrological models is difficult to estimate due to differences in grid averaging and observations at point scale (Wang et al. 2007), but the correlation found in this study and by Calvo-Alvarado et al. (2012b) between canopy parameters such as LAI or normalized difference vegetation index (NDVI) and canopy water losses will facilitate hydrological modeling using remote sensing.

To obtain better precision in rainfall interception studies, there are two basic options: (1) use high numbers of rain gauges (randomly distributed over the plot area) to determine the throughfall or (2) use fewer rainfall gauges having collectors with larger diameters (Lloyd and Marques 1988; Tobon-Marin et al. 2000; Ruiz-Suescún et al. 2005; Moura et al. 2009). The aforementioned studies do not mention sampling errors on interception, throughfall, or stemflow estimations. In our study, if the rain gauge sampling number is increased to obtain a sampling error lower than 5% it is required to have more than 150 rain gauges, which is unfeasible from the field measurement standpoint. Hence, future studies should use similar sampling sizes but increase the diameter of the collector area if such a precision is desired. The 40 rain gauges used in this study only guaranteed an error of 15% for all daily precipitations higher than 20 $mm \cdot day^{-1}$ at all stages (Table 14.5). The sampling design also guarantees a 10% error for all events with more than 5 $mm \cdot day^{-1}$ in the early stage, which represents more than 90% of the total annual precipitation for this plot. In the intermediate stage the error is 15% for the same data set, whereas the late stage with high canopy heterogeneity registered the highest error estimation (20%). As a consequence, the uncertainties associated with these estimations have to be considered with the aim of improving future hydrological modeling in TDFs.

Sampling errors in rainfall interception estimation can be reduced by employing wider funnels for throughfall monitoring, thus avoiding the inefficiency of field data collection. Digital gauges will facilitate stemflow measurements according to the studies by Calvo-Alvarado et al. (2009a, 2012b). Moreover, the plastic ring collector method employed for the present study is an improvement.

14.6 Conclusions

Long-term research is needed for tropical dry ecosystems to better understand their hydrological dynamics and impact, particularly in dry regions where water resources are the key to socioeconomic development. Knowledge of better water flux estimates related to TDF successional stages will improve watershed hydrological modeling, focusing on daily evaporation, interception, transpiration rates, and runoff estimates. Additionally, canopy storage capacity determinations will improve the cross validation of evaporation losses with seasonal patterns of foliage and rainfall events. Rainfall interception of TDFs in leafed periods increases as stand maturity progresses, incrementing canopy storage capacity. This increment depends on VAI, LAI, and tree branch architecture. Additionally, tree bark characteristics of dominant species vary according to forest successional stages, increasing canopy storage capacity, a variable that is not completely quantified in this study and other studies at the species level.

Secondary TDFs increase the forest structure and floristic composition over time. This pattern directly affects water losses from rainfall interception. Some authors describe the importance of this seasonal effect of deciduous forests on rainfall interception, underlining the importance of differentiating between leafed and leafless periods. In this study, we indexed the successional stage effect on forest structure and composition by computing HCI, VAI, and CC, variables that can be classified using remote sensing and be correlated very well with rainfall interception, helping the use of hydrological modeling based on remote sensing land-use input data.

This study had showed that less complex stages such as early forests intercept around 18.0% of gross rainfall and convey 1.7% as stemflow. On the other hand, mature forests such as late stage have the capacity to retain more than 50% of gross rainfall in the canopy. Old tree morphologies make water movement difficult in the trunks in the late stage, flushing less than 1% of gross rainfall. Stand structure in the intermediate stage intercepts 38.2% and drains around 3% of gross rainfall as stemflow. This high value in comparison to early and late stages is a consequence of the increased presence of smooth tree barks, such as *G. americana* and *B. simarouba*.

Inexplicably, the bulk of rainfall interception literature does not report the sampling errors and the statistical exercise required to judge the precision of the results. This study, as well as those conducted by Calvo-Alvarado et al. (2012b) and Carvajal-Vanegas and Calvo-Alvarado (2012), has been driving this effort to improve and calibrate the methodology to estimate rainfall interception under tropical conditions. As demonstrated by our analysis, 40 rain gauges per plot provides a throughfall estimation with an error of 20% for all the stages, which in terms of hydrological studies of this nature can be considered acceptable due to the lack of experience and comparable data. A reduction of this error can be achieved only by placing wider funnels or increasing the sample size.

15

Linkages among Ecosystem Structure, Composition, and Leaf Area Index along a Tropical Dry Forest Chronosequence in Mexico

Yingduan Huang, Arturo Sánchez-Azofeifa, Benoit Rivard, and Mauricio Quesada

CONTENTS

15.1 Introduction

As they are disturbed frequently and severely by human activities and economic pressures (Janzen 1986, 1988a; Mooney et al. 1995; Kalácska et al. 2004b; Quesada and Stoner 2004; Brooks et al. 2009), tropical dry forests (TDFs) are considered to be the most endangered major tropical ecosystem (Janzen 1988a; Portillo-Quintero and Sánchez-Azofeifa 2010). Land cover change in TDFs in the Americas is generally driven by expansion on the agricultural frontier, extensive and intensive cattle ranching, and tourism development (Portillo-Quintero and Sánchez-Azofeifa 2010). Although the conservation of TDFs has been advocated (Sánchez-Azofeifa et al. 2005a-c), little has been done to this end and only a handful of national parks and biological reserves exist in the Americas to counteract fast deforestation processes (Portillo-Quintero and Sánchez-Azofeifa 2010).

In TDFs, phenological characteristics such as leaf shedding, shooting, and flowering present a higher degree of variability than in other tropical ecosystems (Borchert 1994a,b; Burnham 1997). It has been recognized that the principal element controlling the phenology of TDFs is soil moisture availability (Longman and Jenik 1974; Doley 1981; Reich and Borchert 1984), which is highly related to the seasonal variation in rainfall (Schimper 1898; Bullock and Solis-Magallanes 1990). The phenological expression caused by temporal changes of the aforementioned environmental factors can be quantified by temporal changes in leaf area index (LAI). LAI, defined as the total one-sided area of photosynthetic tissue per unit ground surface area (Watson [1947] as cited by Jonckheere et al. [2004]), has been widely used as a convenient way to quantify leaf phenology, gross primary productivity, and carbon and water cycle processes (Chen et al. 1991, 2002; Maass et al. 1995).

Estimation of LAI can be direct or indirect (Jonckheere et al. 2004). Indirect optical techniques are preferred because they are relatively rapid and accurate (Dufrene and Breda 1995). Hemispherical canopy camera and LAI-2000 are two optical instruments that estimate LAI by measuring the canopy gap fraction—the fraction of a field of view that is not obstructed by canopy (Welles and Cohen 1996). Hemispherical canopy camera captures upward-looking photographs through a fish-eye lens positioned beneath the canopy, whereas a LAI-2000 measures the canopy gap fraction through the ratio between the diffuse radiation measured below and that above the canopy (Jonckheere et al. 2004). However, optical instruments are not able to distinguish foliage from trunks and branches (Kucharik et al. 1998) and estimate the overall contribution of leaves and woody components, which is defined as the plant area index (PAI) (Chen et al. 1991). The contribution of woody components is defined as the woody area index (WAI) (Chen et al. 1997). LAI is calculated as follows (Chen et al. 1997; Frazer et al. 2000; Leblanc and Chen 2001; Sánchez-Azofeifa et al. 2009b):

$$\mathrm{LAI} = L_t\left(1-\mathrm{WAI}\right) = L_e\frac{\left(1-\mathrm{WAI}\right)}{\Omega}$$

where L_t is the total PAI; WAI is the ratio of wooded area; L_e is the effective PAI obtained from optical methods (Chen et al. 1991); and Ω is the clumping factor, which is significant for nonrandom foliage such as conifers but can be assumed to be 1 in many studies in tropical environments for practical reasons (Clark et al. 2008; Sánchez-Azofeifa et al. 2009b).

Remote sensing techniques have also been used as tools for indirect LAI estimations from satellites (Chen et al. 2002; Myneni et al. 2002; Cohen et al. 2003; Eklundh et al. 2003; Meroni et al. 2004). The moderate-resolution imaging spectroradiometer (MODIS) provides a 1-km moderate spatial resolution and eight-day temporal resolution LAI product to monitor phenology (Huemmrich et al. 2005). Many researchers have utilized MODIS LAI data in their studies (Fensholt et al. 2004; Wang et al. 2004; Wang et al. 2005; Ahl et al. 2006). Studies report discrepancies between LAI estimates obtained from MODIS and in situ measurements (Cohen et al. 2003; Fensholt et al. 2004; Wang et al. 2004; Ahl et al. 2006). Spatial scale is a factor leading to the discrepancies (Ahl et al. 2006). One MODIS pixel at a 1 km spatial resolution may contain fractions of different land cover types (Tian et al. 2002) that combine to form the spectral reflectance of the pixel (Peddle et al. 1999), affecting the MODIS LAI. Other factors include temporal compositing of MODIS LAI and understory vegetation that could explain the discrepancies between MODIS LAI and in situ LAI (Ahl et al. 2006).

Although many studies have aimed at developing indirect techniques to quantify LAI, most have been conducted for crops (Demarez et al. 2008), temperate forests (Spanner et al. 1990; Gower and Norman 1991; Macfarlane et al. 2007; Sonnentag et al. 2007; Behera et al. 2010), and boreal forests (Deblonde et al. 1994; Chen et al. 1997). For example, Demarez et al. (2008) estimated the LAI of crops in France with hemispherical photographs. Gower and Norman (1991), Behera et al. (2010), and Spanner et al. (1990) estimated LAIs in temperate forests, the first two studies focusing on the use of the LAI-2000 Plant Canopy Analyzer. Sonnentag et al. (2007) applied both a LAI-2000 and a destructive sampling method to measure LAI in a temperate shrub canopy in Canada. Macfarlane et al. (2007) conducted LAI estimation with hemispherical photography in a temperate forest in western Australia. As to studies in boreal forests, Barr et al. (2004) measured LAI with a LAI-2000 in a boreal forest in Canada.

Only a few estimations have been made for tropical forests (Asner et al. 2003). Kalácska et al. (2005b) used both an optical method and litterfall data to estimate seasonal LAIs in different successional stages in a TDF in Costa Rica. Maass et al. (1995) also used both methods to assess LAIs during a three year period in a TDF in west Mexico, whereas Asner et al. (2003) synthesized LAI measurements from more than 1000 published estimates in 15 biomes including arid, temperate, tropical, and boreal ecosystems.

The main objectives of this study were to estimate seasonal changes in LAI in a TDF in Mexico, evaluate the impacts that ecosystem structure and composition have on seasonal changes of LAI by comparing the LAI values obtained from different TDF successional stages, and evaluate the ability of the current MODIS algorithm to estimate LAIs in TDFs.

15.2 Methods

15.2.1 Plot Design

A total of nine plots representing three successional stages (three plots per successional stage) were selected at the Chamela–Cuixmala Biosphere Reserve, Jalisco, Mexico (Chapter 1). The plots were characterized as early (3–5 years old), intermediate (8–12 years old), and late (mature forests) successional stages (Table 15.1). The only sites characterized as old growth, Tejon 1 and Tejon 2, were located inside the Chamela–Cuixmala Biological Station. An inventory of all species with a diameter at breast height (DBH) higher than 0.10 m was compiled with the help of an expert taxonomist.

Spatial autocorrelation of the plots was evaluated using the Ripley's *K* function. This function allowed an assessment of the random spatial distribution of the selected nine sites at different scales. The nine sites were randomly distributed for 100 simulated scales within 99.9% confidence intervals (Kalácska et al. 2005b).

TABLE 15.1

Main Characteristics of Plots Located at the Chamela–Cuixmala Biosphere Reserve

Name	Stage	Latitude, Longitude	Average DBH (cm)	Average Canopy Height (m)	Elevation (masl)	Land-Use History
Caiman 3–5	Early	19° 28′ 40.44″ N, 104° 56′ 5.64″ W	1.91	6	75	Burned in 2001
Ranchitos 3–5	Early	19° 36′ 51.12″ N, 105° 1′ 14.88″ W	1.82	6	195	Burned in 2002
Santa Cruz 3–5	Early	19° 36′ 2.52″ N, 105° 2′ 36.24″ W	2.47	6	108	Burned in 2001
Caiman 8–12	Intermediate	19° 28′ 3.72″ N, 104° 56′ 13.20″ W	3.29	10	49	Burned in 1997
Ranchitos 8–12	Intermediate	19° 35′ 32.28″ N, 105° 0′ 32.76″ W	4.10	10	154	Burned in 1995
Santa Cruz 8–12	Intermediate	19° 35′ 55.32″ N, 105° 2′ 54.6″ W	3.94	10	110	Burned in 1994
Tejon 1	Late	19° 30′ 5.04″ N, 105° 2′ 35.88″ W	4.27	10	71	Undisturbed
Tejon 2	Late	19° 30′ 33.48″ N, 105° 2′ 24.72″ W	4.86	10	78	Undisturbed
Gargollo	Late	19°24′18.36″ N, 104° 58′ 57.72″ W	5.01	20	23	Undisturbed

Three 30 m × 60 m plots were set up for each successional stage. The plot size was the same as that used by Kalácska et al. (2005b) in the same region and designed to allow comparison with their work.

15.2.2 Field Data Collection and Analysis

Data collection was synchronized with the leaf-off leaf-on phenology (Figure 15.1) of the region. WAI information was acquired using a Nikon CoolPix 995 camera with a hemispherical fish-eye lens during the June 2005 dry season when all trees were 100% deciduous. WAI was extracted from color hemispherical photographs by calculating the gap fraction using the software Gap Light Analyzer 2.0 (Simon Fraser University, British Columbia, Canada).

PAI measurements were conducted using a LAI-2000 Canopy Analyzer (Gower and Norman 1991) in July 2005 when the growing season started and in August 2005 when the leaves were fully developed. Measurements were carried under diffuse light conditions, either at twilight or under cloudy conditions, to reduce the radiation scattered by the foliage, which overestimates below-canopy readings, resulting in the underestimation of LAI. A 45°-view cap was used to eliminate the influence of the operator from the field of view. Readings above the canopy were also acquired in canopy gaps close to each one of the plots to calculate PAI. Measurement of WAI and diffuse radiation below the canopy for the calculation of PAI followed a triangular offset-grid sampling scheme developed specifically for the optical

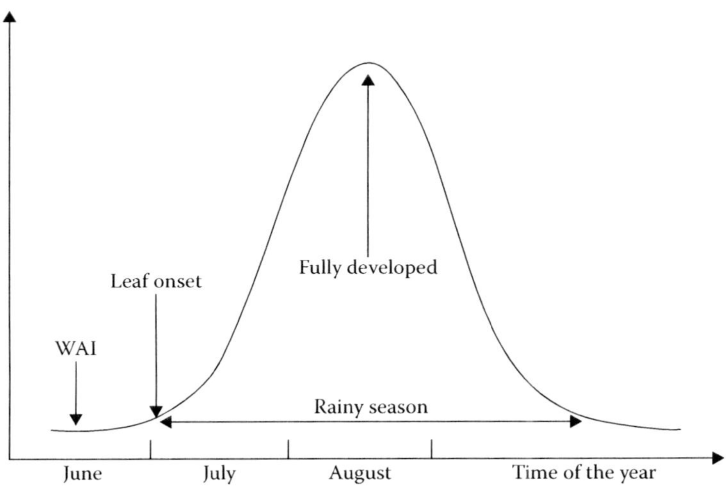

FIGURE 15.1
Illustration of leaves development from the dry season to the rainy season at the Chamela–Cuixmala Biosphere Reserve region.

instrument LICOR-2000 by Kalácska et al. (2005b). PAI was estimated using the following equation (Jonckheere et al. 2004):

$$L = -2\sum_{i} \ln[T_i]\cos\vartheta_i W_i$$

where L is PAI; i varies from 1 to 5, representing the five distinct angular bands of the fish-eye light sensor; T is the light transmission; ϑ is the zenith angle; and W_i are the weighting factors for the five angular bands, which are 0.034, 0.104, 0.160, 0.218, and 0.484 (Jonckheere et al. 2004). When a canopy is clumped, a clumping factor Ω resulting from the nonrandom distribution of foliage is considered (Nilson 1971; Leblanc and Chen 2001; Sánchez-Azofeifa et al. 2009). As Ω is generally unknown, Chen (1991) introduced the term effective PAI (L_e), which is the PAI obtained from optical methods:

$$L_e = \Omega L$$

Ordinary kriging (Luna et al. 2006) was applied to spatially interpolate effective PAI and WAI values. This application was conducted in the software ArcGIS 9.0. Continuous surfaces of PAI or WAI were created for the plots. By subtracting WAI from PAI, the effective LAI was obtained.

Normality of the data was tested using the Kolmogorov–Smirnov test (Lilliefors 1967). A paired t-test was applied to find whether differences existed between LAI measured in July and that in August for each plot, LAI in July and that in August for each successional stage, and LAIs obtained from all plots for the three successional stages in July and in August. A one-way analysis of variance (ANOVA) and an independent t-test were conducted to compare WAI, PAI, and LAI between the three successional stages.

The Morisita–Horn index, which takes into account the differences in total basal area per species among sites, was calculated to examine the differences in species composition that could affect LAI values across sites. The Morisita–Horn index ranges from 0 to 1, with 0 and 1 indicating no similarity and total similarity, respectively, between each pair of sites.

15.2.3 MODIS Data Extraction and Analysis

MODIS provides an eight-day composite LAI product at a spatial resolution of 1 km. MODIS LAI was derived from MODIS spectral reflectance through a three-dimensional radiative transfer model (main algorithm) that provides a good-quality LAI product (Tian et al. 2000). If the main algorithm fails, a backup algorithm is applied to derive the LAI based on the relation between normalized difference vegetation index and LAI (Aragão et al. 2005). To obtain MODIS LAI values for comparison with ground LAI measurements, MODIS LAI images were selected that encompassed the locations and dates of ground measurements. LAI values were extracted from the nine pixels where the nine plots were located and compared with ground LAI values derived using the main algorithm.

15.3 Results

15.3.1 Species Composition and Leaf Area Index

Our results indicate that on average the similarity in species composition was highest within the late stage (Morisita–Horn index = 0.69), moderate within the early stage (Morisita–Horn index = 0.45), and very low within the intermediate stage (Morisita–Horn index = 0.10). Species composition greatly differed between the successional stages, with the Morisita–Horn index varying between 0.09 and 0.19. The largest compositional difference was found between the intermediate and early stages (0.09).

In the early stage, the three sites were moderately similar in species composition with a Morisita–Horn index ranging between 0.35 and 0.64. Similarity was very low in the intermediate stage with a Morisita–Horn index ranging between 0.04 and 0.17. Caiman 8–12 and Ranchitos 8–12 were found to be most different in composition within the intermediate stage (0.04). The three sites in the late stage showed compositional similarity (Morisita–Horn index >0.5) with the highest Morisita–Horn index value of 0.85 between Tejon 1 and Tejon 2, two sites inside the Chamela Biological Station. *Apoplanesia paniculata Presl.* was the dominant species of both sites. LAI values for these two sites were also similar. The comparison within each successional stage indicated that Caiman 3–5 in the early stage, Caiman 8–12 in the intermediate stage, and Gargollo in the late stage had higher effective LAI values. Of these three plots, both Caiman 3–5 and Gargollo had *Piptadenia constricta (Micheli) Macbr.* as the dominant species.

15.3.2 Estimation of Woody Area Index and Plant Area Index

WAI was significantly different between the three successional stages ($F = 48$, $p < 0.0001$). Throughout the comparison, the late stage was found to have the largest amount of woody material followed by the intermediate stage and then the early stage (Figure 15.2).

The comparison of effective PAI values across successional stages indicated that significant difference existed in both July ($F = 37$, $p < 0.0001$) and August ($F = 58$, $p < 0.0001$) (Figure 15.3). In both months, PAI for the intermediate stage was significantly higher than that of the early stage and late stage ($p < 0.0001$). PAI of the late stage was higher than that of the early stage in July ($p < 0.001$) and in August ($p < 0.01$).

For each successional stage, PAI varied between the three plots (Table 15.2). For the early stage, PAI of Caiman 3–5 was higher than that of Santa Cruz 3–5 and Ranchitos 3–5. For the intermediate stage, Caiman 8–12 had a higher PAI than Santa Cruz 8–12 and Ranchitos 8–12. In the late stage, Gargollo had

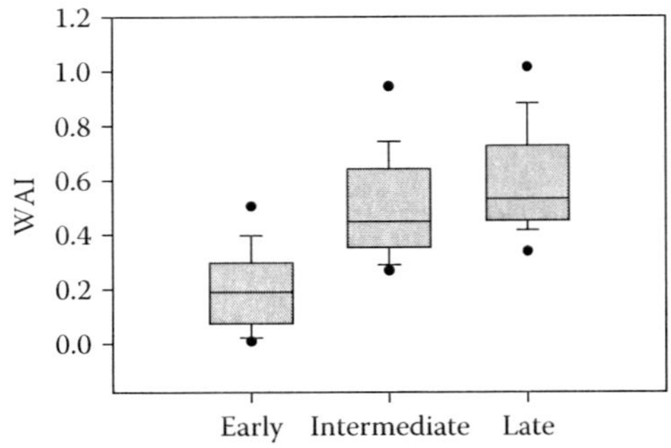

FIGURE 15.2
Comparison of woody area index (WAI) of the three successional stages: each stage contains 36 measurements from three plots. Each box shows the median of WAI values of each stage, and the 25th and 75th percentiles. Black dots are outliers above the 95th and below the 5th percentiles.

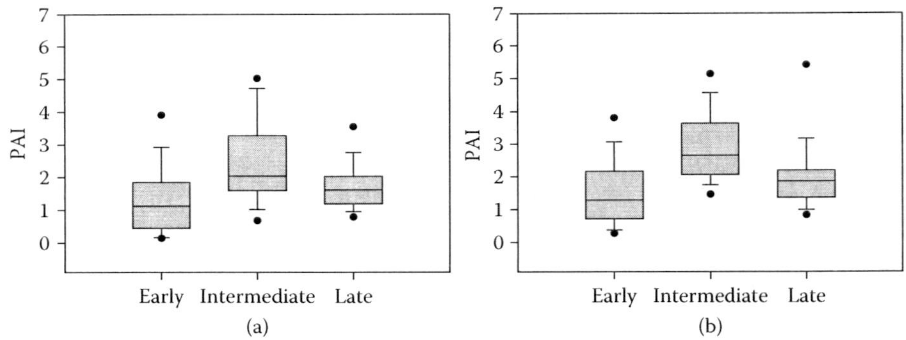

FIGURE 15.3
Comparison of plant area index (PAI) values across successional stages: the intermediate stage had the highest PAI values in July (a) and in August (b) followed by the late stage. The early stage has the lower values in both months. Black dots are outliers above the 95th and below the 5th percentiles.

a higher PAI than the two plots inside the Chamela Biological Station, of which the PAI of Tejon 2 was higher.

15.3.3 Estimation of Leaf Area Index and Seasonality

LAI values were found to vary in a similar manner to PAI values at both the stage level and the plot level (Figure 15.4) Overall, a significant difference was observed between the successional stages in July ($F = 19$, $p < 0.0001$) and in August ($F = 34$, $p < 0.0001$). In July, the intermediate stage had a higher LAI value than the late stage ($p < 0.001$) and the early stage ($p < 0.0001$). In August, the LAI value of the intermediate stage was significantly greater than the LAI value of the late stage and that of the early stage (both p values < 0.0001).

TABLE 15.2

WAI, LAI, and Dominant Species at Sampling Plots Located at the Chamela–Cuixmala Biosphere Reserve

Stage of Succession	Site Name	WAI (June)	LAI (July)	LAI (August)	Family Name	Dominant Species	Dominant Species in the Plot (%)
Early	Caiman 3–5	0.12	1.88	2.17	Leguminosae	*P. constricta (Micheli) Macbr.*	20.39
	Ranchitos 3–5	0.14	0.62	0.74	Polygonaceae	*Coccoloba liebmanii Lindau*	27.12
	Santa Cruz 3–5	0.32	1	1.16	Leguminosae	*Acacia farnesiana (L.) Willd.*	27.21
Intermediate	Caiman 8–12	0.45	3.51	3.64	Euphorbiaceae	*Cnidosculus spinosus Lundell.*	27.90
	Ranchitos 8–12	0.63	0.89	1.21	Euphorbiaceae	*Croton roxanae Croizat*	15.73
	Santa Cruz 8–12	0.47	1.18	2.1	Leguminosae	*P. constricta (Micheli) Macbr.*	12.52
Late	Tejon 1	0.47	0.94	1.14	Leguminosae	*A. paniculata Presl.*	18.13
					Euphorbiaceae	*C. roxanae Croizat*	18.13
	Tejon 2	0.57	1.13	1.22	Leguminosae	*A. paniculata Presl.*	14.22
	Gargollo	0.69	2.75	3.88	Leguminosae	*P. constricta (Micheli) Macbr.*	16.03

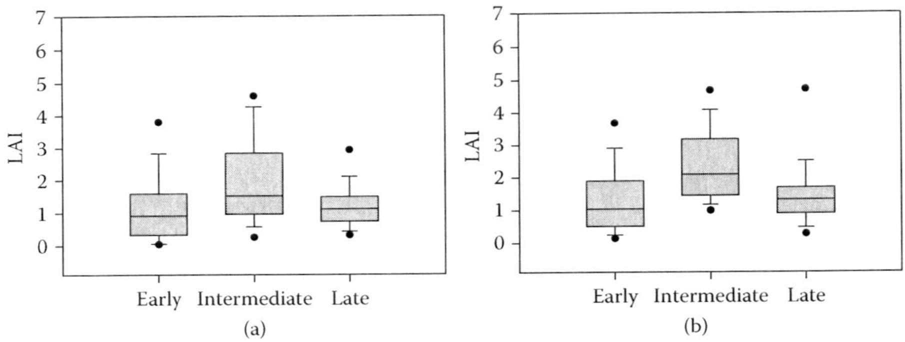

FIGURE 15.4
Comparison of leaf area index (LAI) values across successional stages: patterns are similar to those of PAI after correction for WAI. (a) July collection, (b) August correction.

At the plot level (Figure 15.5; Table 15.2), the highest LAI values were observed for Caiman 3–5 of the early stage, Caiman 8–12 of the intermediate stage, and Gargollo of the late stage when compared with the other two plots within the successional stage. For the early stage, LAI values were ranked in the order of Caiman 3–5 > Santa Cruz 3–5 > Ranchitos 3–5; for the intermediate stage, the order was Caiman 8–12 > Santa Cruz 8–12 > Ranchitos 8–12; and for the late stage, LAI was ranked as Gargollo > Tejon 2 > Tejon 1. The plot-level patterns were observed for both July and August. Since the nine LAI values in July and those in August are not normally distributed ($p < 0.05$), a Spearman rank order correlation was applied to examine the correlation between LAI values and the elevation of the nine plots. LAI values were found to be inversely correlated with the elevation of the nine plots with a correlation coefficient of −0.767 ($p < 0.05$) in July and −0.667 ($p < 0.05$) in August. Segregating LAIs by successional stages makes the linear relation between these two variables more significant with R^2 values ranging from 0.78 to near 1. Murphy and Lugo (1986a,b) found that leaf size decreased with increasing altitude for TDFs, which could be a cause for the lower LAI values observed in plots at higher altitude.

One-way ANOVA was applied to compare the LAIs of the three plots in each successional stage between July and August as a function of elevation. A significant difference was found between plots in the early and intermediate stages ($p < 0.0001$). The difference between plots in the late stage was not significant ($p = 0.28$) because the two plots reside at a similar altitude within the Chamela Biological Station—71 m for Tejon 1 and 78 m for Tejon 2 (Table 15.1).

LAI values showed an increase from July to August for the nine plots (Table 15.2). Results of the *t*-test proved that a significant increase occurred in the LAI values from July to August for most of the plots ($p < 0.05$), except the intermediate plot Caiman 8–12 (increased from 3.51 to 3.64) and the late plot Tejon 2 (increased from 1.13 to 1.22).

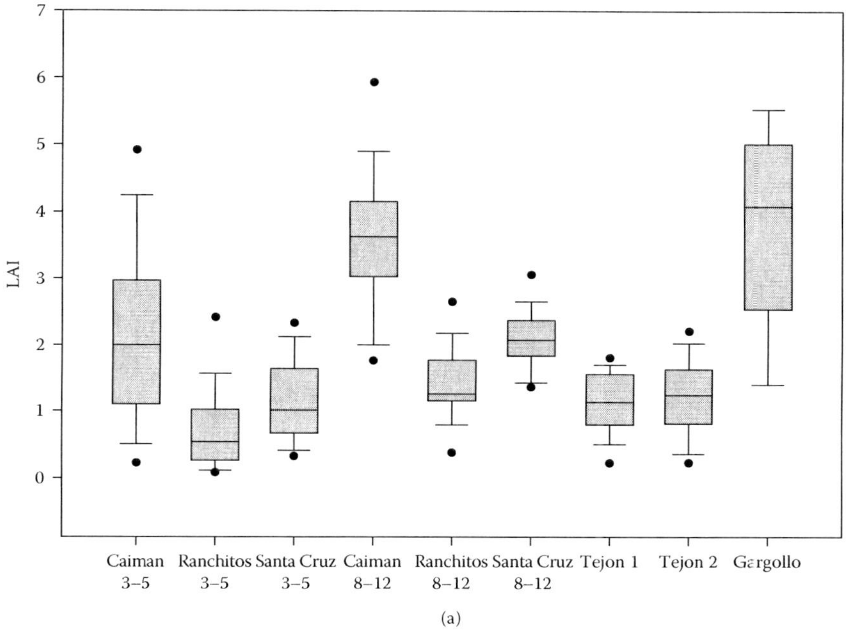

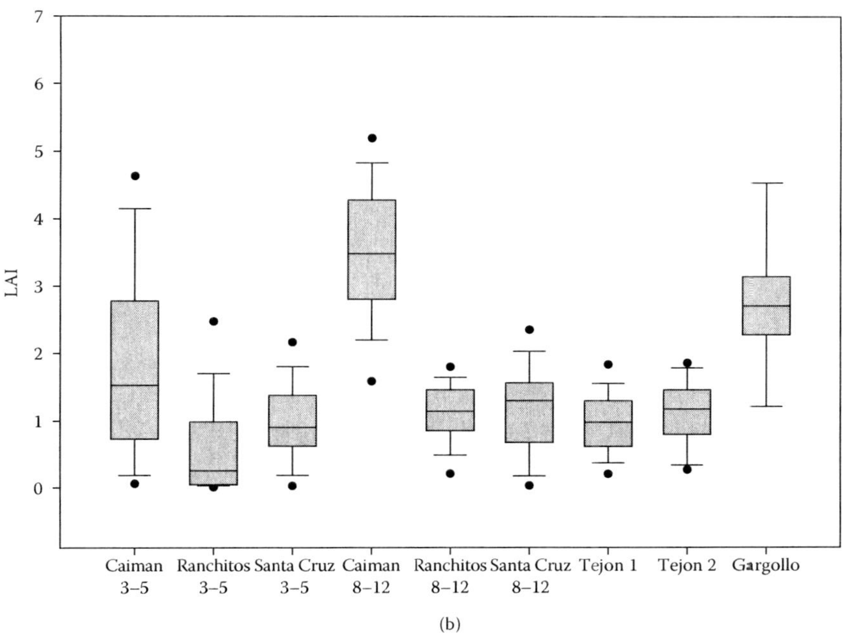

FIGURE 15.5
Comparison of LAI values of the nine plots in (a) July and (b) August: black dots are outliers above the 95th and below the 5th percentiles. Difference can be found between plots in each stage of succession.

TABLE 15.3

Comparison between In Situ LAI and MODIS LAI at the Chamela–Cuixmala Biosphere Reserve

		LAI in July		LAI in August		
Stage of Succession	**Site Name**	**In Situ LAI**	**MODIS LAI**	**In Situ LAI**	**MODIS LAI**	**MODIS Land Cover**
Early	Caiman 3–5	1.88	3.8	2.17	5.3	Savannas
	Ranchitos 3–5	0.62	1.1	0.74	2.1	Mixed forests
	Santa Cruz 3–5	1.00	2.3	1.16	2.1	Woody savannas
Intermediate	Caiman 8–12	3.51	3.7	3.64	6.1	Woody savannas
	Ranchitos 8–12	0.89	1.1	1.21	1.8	Woody savannas
	Santa Cruz 8–12	1.18	1.1	2.10	2.0	Woody savannas
Late	Tejon 1	0.94	1.6	1.14	1.8	Woody savannas
	Tejon 2	1.13	1.2	1.22	1.7	Deciduous broadleaf forests
	Gargollo	2.75	2.2	3.88	0.4	Mixed forests

15.3.4 MODIS Leaf Area Index

MODIS LAI pixel values were extracted and compared with in situ LAI values (Table 15.3). MODIS LAI values were higher than field measurements in July and August except for two sites—Santa Cruz 8–12 and Gargollo. The discrepancy was largest in August, which ranged between −3.48 and 3.13, whereas in July it ranged between −0.55 and 1.92. A linear relationship with an R^2 of 0.55 was found between MODIS LAI and in situ LAI for July. The relationship for August was very poor with $R^2 < 0.1$.

15.4 Discussion

15.4.1 Leaf Area Index and Plant Area Index Patterns

The results of this study indicated that at the Chamela–Cuixmala Biosphere Reserve, the intermediate stage of the TDF had the highest effective PAI and effective LAI values among the three successional stages, followed by the late stage and then the early stage. These comparative results of effective LAI between successional stages are consistent with results presented by Kalácska et al. (2005b), who found the highest effective LAI in the intermediate stage for a TDF during the wet season (September) in Santa Rosa, Costa Rica. The difference in effective LAI is expected because the forest structure and species composition are different across successional stages. Our results also agree with those of Reed et al. (1999) who found a positive relation between LAI and sapling density. According to the census of our nine plots, the intermediate stage has the largest number of stems >1 cm (2812 stems in total), followed by the late stage (2480 stems in total) and then the early stage (1848 stems in total).

We observed a significant increase in LAI across all successional stages, which is controlled mostly by changes in soil moisture regimes. Seasonal changes of LAI in deciduous TDFs are driven by soil moisture related to precipitation (Murphy and Lugo 1986a,b; Maass et al. 1995). Precipitation is strongly seasonal in the Chamela–Cuixmala Biosphere Reserve (Bullock 1986; Maass et al. 1995; Segura et al. 2003). Our records indicated that precipitation measured at the Chamela Biological Station increased from 69.60 mm in July to 127.76 mm in August, which explains the seasonal increase in LAI from July to August found in this study.

15.4.2 Sources of Error in Field Measurements

Estimates of LAI in this study can be impacted by errors in PAI and WAI measurements. The first source of error is clumping, a problem common to all estimates derived from optical instruments since it is assumed that leaves are randomly distributed in space (Dufrene and Breda 1995). Clumping would result in the underestimation of PAI and WAI. To date, the clumping problem has not been satisfactorily solved (Jonckheere et al. 2004). A second error consideration is concerned with the WAI measured in the dry season that includes the contribution of leaves from some evergreens within the plots. This contribution was subtracted from PAI, leading to an underestimation of LAI. Finally, the LAI-2000 aims to minimize the contribution of multiple scattering inside the canopy by collecting diffuse sky radiation at wavelengths below 490 nm where the foliage reflects or transmits little radiation. However, the foliage is not perfectly "black" and scattering still occurs, which caused higher below-canopy readings and, in turn, lower PAI values.

15.4.3 Comparison between MODIS Leaf Area Index and In Situ Leaf Area Index

MODIS LAI was found to overestimate LAI values when compared against field data. Douglas et al. (2006) presented similar results showing MODIS-LAI overestimates during the onset of greenness and maturity. The presence of understory vegetation could be a reason explaining the discrepancy between MODIS LAI and in situ LAI (Ahl et al. 2006). In this study, the greatest discrepancy was found in the early succession stage where canopies at the study site are more open (Sánchez-Azofeifa et al. 2009b).

Other factors leading to the difference between MODIS LAI and in situ LAI include the relatively coarse spatial resolution and temporal resolution of MODIS LAI (Ahl et al. 2006). One MODIS pixel at 1-km spatial resolution may contain several land cover types (Tian et al. 2002). All land cover types in a pixel contribute to the spectral reflectance (Peddle et al. 1999) that is used to derive MODIS LAI (Tian et al. 2000). MODIS land cover misclassification in this area (Table 15.3) indicates that the spectral reflectance of a MODIS pixel is affected by fractions of land cover types. Most of the pixels where the nine plots were located were misclassified by MODIS as savannas. The pixels associated with two plots were classified as mixed forests. Only the pixel where Tejon 2 was located was classified as deciduous broadleaf forests, and better agreement was found between MODIS LAI and field LAI. Lastly, the eight-day temporal resolution of MODIS LAI may not be sufficient to detect phenological changes during the onset of greenness. Ahl et al. (2006) suggested leaf expansion could be better captured with a temporal resolution <1 week.

The overestimation of LAI can also be partially attributed to the MODIS LAI derivation by the backup algorithm and the presence of clouds or aerosols. Among the 18 MODIS LAI values, 4 were retrieved using the backup algorithm. The main algorithm retrieved the remaining 14 values with the presence of some clouds. Partial cloud cover is difficult to avoid during the rainy season and, although the data quality was satisfactory, the conditions were not ideal.

15.5 Final Remarks

LAI plays an important role in calculating carbon, water, and energy exchanges between terrestrial environments and the atmosphere (Sellers et al. 1997) and thus contributes to global climate studies (Myneni et al. 1997). The data and results of this study are important not only in characterizing TDFs but also in developing conservation policies and estimating payments for environmental services for the Chamela–Cuixmala Biosphere Reserve, which also applies to other forest ecosystems (Kalácska et al. 2008).

The examination of the relation between MODIS LAI and in situ LAI in this study did not yield satisfactory results. The underestimation of in situ LAI by optical instruments and the uncertainty of MODIS LAI data (Foody and Atkinson 2003) are two main factors leading to the discrepancy between MODIS and in situ LAI. This study suggests that MODIS LAI data need further validation with field measurements prior to large-scale applications. Deriving LAI from high-resolution remotely sensed imagery and using such estimates to link MODIS and ground LAI may represent an avenue for further studies.

16

Seed Germination of Arboreal–Shrub Species with Different Dispersal Mechanisms in a Brazilian Tropical Dry Forest

Giovana Rodrigues da Luz and Yule Roberta Ferreira-Nunes

CONTENTS

16.1 Introduction

In tropical dry forests (TDFs), anemochory is usually reported as the most common seed dispersal syndrome (Bullock 1995; Nunes et al. 2012). The period of dispersal for these wind-dispersed seeds is the dry season, whereas that for zoochorous plants is usually the rainy season (Lima et al. 2008). Seeds that are dispersed during the dry season and at the beginning of the rainy season germinate immediately after the first rains (Garwood 1983). This strategy seems to be an adaptive characteristic of some plants to maximize the efficiency of water use for seedling establishment (Garwood 1983). However, some seeds are not able to germinate, even when exposed to favorable environmental conditions, due to several factors, such as the tegument impermeability to water and oxygen (Labouriau 1983), immature or undeveloped embryos, special requirements of light or temperature conditions, and the presence of inhibitors of growth, among others (Baskin and Baskin 1998).

Dormancy, which is defined as the inability to germinate even under favorable conditions (Popinigis 1977), is a process that prevents all seeds from germinating simultaneously, increasing their chance of survival and decreasing their risk of species extinction (Carvalho and Nakagawa 2000). Dormant species usually form seed banks in the soil, where the seeds remain until environmental conditions enable seedling establishment (Alves et al. 2006). According to Thompson and Grime (1979), seeds forming the seed bank are small. Conversely, large seeds are usually dispersed by animals and, thus, may be more exposed to predation. However, having a bigger size is advantageous, as it ensures adequate nutrient supply during seedling establishment (Antunes et al. 1998). Therefore, seed size evolved under opposing selective forces (dispersal vs. establishment) (Antunes et al. 1998). In general, anemochoric species produce small, abundant seeds that are dispersed at a particular time of the year; whereas zoochorous species, which use fruits as dispersal units, produce relatively heavy seeds in low quantities that are dispersed over longer periods (Antunes et al. 1998). According to some authors (Antunes et al. 1998; Barbosa 2003), species with autochorous and zoochorous dispersal (dispersal by gravity) present characteristics that are related to the formation of permanent seed banks, whereas anemochorous seeds usually do not form seed banks.

Seeds buried in the soil are exposed to deterioration due to rainfall and high temperatures (Araújo-Neto et al. 2005). The ultimate result of seed deterioration is the loss in ability to germinate (Carneiro and Aguiar 1993). However, some studies have shown that soil-stored seeds can remain viable for long periods and that viability varies among species (Egley and Chandler 1983; Teketay and Granström 1997). Moreover, abiotic factors such as depth, temperature (Egley and Chandler 1983), and soil water potential (Evans and Etherington 1990) can also affect the seed lifespan. On the forest floor, seed dormancy may be reduced by the action of fungi and bacteria that degrade the seed tegument (Fowler and Bianchetti 2000). Laboratory techniques try to simulate such interactions, using a variety of approaches, such as chemical scarification, by using hydrochloric acid or sulfuric acid; thermal scarification by using hot water (70° or 80°C) (Fowler and Bianchetti 2000); and mechanical scarification by using sandpaper, which aims to remove part of or all the tegument of the seed, allowing the entry of water and oxygen (Zaidan and Barbedo 2004). However, the application and efficiency of each method depends on the type and level of dormancy, which varies among species (Alves et al. 2006).

Most studies that focus on plant reproductive processes associate the mechanism of dormancy and storability of the seeds with the characteristics of the ecological groups (pioneer, intermediate, and climax) to which they belong (Cheke et al. 1979; Longhi et al. 2005). Conversely, few studies have investigated the relationship between dormancy, viability, storability, and the process of seed dispersal (see Murali 1997; Araújo-Neto et al. 2005; Vieira et al. 2008). Furthermore, increasing the knowledge of seed dormancy in TDFs and of seed storability in forest soils is extremely important to improve our understanding of ecological processes and to develop conservation, management, and restoration techniques in these environments (Smiderle and Souza 2003; Nunes et al. 2006).

This study aimed to determine whether TDF species that belong to distinct dispersal guilds have different seed germination characteristics. For this purpose, during 4 months, we used different scarification treatments (mechanical, thermal, and chemical) and seed storage types in both a natural environment (forest) and the laboratory. We tested the following hypotheses: (1) seeds of anemochorous species do not present dormancy and germinate faster (higher germination speed) than zoochorous and autochorous seeds that present dormancy and (2) seeds of anemochorous species lose viability faster when stored in the soil than seeds that are primarily (zoochorous) or secondarily (autochorous seeds) dispersed by animals.

16.2 Methods

16.2.1 Study Area

The study was conducted at Mata Seca State Park (MSSP), located in Manga, north of the state of Minas Gerais, Brazil, between coordinates 14°56′38″ S 43°59′11″ W and 14°47′42″ S 44°05′46″ W. The park was created in the year 2000 under the responsibility of the State Forestry Institute of Minas Gerais (IEF 2000) and currently has an area of 15,360 ha (unpublished data). The vegetation is composed of different formations, including evergreen forests in the floodplains of the São Francisco River; semideciduous forests at a higher elevation along the water courses; rupestrian vegetation on lithic and calcareous soils; and deciduous forests on litholic, podzolic, latosol, and cambisol soils (IEF 2000). Among these, seasonally dry deciduous forests (TDFs) can be found at different successional stages and are defined on the basis of forest structure (Madeira et al. 2009). According to the climate classification of Thornthwaite (Thornthwaite 1948), the climate in the study area is semiarid with two marked seasons during the year: a dry season from June to August (Nimer and Brandão 1989) and a rainy season between October and March, with November, December, and January being the wettest months (Antunes 1994). Annual precipitation varies from 733 mm to 1305 mm, and average temperature ranges between 16.8°C and 26.2°C.

16.2.2 Species Studied

Eight species were selected for the experiment on seed dormancy: four zoochorous, two anemochorous, and two autochorous species. However, of these eight species, only five were used to determine germination behavior after soil storage: one anemochorous, one autochorous, and three zoochorous species. Anemochorous seeds are those with wings, plumes, or other means for wind dispersal; whereas autochorous ones are those which are dispersed by free fall or by the explosion of fruits that open abruptly (Vicente et al. 2003). In addition, zoochorous species are those whose primary vector for dispersal is an animal (van der Pijl 1982). The characterization of seed dispersal syndromes into zoochorous, anemochorous, and autochorous species was performed using information from the literature (Britton and Rose 1963; Barbosa et al. 2003; Maia 2004; Lucena 2007; Pereira et al. 2008).

Thus, the species used in the dormancy test were as follows: *Spondias tuberosa* Arruda, *Commiphora leptophleos* (Mart.) J. B. Gillet, *Pereskia stenantha* F. Ritter, and *Pilosocereus pachycladus* Ritter (zoochorous); *Mimosa hostilis Senna spectabilis* (DC.) H.S. Irwin & Barneby & Barneby (autochorous); and *Pseudopiptadenia contorta* (Benth.) and *Myracrodruon urundeuva* Allemão (anemochorous) (Figure 16.1). The species used in the seed storage test were *S. spectabilis* (autochorous), *M. urundeuva* (anemochorous), and three zoochorous species: *P. stenantha, P. pachycladus*, and *S. tuberosa*. Voucher specimens were deposited in the Montes Claros Herbarium (HMC) of the Universidade Estadual de Montes Claros (UNIMONTES), Brazil.

16.2.2.1 Zoochorous Species

16.2.2.1.1 Pilosocereus Pachycladus Ritter (Cactaceae)

This cactus species is found in Brazil from Ceará to Minas Gerais in seasonal dry forests and the Caatinga biome (Oliveira-Filho 2006). *Pilosocereus pachycladus* individuals produce flowers and fruits throughout the year with a peak of fruit production in the rainy season (Lucena 2007). This

FIGURE 16.1
(See color insert.) Details of fruits and seeds, respectively, of *Spondias tuberosa* (a and b), *Commiphora leptophleos* (c and d), *Pereskia stenantha* (e and f), *Pilosocereus pachycladus* (g and h), *Mimosa hostilis* (i and j), *Senna spectabilis* (k and l), *Pseudopiptadenia contorta* (m and n), and details of diaspores in *Myracrodruon urundeuva* (o and p). (Picture: Giovana Rodrigues da Luz)

pattern results in continuous food availability for frugivores, especially birds and bats, which are the main dispersers of Cactaceae species (Lucena 2007). There is no information available on seed germination ecology of this species.

16.2.2.1.2 *Pereskia Stenantha F. Ritter*

This shrub or small tree species, 2–4 m in height, is found in the Caatinga biome on the valley of the São Francisco River, in western, central, and southern Bahia; and in central and northern Minas Gerais (Taylor and Zappi 2004). It is found at elevations below 800 m and is associated with the occurrence of calcareous soils (Zappi 2008). Individuals of this species have rounded areoles that produce straight spines, which increase in number and size in older branches. The leaves vary from obovate to elliptic, and the flowers can be grouped in terminal or solitary inflorescences with orange-red colors (Taylor and Zappi 2004). The fruits (7.3 × 2.6 cm), pyriform or turbinate, may be found as infructescences or solitary fruits, with approximately 30 obovoid seeds (4.5–5.5 mm × 3.2–3.6 mm) (Taylor and Zappi 2004). There is no information available on the seed germination ecology of this species.

16.2.2.1.3 *Commiphora Leptophleos (Mart.) J. B. Gillet (Burseraceae)*

This tree species, 6–9 m in height, is found in calcareous, well-drained, and deep soils (Lorenzi 1998; Maia 2004). The stem of this species is smooth and reddish, and its branches are filled with thorns. Its flowers are small, light green, separate, or in small groups (Lorenzi 1998). Its fruits are green drupaceous capsules that burst open, leaving a single seed exposed and covered by a red aryl in its base, which is dispersed by birds (Maia 2004; Barbosa et al. 2003). Flower production occurs in November and December, along with the emergence of new foliage; fruits mature in March and April, with the onset of leaf fall (Lorenzi 1998; Maia 2004). Seed storage viability is usually short, and radicle emergence occurs after a few weeks, with germination rates below 50% (Lorenzi 1998).

16.2.2.1.4 *Spondias Tuberosa Arruda (Anacardiaceae)*

This species is endemic to the Caatinga; it is 4–7 m in height, has a very short stem covered with smooth bark, a diameter from 40 to 60 cm, a low canopy with profuse branching, is apparently disordered, and has leaves that are composed of three to seven membranous leaflets (Lorenzi 1992). The root system has water and starch storage organs, which provide resistance to prolonged periods of drought (Lorenzi 1992). Fruits are glabrous or slightly hairy drupes, round, 2–4 cm in diameter, and dispersed by animals during the rainy season. Animals consume the fruit and then expel its core, as it is not digested by the digestive tract (Barbosa et al. 2003; Maia 2004). Seed viability after storage at room temperature is high, and percentage germination decreases after 60 months of storage (Cavalcanti et al. 2006). Germination is usually slow and asynchronous (Costa et al. 2001), with radicle emergence after 10–45 days, which may vary according to the scarification method and storage time (Araújo et al. 2001).

16.2.2.2 Anemochorous Species

16.2.2.2.1 *Myracrodruon Urundeuva Allemão (Anacardiaceae)*

This species is a deciduous and xerophytic tree of considerable economic value, mainly due to the durability of its wood (Lorenzi 1992). It often occurs in calcareous soils, and in dry and rocky areas that are typical of dry regions such as the Caatinga and Cerrado biomes (Lorenzi 1992). In the north of Minas Gerais, flowering occurs in July and August, and fruiting occurs from July to November (Nunes et al. 2008). This species has a globular or ovoid drupaceous fruit seed, with a persistent calyx, that is usually dispersed by wind (Barbosa et al. 2003). Seeds are short-lived and deteriorate rapidly, because they are oleaginous (Carneiro and Aguiar 1993; Teófilo et al. 2004). Seed dormancy is not present, and high germination rates and germination speed are recorded (Nunes et al. 2008). Due to indiscriminate deforestation in recent decades, this species was included in the list of endangered species (MMA 2008).

16.2.2.2.2 *Pseudopiptadenia Contorta (Benth.) (Fabaceae-Mimosoideae)*

This species occurs in Atlantic forest formations (Oliveira-Filho 2006). It has inflorescences with gray-pubescent flowers, and follicle fruits, which are straight or curved and are bent in a semicircle with elliptical or oblong, winged seeds (Morim and Barroso 2007) that are dispersed by wind (Pereira et al. 2008). Flowering occurs in November and December, and fruiting takes place from April to October (Pereira et al. 2008). There are no data on the germination rates for this species.

16.2.2.3 Autochorous Species

16.2.2.3.1 *Senna spectabilis DC. H. S. Irwin & Barneby (Fabaceae-Caesalpinoideae)*

This tree species is 6–9 m in height, and it has pinnate compound leaves with 10–20 pairs of leaflets (Lorenzi 1992). It occurs mostly in deep, well-drained, and reasonably fertile soils (Maia 2004). It has large yellow flowers in terminal panicles, which are produced in the rainy season from December to April; whereas the first fruits appear when the tree is still in bloom and mature in the dry season (August–September) (Barbosa et al. 2003; Maia 2004). The fruits are indehiscent legumes, black or dark brown, with large amounts (13–20) of small autochorous seeds (Barbosa et al. 2003; Maia 2004). Seed viability can last longer than 6 months, and radicle emergence occurs after 10–30 days; germination rates are usually lower than 30% (Lorenzi 1992). Therefore, scarification treatments may increase germination rates (Lorenzi 1992).

16.2.2.3.2 *Mimosa Hostilis Benth. (Fabaceae-Mimosoideae)*

This species is frequent in Caatinga formations, 4–6 m in height, with the presence of thorns on the stem and a thin and irregular crown (Lorenzi 1998). Leaves are compound bipinnate; flowers are white, very small, and are arranged in isolated spikes; and fruits are in the form of a small seedcase, with a delayed dehiscence, and with 4–6 small autochorous seeds (Barbosa et al. 2003; Maia 2004). This species is found in secondary floodplain formations with high soil moisture, and deep, alkaline, fertile soils (Lorenzi 1998). The species contributes to the recovery of nitrogen content in the soil, preparing the soil for the appearance of more demanding plant species such as *M. urundeuva* (Maia 2004). Flower production occurs predominantly from September to January, and fruits can be found mainly from February to April (Maia 2004). Radicle emergence takes place between two and four weeks, and germination rates are usually higher in scarified seeds (Lorenzi 1998; Maia 2004).

16.2.3 Seed Collection and Processing

The seeds used for germination tests were collected from 10 to 20 trees of each species. Fruits were collected from May 2008 to June 2009, according to the fruiting period of each species (Table 16.1) and maturity predictors such

TABLE 16.1

Fruit and Seed Collection Dates, Number of Mother Trees Sampled, and Seed Dormancy and Storage Experiment Start Dates from Arboreal Species in Mata Seca State Park, Southeast Brazil

Species	Collection Date	Mother Trees	Fruits Collected	Dormancy Experiment	Storage Experiment	Voucher (HMC)
Commiphora leptophleos	March 2009	15	900	March 2009	–	442
Pereskia stenantha	May 2008	20	100	May 2008	July 2008	443
Pilosocereus pachycladus	February 2009	10	50	February 2009	April 2009	444
Spondias tuberosa	February 2009	17	3250	February 2009	April 2009	445
Myracrodruon urundeuva	August 2008	20	3250	August 2008	October 2008	446
Pseudopiptadenia contorta	June 2009	17	700	July 2009	–	447
Mimosa Hostilis	July 2008	13	700	August 2008	–	448
Senna Spectabilis	July 2008	17	400	August 2008	September 2008	449

Note: Voucher (HMC) = voucher number in the Montes Claros Herbarium.

as changes in fruit color, fall, or beginning of dehiscence. The fruits were placed in paper bags, identified, and taken to the Laboratory of Ecology and Plant Propagation of UNIMONTES (Laboratório de Ecologia e Propagação Vegetal), and, subsequently, a screening was held to select the seeds that would be used. The seeds were separated into homogeneous lots according to color and size, and those with a different appearance from that of normal seeds, attacked by pathogens and predators or with no apparent embryo development (aborted), were eliminated (Nunes et al. 2006). The time between fruit collection and setting up the germination experiment never exceeded 15 days.

16.2.4 Overcoming Dormancy

To check for the presence of seed dormancy, the following procedures were performed: (1) chemical scarification using hydrochloric acid; (2) thermal scarification using hot water at 70° C; and (3) mechanical scarification using sandpaper (n° 80). The three seed treatments were compared with (4) untreated seeds (control). Seeds placed in hot water at 70°C remained under these conditions until water temperature reached 50°C, which took approximately 30 minutes. In the case of mechanical scarification, seeds were manually rubbed with sandpaper until visible tegument wear was observed at the opposite side of the micropyle (Nunes et al. 2006). For chemical scarification, seeds were immersed in hydrochloric acid for five minutes with pH between 3 and 4 to simulate scarification of zoochorous seeds as they passed through the stomach when they were ingested by the dispersers (Bocchese et al. 2007). However, seeds from the three syndromes studied, and not only zoochorous seeds, were used in this treatment.

The seeds were placed in Petri dishes (9 cm in diameter), except the seeds of *S. tuberosa*, which were placed in gerbox boxes (11 cm × 11 cm × 3.5 cm) due to their large size. Seeds were homogenously distributed on filter paper with neutral pH and labeled. Then, 20 mL of distilled water were added to each sample. All equipment used in the experiment was sterilized and cleaned using a detergent, immersed in sodium hypochlorite (2%) for 30 minutes, and subsequently washed with water.

The experiment was conducted in a germination chamber (FANEM; model 347 CDG) with alternating light and temperature conditions (30°C light/12 h: 20°C dark/12 h), and was monitored daily from 1500 to 1700 hours by alternating the tray position in the germination chambers after each evaluation. The experiment was completed after 30 consecutive days of evaluation; it was conducted as a completely randomized design (CRD), with 10 replicates of 20 seeds per treatment, totaling 800 seeds per species. All seeds showing radicle emergence were counted as germinated and were eliminated from the experiment. Water supply was maintained during daily evaluations by adding approximately 0.5 mL of distilled water on filter paper by using disposable syringes when necessary.

Data analysis was performed at the level of dispersal syndromes. Therefore, tree species were separated in groups according to dispersal syndrome: zoochorous, anemochorous, and autochorous. We calculated germinability (G%) and germination speed index (GSI) (Borghetti and Ferreira 2004) for each group. Germinability was determined by the formula: $\% G = Gn/Nn*100$, where Gn represents the total number of germinated seeds in each sampling unit during the evaluation, and Nn indicates the total number of seeds in the sampling unit (Borghetti and Ferreira 2004). GSI is germination speed index (GSI), which measures the speed of seed germination, is the sum of the number of germinated seeds each day divided by the number of days between the beginning of the experiment and radicle emergence, according to the formula described by Maguire (1962): GSI: $G1/N1 + G2/N2 + G3/N3 + G4/N4 + \ldots + Gn/Nn$, where G1, G2, G3,…, Gn are the number of germinated seeds on each count, the first, second, third,…, nth count; and N1, N2, N3,…, Nn are the number of days, day 1, day 2, day 3…, nth day, after the beginning of the experiment.

We used analysis of variance (ANOVA) to determine which syndrome had the highest G% and GSI and the presence or absence of dormancy, with each pregerminative treatment means compared by the Tukey test at 5% significance (Zar 1996). For homogeneity of variance, before the analysis, all G% values were arc-sin transformed using the square root of percentage germination (Santana and Ranal 2004). In addition, to determine whether G% values were similar among species in the same dispersal syndrome group, we performed a cluster analysis with G% using the unweighted pair-group average (UPGA) clustering method with Euclidean distance as the coefficient of association.

16.2.5 Seed Storage Experiments

Seed germination of the five stored species (*S. spectabilis*, *M. urundeuva*, *P. stenantha*, *P. pachycladus*, and *S. tuberosa*) was evaluated through germination tests that were conducted at 30, 60, 90, and 120 days of storage, using the same testing procedure applied to the dormancy experiment described earlier, in a factorial 2 × 4 (two storage locations and four storage periods). Thus, the seeds of each species were categorized into eight groups of 200 seeds each, stored under two different environmental conditions, and divided into two treatments: (1) control in which seeds were placed in nylon bags (12 cm × 18 cm) and kept in the laboratory at room temperature and humidity; and (2) buried, in which seeds were placed in nylon bags and buried 10 cm deep (Souza et al. 2007) in the soil at the collection area, near the parent trees. Thus, germination tests were performed every 30 days, with 200 seeds in the control treatment and another 200 seeds in the buried treatment, until they reached 120 days of storage in the soil and the laboratory. Seeds of the five species were buried in the months subsequent to seed collection, which varied between species. Thus, the seeds of *P. stenantha* were buried in June 2008; *S. spectabilis*, in August 2008; *M. urundeuva*, in September 2008; and *P. pachycladus* and *S. tuberosa*, in March 2009. Twenty more seeds were placed in each container as a margin of safety for the experiment.

Data analysis was also done using dispersal syndromes, and G% and GSI were also determined. To check which syndrome loses viability more rapidly in soil (lower G% and higher GSI values), we performed an ANOVA and, subsequently, a Tukey's test at 5% significance (Zar 1996). Before the analysis, G% values were transformed according to the same methodology described in the dormancy experiment. To perform the ANOVA, in both dormancy and storage tests, when the number of species belonging to the same syndrome was not equal (for example, three zoochorous, one autochorous, and one anemochorous species), the average number of germinated seeds was used for the level (syndrome), and not for each species, which resulted in an average percentage germination for each replicate ($n = 10$). Similarly, in the dormancy experiment, a cluster analysis was performed (UPGA method and Euclidean distance) using the G% of each species to determine the similarity in percentage germination after storage between species with the same syndrome in the soil and the laboratory.

16.3 Results

16.3.1 Dormancy Experiment

16.3.1.1 Germinability

Percentage germination differed significantly among dispersal syndromes (zoochorous, autochorous, and anemochorous) (df = 2, $F = 497.56$, $p < 0.001$, $n = 120$), between scarification treatments (df = 3, $F = 3.85$, $p < 0.05$, $n = 120$), and in the interaction scarification treatment × syndrome (df = 6, $F = 21.29$, $p < 0.001$, $n = 120$). Furthermore, these results show that mechanical scarification resulted in higher seed germination rates in zoochorous ($\overline{X} = 30.12 \pm 3.20\%$) and autochorous ($\overline{X} = 19.75 \pm 8.62\%$) seeds. Conversely, anemochorous seeds were negatively affected by mechanical scarification, with higher germination rates when seeds were intact, without scarification ($\overline{X} = 72.75 \pm 9.89\%$) (Figure 16.2).

16.3.1.2 Germination Speed

The GSI among dispersal syndromes and between scarification treatments was significantly different (df = 2, $F = 975.12$, $p < 0.001$, $n = 120$; df = 3, $F = 6.40$, $p < 0.001$, $n = 120$, respectively). The interaction scarification treatment × dispersal syndrome was also significantly different (df = 6, $F = 3.06$, $p < 0.05$, $n = 120$), with high GSI in the control treatment for the anemochorous syndrome ($\overline{X} = 8.27 \pm 1.57$ days) (Figure 16.3).

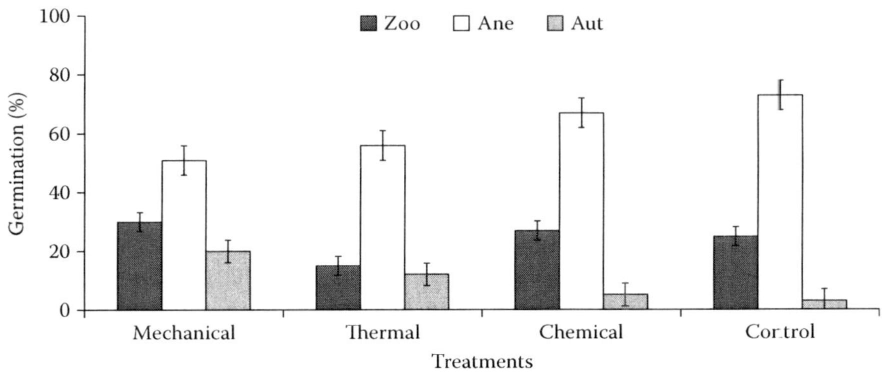

FIGURE 16.2
Average percentage germination of zoochorous (zoo), anemochorous (ane), and autochorous (auto) seeds collected in Mata Seca State Park, north of Minas Gerais, Brazil, submitted to different scarification treatments (mechanical, thermal, and chemical) and controls (intact seeds).

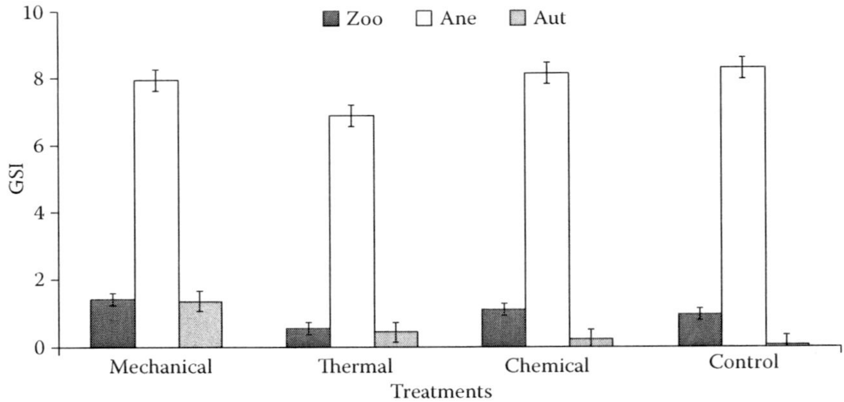

FIGURE 16.3
Germination speed index (GSI) of zoochorous (zoo), anemochorous (ane), and autochorous (auto) seeds collected in Mata Seca State Park, north of Minas Gerais, Brazil, submitted to different scarification treatments (mechanical, thermal, and chemical) and controls (intact seeds).

16.3.1.3 Analysis of Similarity between Species

To evaluate the similarities in germination behaviors among the species, the species were grouped by germinability, and the formation of two distinct groups can be observed (Figure 16.4): (1) a group of dormant, zoochorous, and autochorous species; and (2) a group of nondormant, anemochorous species. *Pilosocereus pachycladus* was the only exception; despite being a zoochorous species, it showed no dormancy, and grouped with the anemochorous species group. Its germination behavior was similar to that of *P. contorta* (anemochorous), as both had low germination rates when treated with hot water, and high germination rates when seeds were intact and treated with acid or sandpaper, with no significant differences between treatments.

In the group of dormant species, there was no apparent separation in the germination of zoochorous and autochorous species. For example, *M. hostilis* (autochorous) and *S. tuberosa* (zoochorous) had high similarity. This fact is evident, because both species had the highest percentage germination in the mechanical scarification treatment (although in *S. tuberosa* this value was not statistically different from other treatments) and the lowest percentage germination in the thermal scarification and control treatments. *Commiphora leptophleos* and *P. stenantha* also responded in a similar manner to pregermination treatments. In both species, percentage germination was significantly higher in the mechanical scarification treatment, and it did not differ between the other treatments. Mechanical scarification also resulted in higher percentage germination in *S. spectabilis* (autochorous), but it was not significantly different from thermal scarification, as observed in other species, indicating its low similarity with zoochorous and autochorous species.

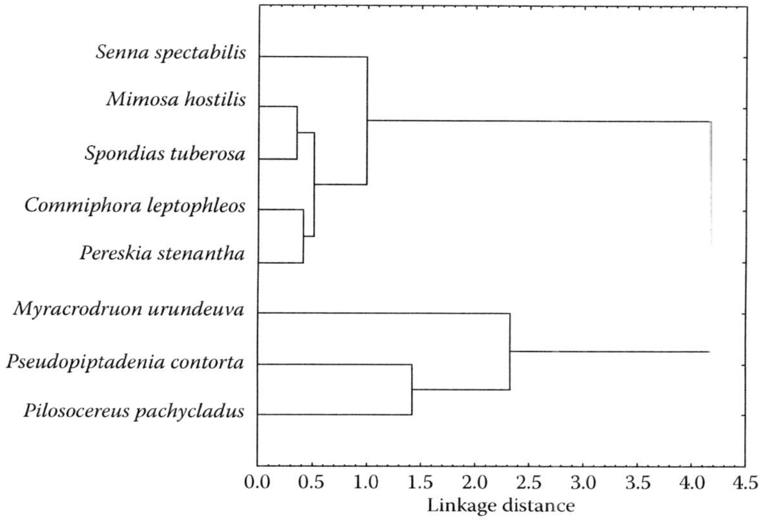

FIGURE 16.4
Dendrogram of percentage germination data from zoochorous (*Commiphora leptophaleos, Pereskia stenantha, Spondias tuberosa,* and *Pilosocereus pachycladus*), autochorous (*Mimosa hostilis* and *Senna spectabilis*), and anemochorous species (*Myracrodruon urundeuva*) using different scarification treatments, and UPGA method with Euclidean distance.

16.3.2 Seed Storage Test

16.3.2.1 Germinability

There were significant differences in germination rates both among dispersal syndromes (df = 2, $F = 272.76$, $p < 0.001$, $n = 240$) and between control and buried treatments (df = 1, $F = 350.74$, $p < 0.001$, $n = 240$). The interactions syndrome × treatment (df = 2, $F = 336.86$, $p < 0.001$, $n = 240$), syndrome × time (df = 2, $F = 13.99$, $p < 0.001$, $n = 240$), and syndrome × treatment × time (df = 6, $F = 6.63$, $p < 0.001$, $n = 240$) were also significant. Percentage germination was not affected by time and the interaction time × treatment only ($p > 0.05$). These results indicate that dispersal syndrome, storage location, and time may affect seed germination in the species evaluated. For instance, anemochorous seeds of the control treatment exhibited higher percentage germination, especially when stored for 60 days under laboratory conditions ($\overline{X} = 84.50 \pm 7.62\%$). Conversely, lower germination rates were observed in buried anemochorous seeds after 90 ($\overline{X} = 0.50 \pm 1.58\%$) and 120 days of storage, with no germinated seeds in the latter period, which was probably due to loss of viability (Figure 16.5).

16.3.2.2 Germination Speed

The GSI differed significantly among dispersal syndromes (df = 2, $F = 1719.75$, $p < 0.001$, $n = 240$), both between the control and buried treatments (df = 1,

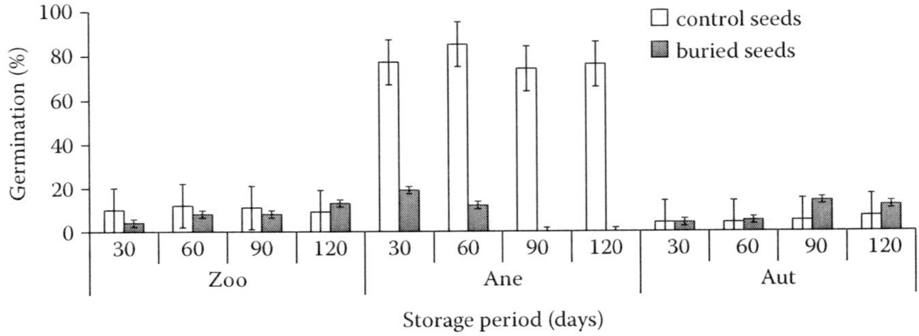

FIGURE 16.5
Average percentage germination of zoochorous (zoo), anemochorous (ane), and autochorous (auto) species stored under laboratory conditions (control) and in forest soils (buried) in Mata Seca State Park, for 30, 60, 90, and 120 days.

$F = 1349.05$, $p < 0.001$, $n = 240$) and between storage periods (df = 3, $F = 10.46$, $p < 0.001$, $n = 240$). The interactions syndrome × treatment (df = 2, $F = 1448.73$, $p < 0.001$, $n = 240$), syndrome × time (df = 6, $F = 4.12$, $p < 0.05$, $n = 240$), and treatment × time (df = 3, $F = 28.85$, $p < 0.001$, $n = 240$), and the relationship between the three variables (syndrome × treatment × time: df = 6, $F = 34.44$, $p < 0.001$; $n = 240$) were also statistically significant, indicating that all factors (dispersal syndrome, storage location, and time), acting alone or in combination, affected the seed germination speed. Thus, seeds of anemochorous species that were stored for 90 days in the laboratory (control) had higher germination rates ($\overline{X} = 11.95 \pm 1.57\%$) than other combinations (Figure 16.6).

16.3.2.3 Analysis of Similarity between Species

The cluster analysis that was performed on the germination rates of seeds subjected to storage in the soil and the laboratory showed that zoochorous and autochorous species had more similarities with regard to germination behavior (Figure 16.7). The zoochorous species *P. pachycladus* and *S. tuberosa* showed high similarity, as the seed germinability of both species increased with increasing storage time when the seeds were buried. Similarly, in both species, seeds from the control treatment exhibited reduced germinability from 30 to 120 days of storage. Although the germination percentage in *Senna spectabilis* (autochorous) was not reduced with the storage time in control seeds, it had higher germination values during the last months of storage in buried seeds, thus showing a certain degree of similarity with zoochorous species *P. pachycladus* and *S. tuberosa*.

Pereskia stenantha (zoochorous) was not very similar to other zoochorous species, probably because its buried seeds exhibited lower germination rates than zoochorous species P. *pachycladus* and *S. tuberosa* and the only autochorous species analyzed (*Senna spectabilis*). Moreover, when each species was

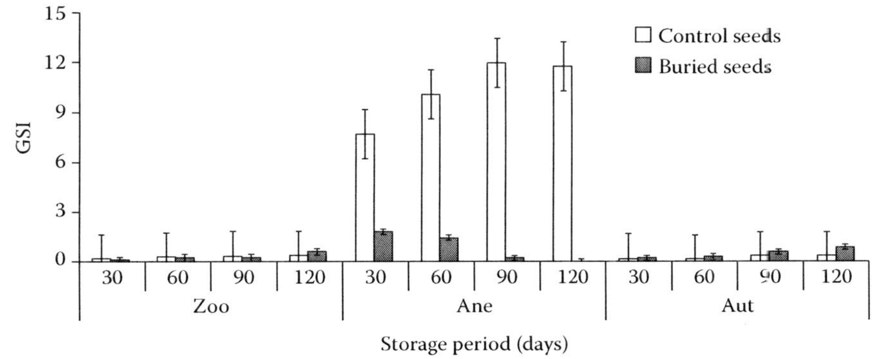

FIGURE 16.6
GSI of zoochorous (zoo), anemochorous (ane), and autochorous (auto) seeds stored under laboratory conditions (control) and in forest soils (buried) in Mata Seca State Park, for 30, 60, 90, and 120 days.

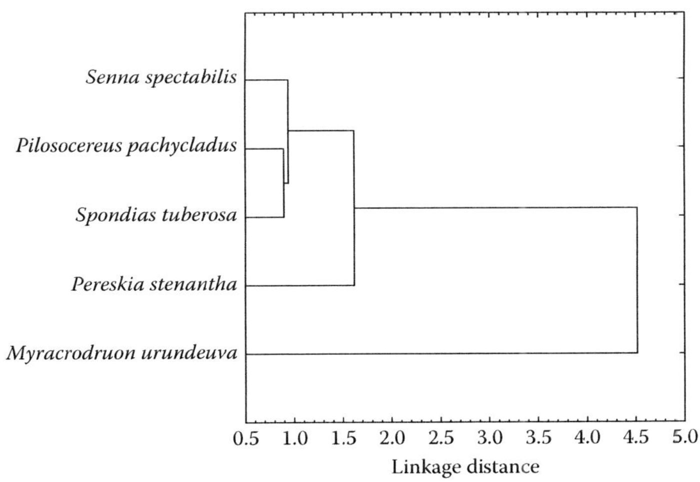

FIGURE 16.7
Dendrogram of percentage germination data from zoochorous (*Pereskia stenantha, Spondias tuberosa,* and *Pilosocereus pachycladus*), autochorous (*Senna spectabilis*), and anemochorous (*Myracrodruon urundeuva*) species stored under laboratory conditions and in forest soils for 120 days (UPGA method and Euclidean distance between species).

analyzed separately, *P. stenantha* had the highest percentage germination when stored for 60 days, contrary to other zoochorous species, which exhibited the highest percentage germination after 120 days of storage.

Myracrodruon urundeuva (anemochorous) was distanced from other species in the cluster because of its different germination behavior. It was the only species whose germination rates remained high (around 80%) without scarification throughout the storage period and the only species whose seeds that were stored in the soil deteriorated in the last two storage periods, with total loss of viability.

In general, we observed similarities at the level of dispersal syndromes when the similarity of species was analyzed after pregerminative or stored treatments (soil and laboratory) for a period of 4 months. Consequently, species from the same dispersal syndrome had similar germinative behavior. For example, zoochorous species had high similarity levels. It should also be noted that the germinative behavior of autochorous species was more similar to that of zoochorous species than that of anemochorous ones.

16.4 Discussion

16.4.1 Dormancy

Seed dormancy can occur due to several factors such as tegument impermeability to water and gases, immature or undeveloped embryos, special light or temperature requirements, and presence of growth inhibitors or promoters (Popinigis 1977; Marcos-Filho 2005). Treatments to overcome seed dormancy have both advantages and disadvantages; so, each treatment should be analyzed carefully for choosing the best method (Passos et al. 2007). The applicability and effectiveness of the treatments depend on dormancy type and intensity, which vary among species (Bruno et al. 2001). In this study, analysis of the occurrence of seed dormancy in different dispersal syndromes indicates that anemochorous species showed no dormancy, as higher seed germinability and germination speed (GSI) values were recorded in intact seeds. Moreover, the germination behavior of autochorous species was more similar to that of zoochorous species, because seeds in both syndromes had tegument dormancy. Nonscarified seeds from these two guilds had very low percentage germination and germination speed, whereas mechanically scarified seeds had higher values. The delay in the germination of nonscarified seeds in autochorous and zoochorous species may have occurred due to the presence of a highly resistant hard tegument, which is an adaptation to prevent seeds from germinating during unfavorable seasons (Floriano 2004). The fact that seeds in autochorous and zoochorous species have a hard tegument may be associated with seed dispersal. Since autochorous seeds may undergo secondary dispersal by animals (Morellato and Leitão Filho 1992a), and endozoochory is one of the forms of seed dispersal by animals (Almeida-Cortez 2004), these seeds need to be dormant and have a hard, thick tegument that protects them from the action of stomach acids and enzymes in the digestive tract of frugivores (Marcos-Filho 2005). This fact can be observed in cactus of the subfamily Opuntiodeae, which has a sclerified aryl that protects the seeds as they pass through the digestive tract of birds (Rojas-Arechiga and Vasquez-Yanes 2000). In addition to vertebrates, many autochorous species may undergo secondary dispersal by ants (Espírito-Santo 2007). Ants contribute to increasing primary seed dispersal distance (gravity or seed weight), which is usually very small in autochorous species; they also transport seeds to their nests, reducing competition between seedlings and seed predation and promoting seed germination, because nest soils are nutrient rich (Espírito-Santo 2007).

Seed dormancy can also be related to seed dispersal season. Thus, anemochorous species, which disperse their seeds in the dry season, are nondormant, probably because seed dispersal begins one or two months before the rainy season (Nunes et al. 2008; Pereira et al. 2008). This behavior decreases postdispersal predation risk and makes dormancy unnecessary because of the proximity to the rainy season (Garwood 1983). The great number of species that germinate at the beginning of the rainy season seems to be an evolved characteristic in TDFs that helps maximize seedling growth in the rainy season, resulting in a higher probability of seed establishment and development (Garwood 1983).

Conversely, because zoochorous and autochorous seeds are dispersed during the rainy season (Nunes et al. 2012), they probably need to remain dormant until the next rainy season, as observed by Garwood (1983) in a semideciduous forest in Panama, where seeds that are dispersed in the rainy season remain dormant for four to eight months, which is the time for the next rainy season. Vieira et al. (2008) also observed that in TDFs, seed dormancy acts to minimize mortality after the first rains, which can sometimes be followed by periods of drought. Consequently, with the increased rates of litter decomposition during the rainy season, soil moisture is higher and there is temporary abundance of nutrients (Morellato 1992), which contribute to initial plant establishment and faster growth before the arrival of the dry season (Nunes et al. 2005). Therefore, seed dormancy is very advantageous for these species and evolves as a survival mechanism during adverse environmental and climatic conditions such as dry seasons (Popinigis 1977).

16.4.2 Seed Storage

As in the dormancy test, anemochorous species also showed better germination performance (G% and GSI) than other syndromes, especially when seeds were stored under laboratory conditions. Germination rates of the anemochorous species, *M. urundeuva*, analyzed were higher in the control treatment after 60 days of storage; whereas seeds lost viability after 120 days of storage in the soil. These results confirm the absence of dormancy in anemochorous species, in particular *M. urundeuva*, and the short storability of its seeds in the soil. The reduction in germinability over time can be considered normal due to the physiological characteristics of seeds in this species: a more fragile tegument and endocarp, which results in faster loss of viability and higher susceptibility to pathogen attack (Ferreira and Cunha 2000). Moreover, this species has oleaginous propagules, and oil accelerates the seed deterioration, giving the seed a short natural longevity (Carneiro and Aguiar 1993). The results for *M. urundeuva* corroborate those obtained by Souza et al. (2007), who also found total loss of viability in seeds of this species after 120 days of storage in a TDF soil.

Germination behavior of the autochorous species analyzed (*S. spectabilis*) was more similar to that of zoochorous species, with higher G% and GSI in the last two storage months, especially in the soil. However, even with higher germination percentage in the last storage month in soil, the germination time of zoochorous seeds varied from 30 to 120 days of storage. These results suggest that some seeds probably had not reached physiological maturity at the start of the germination experiment, which may have been reached after storage (Lima et al. 2008). Because samples were composed of seeds from different parent trees, there may have been seeds in different maturation stages. Piña-Rodriguez and Jesus (1992) argued that seed maturation can continue during storage, which increases germination performance.

Longer germination times (lower germination speed) were observed in zoochorous and autochorous seeds than in anemochorous seeds. This result may be associated with the ability of zoochorous and autochorous seeds to form seed banks in soil, unlike anemochorous seeds that form seedling banks instead (Barbosa 2003; Almeida-Cortez 2004; Nunes and Petrere-Jr 2012), which require greater germination speeds. The increased speed of germination enables anemochorous seeds to escape from predators, resulting in increased chances of seedling survival (Barbosa 2003). Conversely, zoochorous and autochorous species may form a persistent seed bank (Almeida-Cortez 2004), because our results suggest that seeds have great longevity under natural conditions. The persistence of seeds in the soil represents a pool of genetic potential that help maintain species diversity and the future plant community (Grime 1989). Moreover, these seeds can be recruited after disturbance, influencing the direction of forest regeneration and secondary succession (Grime 1989).

The positive relationship between seed dormancy and seed longevity was also supported in this study. In fact, this relationship is essential in nature, because tegument impermeability prolongs the life of the seeds, which enables the formation of a persistent seed bank in the soil and spreads out germination in time and space (Rolston 1978). This may increase the chances of finding favorable conditions for seed establishment in the natural environment (Marcos-Filho 2005).

16.5 Conclusion

The hypotheses tested in this study were confirmed. Anemochorous species showed no dormancy mechanisms and were able to germinate immediately after seed maturation and dispersal. However, zoochorous and autochorous seeds were dormant, with higher percentage germination and germination speed when subjected to mechanical scarification. With regard to seed storability, seeds of the anemochorous species *M. urundeuva* had short life on the forest soil, with loss of seed viability after three months of storage. However, germinability in the control seed treatment in the laboratory was high throughout the storage period, indicating the negative influence of the natural environment on seed viability. The autochorous species *S. spectabilis* showed similar behavior to zoochorous species. Both syndromes showed low percentage germination and low germination speed in the buried and control treatments, with no significant differences in storage period and location. Thus, our results support the relationship between seed dormancy and storability and seed dispersal mechanisms. However, the number of species investigated was too small to make generalizations about the germination behavior of species with different dispersal mechanisms. Finally, this is the first study that investigates patterns on seed storage potential in the soil and seed dormancy in TDFs.

17

Interspecific and Interannual Variation in Foliar Phenological Patterns in a Successional Mosaic of a Dry Forest in the Central Llanos of Venezuela

Soraya del Carmen Villalobos, José A. González-Carcacía, Jon Paul Rodríguez, and Jafet Nassar

CONTENTS

17.1 Introduction

The dynamics of leaf shedding and flushing in tropical dry forests (TDFs) have been strongly associated with the marked seasonality characteristic of these ecosystems, especially in relation to soil water availability and air humidity (Frankie et al. 1974; Reich and Borchert 1984; Medina et al. 1990; Borchert et al. 2002; Kushwaha and Singh 2005). The dry season in these forests can extend from three to six months, during which solar radiation and temperatures are comparatively higher with respect to the rainy season (Frankie et al. 1974; Murphy and Lugo 1986). Because leaf buds require water for their development, hydric stress and reduction of humidity in the soil can significantly limit the time of bud sprouting and leaf expansion during the dry season (Reich and Borchert 1984), and this is the mechanism that determines the deciduous condition of many species of plants, particularly trees, in these ecosystems.

In tropical zones around the world, time of leaf shedding and flushing is significantly variable within and among regions (Bullock et al. 1995; Pennington et al. 2006b). The foliar phenological patterns that TDFs exhibit are affected by environmental factors associated with the sites where these forests are located (Bullock et al. 1995; Pennington et al. 2006). Leaf flushing during the rainy season and leaf abscission during the dry season are the generalized phenological patterns observed in the TDFs of Central and South America (Murphy and Lugo 1995; Albert-Puentes et al. 1993, 2008) and Africa (Menaut et al. 1995). On the other hand, in tropical Asia plants lose their leaves during early dry season and new leaves emerge during late dry season (Rundel and Boonpragob 1995; Maxwell and Elliot 2001; Elliot et al. 2006). Many studies focusing on tropical zones suggest that regulation of the time and intensity of leaf shedding and flushing does not depend exclusively on water availability. The noteworthy interannual variation of foliar phenological patterns observed in some dry forests examined seems to indicate that foliar phenology is also influenced by other factors, such as endogenous plant rhythms (e.g., internal hydric status and concentration of solutes) and external climatic and ecological parameters (i.e., temperature, light, and soil type) (Olivares and Medina 1992; Mooney et al. 1995; Murphy and Lugo 1995). For example, spatial and temporal models of leaf fall have demonstrated that termination of photosynthesis and collapse of foliar tissues are the consequence of a combination of factors including life-form (e.g., succulent, leathery, etc.), age, and nutrients content (Holbrook et al. 1995; Gutiérrez-Soto et al. 2008).

Several authors (Schaik et al. 1993; Rivera et al. 2002; Calle et al. 2010) have proposed that photoperiod and day length increases in a few minutes in tropical zones can act as important triggers for bud breaking and expansion of new leaves before the beginning of the rainy season. Several cases of leaf-flushing events during the dry season, close to the spring equinox, have been reported in Mexico (Calle et al. 2010), Costa Rica (Borchert 1994; Borchert and Rivera 2001; Rivera et al. 2002), Brazil, Australia, Argentina (Rivera et al. 2002), and Thailand (Elliot et al. 2006). The hypothesis of the photoperiod as a leaf-flushing trigger assumes that all individuals of a given species respond in the same way to that stimulus (Rivera et al. 2002). However, despite the fact that a substantial number of species show leaf flushing in response to increments in the photoperiod, several studies point toward a remarkable variation in the bud-breaking pattern as a function of other variables, such as forest type, length of the dry season, and magnitude of precipitation (Rivera et al. 2002; Elliot et al. 2006; Williams et al. 2008; Calle et al. 2010). The photoperiod hypothesis needs further testing, and the dry forests particularly suitable for this purpose include those distributed in areas with substantial interannual variation in the precipitation regime. If the photoperiod triggers massive leaf bud breaking in a set of deciduous species within a dry forest, then we should expect to observe the massive emergence of new leaves in these species during that part of the dry season when the most substantial increase in day length and level of insolation occurs, independently of the interannual variation in the onset of the rainy season.

The Central Llanos of Venezuela represent a mosaic of savannas, riparian, gallery, and seasonal dry forests with an elevated degree of taxonomic and physiognomic differentiation (Huber and Alarcón 1988; Duno de Stefano et al. 2007). Foliar phenology studies have been conducted on plant species inhabiting this region during the last four decades (Medina 1966, 1967; Medina et al. 1969; Monasterio and Sarmiento 1976; Ramia 1977). Most of the available information corresponds to qualitative descriptions showing the date of occurrence of the different phenophases, especially population- and species-level leaf fall. Very few investigations have focused on examining interannual variation in the phenological processes associated with dry forests, determining the dominant factors modulating these patterns and evaluating how much phenological differentiation exists among distinct successional stages within a forest. The variable degrees of anthropogenic intervention observed in the dry forests in Central Llanos, together with a significant interannual variation in the precipitation regime observed during the most recent years in the region, offer an attractive scenario to examine the factors involved in the process of leaf turnover in Venezuelan dry forests.

Foliar phenology can be examined from three different perspectives: (1) leaf longevity, which refers to the time that leaves remain on a plant; (2) time and magnitude of leaf flushing; and (3) leaf habit, which refers to the anatomical traits of leaves (Olivares and Medina 1992; Kikuzawa 1995). In this study, we considered the time and magnitude of leaf flushing and leaf fall as variables to evaluate and compare the foliar phenological patterns of the species of shrubs and trees in a dry forest of the Central Llanos of Venezuela in three different successional stages and during three continuous years of monitoring with significant variations in the rainfall pattern.

17.2 Methods

Foliar phenology was monitored in an area of approximately 50 ha. We evaluated leaf shedding and flushing in three successional stages: (1) "late" mature forest, with more than 50 years after the last known human intervention; (2) "intermediate" forests, which were logged and used for agricultural and cattle breeding purposes and were then protected against human intervention and fire, allowing vegetation recovery for the last 10–15 years; and

(3) "early" forests, which were logged and used for agricultural and cattle breeding purposes and were then protected against human intervention and fire, allowing vegetation recovery for the last 3–5 years. The soil mosaic observed in the early plots corresponds to the loam and sandy-loam types (Villalobos 2010). These soil types allow high water availability and good aeration, which facilitate nutrient absorption. Soils in the intermediate and late stages have high percentages of clays and lime (Villalobos 2010).

Early- and intermediate-stage plots were less separated than late successional plots, because the first two stages were less commonly found inside the study area (Figure 17.1). In the case of early plots, the vegetation did not correspond entirely to the physiognomy of the typical early stage of a dry forest because when those areas were logged, several large trees were left alive to provide shade to cattle. Once abandoned, these trees produced seeds and supplied shade to seedlings of different plant species that established there. This kind of agrosystem helps to speed the process of forest regeneration (Janzen 1988). The overall effect of these conditions on the vegetation of these plots is

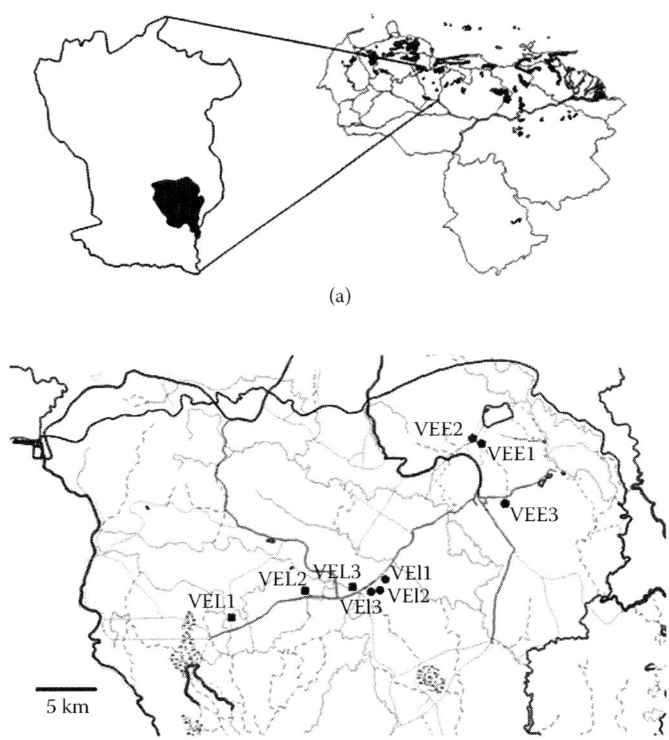

(b)

FIGURE 17.1
(a) Geographic location of phenology plots in UPSA-Piñero and (b) successional plots. Early (VEE), intermediate (VEI), and late (VEL).

the presence of a combination of elements typical of dry forests and elements found in more humid forests. The level of vegetation complexity observed in these early plots was higher than what should be normally expected for early successional dry forests. Table 17.1 summarizes the vegetation information of each of the successional plots examined (Rodríguez and Nassar 2011), including number of individuals, number of families and species of shrubs, trees and lianas with diameter at breast height (DBH) above 5 cm (2.5 cm in the case of lianas), and Holdridge's complexity index (Holdridge et al. 1971). Based on the importance value index (IVI) (Curtis 1959) calculated for each successional stage, the three more important woody species per stage were *Protium heptaphyllum*, *Copaifera pubiflora*, and *Mansoa verrucifera* for the early stage; *Caesalpinia coriaria*, *Machaerium robiniifolium*, and *Cochlospermum vitifolium* for the intermediate stage; and *Pterocarpus acapulcensis*, *Pithecoctenium crucigerum*, and *Ziziphus cyclocardia* for the late stage (Nassar et al., unpublished data).

For each successional stage, we established three 60 m × 30 m plots following the *Manual of Methods of Tropi-Dry* (Nassar et al. 2008), for a total of nine plots. Inside each plot, we delimited a 50 m × 20 m (0.1 ha) subplot. Along the two longest edges of each subplot, considering transects of 1 m width, we marked and labeled all individuals with DBH >2.5 cm.

Every month for 3 years (September 2007–September 2010), we conduced censuses of the status of foliar cover. Foliar cover was classified according to the following foliar phenophases: "new leaves" (small and light green), "young leaves" (average size and bright green), and "senescent leaves" (yellowish or brownish) or "leafless condition." We assigned semiquantitative values for the different foliar cover categories following Fournier (1974): 0 (absence of a particular phenophase), 1 (1%–25%), 2 (26%–50%), 3 (51%–75%), and 4 (76%–100%). For example, for the foliar phenophase leafless condition,

TABLE 17.1

Floristic Composition and Vegetation Structure of Woody Species in UPSA-Piñero

Successional Stage Plots	Number of Individuals	Number of Families	Number of Species	CHCI
Early 1	188	22	34	77.21
Early 2	198	16	20	50.93
Early 3	—	—	—	—
Intermediate 1	127	12	13	16.43
Intermediate 2	175	12	17	38.03
Intermediate 3	227	13	19	78.44
Late 1	121	22	32	99.70
Late 2	115	20	27	157.02
Late 3	157	16	32	200.25

Source: Data from Rodríguez, J.P. and J.M. Nassar, CRN II-021, *Technical* Report Year 5—Venezuela. Centro de Ecología, IVIC, Venezuela, p. 20, 2011.

Note: CHCI, changed Holdridge complexity index.

a value of 0 means that all branches are covered with green leaves and a value of 4 indicates that the plant lost most of its leaves and reached the maximum level of deciduousness possible.

17.2.1 Determination of Deciduous Condition

Deciduousness was established for both species and successional stages based on the semiquantitative values obtained for the leafless phenological stage. This phenophase was evaluated in each individual and averaged across all individuals within each species (species level) and across all individuals of all species within each successional plot (successional stage level). The duration of the deciduous condition was determined for each successional stage, counting the number of months that a particular successional stage remained within a level of deciduousness between 25% and 100%.

17.2.2 Litterfall

Starting in December 2007, we assembled a system of eight 0.5 m × 1.0 m (0.5 m^2) litterfall traps within each of the successional plots to estimate litterfall production at plot level following the Tropi-Dry manual specifications (Nassar et al. 2008). The traps were arranged along two transects that paralleled the phenological sampling transects, and litterfall (leaves and twigs) was collected monthly at the same time that the phenological observations were made. Each transect begins at the 30 m side of the 60 m × 30 m plot. A litter trap was placed 7.5 m from the edge and every 15 m thereafter. Samples from each trap were bagged and dried at <70°C until constant weight was reached. Once dry, the weight of each sample was recorded separately.

17.2.3 Synchrony of Leaf Flushing

We applied circular statistics to analyze the level of flushing synchronization (Morellato et al. 2000; Williams et al. 2008). Dates of flushing were transformed into angle values from 0° to 360°. Records of leaf flushing were represented according to values of foliar coverage (0–4). The mean angle Φ represents the average of dates at which massive leaf flushing occurs and the vector length r corresponds to the temporal concentration of phenological data. Values described were calculated according to the following equations:

$$\Phi = \arctan\left(\frac{x}{y}\right) \text{ if } x > 0 \text{ or } \Phi = 180^\circ + \arctan\left(\frac{x}{y}\right) \text{ if } x < 0$$

where $x = \sum n_i \cos\Phi_i$, $y = \sum n_i \sin\Phi_i$, n is the number of observations of a phenological activity, and Φ is the angle that represents the date at which the phenological activity was recorded (Wright and Calderón 1995). The length of the r vector is a measure of concentration of the phenological activity in time and ranges between 0 and 1.0. Values of r close to 0 indicate that all phenological activities concerning leaf flushing are spread homogeneously or regularly across the year, and r values close to 1.0 indicate that all records of leaf flushing are concentrated in a single month. Even though the r value should be enough to interpret the temporal distribution of phenological data, we applied a Rayleigh test in addition to determine if the distribution of leaf flushing across the entire year was significantly different from a random distribution (Batschelet 1981).

Foliar phenology was analyzed at species and community levels for each successional stage. For each species, individual values of r (r_{ind}) were averaged. Thus, a high r value of a given species is indicative that plants of that species are highly synchronous in producing new leaves and leaf flushing is concentrated for the whole species in a relatively short period of time. On the other hand, a low average value of r indicates that leaf flushing in that species is spread over a broader period of time during the year. At the successional stage level, we estimated the mean angle of all individuals of all species pooled together, which indicates the average date of flushing for the entire community of plants within each successional plot. We also calculated an average r (r_{comm}) value per plot. This value indicates the degree of aggregation of leaf flushing for the entire set of individuals of different species present in a particular successional plot.

17.2.4 Relationship between Leaf-Flushing Phenology, Rainfall Regime, and Solar Radiation

Potential associations between leaf-flushing patterns, rainfall, and daily insolation patterns were explored using a linear regression analysis (Statistic V7), in which average monthly precipitation volumes and daily insolation corresponded to the independent variables and Fournier's records of leaf flushing to the dependent variable. The relationship between rainfall and leaf flushing was examined considering the precipitation records of the month previous to the month under examination for foliar status. Daily isolation is the amount of energy in the form of solar radiation (watt per square meter) that receives a location on earth a given day (Gabler et al. 2008). Annual variations of day length and daily insolation were obtained from the following Internet source: http://aom.giss.nasa.gov/srlocat.html. Daily isolation is determined by the angle at which the sun's rays hit the surface of the earth. This angle varies according to latitude and longitude. In zones near the equator, insolation reaches two maximum annual values that are evident during the two equinoxes in the temperate zones; also, a reduction in the insolation level becomes evident at the start of the summer solstice (http://aom.giss.nasa.gov/srlocat.html).

In the Venezuelan Llanos, rains can vary in intensity and duration during the wet months each year (Figure 17.2). Insolation in the study site shows two important variations in the dates close to the spring equinox (February–March) and the fall equinox (August–September) in the Northern

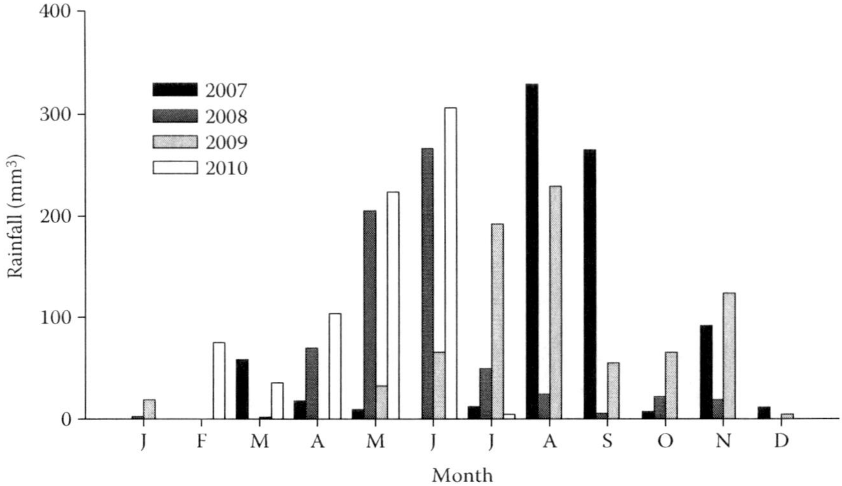

FIGURE 17.2
Monthly rainfall (in cubic millimeter) estimates at the El Baúl Meteorological Station, Cojedes state, the closest location to UPSA-Piñero where rainfall estimates were recorded in the years 2007, 2008, 2009, and 2010.

Hemisphere. If the photoperiod triggers a massive process of leaf bud breaking in some of the studied species, then we should expect to observe the emergence of new leaves during that part of the year when the most substantial changes in day length and level of insolation occur (Rivera et al. 2002; Elliot et al. 2006; Calle et al. 2010). This would imply that leaf-flushing time shows a high consistency across years for that set of species, because the magnitude of changes in day length and level of insolation do not vary among years (http://aom.giss.nasa.gov/srlocat.html). On the other hand, if rainfalls or other local environmental factors related to water availability are the main triggers of leaf flushing in the study area, we should expect to document interannual variation in the time period at which such processes occur, because a substantial level of interannual variation in time and magnitude of precipitations has been documented in the study area.

17.3 Results

17.3.1 Deciduous Condition

A total of 492 individuals corresponding to 75 species of 25 families were labeled and monthly monitored for 3 years (Table 17.2). A single individual represented 23 species, 15 species had less than four individuals, and 37 species had more than five individuals in the sample.

TABLE 17.2

Species Identified and Monitored in the Phenology Plots

Species	Family	Number of Individuals
Acasia glomerosa	Fabaceae	12
A. barinensis	Fabaceae	5
A. niopoides	Fabaceae	9
Annona purpurea	Annonaceae	6
Apeiba tiborbou	Malvaceae	1
A. corallina	Bignoniace	7
A. mollisima	Bignoniace	30
Aspidosperma cuspa	Moraceae	1
Astronium graveolens	Anacardiaceae	5
Bauhinia pauletia	Fabaceae	1
Brosimum alicastrum	Moraceae	1
C. coriaria	Fabaceae	12
C. flexuosa	Capparaceae	5
C. peltata	Cecropiaceae	7
Chomelia spinosa	Rubiaceae	2
Clitoria dendrina	Fabaceae	1
Clitoria falcata	Fabaceae	1
C. caracasana	Polygonaceae	8
C. vitifolium	Cochlospermaceae	1
Combretum fruticosum	Combretaceae	12
C. pubiflora	Fabaceae	10
Copernicia tectorum	Arecaceae	1
Cordia collococca	Boraginaceae	2
Cydista diversifolia	Bignoniace	2
Diospiros inconstans	Ebenaceae	1
Entada polystachya	Fabaceae	2
Erythroxylum orinocense	Erythroxylaceae	3
F. maxima	Moraceae	6
G. americana	Rubiaceae	4
Guapira pubescens	Nyctaginaceae	3
G. ulmifolia	Malvaceae	16
Guettarda divaricata	Rubiaceae	3
Hecatostemon completus	Flacourtiaceae	19
Hirtella racemosa	Chrysobalanaceae	1
Hymenaea courbaril	Fabaceae	1
I. interrupta	Fabaceae	11
Jacaranda obtusifolia	Bignoniaceae	2
Lecythis minor	Lecythidaceae	1
Licania apetala	Chrysolabanaceae	2
Lonchocarpus fendleri	Fabaceae	24

(Continued)

TABLE 17.2 (*Continued*)

Species Identified and Monitored in the Phenology Plots

Species	Family	Number of Individuals
Luehea candida	Malvaceae	4
M. robiniifolium	Fabaceae	17
Maclura tinctorea	Moraceae	2
M. verrucifera	Bignoniaceae	29
Margaritaria nobilis	Euphorbiaceae	2
M. undulata	Apocynaceae	2
Pachira quinata	Malvaceae	1
Passiflora serrulata	Passifloraceae	1
P. cururu	Sapindaceae	5
Paullinia leiocarpa	Sapindaceae	3
Paullinia serrulata	Sapindaceae	1
P. carthagenensis	Fabaceae	5
P. macranthocarpa	Nyctaginaceae	9
P. lanceolatum	Fabaceae	6
P. crucigerum	Fabaceae	1
Platymiscium pinnatum	Fabaceae	3
P. heptaphyllum	Burseraceae	52
Pseudolmedia laevis	Moraceae	1
P. acapulsensis	Fabaceae	18
R. venezuelensis	Rubiaceae	6
S. saponaria	Sapindaceae	7
S. biglandulosum	Euphorbiaceae	7
Securidaca aff. *pubescens*	Polygalaceae	1
Sorocea sprucei	Moraceae	9
Sterculia apetala	Malvaceae	1
Steriphoma ellipticum	Capparaceae	1
Tabebuia ochracea	Bignoniaceae	1
Terminalia oblonga	Combretaceae	2
Trichilia trifolia	Meliaceae	2
Trichilia unifoliola	Meliaceae	13
Vernonanventura brasiliana	Asteraceae	1
Vochysia venezuelana	Vochysiaceae	1
Xylophragma seemanianum	Bignoniaceae	8
Z. cyclocardia	Rhamnaceae	17
Ziziphus venezuelensis	Rhamnaceae	1

The deciduous condition broadly varied among successional stages, from low percentages (<25%) of leafless branches in early-stage plots (Figure 17.3) to almost total loss of leaves (75%–100%) in intermediate-stage plots. In the case of the late successional stage, the percentages of leafless branches per individual varied among years, from 26% to 75% (Figure 17.3).

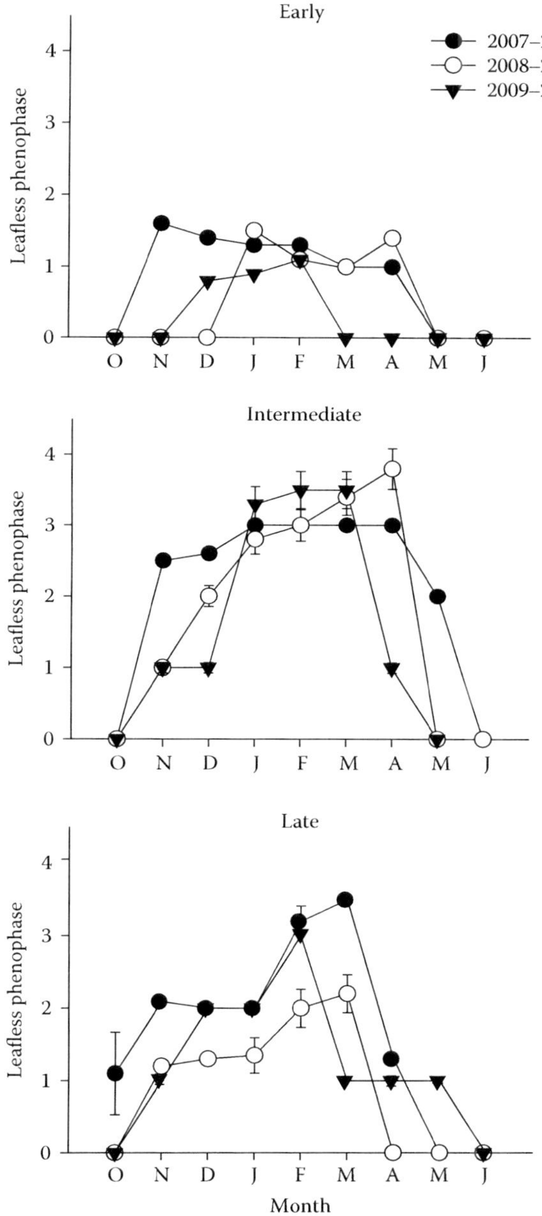

FIGURE 17.3
Phenograms of the leafless condition for the three successional stages examined and for 3 years of observations using Fournier's (1974) semiquantitative scores (monthly average ± 1 SE) for the different foliar cover categories: 0 (absence of a particular phenophase), 1 (1%–25%), 2 (26%–50%), 3 (51%–75%), and 4 (76%–100%).

As shown in Figure 17.3, all successional stages had marked interannual variation in time of leaf-shedding initiation and duration of the leafless condition. During the first year of observations (2007–2008), the leaf shedding initiated earlier (October) in the early successional stage and the leafless condition lasted longer (until May) in the intermediate successional stage. During the second year (2008–2009), leaf shedding initiated almost two months later (December) in the early plots and ended two months earlier in the late successional stage. In terms of duration of the deciduous condition, the shortest record corresponded to the dry season that was initiated at the end of year 2009 in the early plots, when plants lost and remained without leaves during less than three months. The most extended period for the leafless condition lasted seven months and was observed in the intermediate and late successional plots during the dry seasons that were initiated in 2007 and 2009, respectively.

Litterfall production reached maximum values in the late successional plots (Figure 17.4). When we considered the time window when leaf shedding occurs massively, between the end of October and May (only the last two years of sampling were considered), these plots produced averages of 6306.1 ± 279.4 (standard error [SE]) $kg \cdot ha^{-1}$ (2008–2009) and 6869.4 ± 299.2 $kg \cdot ha^{-1}$ (2009–2010). The early successional plots, with a substantial contribution of the arboreal elements present in them, followed second in the production of litterfall, with averages of 5417.5 ± 446.8 $kg \cdot ha^{-1}$ (2008–2009) and 6802.3 ± 603.2 $kg \cdot ha^{-1}$ (2009–2010). Finally, the intermediate successional plots presented the lowest values of litterfall in all the plots examined, with averages of 4109.9 ± 302.9 $kg \cdot ha^{-1}$ (2008–2009) and 4519.5 ± 336.6 $kg \cdot ha^{-1}$ (2009–2010). The most variable interannual patterns of litterfall production were observed in the early plots (Figure 17.4). Extended rains at the end of 2009 and the early onset of rains during the first months of 2010 contributed to move the peak of litterfall production to February. On the other hand, litterfall production extended over a broader time span in the period between October 2008 and April 2009, characterized by low precipitations. In the intermediate and late successional plots, litterfall patterns were quite consistent for the three years of observations, with peaks concentrated between November and December.

At the species level, we were able to consider only plants with sample sizes above five individuals per taxon ($N = 37$ taxa). The deciduous condition varied broadly among the different taxa examined (Figure 17.5), from evergreen plants (Fourier's value ~0) to those that lost their leaves completely (Fourier's value = 4). A few species were evergreen (*P. heptaphyllum*, *Coccoloba venezuelana*, *Cecropia peltata*, and *Z. cyclocardia*) or presented very short periods of leaf shedding (*C. publiflora*, *Pithecelobium lanceolatum*, and *Inga interrupta*). Above 40% of the species examined presented high percentages (75%–100%) of branches without leaves. The duration of the deciduous condition elapsed from three weeks to six months; however, the majority of species (40.5%) remained without leaves between three and five months. Seven species remained without leaves for more than four months (*Capparis*

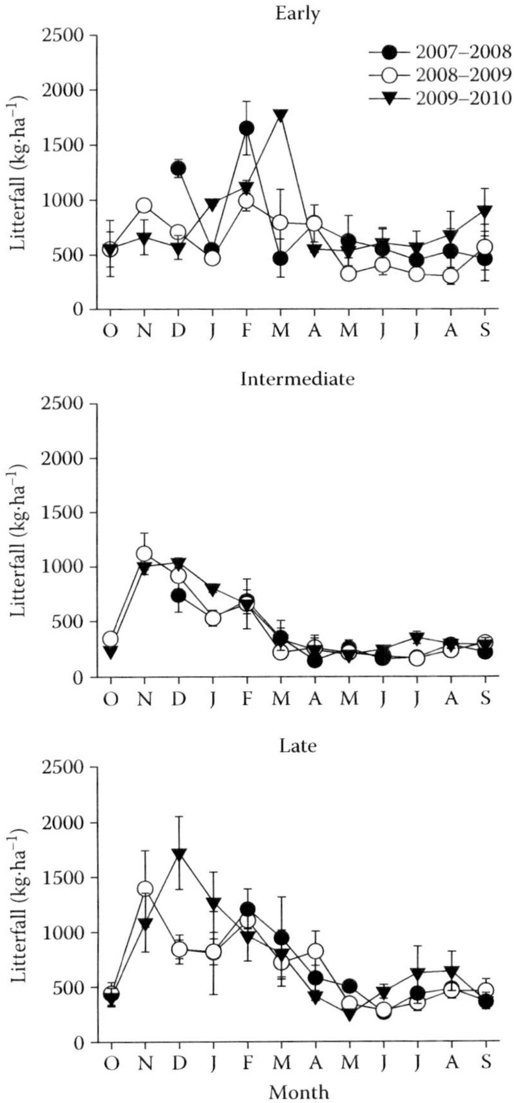

FIGURE 17.4
Leaf litter (kg · ha^{-1}) production (mean ± 1 SE) estimated for each successional stage during the time interval between October and May of years 2008, 2009, and 2010 at UPSA-Piñero.

flexuosa, Marsdenia undulata, Cochlospermun vitifolium, Sapium biglandulosum, Genipa americana, M. robiniifolium, Trichillia unifoliola, and *Piscidia cartaginensis*). Based on the number of months for which the leafless condition lasts, we can differentiate three functional types of plants: (1) leaf-exchanging, (2) brevideciduous, and (3) deciduous species (Figure 17.5).

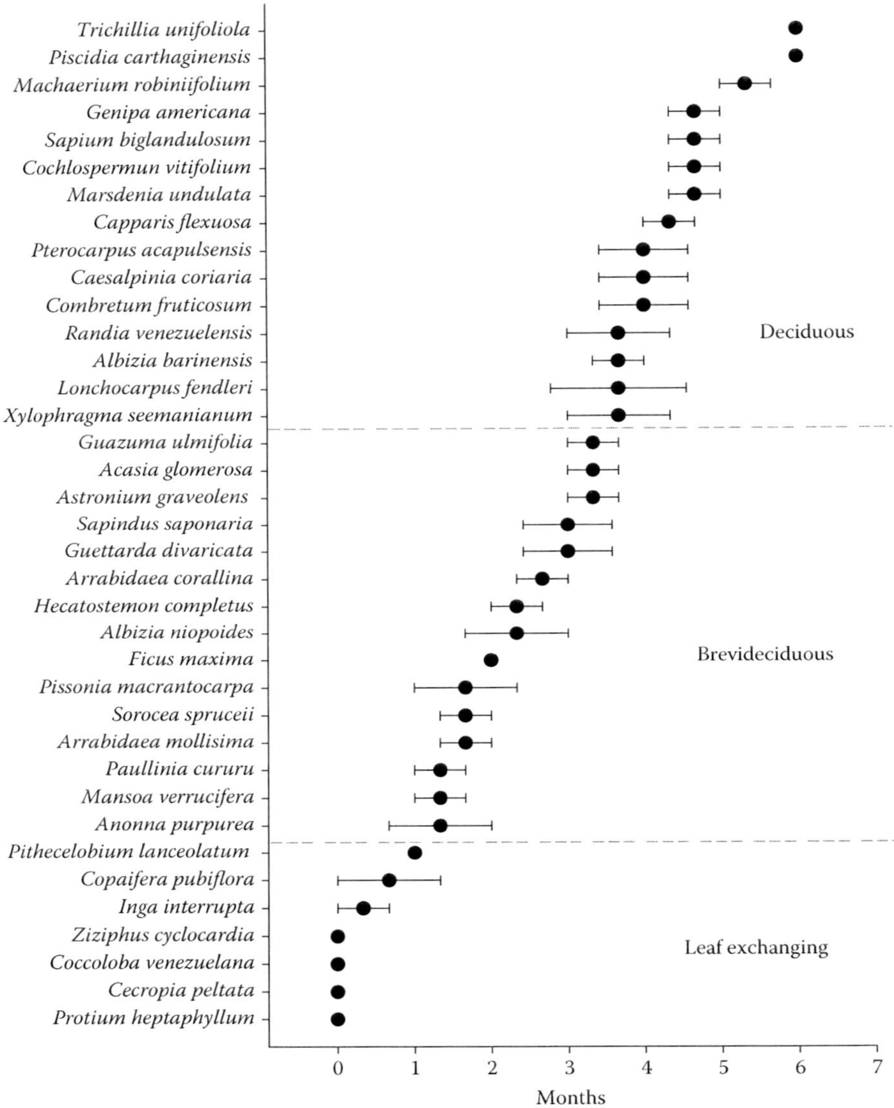

FIGURE 17.5
Classification of 37 species of plants examined in the successional plots at UPSA-Piñero according to the duration of the leafless condition.

Most individuals and species analyzed in the early successional stage corresponded to the leaf-exchanging (46.2%) and brevideciduous (42.3%) categories, and the remaining 11% corresponded to deciduous species. The intermediate state was mainly composed of deciduous taxa (72%), followed by brevideciduous taxa (16.7%), and leaf-exchanging species (11.1%). Finally, the late successional stage was dominated by a combination of deciduous (54%) and brevideciduous (31%) species, with a minor representation (14.2%) of leaf-exchanging taxa.

17.3.2 Leaf Flushing

Leaf flushing does not occur randomly throughout the year in the studied forest (*Rayleigh* test $p < 0.005$) (Table 17.3). For most species, the production of new leaves concentrates during a specific part of the year. The r values estimated for the three successional stages were close to 1.0 during the three years of phenological monitoring (range: 0.790–0.993, Table 17.3), reflecting highly synchronous emergence of leaves. No significant differences in the level of leaf-flushing synchrony were detected among years and successional stages (Table 17.3).

With the exception of the early successional plots during 2010 ($R^2 = 0.436$, $p = 0.037$), no significant association was detected between rainfall and leaf-flushing yearly patterns; however, the initiation of leaf flushing and leaf-flushing peak corresponded closely with the onset of the rainy season during the three years of this study (March–April). Leaf flushing for all species of plants monitored concentrated at the beginning of the rainy season with the exception of the early plots during year 2009 (Figure 17.6), but for that year a few rains occurred in January. In addition, it is important to point out the special circumstances of two of the early plots, which are located close to the hills of Galeras del Pao, Cojedes, Venezuela, and permanent ponds, which

TABLE 17.3

Results of the Circular Statistics Analysis on Leaf Flushing in Three Successional Stages during three Years in UPSA-Piñero

Successional Stage	Year	r	Mean Date of Flushing (Φ)	N	p
Early	2007–2008	0.790	9 May	179	<0.005
	2008–2009	0.987	30 April	155	<0.005
	2009–2010	0.993	10 March	155	<0.005
Intermediate	2007–2008	0.991	12 May	138	<0.001
	2008–2009	0.986	1 May	135	<0.005
	2009–2010	0.993	9 April	132	<0.005
Late	2007–2008	0.985	11 May	184	<0.001
	2008–2009	0.987	1 May	181	<0.001
	2009–2010	0.988	9 April	181	<0.001

Note: r represents the degree of temporal aggregation of leaf-flushing events among individuals within the successional plots, N represents the number of individuals monitored, and p is the significance of the Rayleigh test for deviation from random occurrence of leaf flushing during the year.

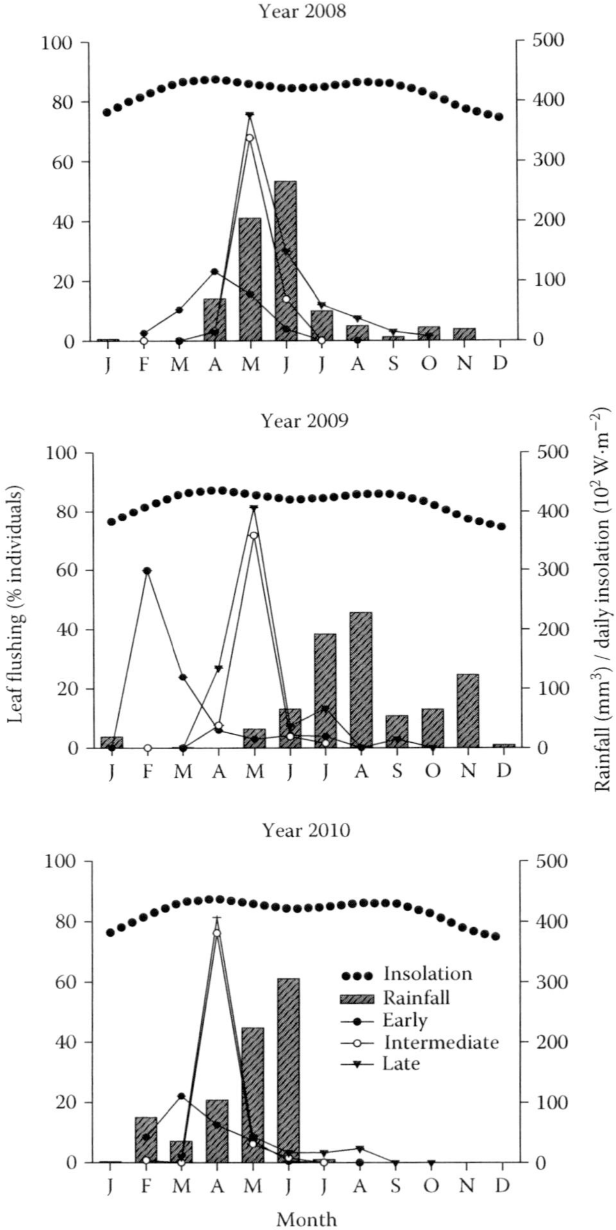

FIGURE 17.6
Leaf-flushing phenograms for the three successional stages examined at UPSA-Piñero during the 3 years of plant monitoring. Together with the phenograms, patterns of average monthly precipitation and insolation are included.

provide good irrigation to the area. For years 2008 and 2009, leaf flushing in the intermediate and late successional stages reached the maximum percentage of occurrence in May, less than a month after the onset of the rainy season. During 2010, leaf flushing for these successional stages reached its maximum percentage of occurrence one month earlier (April) than in the two preceding years. It was in 2010 that the rainy season started unusually early (February) in the Central Llanos of Venezuela.

Overall, leaf flushing was highly synchronous across species (Rayleigh test 6.8×10^{-5}, $p < 0.05$). Species-level values of the r vector ranged between 0.7 and 1.0, which indicates a narrow range of temporal variation for leaf emergence within the species (Figure 17.7; Table 17.4). Most species (86%) showed simultaneous leaf flushing between April and May, when rains began. During the three monitored years, most brevideciduous and deciduous taxa responded predictably to the onset of rains, massively producing new leaves (Table 17.4). Early flushing during 2010 was consistently observed for most species monitored indicating that, independently of the species of plants, rains trigger the production of new leaves in this forest. However, it is important to indicate that a few species repeatedly showed flushing

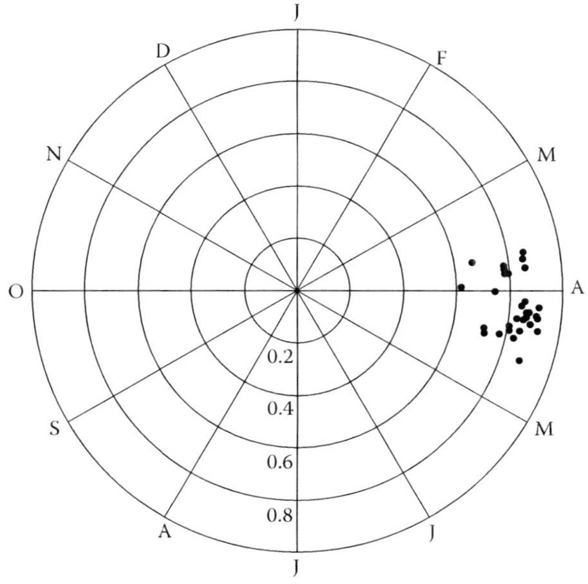

FIGURE 17.7
Species-level values of the r vector for leaf flushing for 34 species of plants examined at UPSA-Piñero, showing the level of synchronicity of leaf production across taxa. Evergreen species were not included in this analysis.

TABLE 17.4

Leaf-Flushing Patterns for 37 Species of Plants Examined at UPSA-Piñero during Three Years of Monitoring

	2007				2008												2009												2010							
Month	S	O	N	D	J	F	M	A	M	J	J	A	S	O	N	D	J	F	M	A	M	J	J	A	S	O	N	D	J	F	M	A	M	J	J	A
Rainfall	266	9	93	13	4	0	0	71	206	267	51	26	7	23	20	0	20	0	2	0	34	67	193	230	56	67	125	6	2	76	37	105	225	307	6	0
Insolation	430	413	387	372	382	409	432	437	429	421	423	430	429	413	387	372	383	410	432	437	429	421	423	430	429	413	387	372	382	409	432	437	429	421	423	430
ACGLO																																				
ALBAR																																				
ALNIO																																				
ANPUR																																				
ARCOR																																				
ARMOL																																				
ASGRA																																				
CACOR																																				
CAFLE																																				
CEPEL																																				
COFRU																																				
COPUB																																				
COVEN																																				
COVIT																																				
FIMAX																																				
GEAME																																				
GUDIV																																				
GUULM																																				
HECOM																																				
ININT																																				
LOFEN																																				

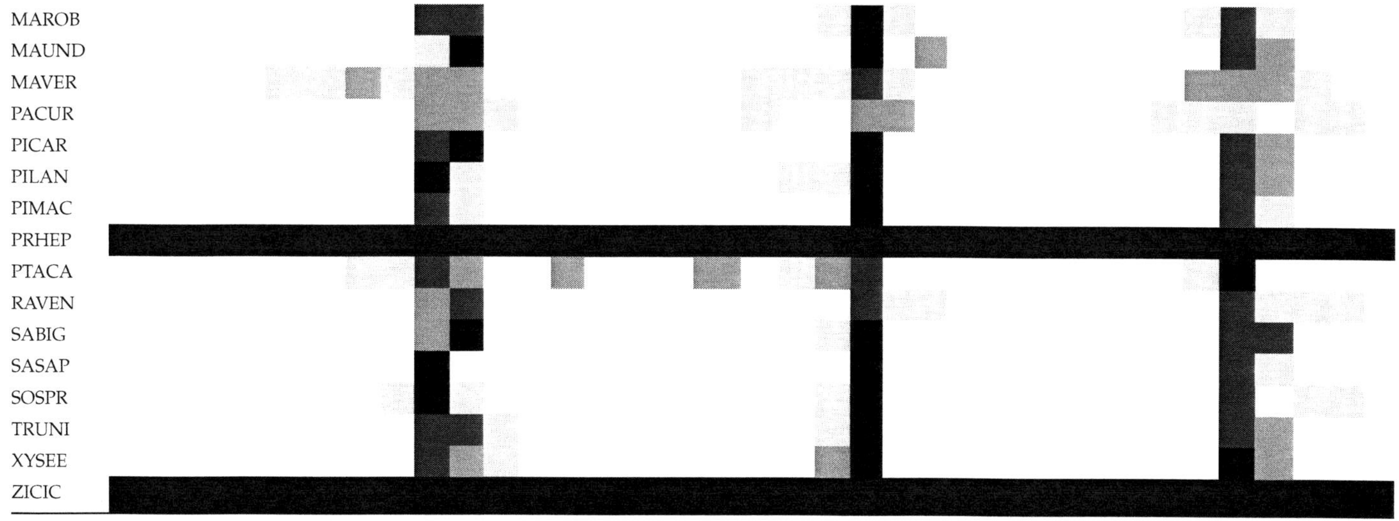

Note: Together with the average Fournier values for the leaf-flushing condition, average monthly precipitation and insolation estimates are included. The species in the table are as follows: *C. pubiflora* (COPUB), *I. interrupta* (ININT), *L. fendleri* (LOFEN), *P. lanceolatum* (PILAN), *P. acapulsensis* (PTACA), *A. barinensis* (ALBAR), *M. robiniifolium* (MAROB), *P. carthaginensis* (PICAR), *A. corallina* (ARCOR), *A. mollisima* (ARMOL), *M. verrucifera* (MAVER), *X. seemanianum* (XYSEE), *A. glomerosa* (ACGLO), *A. niopoides* (ALNIO), *C. coriaria* (CACOR), *A. graveolens* (ASGRA), *P. heptaphyllum* (PRHEP), *T. unifoliola* (TRUNI), *P. cururu* (PACUR), *S. saponaria* (SASAP), *C. peltata* (CEPEL), *F. maxima* (FIMAX), *S. sprucei* (SOSPR), *Z. cyclocardia* (ZICIC), *C. fruticosum* (COFRU), *C. vitifolium* (COVIT), *S. biglandulosum* (SABIG), *C. flexuosa* (CAFLE), *G. ulmifolia* (GUULM), *Pisonia macranthocarpa* (PIMAC), *C. venezuelana* (COVEN), *G. divaricata* (GUDIV), *G. americana* (GEAME), *R. venezuelensis* (RAVEN), *M. undulata* (MAUND), *Hecatostemon completus* (HECOM), and *A. purpurea* (ANPUR). Fournier's new leaves records are as follows: 0 (white), >0–1 (light grey), >1–2 (grey), >2–3 (dark grey), and >3–4 (black). Evergreen species are indicated in the table with a continuous black line across all months.

events during the dry season: *Albizia niopoides, Anonna purpurea, Arrabidaea mollisima, C. pubiflora, M. verrucifera,* and *P. acapulsensis.* It is also notable that even though many species massively produced leaves in a short time interval (less than one month), some presented leaf-flushing activity during several months (Table 17.4), including several months during the rainy season.

No significant relationship was observed between seasonal variation in insolation level and leaf-flushing yearly patterns in any of the successional plots examined; however, for the majority of species evaluated (83%), leaf flushing took place during or after the first peak of insolation that occurs during the year in zones near the equator (Table 17.4). Besides this global trend, it is important to point out several species with leaf-flushing activity taking place during other parts of the year also, including the middle of the dry season (*C. pubiflora, I. interrupta, Paullinia cururu, Z. cyclocardia,* and *S. bigandulosum*) and advanced rainy season (*C. pubiflora, M. robiniifolium, Albizia barinensis, A. niopoides, Arrabidaea corallina, P. cururu, Sapindus saponaria, Ficus maxima, C. venezuelana, Guazuma ulmifolia, Pisonia macranthocarpa,* and *Randia venezuelensis*).

17.4 Discussion

The dry forest examined at UPSA-Piñero, Cojedes state, Venezuela, presents the general foliar phenological pattern described for other TDFs in the region: a dry season, characterized by a massive loss of leaves in part of the woody vegetation, followed by a period of massive leaf flushing. This pattern, however, was affected by the pronounced interannual variation in the precipitation regime, with dry periods as short as two months (2009–2010) to as long as seven months (2008–2009). Continuous monitoring of this forest for three consecutive years allowed us to classify the different plant species according to their degree of deciduousness and to evaluate how they responded to interannual variations in the precipitation regime. Overall, our results support the hypothesis that leaf flushing in the dry forests of the Venezuelan Llanos is primarily driven and influenced by the onset of rains and the level of subsoil water reserves and that, subordinated to these factors, daily insolation could also have some influence on the production and expansion of new leaves in some species.

17.4.1 Deciduous Condition

Like in other dry forests in tropical America, including Costa Rica (Borchert 1994; Rivera et al. 2002) and Mexico (Bullock and Solis-Magallanes 1990), the leafless condition in the studied forest occurred during the dry season and lasted for a variable number of months, depending on the species. Between 2007 and 2010, leaf fall occurred in November, despite the fact that an important amount of rainfall occurred during that month in 2007 and 2009. These two cases illustrate the importance of leaf longevity as one of the endogenous factors triggering leaf shedding, independently of the amount of water available to the plants (Wright and Cornejo 1991; Zalamea and González 2008).

The most marked deciduous condition was observed in the intermediate successional stage, in which trees lost 100% of their foliar cover in many cases. Despite this, total litterfall production in that stage was lower than that in the late successional stage. Taller trees in the latter, with deeper roots, should have better access to phreatic water than smaller trees in the intermediate plots (Sobrado & Cuenca 1979; Holbrook et al. 1995; Singh and Kushwaha 2005), and therefore deciduousness can be less intensive in the latter stage. On the other hand, many of the trees observed in the intermediate plots presented compound leaves, with comparatively smaller leaf areas and sizes with respect to the larger leaves produced by several species in the late plots, which, in addition, presented larger branches. This could explain why litterfall in the late successional plots in our study site was higher than that in the intermediate plots. In the dry forests of Santa Rosa and Palo Verde, Costa Rica, the deciduous condition decreased from earlier to more advanced successional stages (Kalácska et al. 2005a, b); however, these authors found the highest production of litterfall associated with the intermediate successional stage, followed by the late and early stages. On the other hand, many of the trees observed in the intermediate plots presented compound leaves, with comparatively smaller leaf areas and sizes with respect to the larger leaves produced by several species in the late plots, which, in addition, presented larger branches. This could explain why litterfall in the late successional plots in our study site was higher than that in the intermediate plots. In the dry forests of Santa Rosa and Palo Verde, the deciduous condition decreased from earlier to more advanced successional stages (Kalácska et al. 2005a,b); however, Kalácska et al. found the highest production of litterfall associated with the intermediate successional stage, followed by the late and early stages. But the most contrasting difference was observed in relation to the early successional plots. In two of the early plots of our study site, deciduousness was not as marked as expected because water availability in the former plots was comparatively higher due to their proximity to permanent water bodies and the hills of Galeras del Pao, which drain rainwater toward lower elevations in their vicinities. Similar responses to local differences in water availability in dry forests have been observed in the Neotropical region. Ragusa-Netto and Silva (2007) evaluated the canopy phenology of a dry forest in western Brazil, finding that most leaf losses occurred in dry hills, whereas wet valleys with deeper soil layers and shallower water tables remained evergreen.

Different plant species possess different foliar phenological behaviors in TDFs (Reich and Borchert 1982; Murphy and Lugo 1986; Bullock et al. 1995; Pennington et al. 2006; Gutiérrez-Soto et al. 2008). The capacity of water retention and leaf morphological types are two of the plant attributes that play a role in determining the intensity and duration of the deciduous condition in a species (Olivares and Medina 1992). In the current study, we classified the examined species into three functional categories, leaf exchanging, brevideciduous, and deciduous, but only the last two categories were widespread in the local flora. Deciduous species numerically dominated within the intermediate and late successional plots, explaining why in these stages levels of leafless condition reached maximum values. Only a small number of the species monitored presented a green foliage year around. Several of these species were mainly associated with early successional plots. In a dry forest in western Brazil, Ragusa-Netto and Silva (2007) also reported a small proportion (9%) of evergreen species associated with tropical dry vegetation.

17.4.2 Leaf Flushing

The sequence of events from leaf fall to flushing described for many seasonal tropical forests (Morellato et al. 1989) occurs also in the Central Llanos of Venezuela. The role of water availability as a key factor influencing foliar phenology in tropical plants has been the subject of multiple theoretical and experimental studies, which have demonstrated that water is the main triggering factor for production of foliar buds and leaf expansion in many species of plants in dry forests (Frankie et al. 1974; Reich and Borchert 1982; Borchert 1994; Bullock 1995; Reich 1995; Borchert and Rivera 2001; Rivera et al. 2002; Elliot et al. 2006; Williams et al. 2008). Leaf flushing in the study plots concentrated at the beginning of the rainy season, confirming that this factor is the primary one determining the production of new leaves in this forest. For years 2008 and 2009, leaf flushing in the intermediate and late successional plots occurred during the beginning of rains or less than a month after the rains had started. This was the general pattern for the majority of species examined, independently of their taxonomic group.

For the particular case of the early successional plots during 2009, we observed that occurrence of leaf flushing in February followed a few rains that occurred in the region during the previous month. For year 2010, the occurrence of the leaf-flushing peak one month earlier (April) than in the two previous years in intermediate and late successional plots was concordant with the early onset of rains that year in the region; however, rains started as early as February. Then why did leaf flushing not occur immediately after the onset of the rainy season? The maximum level of daily insolation was reached in April, suggesting that this factor could contribute to the determination of leaf-flushing initiation. Production and expansion of new leaves in dates close to the spring equinox have been reported for dry forests in Costa Rica, Colombia, and Panama (Rivera et al. 2002; Calle et al. 2010), monzonic forests in Java, tropical savannas in northern Australia (Rivera et al. 2002), and northern Thailand and India (Elliot et al. 2006). Based on this finding, we propose that even though the dominant factor triggering the production of leaf buds in this forest is precipitation regime, daily insolation could act as a subordinate factor contributing to the stimulation of leaf flushing.

During the dry seasons of the three years monitored in this study, several nonevergreen species presented recurrent leaf-flushing episodes. To understand this phenological behavior, we need to consider the effect of different physiological attributes of the plants, including the level of canopy transpiration, the root system, water storage capacities, and resource requirements, among others (Sarmiento and Monasterio 1983; Reich 1995; Marques et al. 2004). Access to deep soil water through the root system facilitates the survival of many species of tropical trees during the dry season (Medina et al. 1985; Díaz and Granadillo 2005). *A. mollisima* and *M. verrucifera* are two lianas that showed leaf flushing during the dry season. Many lianas are characterized by having a deep radical system, which guarantees physical support to the plants and access to deep soil water (Jackson et al. 1999; Andrade et al. 2005). Three species of legume trees, *C. pubiflora, P. acapulcensis,* and *A. niopoides*, expanded new leaves during the dry season. Many tropical trees in this family possess great tolerance to droughts due to their capacity to extract and store water from the ground (Borchert 1994; Gentry 1996). For example, Jackson et al. (1999) found that the semideciduous species *Dalbergia myscolobium* (Fabaceae) expanded new leaves at the end of the dry season because its roots were able to penetrate deeply into the soil. Rivera et al. (2002) mentioned several species of *Pterocarpus* and *Albizia* as capable of producing new leaves before the onset of rains in dry forests of Costa Rica and the Chaco region, Argentina.

It is important to point out that in the present study we have only considered precipitation and daily insolation as determinant factors of the foliar phenological patterns in the UPSA-Piñero dry forest. A departure from the general leaf-flushing pattern associated with the onset of the rainy season in TDFs probably responds to additional endogenous and exogenous factors not considered in this study, such as soil's water storage and water retention capacities, photosynthetic capacity, and molecular control of foliar bud activity and dormancy, among others. Soil parameters should be considered for a more integral comprehension of these patterns, especially water retention capacities and nutrient availability associated with the different successional stages, following the model of resource investment (Kikusawa 1995), which considers seasonality in resource availability as a primary modulating factor of the leaf exchange rhythm.

17.5 Conclusion

Dry forests in the Central Llanos of Venezuela are exposed to an intense seasonal regime that can vary considerably among years in duration of the dry and rainy periods. The leaf-flushing pattern observed suggests that the foliar phenology of these forests responds primarily to the annual rainfall regime and the level of subsoil water reserves. But, besides this dominant factor, daily insolation and its variation over the year could be playing a secondary role, determining the annual pattern of leaves turnover in these forests. Finally, the particular species composition associated with the different successional stages and the different functional strategies adopted by those species can contribute in a minor extent to variations at species levels in the foliar response of plants to the onset of the dry and rainy seasons.

18

Floristic, Structural, and Functional Group Variations in Tree Assemblages in a Brazilian Tropical Dry Forest: Effects of Successional Stage and Soil Properties

Yule Roberta Ferreira-Nunes, Giovana Rodrigues da Luz, Saimo Rebleth de Souza, Diellen Librelon da Silva, Maria das Dores Magalhães-Veloso, Mário Marcos do Espírito-Santo, and Rubens Manoel dos Santos

CONTENTS

18.1 Introduction

Tropical dry forests (TDFs) in Brazil occur predominantly as forest disjunctions forming natural "islands" amidst all the large Brazilian biomes (Scariot and Sevilha 2005; Espírito-Santo et al. 2009). In these formations, the period of favorable growth is usually restricted to a short rainy season, when seed germination and seedling establishment is expected (Khurana and Singh

2001). Thus, the pronounced seasonal climate determines the patterns of seed production, germination, and seedling survival and development in the plant community (Sampaio 1994).

TDFs are frequently associated with fertile soils, with moderate values of pH and nutrient availability and low levels of aluminum (Ribeiro and Walter 2008). Soils with these characteristics are favorable to agriculture (Ratter et al. 1978), which contributed to the enormous degradation and fragmentation of these forests, resulting in the formation of forest mosaics at different stages of ecological succession (Arroyo-Mora et al. 2005a). Understanding the influence of soil conditions during the successional process is of great relevance, since there is a close relationship between the abiotic conditions and the vegetation (Reatto et al. 1998). According to Mulkey et al. (1996), the geographic distribution of plant communities is influenced by the interaction between the ecophysiological tolerance of each species and the microclimatic and microedaphic variables. In this way, areas of TDFs that had different histories of land use (agriculture and livestock raising), management practices (clear-cut, removal of roots and stumps, fires, use of heavy machinery and fertilizer and pesticide input) (Espírito-Santo et al. 2009), and time since abandonment may present contrasting patterns of regeneration and, consequently, very different vegetation structure.

Areas in different successional stages have distinct environmental complexity, ranging from early forests with simple structure and low diversity to structurally complex late forests with high diversity (Kalácska et al. 2004b; Madeira et al. 2009). Thus, plant species are usually classified, according to their regeneration strategies, into the pioneer, intermediate, and late groups, known as "regeneration guilds" (Swaine and Whithmore 1988). For Rodrigues and Gandolfi (1996), species in each successional group have different biological characteristics, such as different light requirements, growth rates, and duration of life cycle. So, it is observed that the environmental variability that occurs in early stages of succession is higher than that which occurs in more advanced ones, and light availability is one of the main drivers of species replacement during succession (Ronce et al. 2005). As the arboreal components of the community increases in frequency, pioneer species are replaced with intermediate and late species, with progressive canopy connectivity causing a decrease in light availability in the understory (Madeira et al. 2009). So, the range of plant species' shade tolerance and the canopy architecture are important in determining successional sequences in TDFs (Bazzaz 1979).

In addition to the physiological needs of each species to be established during forest succession (regeneration guilds) (Rodrigues and Gandolfi 1996), the process of seed dispersal (dispersal guilds) (Ibarra-Manríquez and Oyama 1992) and tree height (stratification guilds) (Nunes et al. 2003) are important characteristics that can influence the plant dynamics throughout succession. Seed dispersal, for example, is an important trait for understanding the structure and function of forest communities. In the case of natural regeneration of communities, seed dispersal plays a crucial role in the events that accelerate the recovery of degraded forests (Wunderlee 1997). The management and restoration of disturbed forests depend on the efficiency of the processes of propagule dispersion and on the establishment of species from different successional stages (Rondon-Neto et al. 2001). The anemochoric guild (wind dispersal) is fairly common in TDFs (van Schaik et al. 2003) and, in general, the wind speed increases with canopy height (Oliveira and Moreira 1992). Pezzini et al. (2008) studying plant phenology in a TDF found that anemochory was more representative in the intermediate and late stages, the ones that had taller individuals. According to Odum (1998), studies that deal with guilds are very important because these are convenient units to determine interactions between species and functional aspects of the community. Therefore, the grouping of species into guilds and their association with quantitative studies of vegetation structure provides a better knowledge of the interplay between species and their physical environment, which can guide more appropriate practices for recovering degraded areas.

The knowledge on floristic composition, structure and function, and soil attributes of a forest in different successional stages not only contributes to a better understanding of the community regeneration but also reveals some important aspects of species successional strategies. This information is essential when planning activities for sustainable management of forest resources, monitoring permanently preserved areas, and developing commercial forests with native species (Vaccaro 1997). In this way, the present study aimed to evaluate and compare the floristic composition, the structural and functional groups (dispersal, regeneration, and stratification guilds) of plant species along a successional gradient in a Brazilian TDF, and the soil composition and its influence on these plant communities.

18.2 Methods

18.2.1 Vegetation Sampling

Vegetation sampling took place at the Mata Seca State Park located in the State of Minas Gerais (Figure 1.4, see Chapter 1 for a complete description). The floristic and structural sampling of arboreal–shrub vegetation was conducted in March 2009, within six plots of 1000 m^2 (20 m × 50 m) in each successional stage, totalling a sample area of 1.8 ha. We sampled all shrub and tree individuals with diameter at breast height (DBH, measurements taken at 1.30 m above the ground) ≥5 cm. Plants were marked with numbered aluminium tags and identified at the species level. The height of each individual was estimated using the projection of the poles of the tree trimmer. The collected plant material was treated according to conventional herborization techniques and deposited in the Montes Claros Herbarium (HMC) at the Universidade Estadual de Montes Claros. Plant identification was conducted by consulting specialists, using literature and comparing samples with the specimens available in the HMC. For plant family classification, we used the Angiosperm Phylogeny Group III (APG III 2009).

18.2.2 Ecological Characterization of Plant Species

To compare the ecological strategies of plant species occurring at different successional stages, we classified each species into dispersal, regeneration, and stratification guilds (see Nunes et al. 2003). Based on the criteria and categories proposed by van der Pijl (1982), the dispersal guild was classified into three basic groups: (1) anemochoric species, which have mechanisms that facilitate dispersal by wind, (2) zoochoric species, whose dispersal is performed by animals, and (3) autochoric species, which disperse the diaspores by gravity or mechanisms of self-dispersal such as explosive dehiscence (ballistic dispersal). For the regeneration guilds, plant species were classified according to the definition of Swaine and Whitmore (1988) as follows: (1) pioneers, the ones that need direct light to germinate and develop; (2) light-demanding climax species, whose germination occurs under shade conditions, although immature individuals require abundant light for its development until they reach the canopy, and (3) the shade-tolerant climax species, whose germination and development occur in the shaded understory, with plant establishment and growth under the forest canopy. The classification of species into stratification guilds was based on the adult's average height, using the same categories and intervals defined by Oliveira-Filho et al. (1997): (1) short species, with heights between 2.0 and 7.9 m; (2) medium species, ranging from 8.0 to 17.4 m in height; and (3) tall species, which can exceed 17.5 m in height. For seed dispersal and regeneration syndromes, the species were classified based on field observations and literature information (Lorenzi 1992; Morellato and Leitão-Filho 1992a; Lorenzi 1998; Barroso et al. 1999; Barbosa and Lima 2003; Carvalho 2003, 2006, 2008).

18.2.3 Soil Sampling

Soil analyses were conducted with samples of superficial soil (0–20 cm) from each of the 18 plots used in the present study. The collected material was brought to the Laboratory of Soil Analyses in the Instituto de Ciências Agrárias of Universidade Federal de Minas Gerais, where the chemical and the textural analyses were performed, according to the EMBRAPA protocol (1997). From these analyses, we obtained the following variables: pH in water (pH); potassium (K); phosphorus content (P-Mehlich); remainder phosphorus (P-rem); calcium (Ca); magnesium (Mg); aluminum (Al); hydrogen + aluminum (H + Al); sum of bases (SB); base saturation (V); effective cation exchange capacity (t); aluminum saturation (m); cation exchange capacity at pH 7.0 (T); organic matter (OM); and proportions of thick sand (2–0.2 mm), fine sand (0.2–0.05 mm), silt (0.05–0.02 mm), and clay (<0.02 mm).

18.2.4 Data Analyses

Differences in vegetation structure between successional stages were evaluated by calculating the classical quantitative parameters of density, dominance and relative frequency, and importance value index (IVI) (Mueller-Dombois and Ellenberg 1974) for each species at each stage. Moreover, comparisons of diversity, distribution of individuals, and species richness between successional stages were performed by calculating the Shannon diversity index (H'), Pielou's evenness index (J') (Krebs 1989), and the Jackknife estimator 1 (Burnham and Overton 1979), respectively. The H's of the different successional stages were compared by the t-test of Hutcheson (Zar 1996). The analysis was made between pairs of successional stages (Nunes et al. 2003). To evaluate whether the classification of the forest fragments as early, intermediate, and late successional stages was compatible with the level of complexity of the vegetation, we calculated the Holdridge complexity index (C_{HCI}) (Holdridge 1967; Holdridge et al. 1971). The three successional stages were compared for the following structural characteristics: height, basal area, density, richness, and C_{HCI} by analysis of variance (ANOVA) in the GLM (generalized linear model) procedure. The species similarity between the stages was verified both in relation to floristic composition and abundance of each species. For the first analysis, we used the Sørensen similarity index (Sørensen 1948) considering only the presence and absence of the each species; for the second analysis, we used the Morisita index (Morisita 1959), considering the number of individuals per species in each plot.

To compare the frequency distribution of individuals of shrub and tree species into the guilds (regeneration, stratification, and dispersion guilds) between successional stages, we performed a chi-square test using a contingency table (Zar 1996; Dytham 2011). The same analysis was used to compare the frequency of individuals between regeneration × dispersal, regeneration × stratification, and stratification × dispersal guilds (Zar 1996; Dytham 2011).

Differences in the chemical and physical soil attributes between the successional stages were evaluated by ANOVA (one-way ANOVA) followed by Tukey's HSD post hoc tests ($p < 0.05$). To verify the correlations between the soil characteristics and vegetation composition in each sample unit (plot), we performed a mixed-gradient analysis using a canonical correspondence analysis (CCA) (ter Braak 1987). For this purpose, we constructed a matrix with presence and absence data for each species, and a matrix with soil variables. We performed a Monte Carlo permutation test to evaluate the significance level of the main axis and confirm the odds on correlations between environmental variables and vegetation (ter Braak and Smilauer 1998). After the preliminary analyses, edaphic variables that were nonsignificant or that had high redundancy ($p > 0.05$) were eliminated from the model, leaving only the following soil properties, which were closely related to the floristic variables: pH, K (potassium), H + Al (hydrogen + aluminum), SB (sum of bases), *t* (effective cation exchange capacity), OM (organic matter), and *V* (base saturation).

18.3 Results

18.3.1 Floristic Composition and Structure

We sampled 1712 individuals, representing 89 species from 25 botanical families (see appendix to this chapter), and only one species was not identified. In the early stage, 517 individuals from 35 species and 14 families were recorded. In intermediate stage, we sampled 520 individuals belonging to 46 species and 16 families, and in the late stage, 675 individuals from 55 species and 18 families were sampled. In total, the five families with greatest richness were Fabaceae, Bignoniaceae, Euphorbiaceae, Malvaceae, and Anacardiaceae, which together encompassed 61.5% of all species sampled. Fabaceae was the richest family, with 35 species, representing 38.4% of the total species sampled.

The species accumulation curves based on the number of individuals (Figure 18.1a) and the number of plots sampled (Figure 18.1b) showed that species richness was higher in the late stage and lower in the early stage. However, only the richness in the early stage differed statistically from the other stages (Table 18.1). Besides, both the species diversity and the equitability were higher in the intermediate stage ($H' = 2.87$, $J' = 0.74$), followed by the late ($H' = 2.71$, $J' = 0.68$) and early ($H' = 2.27$, $J' = 0.64$) stages. The H' statistically differed between the stages (early × intermediate: $v = 1032$, $t = 7.291$, $p < 0.001$; early × late: $v = 1170$, $t = 5.367$, $p < 0.001$; intermediate × late: $v = 1169$, $t = 1.918$, $p < 0.05$).

The floristic similarity between the three successional stages was low, with only seven species in common: *A. lentiscifolium*, *A. excelsa*, *C. duarteanum*, *D. cearensis*, *M. acutifolium*, *M. Urundeuva*, and *P. pluviosa*. There were 20, 10, and 20 species exclusive to the early, intermediate, and late stages, respectively. The Sørensen similarity (Figure 18.2a) and Morisita indices (Figure 18.2b) demonstrated a greater similarity between the intermediate and late stages (0.47), with 25 species in common, and a low similarity between the early and the late stages (0.04), with only three shared species. The early and intermediate stages also showed low similarity (0.12), with only five common species.

Individuals with 5–8 cm in DBH predominated in all successional stages. However, we observed a higher frequency of individuals with DBH above 8 cm in the late and intermediate stages (Figure 18.3). Height, basal area, and density values were lower for individuals in the early

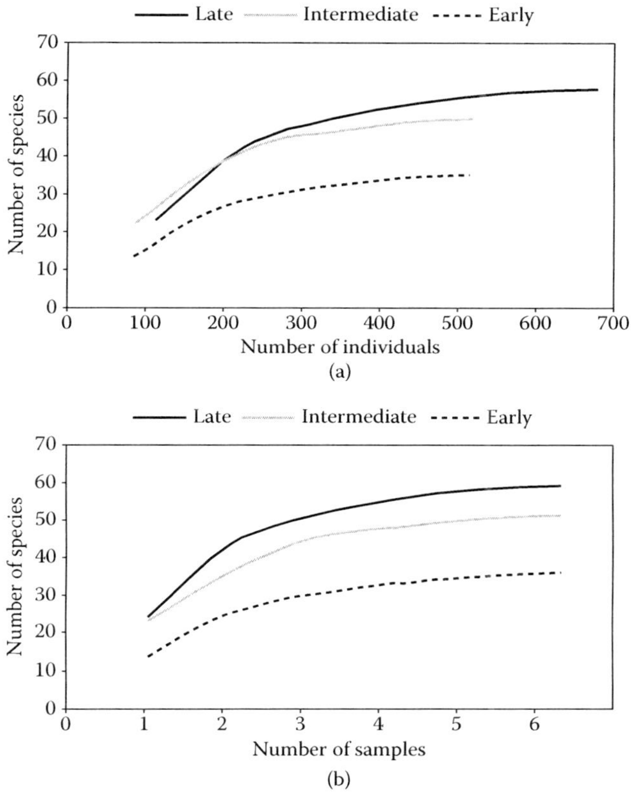

FIGURE 18.1
Rarefaction curves of diversity estimator (Jackknife estimator 1), number of species versus number of individuals, and number of species versus the samples of 18 plots of the three stages of ecological succession in the Mata Seca State Park (MSSP).

TABLE 18.1

Attributes (Mean Values with Standard Deviation) of Tree and Shrub Communities Sampled at Different Successional Stages in a Tropical Dry Forest in Southeastern Brazil (Mata Seca State Park [MSSP], Minas Gerais, Brazil)

Attributes	Early	Intermediate	Late	F	p
H	5.93 ± 2.92^{b}	9.53 ± 5.33^{a}	11.33 ± 5.45^{a}	24.789	<0.001
BA	4.00 ± 0.02	9.50 ± 0.04	15.88 ± 0.05	0.826	0.4566
D	863.33 ± 398.46	866.66 ± 143.48	1125.00 ± 168.73	2.972	0.082
S	13.67 ± 3.56^{b}	22.00 ± 2.83^{a}	22.83 ± 4.36^{a}	11.669	<0.001
C_{HCI}	4.97 ± 3.44^{b}	37.61 ± 15.71^{ab}	111.27 ± 64.05^{a}	5.239	0.051

Note: H = height (m); BA = basal area ($m^2 \cdot ha^{-1}$); D = density (indiv. ha^{-1}); S = richness; C_{HCI} = Holdridge complexity index. Different letters indicate statistical differences at 5% of probability.

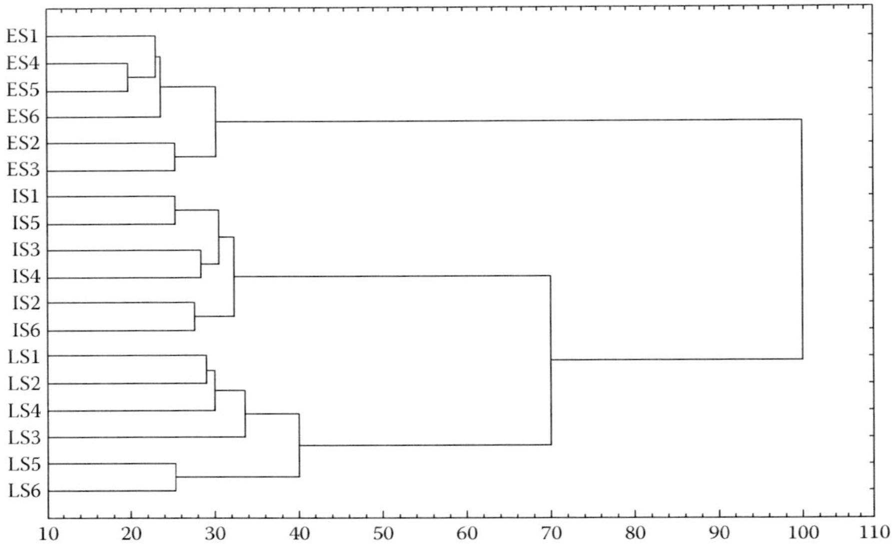

$(D_{link}/D_{max})\cdot 100$

(a)

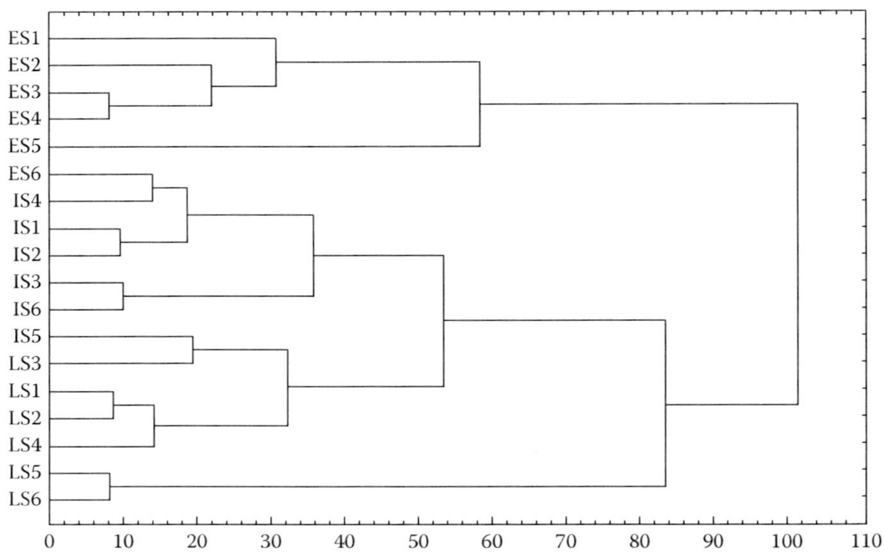

$(D_{link}/D_{max})\cdot 100$

(b)

FIGURE 18.2
Dendograms of similarity based on (a) abundance of species (Sørensen) and (b) presence or absence of species (Morista) of the 18 plots sampled in the MSSP. ES = early stage; IS = intermediate stage; LS = late stage.

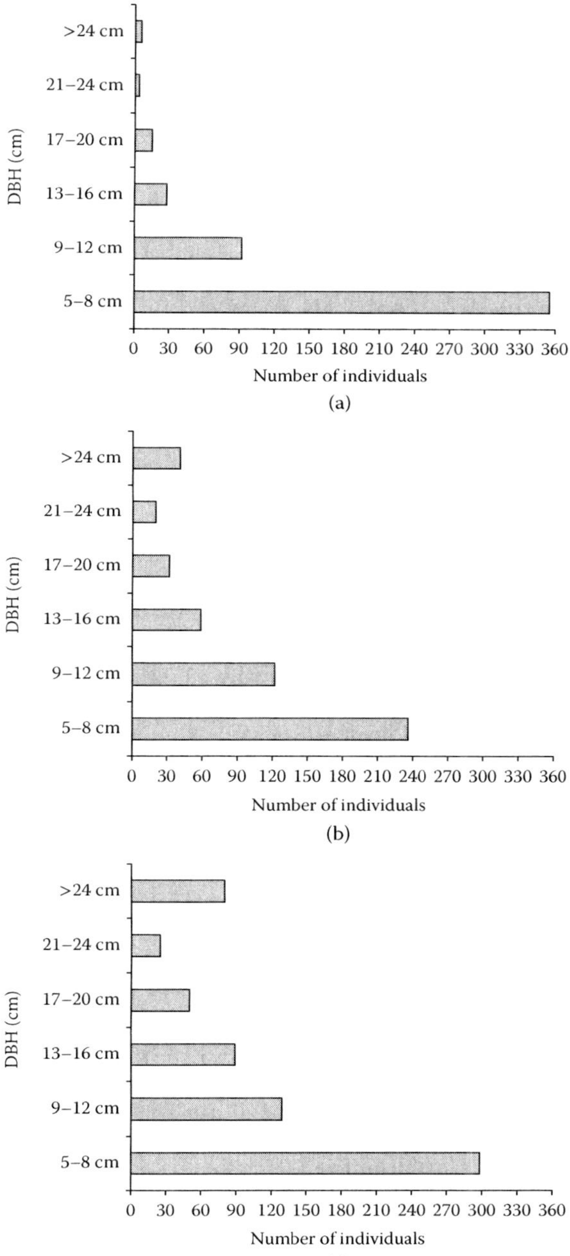

FIGURE 18.3
Number of trees in the (a) early, (b) intermediate, and (c) late stages in the diameter classes of 5–8, 9–12, 13–16, 17–20, 21–24, and >24 cm.

stage and higher in the late stage, but basal area and density values did not statistically differ between stages (Table 18.1). Only the height of the individuals in the early stage differed from the other stages. The C_{HCI} also followed the pattern observed of lowest value for the early stage and higher for the late stage, with statistically significant difference between the two.

There was a significant change in the species IVI along the successional gradient (Figure 18.4). Three of the five most important species (IVI) were shared between the intermediate and late stages (*C. duarteanum*, *H. reticulatus*, and *C. leptophloeus*). *M. urundeuva* was among the five most important species in the early and late stages. However, none of the top five species of highest IVI were common to early and intermediate stages.

18.3.2 Guild Composition

We were able to classify 88 shrub and tree species according to their dispersal syndromes (Appendix). We observed a significant predominance of anemochoric species (50%), followed by zoochoric (33%) and autochoric species (17%). The frequency of individuals per syndromes varied between successional stages: 56%, 50%, and 54.5% of the species were anemochoric in early, intermediate, and late stages, respectively, but this difference was not statistically significant (see Table 18.2 for the observed and expected cell counts). The proportion of zoochoric species varied significantly between the early, intermediate, and late stages (23.5%, 37%, and 33%, respectively), the same being observed for autochoric species (20.5%, 13%, and 12.5%, respectively). Guild frequency also differed significantly within stages: the early stage had more autochoric and less zoochoric individuals than expected by chance. The intermediate and late stages had fewer autochoric and more zoochoric individuals than expected. However, the distribution of anemochoric individuals did not differ from the expected in all stages of succession.

Regarding the regeneration guilds, there was a general predominance of shade-tolerant species (52%), followed by light-demanding (25%) and the pioneer species (23%). The frequency distribution of individuals into regeneration guilds also varied between successional stages: 23.5%, 26%, and 25.5% were light-demanding climax in early, intermediate, and late stages, respectively, but this difference was not statistically significant (Table 18.2). The proportion of shade-tolerant climax species differed significantly between the early, intermediate, and late stages (32.5%, 56.5%, and 63.5%, respectively), the same being observed for pioneer species (44%, 17.5%, and 11%, respectively). Guild frequency also differed significantly within stages: in the early stage, the number of pioneer individuals was higher than expected by chance and the opposite was observed for shade-tolerant individuals. A reversed pattern was observed for the intermediate and late

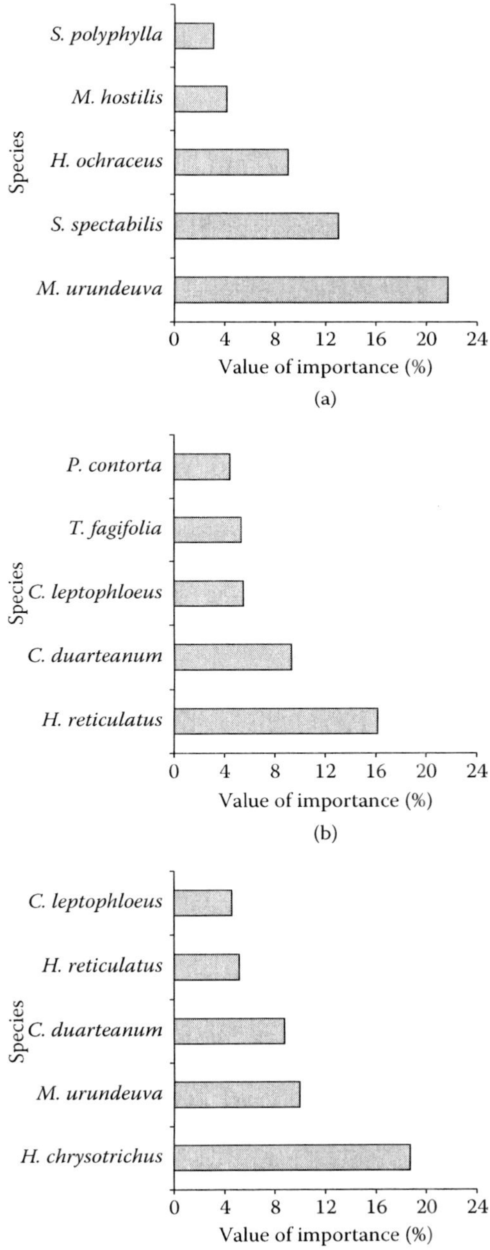

FIGURE 18.4
Tree species with the highest importance value index (IVI) in each successional stage: (a) early, (b) intermediate, and (c) late, sampled in the MSSP.

TABLE 18.2

Contingency Tables with the Observed and Expected (in Parentheses) Frequencies of Trees for Regeneration, Stratification, and Dispersal Guilds in the Three Stages of Ecological Succession Sampled in the MSSP

			Forest Stages			Chi-Square	
			ES	IS	LS	χ^2	p
Dispersal guilds	Anemochoric		348 (376.5)	375 (379.4)	525 (492.1)	4.39	>0.05
	Autochoric		143 (75.4)	59 (76.0)	48 (98.6)	90.34	<0.001
	Zoochoric		23 (62.1)	84 (62.6)	99 (81.3)	35.83	<0.001
Chi-square		χ^2	87.38	11.15	32.03	130.56	
		p	<0.001	<0.005	<0.001		<0.001
Regeneration guilds	Pioneer		193 (103.4)	86 (104.3)	61 (135.3)	118.23	<0.001
	Light-demanding climax		130 (124.8)	120 (125.5)	164 (162.9)	0.41	>0.75
	Shade-tolerant climax		191 (286.0)	312 (288.2)	444 (373.8)	46.00	<0.001
Chi-square		χ^2	109.05	5.41	50.72	164.64	
		p	<0.001	>0.05	<0.001		<0.001
Stratification guilds	Tall		204 (265.1)	267 (267.2)	408 (346.7)	24.96	<0.001
	Medium		249 (192.2)	181 (193.6)	207 (251.2)	25.43	<0.001
	Short		61 (56.7)	70 (57.2)	57 (74.1)	7.18	<0.05
Chi-square		χ^2	31.25	3.71	22.60	57.56	
		p	<0.001	>0.10	<0.001		<0.001

Note: The results of the chi-square test are provided for each table as a whole and for its rows and columns separately. ES = early stage; IS = intermediate stage; LS = late stage.

stages, where the number of pioneering individuals was lower and the number of shade-tolerant plants was higher than expected. The frequency of light-demanding individuals did not differ from the expected in any successional stage.

Among the stratification guilds, in all successional stages there was a predominance of tall species (43%), followed by medium (33%) and short species (24%). The frequency distribution of individuals into all stratification guilds varied significantly between successional stages (Table 18.2): tall individuals comprised 32.5%, 45.5%, and 49% of individuals in the early, intermediate, and late stages, respectively. The frequencies were 41%, 35%, and 33% for medium-sized plants, and 26.5%, 19.5%, and 18% for short individuals, along the successional sequence (early, intermediate, and late stages, respectively). Stratification guild frequency also differed significantly within stages: tall individuals were more frequent than expected in the late stage and less frequent in the early stage. The observed frequency of individuals of medium size was higher than expected in the early stage and lower in the intermediate and late stages. For short trees, the number of individuals was higher than expected in the early and intermediate stages and lower in the late stage.

18.3.3 Interactions between Guilds

We found significant differences in the observed and expected distribution frequencies of individuals between guilds (seed dispersal, regeneration, and stratification) (Table 18.3). The frequency of pioneer individuals with anemochoric and zoochoric syndromes was lower than expected, and the opposite was observed for the autochoric syndrome. Shade-tolerant individuals showed a reverse pattern. For light-demanding individuals, anemochory was more frequent compared to autochory and zoochory. In the pioneer guild, the frequencies of medium and short individuals were higher than expected, and the opposite was observed for tall individuals. The reverse pattern was observed for shade-tolerant individuals. For the light-demanding functional group, the frequency of medium-sized individuals was higher than expected, while both tall and short individuals were less frequent than expected. Tall and medium individuals had a higher frequency of anemochory and a lower frequency of zoochory than expected. A reverse pattern was observed for short individuals. Autochory was more common than expected for short and medium individuals, the opposite being observed for tall individuals.

18.3.4 Soil–Vegetation Interactions

We identified two soil types in the plots established at the Mata Seca State Park (MSSP), Minas Gerais, Brazil: (1) Red Latosol, an eutrophic soil originated from the Bambuí group and colluvial–eluvial soils from Quaternary

TABLE 18.3

Contingency Tables across the Frequency Distribution of Observed and Expected (in parenthesis) Individuals of the Species Sampled in the MSSP between the Regeneration, Stratification, and Dispersal Guilds

		Dispersal Guilds			Chi-Square	
Regeneration Guilds		**Anemochoric**	**Autochoric**	**Zoochoric**	χ^2	p
Pioneer		93 (251.2)	215 (50.3)	35 (41.5)	135.02	<0.001
Light-demanding climax		407 (302.5)	2 (60.6)	5 (50.4)	639.54	<0.001
Shade-tolerant climax		748 (694.3)	33 (139.1)	167 (114.6)	109.02	<0.001
Chi-square	χ^2	139.91	676.47	67.21	883.59	
	p	<0.001	<0.001	<0.001		<0.001
Stratification guilds						
Regeneration guilds		*Tall*	*Medium*	*Short*		
Pioneer		117 (176.9)	159 (128.2)	67 (37.9)	521.95	<0.001
Light-demanding climax		26 (213.0)	379 (154.4)	8 (45.6)	50.16	<0.001
Shade-tolerant climax		736 (489.0)	99 (354.4)	113 (104.5)	309.46	<0.001
Chi-square	χ^2	309.26	518.2	54.11	881.56	
	p	<0.001	<0.001	<0.001		<0.001
Dispersal guilds						
Stratification guilds		*Anemochoric*	*Autochoric*	*Zoochoric*		
Tall		750 (643.8)	73 (129.0)	56 (106.3)	65.59	<0.001
Medium		480 (466.5)	101 (93.5)	56 (77.0)	6.73	<0.05
Short		18 (137.7)	76 (27.6)	94 (22.7)	412.54	<0.001
Chi-square	χ^2	121.96	109.89	253.01	484.86	
	p	<0.001	<0.001	<0.001		<0.001

Note: The results of the chi-square test are provided for each table as a whole and for rows and columns separately.

deposits; (2) Haplic Cambisol, an eutrophic–petroplintic soil also derived from the Bambuí group. The plots in the early successional stage had soils with mollic (chernozemic) horizons, intermediate texture in flat terrain. In the intermediate stage, we found soils with moderate horizons, clay texture in flat terrain. In the late stage, we detected both soil types (1 and 2) to moderate, with a clay texture and ranging from flat to soft undulating terrain.

The edaphic variables evaluated in the present study differed significantly between the three successional stages, except for P-Mehlich and thick and fine sand (Table 18.4). The Tukey's HSD post hoc test showed that the late and early stages had higher similarity in their chemical composition, specifically for pH, Mg^{2+}, Al, H + Al, *m*, *V*, silt, and clay.

According to the CCA (Figure 18.5), the total variance of the data explained 3.8%, being that axis 1 explained 18.5%, axis 2, 9.8%, and axis 3, 5.4%. Eigenvalues were 0.71, 0.37, and 0.29, for the axes 1, 2, and 3, respectively, indicating strong gradients for axis 1 and a short gradient for the axes 2 and 3. Strong correlations were observed for pH (0.766), potassium (0.821), hydrogen + aluminum (−0.656), total bases (0.926), effective cation exchange capacity (0.936), OM content (0.764), and bases saturation (0.668). The remaining variables had low correlations or were redundant.

TABLE 18.4

Analysis of Variance (ANOVA) and Characterization of Soil Attributes with Mean Values and Standard Error for the Three Stages of Ecological Succession in the MSSP

	Stages			ANOVA	
Edaphic Variables	**ES**	**IS**	**LS**	**F**	***p***
pH in H_2O	7.1 ± 0.18^a	5.42 ± 0.19^b	6.67 ± 0.03^a	34.01	***
P-Mehlich ($mg \cdot kg^{-1}$)	12.82 ± 5.71	1.55 ± 0.23	2.36 ± 0.38	3.62	ns
P-rem ($mg \cdot L^{-1}$)	30.08 ± 1.52^a	25.4 ± 0.72^b	30.29 ± 0.84^a	6.52	**
K^+ ($mg \cdot kg^{-1}$)	303.78 ± 63.57^a	58.35 ± 4.55^b	147.74 ± 31.66^b	9.14	**
Ca^{++} ($cmolc \cdot dm^{-3}$)	11.03 ± 4.50^a	2.78 ± 1.14^c	7.68 ± 3.14^b	51.05	***
Mg^{2+} ($cmolc \cdot dm^{-3}$)	2.77 ± 0.17^a	1.25 ± 0.12^b	2.18 ± 0.19^a	22.16	**
Al ($cmolc \cdot dm^{-3}$)	0^b	0.4 ± 0.12^a	0^b	10.43	**
H + Al ($cmolc \cdot dm^{-3}$)	0.79 ± 0.05^b	1.74 ± 0.17^a	0.89 ± 0.02^b	25.2	***
SB ($cmolc \cdot dm^{-3}$)	14.58 ± 0.79^a	4.18 ± 0.57^c	10.25 ± 0.65^b	59.57	***
t—cation exchange capacity ($cmolc \cdot dm^{-3}$)	14.58 ± 0.79^a	4.58 ± 0.47^c	10.25 ± 0.65^b	59.3	***
m—aluminum saturation (%)	0^b	9.96 ± 3.20^a	0^b	9.7	**
T ($cmolc \cdot dm^{-3}$)	15.36 ± 0.75^a	5.92 ± 0.44^c	11.14 ± 0.65^b	56.84	***
V—base saturation (%)	94.76 ± 0.54^a	69.4 ± 4.14^b	91.84 ± 0.55^a	32.52	***
Organic matter ($dag \cdot kg^{-1}$)	4.99 ± 0.26^a	3.76 ± 0.21^b	4.2 ± 0.27^b	6.49	**
Coarse sand ($dag \cdot kg^{-1}$)	14.2 ± 0.75	15.82 ± 0.69	11.23 ± 0.59	1.93	ns
Fine sand ($dag \cdot kg^{-1}$)	31.13 ± 1.91	28.52 ± 2.32	35.77 ± 3.72	1.76	ns
Silt ($dag \cdot kg^{-1}$)	34.68 ± 1.98^a	25.33 ± 1.33^b	35.33 ± 2.67^a	7.37	**
Clay ($dag \cdot kg^{-1}$)	20 ± 2.30^b	30.33 ± 2.50^a	17.67 ± 1.75^b	9.32	**

ES = early stage; IS = intermediate stage; LS = late stage.

$^*p < 0.05$, $^{**}p < 0.01$, $^{***}p < 0.001$, ns = not significant. Different letters indicate statistical differences at 5% of probability.

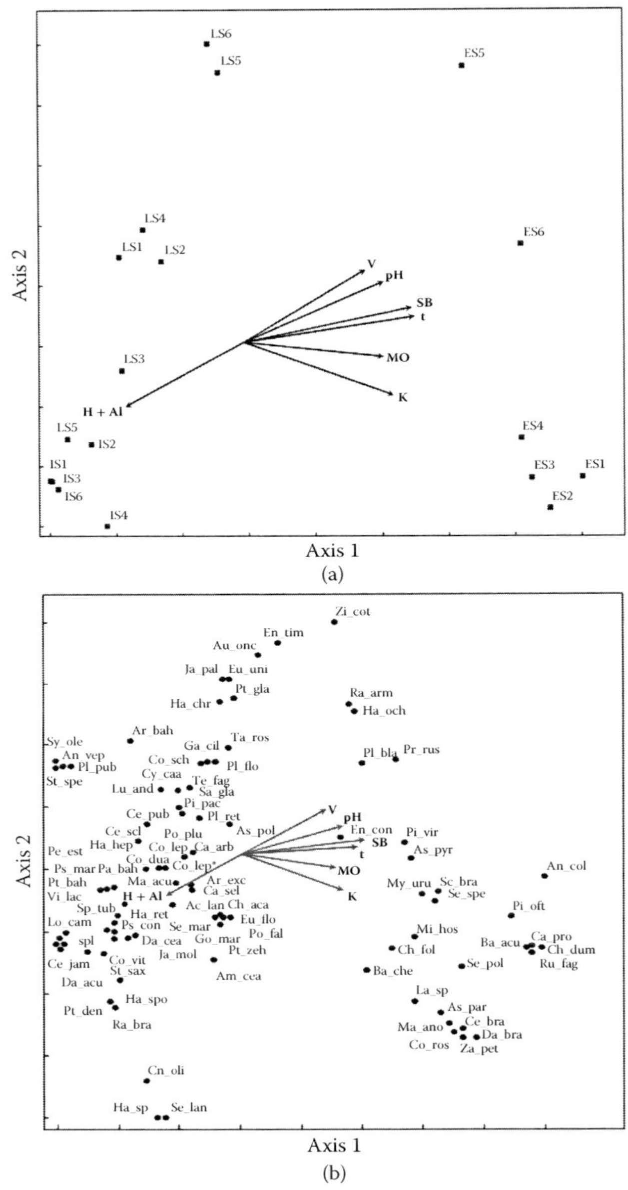

FIGURE 18.5

Diagrams of canonical correspondence analysis of (a) plots and (b) species based on the distribution of the number of individuals of 89 species in 15 plots and its correlation with the eight environmental variables. ES = early stage; IS = intermediate stage; LS = late stage; K = potassium; H + Al = aluminum + hydrogen; SB = sum of bases; t = effective cation exchange capacity; OM = organic matter content; V = base saturation.

The Monte Carlo permutation test indicated that the species and environmental variables correlated significantly ($p < 0.05$) and correlations were high (>0.91). We observed a clear aggregation of plots from the same successional stage in the ordination space (Figure 18.5a). The first group was composed of plots from the late stage, where we found the species *C. schwackeana*, *H. chrysotrichus*, *P. glabra*, *S. oleracea*, *S. speciosa*, and *T. fagifolia* (Figure 18.5b). The second group was composed of early stage plots, with the occurrence of *A. parvifolium*, *B. acuruana*, *M. anomala*, *M. urundeuva*, *S. brasiliensis*, and *S. spectabilis*. A third group was formed by plots from the intermediate stage, characterized by the presence of *C. vitifolium*, *D. acuta*, *D. cearensis*, *H. reticulatus*, *L. campestris*, and *P. contorta*.

Most of the species in the first group (late stage plots) are shade tolerant and associated with slightly acidic soils with low concentration of H + Al. The second group (early stage plots) is composed of mostly pioneer and light-demanding species associated with a slightly basic soil with high concentrations of OM and potassium. The species aggregated in the third group (intermediate stage plots) are distributed among the three regeneration guilds, with a slight tendency for shade tolerance, and are associated with acidic soils with high concentrations of H + Al.

18.4 Discussion

In general, Fabaceae is the most speciose botanical family in fragments of TDFs situated in the north of Minas Gerais, Brazil (Santos et al. 2007; Arruda et al. 2011; Santos et al. 2011b, 2012). This was observed in the present work at the MSSP, where Euphorbiaceae, Malvaceae, and Apocynaceae were also common, corroborating other studies in TDFs in different parts of Brazil (Silva and Scariot 2003; Salis et al. 2004; Santos et al. 2007; Madeira et al. 2009) and of the Americas (Gentry 1995; Kalácska et al. 2004; Oliveira-Filho et al. 2006). Moreover, the genera *Combretum*, *Commiphora*, and *Handroanthus* are recurrently mentioned in the floristic samples conducted in Brazilian TDFs (Werneck et al. 2000; Silva and Scariot 2003; Santos et al. 2007), probably because they are heliophytes and xerophytes and occur on well-drained soils (Maia 2004). Also, *M. urundeuva* is abundant in this type of vegetation due to its tolerance to a wide range of environmental conditions (Brandão 2000), which allows its occurrence both in the early and late successional stages (Lorenzi 1992; Carvalho 1994).

The five most important (IVI) tree species in the early and intermediate stages are characterized mostly by their high density, whereas in the late stage the most important species have high basal areas. In the early stage, three of the most common species, *M. hostilis*, *S. spectabilis*, and *S. polyphylla*, are typically pioneer species from the Caatinga biome that occur mainly in early secondary forests with deep, well-drained, and fertile soils (Lorenzi 1992; 1998). Other common TDF species from MSSP, such as *P. contorta*, are representatives of the Atlantic rainforest (Ribeiro and Barros 2006), corroborating the existence of a strong floristic relationship between the Atlantic rainforest and TDFs (Santos et al. 2011, 2012) and reinforcing the need for conservation of these highly threatened ecosystems.

The diversity and equitability observed for each stage were similar to those found in other Brazilian TDFs, such as in Goiás ($H' = 2.99$, $J' = 0.83$) (Silva and Scariot 2003) and São Paulo states ($H' = 3.00$, $J' = 0.70$) (Ivanauskas and Rodrigues 2000). Moreover, the successional pattern observed here confirmed the intermediate disturbance hypothesis, which predicts that the diversity is greater in environments with intermediate levels of disturbance (Connell 1978). Thus, in the early stage, the diversity would be lower because the succession did not progress much beyond the pioneer species. On the other hand, in the late stage, the climax species are dominant and likely excluded intermediate species through competition, reducing plant diversity. Therefore, pioneer, intermediate, and climax species co-occur in the intermediate successional stages (Molino and Sabatier 2000), increasing their total diversity.

We observed striking differences between the successional stages in forest structure (basal area, height, density, frequency, dominance, equitability, and diversity), species composition, and ecological functions (proportions of dispersal, regeneration and stratification guilds). Changes in structure are demonstrated by the C_{HCI}, evidencing a gradual increase in the values of height and basal area, which corroborates other studies in TDFs (Guariguata and Ostertag 2001; Ruiz et al. 2005b). However, the density of individuals also increased along the chronosequence in the MSSP, contrary to the most conventional successional patterns (Saldarriaga et al. 1988; Kennard 2002; Ruiz et al. 2005). The highest density of individuals in the late stage may be related to the sprouting rate of some species and to the sampling method used here, which excluded many shrubs and treelets that represent most individuals in the early stage. According to Dislich et al. (2001), the difference in density values among successional studies is influenced by inclusion criteria because the density of larger trees increased, while that of the smaller trees decreased in response to the reduction of light incidence under the canopy. So, a change in format of the diameter frequency distribution can occur, masking possible relationships between time succession and density (Dislich et al. 2001).

The frequency distribution of individuals in the guilds demonstrates strong effects of forest disturbance in ecological functions. As predicted by classical successional models (Egler 1954), the relative importance of shade-tolerant species (such as *S. brasiliensis*, *A. polyneuron*, *A. excelsa*, *E. timbouva*, and others) increased along the chronosequence, whereas the opposite is observed for pioneer species (*C. leptophloeus*, *M. anomala*, *S. spectabilis*, among others). It is, therefore, inferred that the distribution pattern of regeneration guilds found in this study is common; the number of pioneer trees was higher in early stages and lower in later stages where the shade-tolerant individuals were more frequent. This also corroborates the initial floristic model (Egler 1954), which predicts that all groups of regeneration are present since the beginning of the succession and remain in the later stages, changing only the dominance between groups during the sequence. However, the recovery period of a forest depends not only on time but also on the intensity of disturbance and on environmental conditions (Klein 1980; Tabarelli and Mantovani 1999; Nunes et al. 2003).

It is expected that the stratification complexity of the forest increase along the succession with the increase of larger species in the forest (Nunes et al. 2003). According to Sterck et al. (1999), the changes in architecture, physiology, and growth are determined by light intensity variation. Studies demonstrated that species change can be driven by the variation in light penetration through the canopy along the gradient of succession (Madeira et al. 2009). Besides, the light quality and quantity determine the survival of pioneer and shade-tolerant species in the environment, and therefore the variation in floristic composition among successional stages (Madeira et al. 2009). For instance, *M. acutifolium*, a shade-tolerant climax species, was found in the three successional stages but was more abundant in the late stage. On the other hand, the pioneer *P. pluviosa* was also found in the three stages, but likewise, it was more abundant in the late stage. This result can be related to the large size and rapid growth of this species, ensuring the canopy exposition and, consequently, higher light incidence in their crowns.

In general, anemochoric was the predominant dispersal syndrome in the studied TDF, regardless of the successional stage. According to Gentry (1995) and Pennington et al. (2000), this is a characteristic of the deciduous formations, which are dominated by the families Bignoniaceae and Fabaceae that are mainly wind dispersed. Furthermore, anemochory is favored in tropical deciduous forests by the low relative humidity (Janzen 1967) and the leaf loss during the dry season, which allows greater air circulation, both in the canopy and in the lower strata (Morellato et al. 1989; Griz and Machado 2001). Gentry (1982b) also suggests that the frequency of anemochoric species increases from humid toward dry areas, the opposite occurring with zoochoric species, which predominate in the humid forests (Howe and Smallwood 1982; Gentry 1983, 1995). In addition, when evaluating the frequency distribution of dispersal guilds in successional stages, we observed a gradual increase of zoochoric species as the forest progressed toward maturity. For

Fenner (1985) and Mikich and Silva (2001), this change in the plant community toward greater animal–plant interactions is promoted by the increase in forest complexity, allowing the occurrence of dispersing agents such as birds and mammals.

Throughout succession, several ecological mechanisms determine the directional changes in plant composition, which result from the relationship between the regeneration, dispersal, and stratification guilds (Tabarelli and Mantovani 1999; Nunes et al. 2003). Wikander (1984) studied a TDF in Venezuela and observed that dispersal mechanisms vary according to forest stratification: in the upper and middle strata, three modes of dispersal were recorded (zoochoric, autochoric, and anemochoric), whereas only zoochoric occurred in the lower stratum. Similar results were obtained in a semideciduous forest (Morellato and Leitão-Filho 1992), in the Brazilian Cerrado (Batalha et al. 1997; Batalha and Mantovani 2000), and in other TDFs (Frankie et al. 1980; Lieberman 1982; Lampe et al. 1992; Borchert 1996). In this study, all dispersal syndromes were encountered along the stratification guilds, but our results partially corroborate the above-mentioned studies, with a higher than expected frequency of zoochoric for short individuals. Thus, there is a strong interaction between ecological functions (i.e., stratum position versus dispersal syndrome) along forest natural regeneration that accentuate the differences in ecosystem functioning between successional stages.

The soil type in the intermediate stage showed striking differences from soils at the early and late stages. Although such contrasts in edaphic characteristics may occur due to distinct previous land use at each site, it is likely that the intermediate forest at the MSSP grows on a naturally different soil type. According to the chemical classification of Álvares et al. (1999), only the intermediate stage has a medium acidity and may be harmful to plants. The early and late stages, respectively, show weak acidity and weak alkalinity, providing a pH range where most of the nutrients (K, Ca, Mg, N, S, P, and B) are more available (Silva and Souza 1998), benefiting plant growth. These results corroborate those of Ratter et al. (1978), who stated that TDFs typically occur on fertile soils, with moderate values of pH, moderate nutrient availability, and low levels of aluminium. In spite of that, the species composition, forest structure, and ecological functions are more similar between intermediate and late stages, suggesting that, in this particular case, time since abandonment plays a more important role than soil type in defining the local communities.

The CCA results demonstrate that there is a species replacement (axis 1) and a variation in the relative abundances of species (axis 2) between the areas. The axis 1 was responsible for segregating the early stage plots from both intermediate and late stages, mainly due to OM, pH, base saturation, SB, effective cation exchange capacity, and potassium. The axis 2 separated the intermediate from late plots, based on H + Al levels. In the case of plant species, two groups were formed along the axis 1 influenced by the availability of mineral elements in the soil (Oliveira-Filho et al. 2001; Pinto et al. 2005), which in turn depends on several factors such as pH, concentration of the element in the soil, OM, moisture, aeration, competition between ions for sites, among others (Meurer 2007). The soil pH can simultaneously influence many chemical and biological processes, such as microbial activity, seed dormancy, predation, and responses to chemical compounds such as nitrate and nitrite (Gianello et al. 1995; Favreto and Medeiros 2006), affecting the distribution of plant species.

18.5 Conclusion

The most important arboreal species found in the different successional stages stood out by the density in the early and intermediate stages and by dominance in the late stage, being that the communities studied had similar diversity and equability to studies developed for this type of vegetation in Brazil. As expected, the diversity in early successional stage is lower. However, the intermediate stage has a higher diversity than the late stage, showing a higher occurrence of species from different guilds of succession.

The successional stages presented differences in structural and functional groups and, as the forest advances to maturity, there is a tendency of increasing structural parameters. However, in this study, the density of individuals did not follow the expected pattern. The shade-tolerant species increase their level of importance toward the late stage, while the pioneers decline. Moreover, smaller trees are more frequent in the early stage and larger trees in the late stage. Also, as expected, anemochoric dispersal is dominant in the community studied, but there is a gradual increase of species with zoochoric propagation, as the forests progress toward maturity.

The significant variations of the soil characteristics cause substitution and changes in the relative abundances of species between areas and reinforces a change in vegetation complexity as it progresses in the ecological succession. The soil properties that demonstrated a close relationship with floristic variables were pH, potassium, hydrogen + aluminum, SB, effective cation exchange capacity, OM, and base saturation. The soil fertility and characteristics are related to the diversity of trees in the different stages, limiting or favoring different species. Nevertheless, it is considered for this study that the disturbance promotes much greater local influence on the distribution of species than the edaphic characteristics. Therefore, the time for regeneration after the disturbance is the most relevant factor in the distribution of species in this study.

Appendix

List of sampled species in three stages of ecological succession in the MSSP

	Stages			Guilds		
Family/Species	**ES**	**IS**	**LS**	**Dis.**	**Reg.**	**Str.**
Anacardiaceae						
Cyrtocarpa caatingae J.D.Mitch. & Daly	—	3	6	Zoo	STC	T
Myracrodruon urundeuva Allemão	170	5	36	Ane	STC	T
Schinopsis brasiliensis Engl.	8	—	—	Ane	LDC	T
Spondias tuberosa Arruda	—	6	1	Zoo	STC	T
Annonaceae						
Annona vepretorum Mart.	—	—	1	Zoo	STC	M
Apocynaceae						
Aspidosperma parvifolium A.DC.	19	—	1	Ane	LDC	M
Aspidosperma polyneuron Müll.Arg.	—	—	10	Ane	STC	T
Aspidosperma pyrifolium Mart.	2	—	—	Ane	LDC	M
Araliaceae						
Aralia excelsa (Griseb.) J.Wen	1	5	7	Zoo	STC	T
Arecaceae						
Syagrus oleracea (Mart.) Becc.	—	—	5	Zoo	STC	T
Calotropis procera R. Br.	1	—	—	Ane	P	S
Bignoniaceae						
Arrabidaea bahiensis (Schauer) Sandwith & Moldenke	—	—	5	Ane	LDC	M
Handroanthus chrysotrichus (Mart. ex A.DC.) Mattos	—	5	206	Ane	STC	T
Handroanthus heptaphyllus (Martius) Mattos	—	5	2	Ane	LDC	M
Handroanthus ochraceus (Cham.) Mattos	95	3	—	Ane	LDC	M
Handroanthus reticulatus A.H.Gentry	—	135	41	Ane	STC	T
Handroanthus sp.	—	1	—	Ane	LDC	T
Handroanthus spongiosus (Rizzini) S.O.Grose	—	15	1	Ane	STC	T
Tabebuia roseoalba (Ridl.) Sandwith	—	—	4	Ane	STC	T

	Stages			Guilds		
Family/Species	ES	IS	LS	Dis.	Reg.	Str.
Boraginaceae						
Auxemma oncocalyx (Allemão) Taub.	—	—	6	Ane	LDC	S
Patagonula bahiensis Moric.	—	17	9	Ane	STC	T
Burseraceae						
Commiphora leptophloeus (Mart.) J.B.Gillet	—	14	13	Zoo	P	M
Cactaceae						
Cereus jamacaru DC.	—	1	—	Zoo	LDC	S
Pereskia stenantha F. Ritter	—	8	3	Zoo	STC	S
Pilosocereus pachycladus Ritter	—	3	4	Zoo	STC	M
Cannabaceae						
Celtis brasiliensis (Gardner) Planch.	2	—	—	Zoo	P	S
Combretaceae						
Combretum duarteanum Cambess.	2	91	126	Ane	LDC	M
Combretum leprosum Mart.	—	—	2	Ane	LDC	M
Terminalia fagifolia Mart.	—	9	13	Ane	STC	M
Euphorbiaceae						
Cnidoscolus oligandrus (Mull.Arg.) Pax	—	10	—	Aut[a]	STC	T
Jatropha mollissima (Pohl) Baill.	—	17	3	Aut[a]	STC	S
Jatropha palmatifolia Ule	—	—	1	Aut[a]	STC	S
Manihot anomala Pohl	7	—	—	Aut[a]	P	S
Sapium glandulosum (L.) Morong	—	5	13	Zoo	STC	S
Stillingia saxatilis Müll.Arg.	—	23	2	Zoo	STC	S
Fabaceae–Caesalpinoideae						
Goniorrhachis marginata Taub.	—	—	2	Aut	LDC	T
Poincianella pluviosa DC.	1	21	37	Aut	P	T
Senna spectabilis (DC.) H.S.Irwin & Barneby	77	—	—	Aut	P	M
Bauhinia acuruana Moric.	3	—	—	Aut	P	S
Bauhinia cheilantha (Bong.) Steud.	1	—	—	Aut	P	S
Fabaceae–Faboideae						
Acosmium lentiscifolium Schott	1	4	5	Ane	LDC	M
Amburana cearensis (Allemão) A.C.Sm.	1	1	—	Ane	LDC	T

(Continued)

Family/Species	Stages			Guilds		
	ES	IS	LS	Dis.	Reg.	Str.
Centrolobium sclerophyllum H.C.Lima	—	—	2	Ane	LDC	T
Coursetia rostrata Benth.	1	—	—	Ane	STC	S
Dalbergia acuta Benth.	—	13	1	Ane	STC	M
Dalbergia cearensis Ducke	1	9	4	Ane	LDC	M
Lonchocarpus campestris Mart. ex Benth.	—	2	—	Ane	LDC	T
Luetzelburgia andrade-limae H.C.Lima	—	5	7	Ane	STC	T
Machaerium acutifolium Vogel	2	11	13	Ane	STC	M
Platymiscium floribundum Vogel	11	—	1	Ane	P	T
Platymiscium pubescens Micheli	—	—	3	Ane	LDC	T
Poecilanthe falcata (Vell.) Heringer	—	—	1	Aut	STC	T
Pterocarpus zehntneri Harms	—	—	1	Ane	STC	T
Fabaceae–Mimosoideae						
Anadenanthera colubrina (Vell.) Brenan	1	—	—	Ane	P	T
Chloroleucon acacioides (Ducke) Barneby & J.W.Grimes	—	—	1	Zoo	STC	T
Chloroleucon dumosum (Benth.) G.P.Lewis	1	—	—	Zoo	STC	M
Chloroleucon foliolosum (Benth.) G.P.Lewis	7	2	—	Zoo	STC	M
Enterolobium contortisiliquum (Vell.) Morong	4	2	—	Zoo	P	T
Enterolobium timbouva Mart.	—	—	1	Zoo	STC	T
Piptadenia oftalmocentra Mart.	17	—	—	Ane	P	M
Piptadenia viridiflora (Kunth) Benth.	4	—	—	Ane	P	T
Plathymenia reticulata Benth.	—	1	8	Ane	STC	M
Prosopis ruscifolia Griseb.	10	—	—	Ane	P	S
Pseudopiptadenia contorta (DC.) G.P.Lewis & M.P.Lima	—	26	8	Ane	P	T
Senegalia langsdorffii (Benth.) Bocage & L.P.Queiroz	—	1	—	Aut	P	S
Senegalia polyphylla (DC.) Britton & Rose	30	—	—	Aut	P	S
Lamiaceae						
Vitex laciniosa Turcz.	—	2	—	Zoo	STC	T
Malpighiaceae						
Ptilochaeta bahiensis Turcz.	—	2	—	Zoo	STC	M

	Stages			Guilds		
Family/Species	**ES**	**IS**	**LS**	**Dis.**	**Reg.**	**Str.**
Ptilochaeta densiflora Nied.	—	1	—	Zoo	STC	M
Ptilochaeta glabra Nied.	—	—	4	Zoo	STC	M
Malvaceae						
Cavanillesia arborea (Willd.) K.Schum.	—	1	3	Ane	LDC	T
Ceiba pubiflora (A.St.-Hil.) K.Schum.	—	1	1	Ane	LDC	T
Pseudobombax marginatum (A.St.-Hil.) A.Robyns	—	3	2	Ane	STC	T
Sterculia speciosa K.Schum.	—	—	2	Ane	STC	T
Myrtaceae						
Eugenia florida DC.	—	—	1	Zoo	STC	M
Eugenia uniflora L.	—	—	2	Zoo	LDC	M
Nyctaginaceae						
Ramisia brasiliensis Oliver	—	1	—	Zoo	LDC	S
Polygonaceae						
Coccoloba schwackeana Lindau	—	1	10	Zoo	STC	T
Cochlospermum vitifolium (Willd.) Spreng.	—	12	2	Ane	P	M
Ruprechtia fagifolia Meisn.	1	—	—	Ane	STC	T
Rhamnaceae						
Ziziphus cotinifolia Reissek	1	—	—	Zoo	LDC	M
Randia armata (Sw.) DC.	6	—	2	Zoo	STC	S
Rutaceae						
Galipea ciliata Taub.	—	—	1	Aut	STC	S
Zanthoxylum petiolare A.St.-Hil. & Tul.	1	—	—	Zoo	STC	T
Salicaceae						
Casearia selloana Eichl.	—	5	23	Zoo	STC	S
Verbenaceae						
Lantana sp.	2	—	—	Zoo	P	S

Note: List of sampled species is arranged in alphabetical order by family, followed by the number of individuals sampled in each stage of ecological succession: ES = early stage, IS = intermediate stage, LS = late stage and their respectively guilds, dispersal (Dis.), regeneration (Reg.), and stratification (Str.). Guilds: Ane = anemochoric; Aut = autochoric; Zoo = zoochoric; P = pioneer; LDC = light-demanding climax; STC = shade-tolerant climax; T = tall; M = medium; S = short.

[a] Occurrence of secondary dispersal.

19

Tree Diameter Growth of Three Successional Stages of Tropical Dry Forests, Santa Rosa National Park, Costa Rica

Dorian Carvajal-Vanegas and Julio Calvo-Alvarado

CONTENTS

19.1 Introduction

The northwest Pacific coast of Costa Rica was dedicated 100 years ago to cattle farming, which jeopardized fragile ecosystems within the region. In the 1970s, the Guanacaste Conservation Area (ACG) was established to protect forest ecosystems such as the tropical dry forest (TDF), transforming this region into an open laboratory for research on natural forest restoration (Fedigan et al. 1985; Janzen 2000; Molina 2002; Calvo-Alvarado et al. 2009a). The Parque Nacional Santa Rosa (PNSR) is one of these conservation areas dedicated to safeguarding the TDFs within the ACG; but this protection also encompasses mangrove ecosystems; turtle nesting sites; and a large amount of microhabitats for mammals, reptiles, and bird species (Boza 1993; Carrillo et al. 1994; Campbell et al. 2005).

TDF is one example of a tropical ecosystem that has been threatened for centuries by human development (Fuchs et al. 2003; Marin et al. 2005; Calvo-Alvarado et al. 2009a; Powers et al. 2009a). This pressure is a consequence of the high soil fertility and favorable climatic conditions of the area that enhanced anthropogenic activities such as agriculture and cattle farming, increasing environmental degradation (Murphy and Lugo 1986a; Gillespie et al. 2000); forest fires, and alien species invasion (Johnson and Wedin 1997; Vega 2002). Because of this high degradation most TDFs can no longer be considered as pristine old-growth forests; rather, they are a mosaic of successional stages (Janzen 1986b, 1988b; Arroyo-Mora et al. 2005b; Calvo-Alvarado et al. 2009a; Quesada et al. 2009), which justifies the study of the ecology of different TDFs across successional stages and not just old growth (Quesada et al. 2009). It is well known that successional processes in TDFs are driven by tree densities and species composition, which, in turn, are determined by seed availability (Wijdeven and Kuzee 2000), seed dispersal and seed predation (Fedigan et al. 1985; Janzen 1987; Corlett 1995; Curran and Leighton 2000); previous land use (Alfaro et al. 2001; Leiva et al. 2009), and climate and soil characteristics (Murphy and Lugo 1986b; Enquist and Leffler 2001).

Much of the emphasis on successional forest dynamics has been focused on characterizing the variation in species composition and forest structure among successional stages (Guariguata and Ostertag 2001). Most of these studies have been conducted in humid or wet forests and very few in TDFs (Mooney et al. 1995; Kennard 2002). This bias has allowed a relatively well-defined characterization of how these stages compare with pristine old-growth forests (Clark and Clark 1992; Guariguata et al. 1997; Martin et al. 2004). Despite the interest on forest succession studies, there is still a tremendous need to collect information about tree growth rates, tree recruitment, and mortality among TDF successional stages to help better understand forest dynamics (Marin et al. 2005; Quesada et al. 2009; Piotto et al. 2010). Information about tree growth rates will improve the knowledge on forest restoration processes, organic matter production, soil carbon stocks, and net primary productivity (Brown and Lugo 1990; Lugo and Helmer 2004; Lawrence 2005; Chave et al. 2009b; Becknell et al. 2012). Moreover, forest growth rates can be used as a practical tool for planning forest restoration and silvicultural management programs of degraded TDF areas (Aide et al. 2000; Guariguata and Ostertag 2001; Lugo and Helmer 2004).

Several studies on the dynamics of species composition and forest structure associated with different successional stages had been carried out in TDFs of Costa Rica, specifically in PNSR. The studies by Kalácska et al. (2004b) and Powers et al. (2009a) are some examples of studies that considered geographical and environmental patterns, whereas those by Alfaro et al. (2001) and Leiva et al. (2009) emphasized edaphic changes associated with successional stages in a TDF. However, there are very few studies on tree diameter increment in Costa Rican TDFs, and there are even lesser if successional gradients are considered. Monge et al. (2002) studied the dynamics of a TDF in the Palo Verde National Park, Guanacaste (10°20′N, 85°20′W), and reported an annual diameter increase of 4.2 mm · year^{-1}. Meza and Mora (2002) found an average annual increment of 2.2 mm · year^{-1} in the Guanacaste National Park. Both studies were conducted on late successional stages.

The aforementioned studies describe forest dynamics in terms of species composition and growth rates. However, except for the review by Becknell et al. (2012), studies of total carbon storage in TDF successional stages are lacking. The knowledge in this field can be improved following the methods outlined by Curtis et al. (2002) and Gough et al. (2008) in temperate forests. They used biometric measurements and diameter increment values to develop carbon storage models.

The aim of this study is to determine and compare the structure, species composition and annual tree diameter increment, tree recruitment, and mortality of the three successional stages of the TDF in PNSR. The results generated in this study address the current lack of key information to better understand and model forest restoration and carbon sequestration processes of TDF (Jiang et al. 1999; Thürig et al. 2005; Tan et al. 2010). The study is novel because it evaluates tree growth changes among three successional stages, which are hardly reported in the literature on TDFs.

19.2 Methods

19.2.1 Experimental Design

In 2006, nine permanent plots (Figure 19.1) of 1000 m^2 (20 m × 50 m) were established (three per successional stage, early, intermediate, and late). The plots followed the research protocols of the Tropi-Dry project described by Nassar et al. (2008). All trees with a diameter at breast height (DBH) greater than 5 cm (1.3 m) were measured yearly to estimate tree diameter increment, tree recruitment, and mortality. Field data were collected every dry season during the month of February and for the period of 2006–2011.

Kalácska et al. (2004b, 2005c) described the structure, floristic composition, leaf area index, and biomass of three successional stages of the TDF in PNSR and summarized their historic land-use change. According to Kalácska et al. (2004b), the early-stage forests grew after several intensive pasture fires that took place late in the 1980s; plots in the intermediate forest stage were affected by logging and less intense fires early in the 1970s. The late stages are located in areas where the last reported selected timber harvesting disturbance took place in the 1920s, but authors such as Leiva et al. (2009) indicate that the late forest stage of PNSR has more than 100 years (Figure 19.2).

19.2.2 Data Analysis

Characterization of floristic composition (number of families and species) was performed by taking into account all tree individuals larger than 5 cm in diameter, according to the Tropi-Dry methodology (Nassar et al. 2008). We also calculated the vegetation importance value index (IVI), which was

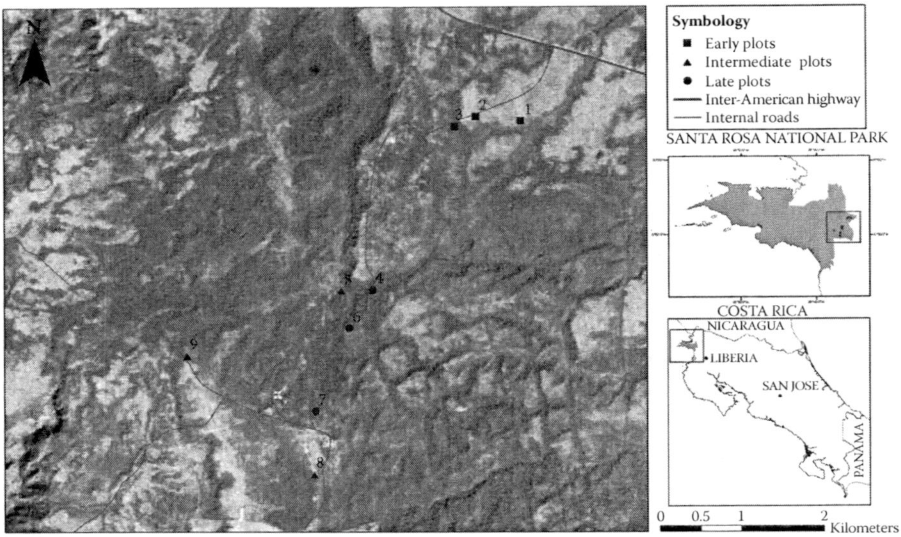

FIGURE 19.1
Distribution of permanent plots in the Santa Rosa National Park.

FIGURE 19.2
(See color insert.) Photographs illustrating selected tropical dry forest (TDF) successional stages in the Santa Rosa National Park: (a) early, (b) intermediate, and (c) late stages and (d) the early stage with a remaining tree.

proposed by Curtis and McIntosh (1951) and is one of the most used indexes to obtain a classification hierarchy using abundance, frequency, and relative dominance values within stages (Monge et al. 2002; Spittler 2002).

The demographic parameters including tree mortality (*m*), recruitment (*r*) and annual rates of basal area growth loss (*l*), gain (*g*), and ingrowth (*i*) were calculated using a logarithmic model for a period of 5 years (2006–2011). The logarithmic model was used following the analysis of Sheil et al. (1995) and other studies (Lieberman and Lieberman 1987; Condit et al. 1999; Hoshino et al. 2002; Marín et al. 2005), which argued that interpretation of mortality rates does not follow a constant probability. All the aforementioned parameters, including the Holdridge complexity index (HCI), dominant tree height, basal area, species, and tree density, were calculated using all tree individuals with DBH over 10 cm. We fixed this diameter to facilitate the comparison of results, since most literature results are based on permanent plots that measure only trees greater than 10 cm in diameter (Holdridge 1967; Lewis et al. 2004). Mortality and recruitment in the number of trees per hectare were calculated by the following formulas:

$$m = \frac{\ln N_{06} - \ln N_S}{T} \quad (19.1)$$

$$r = \frac{\ln N_{11} - \ln N_S}{T} \quad (19.2)$$

where N_{06} is the number of individuals present in 2006, N_S is the number of individuals surviving in 2011 ($N_S = N_{06}$ – number of individuals dying during the period), and $N_{11} = N_S$ + number of individuals recruited during the time period *T*.

$$l = \frac{\ln \mathrm{BA}_{06} - \ln \mathrm{BA}_{S06}}{T} \quad (19.3)$$

$$g = \frac{\ln \mathrm{BA}_{11} - \ln \mathrm{BA}_{S06}}{T} \quad (19.4)$$

$$i = \frac{\ln \mathrm{BA}_{S11} - \ln \mathrm{BA}_{S06}}{T} \quad (19.5)$$

where BA_{06} is the basal area of live stems in 2006, BA_{S06} is the basal area in 2006 of the surviving stems in 2011, BA_{S11} is the basal area in 2011 of the live stems in 2011, and $\mathrm{BA}_{11} = \mathrm{BA}_{S11}$ + the basal area recruitment stems in 2011.

Since mean annual increments (MAIs) in TDFs are very small ($<2\text{mm} \cdot \text{year}^{-1}$), the recommended procedure to measure it is through the estimation of diameter difference registered during a long period of years (Murphy and Lugo 1986b). Consequently, in this study MAI was estimated with a database of 5 years (2006–2011) with the information of all trees with

DBH ≥ 5 cm that survived during the same period. The MAI of each individual was estimated using the following equation (Marín et al. 2005):

$$\text{MAI} = \frac{\text{DBH}_{06} - \text{DBH}_{11}}{T} \tag{19.6}$$

Increment data inconsistencies (due to measurement or annotation errors) were not considered for this analysis. Large trees left after fires (Figure 19.2) were not considered in this analysis, as they misrepresent the MAI values for the early successional stages. For individuals with several stems below the DBH measuring point, an average increment for all stems was calculated.

Statistical analyses were performed with the software Statistica 6.0 (StatSoft Inc. 2003). As MAI did not show a normal distribution, we used a Kruskal–Wallis test (Kruskal and Wallis 1952). The median, rather than the average, was used to compare the MAI values by diameter classes, species, and successional stages, considering that this parameter reduces the bias caused by extreme values (Finegan and Camacho 1999; Ortiz and Carrera 2002; Marin et al. 2005).

19.3 Results

19.3.1 Forest Structure and Species Composition

We identified a total of 1001 tree individuals distributed in 36 families and 108 species among all plots. The number of families and species per succession stage is illustrated in Table 19.1. Individual species dominance changed along the successional stage gradient. In the early stage, there were 295 individuals

TABLE 19.1

Families with a Percentage Composition >5% in the Three Successional Stages for 2011

Stage	Family	Composition (%)	Number of Species
Early	Cochlospermaceae	14.9	1
	Fabaceae	24.4	7
	Hippocrateaceae	5.1	1
	Verbenaceae	31.2	2
	Other families (13)	24.4	21
Intermediate	Bignoniaceae	5.6	4
	Boraginaceae	6.6	3
	Fabaceae	16.5	12
	Flacourtiaceae	6.3	3
	Malvaceae	17.8	7

TABLE 19.1 (*Continued*)

Families with a Percentage Composition >5% in the Three Successional Stages for 2011

Stage	Family	Composition (%)	Number of Species
	Meliaceae	7.3	6
	Moraceae	5.9	3
	Rubiaceae	8.9	4
	Other families (20)	25.1	27
Late	Bignoniaceae	7.7	5
	Burseraceae	5.6	3
	Euphorbiaceae	18.3	3
	Fabaceae	8.4	8
	Hippocrateaceae	12.6	1
	Malvaceae	5.4	5
	Rubiaceae	15.8	7
	Other families (26)	73.8	34

and 17 families represented by 32 species. Four of these families (11 species) cover 75.6% of the sampled individuals. In the intermediate stage, there were 303 individuals (69 species) belonging to 28 families. Eight of these families (42 species) cover 74.9% of the sampled trees. From a total of 33 families and 403 individuals (66 species) in the late stage, there are 7 families (32 species) that cover 73.8% of the sampled individuals.

As expected, all forest structural parameters included in Table 19.2 show increased values as stand age increases. However, high heterogeneity is displayed by the high standard deviation, implying differences among sampled plots per stage. HCI for the late stage is 74.2 and according to Holdridge (1967) the central value for pristine old-growth TDFs should be around 45, whereas the value for a humid tropical forest is 270. As expected, early and intermediate stages have less complex HCIs.

19.3.2 Tree Recruitment, Mortality, and Basal Area

During the period 2006–2011, the following three results were obtained: (1) in the early stage with a total of 207 individuals per hectare (≥10 cm DBH) living in 2006, 13 trees died and 103 new trees were recruited; (2) in the intermediate stage with 520 individuals per hectare, 63 trees died and 123 were recruited; and, finally, (3) in the late stage with 660 individuals per hectare, 47 trees died and 63 were recruited. The percentages of mortality rates calculated for the same period (logarithmic model) were 1.3%year^{-1} (early), 2.6%year^{-1} (intermediate), and 1.5%year^{-1} (late). The average recruitments were 8.8%year^{-1}, 4.8%year^{-1}, and 1.5%year^{-1}, respectively, for early, intermediate, and late stages (Table 19.3). As for mortality rates, the highest basal area loss corresponded with the intermediate stage (1.5%year^{-1}). The gain in

TABLE 19.2

Mean Values with Standard Deviation of Forest Structural Characteristics (Height, Basal Area, Density, and Number of Species) and the HCI for the Year 2011 among Successional Stages in Santa Rosa National Park

Stage	Height (m)	Basal Area (m^2 0.1 ha^{-1})	Density (Stems 0.1 ha^{-1})	Species Density (Species 0.1 ha^{-1})	HCI
Early	9.9 ± 0.9	0.6 ± 0.5	24 ± 7	6 ± 4	1.6 ± 2.4
Intermediate	17.1 ± 3.5	2.5 ± 1.6	49 ± 12	20 ± 6	50.9 ± 39.7
Late	22.4 ± 2.6	2.7 ± 0.3	62 ± 13	19 ± 4	74.2 ± 28.1

TABLE 19.3

Mortality and Recruitment Rates and Loss, Gain, and Ingrowth Rates in Basal Areas for Stems ≥10 cm DBH among Successional Stages in the Santa Rosa National Park

Stage	Mortality (%year $^{-1}$)	Recruitment (%year $^{-1}$)	Loss (%year $^{-1}$)	Gain (%year $^{-1}$)	Ingrowth (%year $^{-1}$)
Early	1.3	8.8	1.0	5.5	4.1
Intermediate	2.6	4.8	1.5	4.8	3.2
Late	1.5	2.0	0.9	3.3	2.3

basal area was higher in the early stage (5.5%year^{-1}), and the highest increase in basal area was also in the early stage (4.1%year^{-1}).

19.3.3 Diameter Increments by Stage

The greatest MAI was observed in intermediate (2.2 mm · year^{-1}) and early (1.6 mm · year^{-1}) stages with no statistical differences between them ($p > 0.05$). This similarity could be the result of the high MAI variability among species, particularly in the intermediate stage. On the other hand, early and intermediate stages are statistically distinct from the late stage (1.2 mm · year^{-1}) ($p < 0.001$) (Table 19.4).

TABLE 19.4

Current Annual Diameter Increment MAI and Tree Sample Size in Three Successional Stages

Stage	MAI (mm · year^{-1})	Interval (mm · year^{-1})	Number of Trees
Early	1.6 (a)	0.0–8.6	200
Intermediate	2.2 (a)	0.0–19.0	213
Late	1.2 (b)	0.0–12.4	311

Note: Groups with same letters are statistically similar ($p < 0.05$).

19.3.4 Increments by Diameter Class

In the early stage, the highest diameter increment is concentrated in the diameter class 15–20 cm (5.20 mm · year^{-1}) and there is only a significant difference ($p < 0.001$) between this class and the 5–10 cm class (1.40 mm · year^{-1}). As mentioned earlier, the intermediate stage shows the highest increment in the class 20–25 cm (4.50 mm · year^{-1}) and the class 40–45 cm (4.50 mm · year^{-1}). However, the last class is statistically different from the 20–25 cm class even if both have the same MAI value. Last, the late stage concentrates the highest increment in the class 35–40 cm (4.90 mm · year^{-1}) and the lowest value in the class 5–10 cm (0.70 mm · year^{-1}) (Figure 19.3).

Figure 19.4 shows the three species with the highest IVI for the early stage. In the lower diameter class (5–10 cm) increments did not exceed 3 mm · year^{-1}, and in the class 15–20 cm the species *Rehdera trinervis* and *Byrsonima crassifolia* reached their highest increment surpassing the increment rate of 5 mm · year^{-1}. The behavior of the two species was very similar to the results shown in Figure 19.3 for early stage; hence, it can be assumed that these two species are the ones that determine the diameter increment pattern of the early stage.

In the intermediate stage (Figure 19.5), R. trinervis again has the highest IVI value; however, the MAIs in this stage were more than double (13.2–18.0 mm · year-1) those of the early stage (1.54–5 mm · year-1). A particular case is shown by Semialarium mexicanum; although it is the second species in IVI value, it is not the most abundant species. Hence, increments in different diameter classes are given only by very few individuals 15–20 cm (1 in Figure 19.5) and 20–25 cm (2 in Figure 19.5). Then, it would be misleading to take these values as representative for this species. The aforementioned

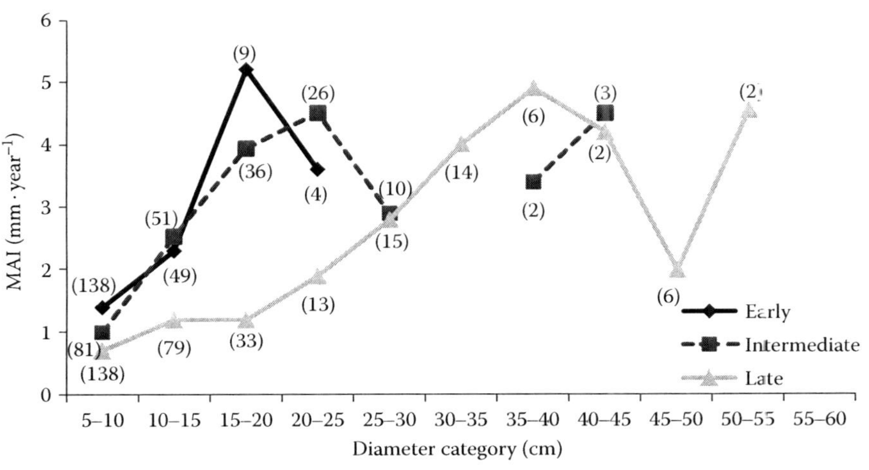

FIGURE 19.3
Current annual diameter increment MAI (in millimeters per year) by diameter classes in three TDF successional stages in the Santa Rosa National Park. Numbers in parentheses indicate the numbers of trees included in each diameter class.

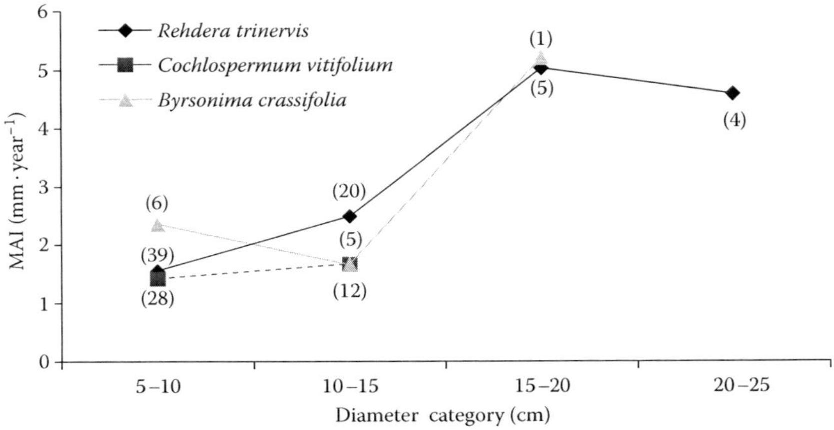

FIGURE 19.4
Current annual diameter increment MAI (in millimeters per year) by diameter classes for three species in the early stage in the Santa Rosa National Park: numbers in parentheses indicate the numbers of trees included in each diameter class.

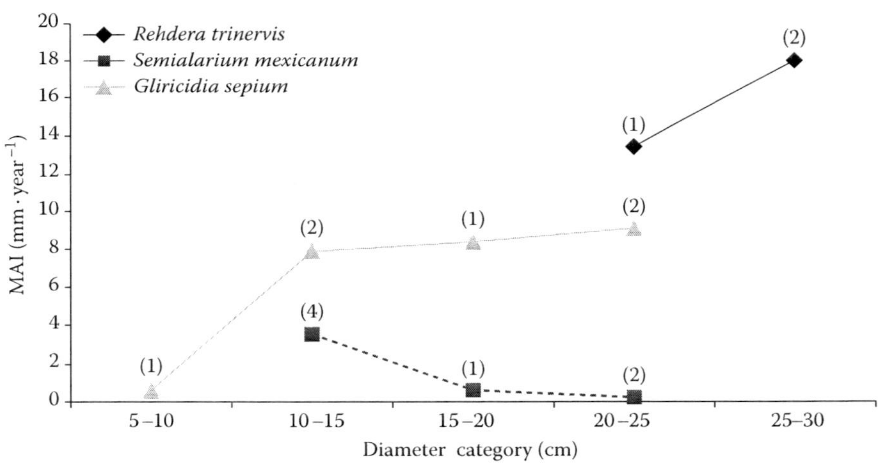

FIGURE 19.5
Current annual diameter increment MAI (in millimeters per year) by diameter classes for three species in the intermediate stage in the Santa Rosa National Park: numbers in parentheses indicate the numbers of trees included in each diameter class.

statement is also exemplified by Gliricidia sepium in the intermediate stage, this species is not the most abundant either, and the increase in the class 5–10 cm is given by a single individual.

In the late stage, the species *Exostema mexicanum* y and *Sebastiania pavoniana* do not outweigh the value of 4 mm · year^{-1}and only *Lysiloma divaricatum* double this value in the diameter classes of 30–35 cm and 50–55 cm (Figure 19.6). Moreover, canopy dominant species with DBHs greater than

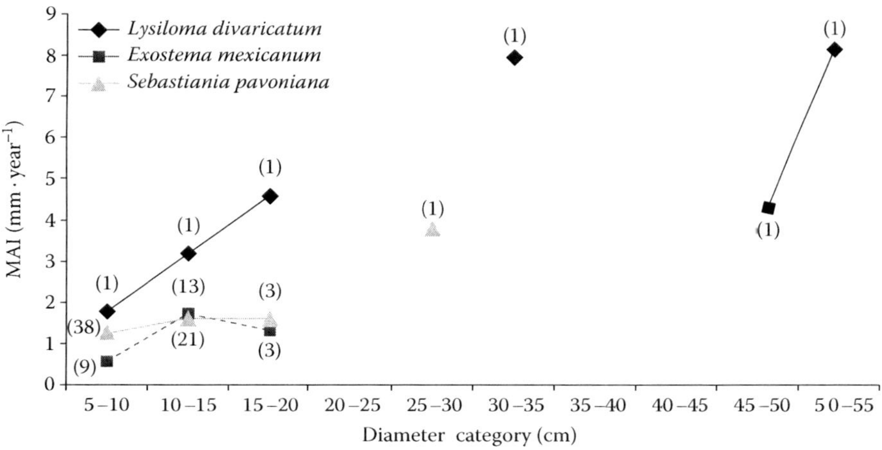

FIGURE 19.6
Current annual diameter increment MAI (in millimeters per year) by diameter classes for three species in the late stage in the Santa Rosa National Park. Numbers in parentheses indicate the numbers of trees included in each diameter class.

30 cm show higher growth rates. *Thouinidium decandrum* (7.2–9.2 mm · year^{-1}), *Guettarda macrosperma* (8.0 mm · year^{-1}), and *Quercus oleoides* (6.8 mm · year^{-1}) are good examples of dominant species with high light availability.

19.3.5 Species with Higher and Lower Increments

Seven species from the intermediate stage owned the highest growing rates among all the species (Table 19.5). During 2006–2011, *R. trinervis* and

TABLE 19.5
Tree Species with the Highest Mean Annual Diameter Increment MAI in Three Successional Stages in the Santa Rosa National Park

Stage	Species	MAI (mm · year^{-1})	Interval (mm)	Number of Trees
Intermediate	*R. trinervis*	17.0	13.4–19.0	3
Intermediate	*S. macrophylla*	12.2	12.0–12.4	2
Intermediate	*Cochlospermum vitifolium*	9.8	7.8–11.8	2
Intermediate	*Trophis racemosa*	9.1	2.0–14.2	10
Intermediate	*B. crassifolia*	8.3	1.8–13.8	8
Late	*T. decandrum*	8.2	7.2–9.2	2
Intermediate	*G. sepium*	7.8	0.6–11.2	6
Intermediate	*Acosmium panamense*	6.9	5.2–8.6	2
Early	*Pisonia aculeata*	6.0	2.4–8.6	6
Late	*Q. oleoides*	4.7	2.6–6.8	16

TABLE 19.6

Tree Species with the Lowest Mean Annual Diameter Increment MAI in Three Successional Stages in the Santa Rosa National Park

Stage	Species	MAI ($mm \cdot year^{-1}$)	Interval (mm)	Number of Trees
Intermediate	*Sterculia apetala*	0.3	0.2–0.4	2
Late	*Hymenaea courbaril*	0.3	0.2–0.4	2
Intermediate	*Piptadenia flava*	0.3	0.2–1.4	4
Late	*Chrysophyllum brenesii*	0.3	0.2–0.4	2
Late	*Lonchocarpus minimiflorus*	0.3	0.0–3.8	12
Intermediate	*Sapranthus palanga*	0.4	0.0–2.4	5
Late	*S. palanga*	0.4	0.2–2.8	63
Late	*Tabebuia ochracea*	0.4	0.0–5.8	2
Intermediate	*Anona reticulata*	0.5	0.2–0.8	2
Intermediate	*Cordia panamensis*	0.6	0.2–3.0	11

Swietenia macrophylla grew the most (17.0 and 12.2 $mm \cdot year^{-1}$, respectively). On the other hand, five species have growth rates of less than 0.3 $mm \cdot year^{-1}$, being associated with the intermediate and the late stages (Table 19.6).

19.4 Discussion

The results of this study are comparable to those of the study by Kalácska et al. (2004b) in the Santa Rosa National Park, confirming successional stages differences and a progressive recovery trend of forest structure and species composition characteristics from the early stage to the late stage. Basal areas estimated for the intermediate (25 $m^2 \cdot ha^{-1}$) and the late (27 $m^2 \cdot ha^{-1}$) stages in this study are within the range of 17–40 $m^2 \cdot ha^{-1}$ for old-growth TDFs across the world reported by Murphy and Lugo (1986a). These results are even higher than the average basal area reported by Gillespie et al. (2000) for six old-growth TDFs in Nicaragua (22 $m^2 \cdot ha^{-1}$). Spittler (2002) found a basal area of 28 $m^2 \cdot ha^{-1}$ in secondary TDFs in the Chorotega region (Guanacaste); this value of a 50-year-old forest is similar to the late-stage value reported in this study (Table 19.2). The author also mentions that such basal area values are not different from old-growth TDFs.

According to Marin et al. (2005), the mortality rates for tropical old-growth forests vary considerably, from 0.9% in a TDF in Ghana to 2.91% in a mixed deciduous forest in Thailand. Mortality rates for the intermediate (2.6%) and late (1.5%) stages estimated in this study are in this expected world range for old-growth forests and very close to the rate reported for TDFs by Sabogal and Valerio (1998) in Nicaragua (2.1%). The high mortality rate for the intermediate stage is probably due to the competition for resources (Louman et al. 2001) and the mortality of pioneer species reaching the end of their life cycle. The highest recruitment values corresponded to the early stage due to the fact that this stage has greater space and availability of resources for individuals to reproduce and grow (Connell et al. 1984) in comparison with the other two stages. Finally, the late stage showed the lowest values for mortality and recruitment, an expected tendency for a mature successional stage. Lieberman and Lieberman (1987) found that tree recruitment and mortality in old-growth tropical wet forests display the same pattern.

Murphy and Lugo (1986a) reported that diameter increments for TDFs are usually lower than 2 mm · year^{-1}, which corresponds to the results obtained for the three stages studied in this chapter. The low MAI value for the late stage is associated with the aging of established trees and the stage-specific floristic composition. The late stage is dominated by shade-tolerant plant (sciophytes) species present in the understory that have low growth rates than the pioneering species and other plants that grow in full sunlight conditions (heliophytes) found in the early and intermediate stages (Valerio and Salas 1998; Louman et al. 2001). In Chacocente Wildlife Reserve (Nicaragua), Marin et al. (2005) reported 1.4 mm · year^{-1} of diameter increment in a deciduous forest and 2.4 mm · year^{-1} in a gallery forest. The aforementioned study depicts diameter increments similar to those found in PNSR ranging from 1.6 to 2.2 mm · year^{-1} (present study).

The late-stage results are similar to the ones reported by Meza and Mora (2002) in their study in the Guanacaste National Park, where they found that MAI peaks in the class 35–45 cm (2.62–2.78 mm · year^{-1}). Meanwhile, in the Palo Verde National Park, Monge et al. (2002) found the highest increment in the class 50–60 cm (5.85 mm · year^{-1}). Presumably, in late successional stages when the trees reach the canopy and even surpass it they will have higher MAIs as they have greater resource availability, especially access to light. However, it must be pointed out that mature dominant trees do not grow indefinitely because their roots and foliage cannot fulfill their physiological demands and there is some point at which their growth stops (Louman et al. 2001). Ghazoul and Sheil (2010) also mentioned that canopy and emergent trees generally have a greater diameter increase when the trees reach larger diameters, which is the case for *R. trinervis* trees found in the intermediate stage.

In the late stage, *L. divaricatum* has the highest IVI value; this species comprises 15% of the total basal area for this stage (Jiménez-Rodríguez 2010). The discontinuous line in Figure 19.6 is due to the lack of individuals in all the diameter classes. This pattern shows the expected behavior for late-stage species registered also in tropical rain forests by Finegan et al. (1999), Clark and Clark (1992), and Lieberman and Lieberman (1987). Finegan et al. (1999) indicated that higher increments occur in groups of emergent and upper-canopy species. In the La Selva Biological Station, Clark and Clark (1992) recorded diameter increments between 5 and 14 mm · year^{-1} in emergent tree species with DBHs >30 cm. Finegan et al. (1999) and Clark and Clark (1992) argue that trees with well-exposed crowns located in the upper canopy have more possibility of growth than other species. This argument is supported in this study by the fact that canopy dominant species such as *T. decandrum*, *G. macrosperma*, and *Q. oleoides* had growth rates > 6 mm · year^{-1}.

The species *E. mexicanum* and *S. pavoniana* illustrate what Louman et al. (2001) had described for species growing under the canopy, resulting in flatter growing curves. Both species are sciophytes and are frequent in the late stage, particularly *S. pavoniana*, which alone represents 36% of the total late-stage individuals. It could be concluded that the only individual of *S. pavoniana* species in the 25–30-cm-diameter class is the tree seed source for many individuals regenerating in the lower classes.

Worbes (1999) reported increments between 1 and 10 mm · year^{-1} for 37 tree species in a semideciduous forest in Venezuela. Finegan et al. (1999) reported increments of 0 to 19 mm · year^{-1} for species of the rain forest in Costa Rica. In this study, the species increment interval was found to be between 0.3 and 17 mm · year^{-1} for 63 tree species, very close to the figures reported by Finegan et al. (1999). *R. trinervis* and *S. macrophylla* have the highest diameter increments (Table 19.3); both were found in the upper canopy of the intermediate stage: *R. trinervis*, 12.3 ± 0.6 m; and *S. macrophylla*, 13.5 ± 0.7 m.

All the 10 species reported in Table 19.4 with the lowest MAI values are found in the intermediate and late stages and none of these trees belong to the upper canopy. Vieira et al. (2005) found increments similar to these species in Manaus, Brazil, where some of the longer lived species (up to 1000 years, radiocarbon dating) have annual increments of 0.1–0.3 mm.

19.5 Conclusions

This study confirms the successional stages differences and the progressive ecological recovery trend of forest structure and species composition characteristics of TDFs from early to late stages. The estimated figures from this study for the intermediate and late stages are within the reported values for old-growth TDFs across the word.

Results obtained in this study indicated that the highest diameter increments were found in the intermediate and early stages mainly due to their floristic composition and age. The highest increments in the three stages were found in diameter classes where upper canopy trees are found. This is explained by their greater access to resources such as light and soil nutrients. In general, species with the highest increments are mostly found in the intermediate stage as are species that belong to the upper canopy. Species with lower increments are found in the understories of intermediate and late stages, which face a higher level of competition for resources.

Aside from the difference in growth rates along the stages (Clark and Clark 1992; Finegan et al. 1999), there could also be a variation in the diameter increment during the year influenced by unfavorable environmental conditions such as a prolonged dry season that induces cambial dormancy periods (Rozendaal 2010). Worbes (1999) found in the Venezuelan Caparo Forest Reserve that evergreen species only stop growing at the end of the dry season, whereas deciduous species stop their growth with the end of the rainy season. These and other variations in the diameter increment of TDF species make it even more difficult to obtain accurate values if one year or even shorter periods are used to estimate diameter increment rates (Murphy and Lugo 1986b). Hence, to compensate for these variations in TDFs, it is necessary to take measurements between longer periods of time to average out variations among years.

This study is one of the few that report estimates and compare MAI values among TDF successional stages. Even though our study might have some sampling limitations, our results are useful to understand the restoration processes of TDFs and can be of use in improving the prediction of forest landscape dynamics, carbon storage, and sequestration processes. The study is novel because it evaluates tree diameter growth changes among three successional stages of TDFs, which are hardly reported elsewhere in the literature.

Further work will consist of determining and comparing the structure, species composition and annual tree diameter increment, tree recruitment, and mortality of the three successional stages of TDF permanent plots established in supplementary Tropi-Dry sites (Chamela in México and Mata Seca in Brazil). This effort will provide a better analysis to draw robust conclusions about Neotropical dry forest successional processes, with the aim of assisting decision makers and practitioners in developing conservation and management strategies for these important and unique forest ecosystems. Because of this, the measurement of Tropi-Dry permanent plots must be continued every year to create a vigorous database that will allow the study of the relationship of successional processes with climate variability as well as the generation of diameter increment information for key selected tree species.

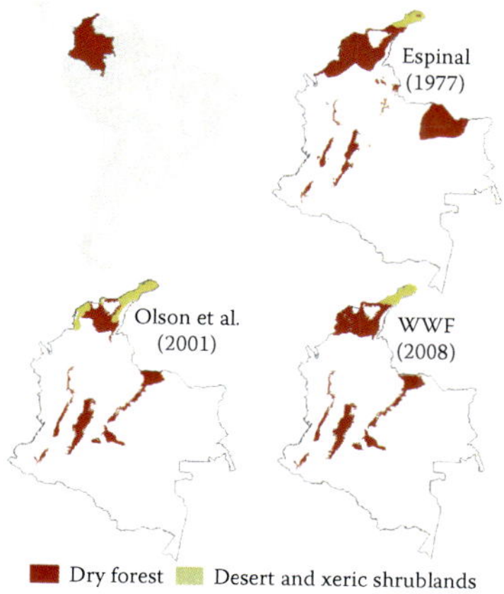

FIGURE 3.1
Maps of dry forest distribution in Colombia drawn by several authors/organizations (Espinal 1977) grouped the dry forest into different life zones (TDF, PMDF, TVDF, and STDF), as well as desert and xeric scrubland (TSTM and STDS) (see Table 3.1 for an explanation of the acronyms).

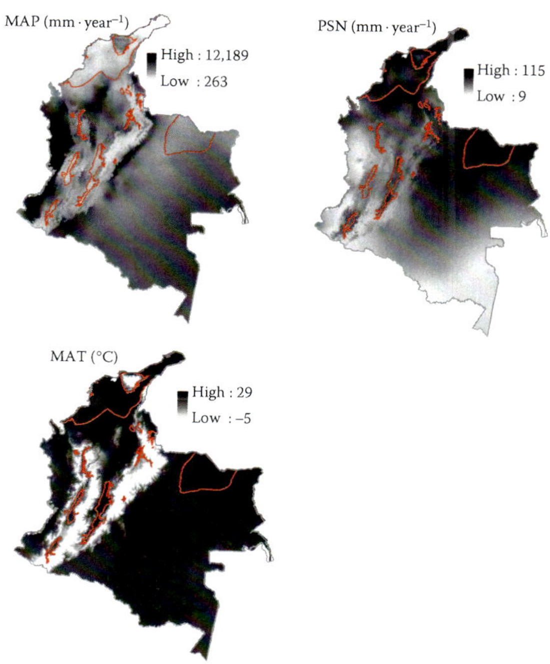

FIGURE 3.2
Climatic characteristics of the dry formations in Colombia: the figure presents the ranges for Colombia in terms of mean annual precipitation (MAP); seasonality of precipitation (PSN), defined as the coefficient of variation of mean monthly precipitation; and mean annual temperature (MAT). TDF ranges are MAP = 21–29 mm · year^{-1}, PSN = 263–2300 mm · year^{-1}, and MAT = 55°C–115°C. (From Hijmans, R.J., et al., *Int. J. Climatol.*, 25, 1965–1978, 2005.)

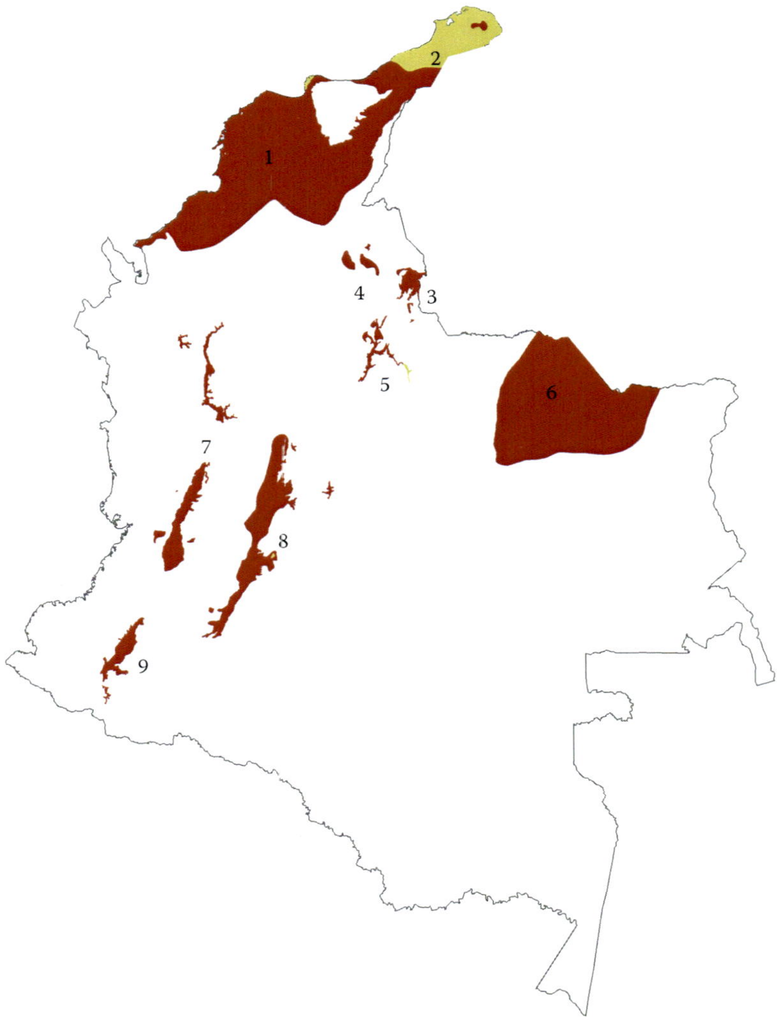

FIGURE 3.3
Regions having dry forests (tropical, subtropical, and mountainous) in Colombia: (1) the Caribbean coast, (2) the Guajira, (3) the Catatumbo region, (4) Aguachica, (5) the Chicamocha canyon, (6) Arauca, (7) the Cauca valley and river canyon, (8) the upper Magdalena River valley, and (9) the dry enclave of the Patía and the Dagua Rivers.

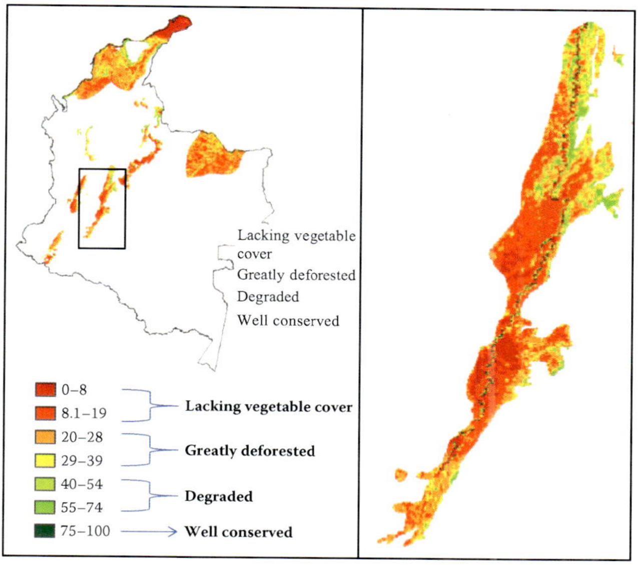

FIGURE 3.6
The status of dry forest conservation in Colombia and the upper Magdalena valley.

FIGURE 7.S2
Photographs of the TDF of Chamela, Mexico, taken from the same position on 6 consecutive days at the start of the rainy season in 2009. Leaf expansion of the closest trees happening within a few days can be appreciated; in a week's time, the forest turned green.

FIGURE 16.1
Details of fruits and seeds, respectively, of *Spondias tuberosa* (a and b), *Commiphora leptophleos* (c and d), *Pereskia stenantha* (e and f), *Pilosocereus pachycladus* (g and h), *Mimosa hostilis* (i and j), *Senna spectabilis* (k and l), *Pseudopiptadenia contorta* (m and n), and details of diaspores in *Myracrodruon urundeuva* (o and p). (Picture: Giovana Rodrigues da Luz)

(a) (b) (c) (d)

FIGURE 19.2
Photographs illustrating selected tropical dry forest (TDF) successional stages in the Santa Rosa National Park: (a) early, (b) intermediate, and (c) late stages and (d) the early stage with a remaining tree.

FIGURE 22.1
Poster disseminated by rural associations in the north of Minas Gerais, inviting people for a debate about the legal dispute over the Tropical Dry Forests in the region. Preserving the forests means "poverty," whereas converting them to irrigated agriculture means "progress."

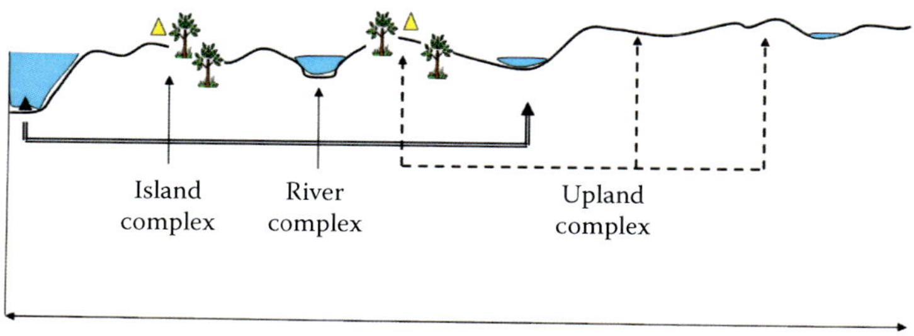

FIGURE 22.2
Schematic definition of the territorial organization of the vazanteiros at the margins of the São Francisco River, whose cycle of floods and ebbs defines the ecological dynamics and, therefore, the management strategies of each complex. (Modified from Fernandes, L.A. et al., Proposta de ocupação e uso dos ambientes pelos vazanteiros de Pau Preto: Novos indicativos à proposta da RDS no context do diálogo dos Vazanteiros de Pau Preto com o Instituto Estadual de Florestas. Technical Report, Montes Claros, 2010.)

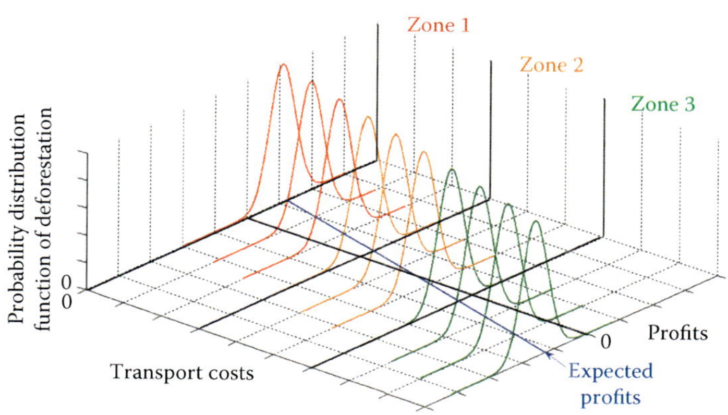

FIGURE 23.1
Variation in private deforestation pressure across the landscape.

FIGURE 24.3
The ecosystems used by the local communities in Serra do Cipó. At the top right the cerrado ecosystem (first plan), followed by rock outcrops lapeiros and mountain ecosystems. A significant number of species are used as firewood (photographs at the top left). The palm *Acromia aculeata* (at the bottom left) is one of the most abundant species and is used by the local communities in constructions and for oil extraction from its fruits (at the bottom right).

20

Carbon Dioxide Enrichment Effects in the Spectral Signature of Lianas and Tree Species from Tropical Dry Forests

Yumi Oki, Arturo Sánchez-Azofeifa, Carlos Portillo-Quintero, Payri Yamarte-Loreto, and Geraldo Wilson Fernandes

CONTENTS

20.1 Introduction

Carbon dioxide (CO_2) increases in the atmosphere (currently around 381 ppm) have been documented since the 1950s with an expectation that these values will surpass 700 ppm by 2100 (IPCC 2007b). Previous studies suggest that atmospheric CO_2 accretions and differential responses of plant species have led to changes in the structure and dynamics of natural ecosystems in the tropics (Körner 2003, 2004; Schnitzer and Bongers 2011). One of the changes in ecosystem structure involves an increase in liana abundance (Schnitzer and Bongers 2011). Considering the importance of tropical forests in regulating the global carbon (C) cycle (contributing to global primary productivity and climatic regulation of the planet), increased CO_2 concentrations in tropical forests may generate a cascade of strong, yet unpredicted, global ecological consequences.

Elevated CO_2 concentrations enhance C fixation and produce direct effects on photosynthetic processes (Owensby et al. 1999). Generally, the effects of elevated CO_2 include increase in biomass (Erice et al. 2006; Housman et al. 2006), change in growth rates (Soulé and Knapp 2006) and plant structure

(Pritchard et al. 1999), alteration in the patterns of allocation of nutrients (Nagel et al. 2005), change in the efficiency of water use (Eamus 1991; Li et al. 2003), and energy assimilation (Nagel et al. 2005). However, the magnitude of these effects varies depending on the plant species or the plant functional type (Körner 2004). Temperate lianas have responded more rapidly than trees (increase in leaf area and plant biomass) to enriched CO_2 concentrations (see the studies by Mohan et al. [2006] and Zotz et al. [2006]), suggesting that the increase in lianas observed in some ecosystems may be associated with elevated levels of CO_2 (Phillips et al. 2002). While investigating tropical species in China, Zhu and Caos (2010) postulated that lianas have some leaf characteristics (such as high ratio of leaf area to total plant mass) that allow greater fixation of C and a faster response to CO_2 enrichment. The authors observed a higher specific leaf area and resulting amplified photosynthetic rates in lianas in comparison with trees in response to atmospheric CO_2 enrichment.

Biochemical and biophysical plant alterations accompanying increased CO_2 concentration can also be observed in plant reflectance (Peñuelas and Filella 1998). Leaf spectral reflectance responses have been associated with ontogenetic alterations in leaf structure and physiology that occur as a response to changes in environmental conditions (Sims and Gamon 2002), although few experiments have assessed doubled CO_2 concentrations. Three tropical tree species under CO_2 enrichment treatment responded with a 9%–23% increase in spectral leaf reflectance to light in the visible (400–700 nm) wave band (Thomas 2005). According to Sims and Gamon (2002), leaf spectral reflectance changes at 445 nm are related to leaf surface reflectance, whereas reflectance changes at 800 nm are related to light scattering as a function of reflectivity and changes in internal leaf structure. Reflectance-based spectral indices also provide useful metrics when assessing plant physical and physiological processes (Griffith et al. 2010). Spectral reflectance indices such as the normalized difference vegetation index (NDVI) have also been shown to be sensitive to potential changes of the photosynthetic rate of plants (Whiting et al. 1992). Simple ratio (SR) is another spectral index that has been widely used to estimate photosynthetic pigments and green biomass (Peñuelas and Filella 1998; Sims and Gamon 2002).

The aim of this study was to evaluate spectral changes as they relate to the development and structure of dry tropical lianas (*Amphilophium paniculatum* [Bignoniaceae], *Callichlamys latifolia* [Bignoniaceae], and *Mascagnia nervosa* [Malpighiaceae)] and tree species (*Chrysophyllum cainito* [Sapotaceae], *Cecropia insignis* ([Urticaceae], *Luehea seemannii* [Malvaceae], and *Tabebuia rosea* [Bignoniaceae]) from Panama under conditions of enriched CO_2 concentration. Based on the aforementioned studies, we examine the hypothesis that plant structural changes occur with the increase of CO_2 and consequently affect leaf reflectance. To examine this hypothesis and to understand the effects of CO_2 enrichment on tropical plant species, we addressed the following question: what spectral changes and associated structural changes are observed in the leaves of lianas and tree species under CO_2 enrichment?

20.2 Carbon Dioxide Effects on Leaf Spectral Reflectance of Dry Forest Species

The effects of changes induced by augmented CO_2 have been effectively studied in many temperate species, although information regarding their effects on tropical forest species is still not fully understood. The assumption of a single response of all plant species to increased CO_2 may be misleading as phylogeny, growth form, preadaptations, habitat interaction, and even interactions with symbiotic microorganisms may mediate and change such responses (see the studies by Fernandes et al. [2011] and Sánchez-Azofeifa et al. [2012]). The relevance of this knowledge then becomes even greater when investigations indicate that tropical rain forests may turn into dry tropical forests due to global warming.

The responses of plants to enriched CO_2 levels vary according to the physiological features and structural adaptation of each plant species (Diaz 1995). Moreover, for each species or plant functional type (lianas, shrubs, and trees) the time of physiological responses under CO_2 enrichment can be different (Körner 2004). Despite there being various studies showing the effects of plant development under enriched CO_2 treatments, investigations about the structural changes that can be revealed from spectral analyses are few. In such studies, the spectral responses to CO_2 increase were clearer than the developmental parameters used for both lianas and trees. The spectral responses to CO_2 increase were clearer for both tropical dry forest lianas and trees after 90 days. The spectral responses indicate that biochemical and biophysical changes occurred under the CO_2 enrichment treatment, although the short time span of the experiment (90 days) did not allow for macrophysiological effects (such as growth) to be noticed.

20.2.1 Spectral Responses on Growth Habitat: Trees versus Lianas

A clear variation in the leaf reflectance for the group of trees (almost 4%) and lianas (around 1%) occurred at 800 nm (Figure 20.1) with the increase in CO_2 concentration. The leaf reflectance of the trees diminished along 500–700 nm. Spectrally, the trees presented lower values of SR 750/705 (21% lower), SR 800/680 (20% lower), and Modified normalized difference index (mND) 750 (20% less) at higher CO_2 concentrations (Table 20.1). Another trend observed was that the NDVI $(800-680)/(800+680)$ of tree leaves decreased with the increase of CO_2 ($P = 0.06$). All leaf spectral indices of the lianas significantly

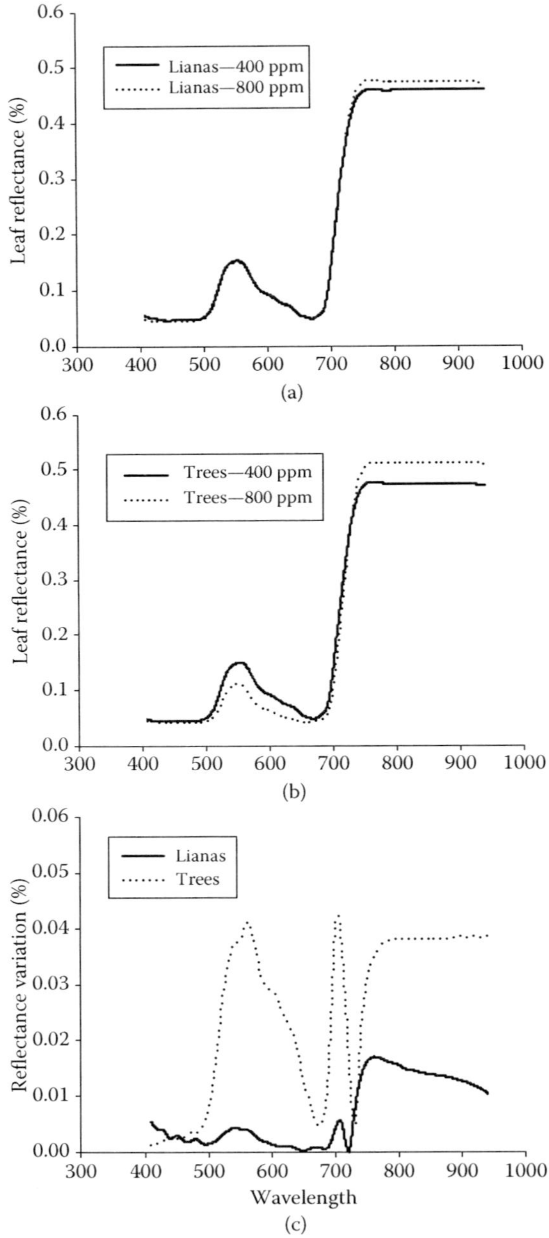

FIGURE 20.1
Leaf reflectance (%) found in lianas (a) and trees (b) under a carbon dioxide (CO_2) environment (400 ppm) and under a CO_2-enriched environment (800 ppm), and reflectance variation (%) of lianas and trees (c) between the CO_2 environment (400 ppm) and the CO_2-enriched environment (800 ppm).

TABLE 20.1

Leaf Spectral Indices [NDVI (800 − 680)/(800 + 680), SR 750/705, SR 800/680, and mND 750] of Tropical Plants Separated by Growth Habitat (trees—4 species and lianas 3 species) under CO_2 Environment (400 ppm) and CO_2 Enrichment (800 ppm)

Growth Habitat	CO_2 Concentration (ppm)	NDVI (800–680)/(800+680)	SR 750/705	SR (800/680)	mND 750
Trees	400	0.82 ± 0.0058 a	3.22 ± 0.0814 a	10.4 ± 0.361 a	0.60 ± 0.0141 a
	800	0.80 ± 0.0041 a	2.67 ± 0.115 b	8.70 ± 0.19 b	0.50 ± 0.0226 b
Lianas	400	0.74 ± 0.0095 a	2.63 ± 0.132 a	6.72 ± 0.261 a	0.56 ± 0.0256 a
	800	0.70 ± 0.0151 a	2.18 ± 0.141b	5.94 ± 0.274 b	0.45 ± 0.0306 b

Note: Mean values followed by different letters in the same column indicate significant differences between CO_2 treatments ($P < 0.05$).

diminished with the doubling of the concentration of CO_2. The NDVI reduced by about 6%, whereas SR 750/705 decreased by 21% and SR 800/680 and mND 750 were, respectively, 13% and 24% lower in the treatment with enriched CO_2.

20.2.2 Spectral Responses per Tree and Liana Species

Among the trees, *T. rosea* was the species that presented differences in all spectral indices between CO_2 treatments. The values of NDVI 800, SR 750/705, SR 800/680, and mND 750 found in the leaves of this species under CO_2 enrichment treatment were significantly lower, 4.5%, 26.7%, 22.54%, and 15.75%, respectively, than those in the CO_2 environment (Table 20.2). On the other hand, *C. cainito* trees did not show differences in the values of leaf spectral indices ($P = 0.850$ for NDVI (800–680)/(800+680), $P = 0.944$ for SR 750/705, $P = 0.881$ for SR 800/680, and $P = 0.815$ for mND 750) with increases in CO_2. For *C. insignis*, only the spectral index SR 800/680 values varied between the CO_2 treatments (28% lower in enriched CO_2; $P = 0.014$). Moreover, for *L. seemannii* the values of NDVI and SR 800/680 did not differ between the two CO_2 treatments ($P = 0.476$ for NDVI; $P = 0.491$ for SR 800/680) and the values of SR 750 and mND 750 decreased by about 66% and 29%, respectively, with the enrichment of CO_2.

A decrease in the values of leaf spectral indices was also observed in some liana species under CO_2 enrichment treatment compared to those under ambient treatment (Table 20.2). For *A. paniculatum* the leaf spectral indices did not vary between the CO_2 treatments, whereas for *C. latifolia* the values of SR 750/705 and mND 750 diminished when the concentration of CO_2 was high, although the values of NDVI and SR800/680 of the leaves did not differ between the CO_2 treatments. The leaves of *M. nervosa* presented lower values of NDVI (4% lower, $P = 0.008$), SR 800/680 (17% less, $P = 0.007$), and mND 750 (32% lower, $P = 0.001$) and no variation for the values of SR 750 when the concentration of CO_2 was doubled.

TABLE 20.2

Leaf Spectral Indices (NDVI, SR 750/705, SR 800/680, and mND 750) of Tree and Liana Species under CO_2 Environment (400 ppm) and CO_2 Enrichment (800 ppm)

Plant Species	CO_2 Concentration (ppm)	NDVI (800–680)/ (800+680)	SR (750/705)	SR (800/680)	mND750
Tree Species					
C. insignis	400	0.84 ± 0.006 a	3.24 ± 0.240 a	11.56 ± 0.470 a	0.57 ± 0.023 a
	800	0.81 ± 0.009 a	2.47 ± 0.203 a	9.06 ± 0.301 b	0.44 ± 0.040 a
C. cainito	400	0.79 ± 0.005 a	3.27 ± 0.162 a	8.59 ± 0.211 a	0.64 ± 0.030 a
	800	0.79 ± 0.0049 a	3.26 ± 0.181 a	8.54 ± 0.248 a	0.63 ± 0.029 a
L. seemannii	400	0.85 ± 0.008 a	3.32 ± 0.18 a	12.15 ± 0.65 a	0.60 ± 0.020 a
	800	0.84 ± 0.0041 a	2.00 ± 0.091 b	11.68 ± 0.313 a	0.46 ± 0.013 b
T. rosea	400	0.83 ± 0.006 a	3.11 ± 0.142 a	11.11 ± 0.44 a	0.58 ± 0.017 a
	800	0.80 ± 0.006 b	2.46 ± 0.081 b	9.06 ± 0.295 b	0.47 ± 0.016 b
Liana Species					
A. paniculatum	400	0.80 ± 0.0056 a	2.31 ± 0.30 a	9.00 ± 0.292 a	0.46 ± 0.061 a
	800	0.79 ± 0.005 a	2.07 ± 0.12 a	8.84 ± 0.222 a	0.40 ± 0.030 a
C. latifolia	400	0.79 ± 0.0071 a	2.26 ± 0.095 a	8.93 ± 0.329 a	0.43 ± 0.020 a
	800	0.78 ± 0.0072 a	1.81 ± 0.053 b	8.04 ± 0.280 b	0.32 ± 0.014 b
M. nervosa	400	0.80 ± 0.0050 a	3.22 ± 0.072 a	9.21 ± 0.262 a	0.60 ± 0.008 a
	800	0.77 ± 0.0103 b	2.44 ± 0.143 a	7.78 ± 0.372 b	0.48 ± 0.032 b

Note: Mean values followed by different letters in the same column indicate significant differences between CO_2 treatments ($P < 0.05$).

20.2.3 Considerations of Spectral Responses of Tropical Dry Forests

The leaf reflectance changes in the visible light spectrum (400–700 nm) found in the tree species of our study were also observed in the three tree species studied by Thomas (2005). A change in this wavelength spectrum indicates an alteration in the concentration of leaf photosynthetic pigments (Ormrod et al. 1999) and/or cuticular wax deposition (North et al. 1995). The increase in leaf reflectance along the near-infrared spectrum (700–900 nm), observed in the lianas and trees under enriched CO_2 treatments, may represent leaf structural changes (Xu et al. 2007). Slaton et al. (2001) observed that the responses of reflectance at 800 nm in 48 species were related to their coloration of leaves, thicker cuticle, and higher proportion of mesophyll cell surface area exposed to intercellular air space per unit leaf surface area. Surely, leaf reflectance variations of lianas and trees between CO_2 treatments (environment and enrichment CO_2 treatments) may be a result of changes in leaf structure and composition (photosynthetic pigment content), cell size, and cell wall composition and structure.

Vegetation indices have been efficient in diagnosing changes in plant metabolism and growth due to environmental conditions (Carter et al. 1994; Delalieux et al. 2009). The indices used in our study, NDVI, SR and mND, are linked to the absorption and reflectance of photosynthetic pigments and the biomass (Sims and Gamon 2002). Here, the lowest values of NDVI, SR, and mND observed in some liana (*M. nervosa* and *C. latifolia*) and tree (*T. rosea*, *L. seemannii*, and *C. insignis*) species under enriched CO_2 treatments indicate the occurrence of physiological and structural alterations. It is possible that such alterations are associated with metabolic disturbances observed under stress conditions (Knipling 1970). Generally, lower values of NDVI, SR, and mND have been found when plant species are subjected to biotic or abiotic stress (Carter 1994; Delalieux et al. 2007, 2009). Thus, our results suggest that the tree *C. cainito* and the liana *A. paniculatum* are tolerant to atmospheric enrichment of CO_2 as we failed to identify any difference in the values of all spectral indices measured. The results suggest that spectral indices represent important and reliable parameters to understand the response of tropical dry forest species to increasing atmospheric CO_2.

20.3 Final Remarks

Tree and liana plant species respond differentially to CO_2 enrichment, and the effects of such responses may affect the structure and biodiversity of ecosystems. The results of this study indicate that enrichment of CO_2 concentration has the potential to modify the structure of tropical liana and tree species and can lead to other ecological impacts such as species loss and changes in carbon storage and can perhaps affect a forest form that is totally different from what is known today. Knowledge about these effects on trees and lianas may help in understanding the complex changes taking place in tropical ecosystems with the current and future rise in ambient CO_2 concentrations. More detailed studies on the CO_2 effects on tropical species are called for in an attempt to augment our knowledge on the differential effects of CO_2 on plant performance and success and resulting community changes. Albeit with a limited species sample, we hope we have provided the fuel in the search for trends on the impact of CO_2 on tropical dry forest ecosystems.

21

Tropical Dry Forests in Latin America: Analyzing the History of Land Use and Present Socio-Ecological Struggles

Alicia Castillo, Mauricio Quesada, Francisco Rodriguez, Felisa C. Anaya, Claudia Galicia, Francisco Monge, Rômulo S. Barbosa, Andréa Zhouri, Julio Calvo-Alvarado, and Arturo Sánchez-Azofeifa

CONTENTS

21.1 Introduction

Tropical dry forests (TDFs) cover 42% of the world's tropical ecosystems (Murphy and Lugo 1995), with extensive coverage in Latin America. More than 50% of all TDFs occur in South America, and about 15% occur in Mexico and Central America (Miles et al. 2006). Due to their high biological diversity and rapid loss through deforestation, TDFs have evinced increased scientific interest in the last decades (Murphy and Lugo 1986, 1995; Janzen 1988;

Mooney et al. 1995; Sánchez-Azofeifa et al. 2005, 2009; Portillo-Quintero and Sánchez-Azofeifa 2010). In order to construct management strategies that secure TDF's long-term maintenance and to ensure a continuity of ecosystem services provided to human societies, an interdisciplinary analysis of socio-ecological processes is needed (Berkes and Folke 2000). However, a comprehensive understanding of societal/TDF interactions has not been achieved, particularly when addressing the complex social processes that occur when different stakeholders vary in their needs, views, economic and political interests, and power regarding the appropriation of natural resources (Adams et al. 2003; Waltner-Toews et al. 2003).

Such interactions can be better understood when viewed through more than one lens (Berkes and Folke 2000). Ethno-botanical studies, for example, have documented indigenous and local communities' various uses of common TDF plant species (Casas et al. 2001; Bye 2002; González-Rivas et al. 2006; Barrance et al. 2009). Studies of land-use change provide relevant information for conservation (Trejo and Dirzo 2000, 2002; Miles et al. 2006) and describe how lands have been transformed, but, in general, few such studies consider the social histories that explain today's fragmented landscapes. Rare examples of social influence-inclusive studies elucidate the importance of incorporating such particulars. A detailed account of historical events given by González-Rivas et al. (2006) described the effect of sociopolitical events on TDF ownership, use, and conservation in Nicaragua. Similarly, Roth (1999) evaluated TDF change occurring during a century-long period in the Dominican Republic and described the transformation of TDFs into pastures that were later abandoned and invaded by thorny woody species, which were subsequently used for charcoal production. Another example of socio-ecological histories was conducted in Western Mexico, where both government and peasant strata consider TDFs to have little value, and have consequently promoted TDF transformation into pasture lands, and, more recently, into large-scale tourist center developments, threatening the conservation of TDF remnants (Castillo et al. 2005; Pujadas and Castillo 2007; Castillo et al. 2009).

Relating scientific information about TDFs to the needs of the human population residing nearby and using these TDFs is a crucial element in developing effective land-use decision-making processes (Murphy and Lugo 1986). Tropi-Dry, an international project, was launched in 2006 with the main goal of generating information regarding TDF status in the Americas by examining socio-ecological processes in five countries: Brazil, Costa Rica, Cuba, Mexico, and Venezuela. Emphasis was given to investigating the ways in which societies have traditionally and recently benefited from and influenced this ecosystem, and to identifying the socioeconomic, political, and cultural drivers that cause TDF transformation. This study presents the results of case studies conducted in Mexico, Costa Rica, and Brazil (Figures 1.2 through 1.4 of Chapter 1, respectively). A comparative analysis is made based on the following research questions: (1) How have various societies interacted with TDFs over a given period of time? (2) What are the driving forces that explain TDF changes? (3) How do local actors perceive TDF use, transformation, and conservation? (4) What have been the responses of different stakeholders to TDF changes?

The aim of this chapter is to examine the dynamic human/ecosystem interactions and economic processes that influence TDF conversion at selected research sites in the three countries mentioned earlier. A comparative analysis of land-use histories and policies specific to each site (agricultural, forestry, development, and environmental) reveals strategies that can be used to safeguard TDFs for sustainable, long-term use while concurrently maintaining peoples' livelihoods through and preserving the dignity of the indigenous people and local communities who form an integral part of these TDFs.

21.2 Comparative Case Studies: Conceptual Structure, Research Methodology, and Results of Human Dimensions

As an international collective initiative, unified research protocols were prepared (Nassar et al. 2008) for the different components of the project Tropi-Dry. The social component's conceptual framework focused on the collection of data regarding environmental history with an emphasis on understanding the influence of governmental policies on TDF use and conservation, and on the documentation of local people's perspectives on TDF use, transformation, and maintenance. Considering the fact that each country and research site has unique features, protocols have served as guidelines for conducting case studies in order to standardize comparisons among countries (Yin 2003). The main research techniques included (a) document assessment (archival records, historical accounts, and published research; Creswell 2003); (b) direct and participant observation recording (Patton 2002); and (c) semi-structured interviews (Robson 1994). Since previous research and accessible information varied from one country to another, case studies were constructed that maximized commonly occurring characteristics in order to facilitate a comparative analysis. Specific information regarding the methods used in each case study is presented in Table 21.1; Table 21.2 summarizes the main issues examined in this chapter: history, drivers of change, social perceptions, and consequences of policies on TDF.

TABLE 21.1

Research Methods Used in Each Case Study

	Brazil	Costa Rica	Mexico
Document review (number of documents)	70	24	20
Direct interactions with participants (fieldwork days)	35	23	43
Semi-structured interviews (number of interviews)	25	36	45

TABLE 21.2

Historic Time for Tropi-Dry Sites in Mexico, Costa Rica, and Brazil

	Mexico (Chamela–Cuixmala)	Costa Rica (Guanacaste)	Brazil (Minas Gerais)
Pre-Hispanic times	- Human presence dating from 3,000 years ago - Nahuas indigenous peoples	- TDF dwellers influenced by Mesoamerican culture - Slash and burn methods used to clear TDF lands	Indigenous people subsisting from fishery, hunting, and fruit collection
1492–1900	- Spanish conquest: Consequences included a decrease in indigenous populations (disease and human exploitation) - TDF land with low agricultural productivity. There were no mines, so the region was not seen as important to Spaniards - Establishment of the hacienda system (large pieces of land owned by a few families) Indigenous people's lands were claimed by European settlers. Conflicts between landowners and indigenous people arose under this form of land tenure - Main productive activities were agriculture and cattle raising	- Spanish conquest: Exploitation of economic minerals (mining) and cattle raising formed the dominant exports - Haciendas (large landowners) initiated deforestation	- European colonizers (mainly Portuguese), - Africans (slaves) - Black *aquilombado* subsisting from agriculture (cassava, beans, rice, corn, banana, etc.), cattle raising in natural habitats, fishery, hunting, and fruit collection - *Bandeirantes* (pioneers): farm-centered society, with the establishment of large farms for extensive cattle raising and agriculture for local provision (northeastern region first, then the mining region of Minas Gerais state)
1900–1950	- Large haciendas were present until 1940–1950. - Apart from agriculture and cattle raising, extraction of fine woods by private owners became relevant - Start commercial agriculture (coconut palms; bananas) - Policies to colonize the Jalisco coast because of its low population - Haciendas dismantled to distribute land to peasants (begins period of creation of most ejidos)	- Coffee and banana cultivation - Cattle commerce for the Americas - Wood exploitation (Guanacaste region)	Cattle raising and agriculture for regional provision (center-north of Minas Gerais and south-southeastern Bahia) - Consolidation of large farms for extensive cattle raising and agriculture (corn, beans, and cotton) - Small agriculturists still planting diversified crops (cassava, cereals, sugarcane, and grape) and practicing extractivism, which is associated to cattle raising

1950–2010	- First colonizers (bought lands) - 1959: hurricane - Agrarian Reform: 1975 creation of ejido lands, which were distributed without regard to previous claims by colonizers. Conflict. - From 1970 to 1980 National program for clearing forests for cattle raising - 1971: creation of UNAM Chamela research station - 1993: decree of Chamela–Cuixmala Biosphere Reserve - 1999: decree of Land-Use Planning Program	- 1980s: decline of cattle raising and promotion of exportation crops (cotton and sugar) - 1970s: creation of conservation areas - 1972: Santa Rosa National Park - 1974: Rincón de la Vieja National Park - 1980s: reforestation programs - 1990s: environmental services payments - Tourism development	- Government policies for industrialization - Programs in rural areas for: - Irrigated cash crops (fruit trees) - Extensive cattle ranching - Eucalyptus cotton plantations
Stakeholders identification	- Government (local, Jalisco state, municipality) - Peasant ejidatarios - Avecindados (landless people) - Academic community	- Local leaders - Local population	- Government - Environmentalists - Agribusiness enterprises - Traditional people and local population
TDF transformation drivers	- Governmental colonization program (1940s and 1950s) - Extensive cattle raising - Tourism development	- Extensive cattle raising - Tourism development (massive and big tourist and real-state enterprises gain control over resources)	- Extensive cattle raising - Agriculture for food production
TDF use	- Extraction of fine woods by private owners - TDF as land to be transformed for agriculture or cattle raising in ejidos - Low use of TDF species (wood and non-wood products)		- Traditional populations and communities: TDF as a territorial patrimony to be preserved for cultural, social, and economic reproduction; its use is related to the notion of sustainable extractivism - Large farmers: TDF as an economic asset to be transformed in pastureland or monoculture and charcoal production - Environmentalists: asset to be preserved in isolation from human presence

(Continued)

TABLE 21.2 ***(Continued)***

Historic Timeline for Tropi-Dry Sites in Mexico, Costa Rica, and Brazil

	Mexico (Chamela–Cuixmala)	Costa Rica (Guanacaste)	Brazil (Minas Gerais)
TDF maintenance	- Recognition of TDF as a provider of ecosystem services by local people: provision (water, food, and fuel wood); regulation ("fresh" climate); cultural (recreation, enjoyment of natural landscapes) - Recognition that animals in TDF have diminished through time - Scientists providing information for TDF conservation; more open to interact with local people but lacking abilities and the necessary time for building interactions		- Traditional populations: environmental preservation associated to agricultural production and extractivism (fishery, fruit collection) - Large farmers or agribusiness: deforestation to increase pasturelands and agriculture (fruit production for exportation) - Environmentalists: environmental production through conservation units (CUs) of restricted use, especially parks
For ecosystem conservation	- Old notion of Biosphere Reserve: conservation of pristine ecosystem - No raising of conservation awareness	- Conservation areas - Reforestation programs - Environmental services payments (private owners) Conservation tourism (Guanacaste) - Biological Conservation Education Program	- Conservation areas for pristine ecosystem maintenance (7)
Consequences of conservation policies	- Ejidos and reserve just as neighbors; no mutual interest - Opposition to Land-Use Planning Program (POET): conservation as imposition	- Reserves created expropriating haciendas and smaller land owners - Parks need to provide more jobs to locals	- Local peoples are expelled from their territories - Symbolic dispute over the definition of TDF (Cerrados and Caatinga for agriculture)

Consequences of agricultural, social development, and other policies	- Government is not interested in supporting agriculture - Programs designed for those already in better positions (more lands or financial capacity)		- The creation of CUs of restricted use constrains the agricultural options for food production of traditional populations
Social control and power	- Government creates conflicts for land (from creation of ejido in the 1970s) - Bad administrative practices among ejidatarios produce mistrust - Rich people (national but importantly foreign) as having best lands	- Conflicts between the protected areas and the human communities in their influence areas - Regulation to the exploitation of marine resources (fish and forests) - Tourism: locals are also excluded	- State contradictions: as compensation for agribusiness, expropriate territories (traditional) for TDF conservation - Subsidies to agribusiness and "green certifications" - Environmental policies legitimize deforestation (for agribusiness) by protecting small TDF areas
	- High levels of migration (the Americas and other states in the country) - At municipality level; District Council for Rural Sustainable Development (five municipalities) Collective construction of analysis of problems and search for solutions		- Creation of traditional peoples associations and unions - TDF relevant for its symbolic meaning: not only land; it also represents cultural identity

21.3 Human Dimensions of the Chamela–Cuixmala Biosphere Reserve, Jalisco, Mexico

In Mexico, the TDF is the predominant tropical vegetation type, comprising more than 60% of total tropical vegetation by area, and it represents the most threatened and least protected ecosystem (Trejo and Dirzo 2000). The Chamela–Cuixmala Biosphere Reserve, located on the Jalisco coast between the ports of Manzanillo and Puerto Vallarta, includes one of the very few protected TDF segments in Mexico (Figure 1.2). Conservation for this area was decreed in 1993, establishing 13,142 ha of protected TDF lands (Diario Oficial de la Federacion 1994). The Chamela Biological Research Station, created in 1971 by the Universidad Nacional Autónoma de México, protects 3,319 ha of TDF with the main purpose of conducting research on the biology and ecology of TDFs. In 1988, a private owner agreed to donate roughly 10,000 ha more, making the parcel eligible to become a natural protected area (Figure 1.2 in Chapter 1). No human settlements exist inside the reserve, but influence zones at the area's margins are largely controlled by Ejidos (a form of land tenure created after the 1910–1917 Mexican Revolution that at present combines private and communal land management) and by private landowners. Ejidos can have one or more settlements where ejidatarios live and work the land, but may also support a number of families who are employed at private ranches, in commerce or in the tourist industry. Because of the beauty of the beaches and the landscape, government authorities have recently approved massive hotel developments, with socio-ecological impacts and implications that remain unknown (Castillo et al. 2009).

21.3.1 Methods and Analyses

Since previous research had been conducted in Ejidos, located in the northern part of the reserve's influence area (Castillo et al. 2005), the *Ejido Ley Federal de Reforma Agraria* was selected for analysis in the present study. This Ejido, which is located in the southern part of the Reserve and owned by 486 landowners who live in six settlements across its extent, covers an area of more than 17,000 ha (more than the area protected by the reserve). Investigators reviewed local, regional, and federals laws; Ejido legal papers; and related documents regarding governmental policies and programs. Ejido leaders approved permits to conduct the study, and 20 pilot interviews were initially conducted, including the primary local Ejido authority. A total of 45 interviews with landowners were carried out in the six settlements using the snowball sample method (Patton 2002). All interviews were electronically audio recorded and fully transcribed. An analysis was conducted by a line-by-line examination of transcripts using the Atlas.ti software program (version 5.2).

21.3.2 Historical Dimension of Society

Human settlements on the coast of Jalisco go back 3000 years (Mountjoy 2008). For centuries, indigenous communities lived in sparsely dispersed communities and exerted a low impact on their host ecosystems. Following the European conquest, indigenous populations drastically declined due to disease and exploitation (Rodríguez 1989). Spaniards found the TDF environment to be a wild and difficult frontier, and because no economic minerals attracted settlement, these areas remained largely undeveloped (Rodríguez 1991). European settlers made land claims in spite of long habitation by indigenous people, and created the extensive hacienda system to promote agriculture and cattle ranching. Continuous disputes for territory characterized more than 4 centuries of Mexican history. It was not until the twentieth century that haciendas lost their territorial, economic, and political power.

On the Jalisco coast, several haciendas occupied areas as large as 90,000 ha each (Castillo et al. 2005). Lands of the *Ejido Ley Federal de Reforma Agraria* were part of the Hacienda Apazulco, which existed in 1950 with an extent of 35,000 ha. This hacienda belonged to Rodolfo Paz Vizcaino, who is recognized by current landowners as being responsible for cutting most of the native fine woods (mahogany and cedar) and for opening lands for coconut palms and bananas. During the 1940s and 1950s, a federal law that promoted the colonization of Mexican coastlines allowed for the creation of new settlements by expropriating the lands of Rodolfo Paz Vizcaino (Castillo 1991). This federal law can be recognized as the first important driver of TDF transformation in the Jalisco coast region. Because demand for land distribution remained strong, in 1960, the Agrarian Reform Office expropriated the lands given to the first landowners. After 15 years of conflict among different groups, in 1975, the *Ejido Ley Federal de Reforma Agraria* was decreed. The main productive activities included subsistence agriculture (maize and beans); coconut plantations; and the production of other commercial crops such as lemon, papaya, and mangoes. In the 1980s, federal credits supported irrigation systems to increase cultivation of watermelon, tomatoes, and chilies. Policies that declared forests to be "useless lands" resulted in increased conversion of TDFs to agricultural lands bearing produce that could contribute to the national gross domestic product. During the 1970s, government credits were issued to help private landowners purchase machinery to cut down trees, and technical assistance was provided to explicitly describe to landowners how to transform tropical forests into pasture lands (Paré 1995). With regard to forest use, interviewees explained that fine wood exploitation was common four or five decades earlier, but at present, such species are difficult to find.

Despite the fact that the Mexican government has conducted land evaluations since the 1940s and was aware of erratic rain patterns in the Chamela–Cuixmala region, which make it inhospitable for large-scale agriculture and cattle ranching (Rodríguez 1991), government officials have continued to institute policies that encourage the production of agricultural goods and cattle. As a result, TDF lands have been cleared at high rates, and now, these lands exist only in fragments. Many of the families that followed the guidance of the Mexican government and cleared their lands for cattle and/or agriculture have not profited, and as a consequence, many heads of households have been forced to migrate to support their families. According to local authorities of the Ejido Ley Federal de Reforma Agraria, it is estimated that 70% of men between the ages of 18 and 25 from this region of Mexico work in the Americas.

Constitutional changes made in 1992 modified the way in which Ejidos and indigenous communities owned their lands. Initially, each individual landowner was assigned a parcel and communal rights to jointly owned land. Current changes in the present legislation suggest the possibility that each landowner may receive ownership titles for their parcels. People expressed the opinion that as "legitimate owners" they can ask for individual credits and make decisions that once needed to be approved by the Ejido assembly. Some peasants complained, however, about governmental programs for Ejidos. The main perception is that the Mexican government is not interested in supporting primary activities such as subsistence agriculture. Peasants also argued that support programs are designed for those who are already in better economic positions, because they either have more land area or have financial capacity to invest in projects for which the government provides complementary support. Poorer peasants do not have access to such programs. Finally, landowners also acknowledged that because the area has been recognized as important for tourism, politicians may choose to further displace indigenous and local peoples to accommodate more tourist establishments.

21.3.3 Cultural, Political, and Economic Dimensions

The establishment of the Chamela–Cuixmala Biosphere Reserve in 1993 and the decree of the Coast of Jalisco Land Use Planning Program (Spanish acronym POET) in 1999 represent the primary conservation activities of the region. As explained by Pujadas and Castillo (2007), the reserve is atypical when compared with most biosphere reserves: Most of its lands are private (owned by a foundation created by a European family) with some portions owned by the Universidad Nacional Autónoma de México (UNAM). Human communities are not present inside its boundaries, and very little interaction takes place with adjacent Ejido communities. It is interesting to note that a recent study using remote sensing (Sánchez-Azofeifa et al. 2009) reveals that Ejidos lands retain large forested areas, and that those closest to the reserve present the highest TDF cover percentages but are not protected. Only the Chamela–Cuixmala Biosphere Reserve is well conserved (>13,000 ha). On the other hand, interviews conducted with the Ejidos in the northern part of the reserve revealed that most people do not have a clear idea of the purpose of the biosphere reserve and its functions (Castillo et al. 2005, 2007). Historically, familiarity with the UNAM research station is derived from guided tours offered to school groups. More recently, public awareness of the station has been bolstered by a one-day open house called "Open Doors." This event commenced in December 2007, when, for the first time in decades, the station invited local people to enter the premises and allowed researchers and students to explain their work.

In contrast, POET, an instrument for planning social development and environmental conservation in the 10 municipalities of the Jalisco coast (almost 1.5 million ha), favors private entrepreneurs (Pujadas-Botey 2003), though it prohibits the opening of new productive areas. In the past five years, several tourism project proposals have been submitted to environmental authorities. Until 2011, all were rejected, because they would drastically alter ecosystem functioning, particularly water availability for the same ecosystem and for human use. One project has recently been approved despite the opinion presented by biosphere reserve scientists based on a scientific examination of the proposal (Boege-Pare et al. 2010).

21.4 Santa Rosa National Park, Guanacaste, Costa Rica

Eleven different conservation areas comprising 10 national parks and reserves are recognized in Costa Rica today, with protected TDFs in only two of these areas (Sánchez-Azofeifa et al. 2003). TDF-rich conservation areas include Guanacaste Conservation Area, specifically the Guanacaste National Park, which protects approximately 50,000 ha of TDFs, and the Tempisque Conservation Area, specifically the Palo Verde National Park and the Lomas Barbudal Biological Reserve, which together protect an additional 20,000 ha of TDFs. Although most of the original TDF cover has disappeared in Mesoamerica, Guanacaste possesses the greatest cover of TDFs among all the Costa Rican provinces. Our study site was the Guanacaste Conservation Area and its surroundings (Figure 1.3 of Chapter 1).

21.4.1 Methods and Analyses

Two communities located within the influence area of the Guanacaste National Park were selected: (1) Colonia Bolaños, located in the eastern portion of Guanacaste, in the Cantón de la Cruz; and (2) Quebrada Grande in southern Guanacaste, within the Cantón Liberia (see Figure 1.3 of Chapter 1). In each community, two types of social actors were selected: local leaders and local inhabitants, a group that comprises the primary users of TDF resources. Two pilot interviews were conducted in each community in order to make the interview template appropriate to local conditions. In Colonia

Bolaños, 16 interviews were conducted, including 6 interviews of leaders and 10 interviews of local people. Similarly, in Quebrada Grande, 18 interviews were conducted, including 8 interviews of leaders and 10 interviews of local users. Investigators electronically audio recorded and fully transcribed all interviews. A detailed examination of all texts was performed, from which a narrative was constructed that presents the collective views of participants.

21.4.2 Historical Dimension of Society

From pre-Hispanic times, indigenous communities in Costa Rica were strongly influenced by Mesoamerican cultures with some influence from South American cultures (Quesada 1980). Mesoamerican cultures generally rely on slash and burn methods to clear forests, and then use these lands to cultivate maize, beans, and squash (González 1985). During the Spanish conquest, mining, timber extraction, and cattle ranching became the main economic activities. For more than four centuries, the hacienda system comprised the principal land tenure structure and provided the framework for most social interactions. The actions of small, elite groups of landowners ("terratenientes") contributed to the poor living conditions for a majority of peasants and landless people. These conditions resulted in high rates of migration, circumstances that continue to this day (González 1985). Extensive plains and abundant rivers enabled the expansion of cattle raising in the haciendas; producing massive deforestation until the 1950s. Intensive campaigns for hunting puma (*Puma concolor*) and jaguars (*Pantera onca*) existed under the auspices of reducing cattle predation (Cabrera 2007). Beginning in the 1950s, cattle trade expanded to include exports to North America, diversifying economic activities that had previously been concentrated in coffee and banana cultivation (Vega 1985). After 1980, the cattle industry became the primary contributor to deforestation in Costa Rica, accounting for more deforestation than all other economic activities combined, including commercial logging (Lehmann 1992). The construction of the Pan-American highway in the 1950s opened previously inaccessible locations by providing easy access to the entire country, facilitating both settlement and marketing in Guanacaste (Williams 1986). Development of the cattle industry became a major goal of the Costa Rican government in the 1950s, beginning with the exportation of live cattle in 1954, a practice that was soon replaced by the exportation of refrigerated beef (Hall 1985). The government contributed with national and international credits to financially support the growing cattle industry. It has been estimated that approximately 50% of all agricultural credit went to cattle production in the early 1970s (Leonard 1987). On a countrywide basis, by 1973, more than one-third of the country's territory consisted of cattle pastures (Lehmann 1992). The area of the country dedicated to pasture increased from 630,000 ha in 1950 to more than 2,000,000 ha in 1994 (IICA 1995), which was a rise of almost 400%. The increase in area dedicated to pastureland corresponded to the increase in cattle production, both of which followed a pattern of exponential growth until 1990; whereas forest cover decreased linearly between 1950 and 1998 (Calvo-Alvarado et al. 2009a). In addition, other economic activities related to wood exploitation and expansion of cotton and sugar crops increased during the period between 1950 and 1985, and these increases also contributed to TDF deforestation in the Guanacaste region (Edelman 1998; Arroyo-Mora et al. 2005; Calvo-Alvarado et al. 2008, 2009). The cattle-ranching industry decayed in the mid-1990s after a decline in the international market price for beef, and changes in agricultural policy geared to promote the exportation of crops (Edelman 2005).

Environmental conservation policies began in the 1970s with the creation of conservation areas. Because the TDF land type was identified as a degraded and vulnerable ecosystem, several protected areas were decreed: In 1972, the Santa Rosa National Park was established with 37,117 ha on land and 78,000 ha on sea; in 1974, the Rincón de la Vieja National Park was established with 14,161 ha; and in 1989, the Guanacaste National Park was established with 34,651 ha that aimed at protecting TDFs on the Orosí and Cacao volcanoes (Sistema Nacional de Areas de Conservacion 2008). The expropriation of hacienda lands not only created most reserves, but also subsumed lands of small- or medium-sized properties, causing some conflicts between the government that developed the protected areas and the human communities surrounding these areas (Jiménez 2006).

At the end of the 1980s, the Costa Rican government initiated reforestation policies and supported conservation efforts by private owners in order to encourage payments for environmental services programs (Sánchez-Azofeifa et al. 2007). Costa Rica is one of the first countries to establish such an extended environmental policy. Ecotourism, which became viable in the 1990s, has ironically produced new environmental problems. When tourist and real-estate enterprises realized the lucrative nature of ecotourism, these businesses sought to accommodate the massive influx of visitors. Paradoxically, ecotourism has had a number of ecologically destructive consequences, including deforestation, river-course alterations, damage to coastal and mangrove areas, improper sewage and garbage disposal, disturbance of nesting areas of endangered sea turtles, and increased underground water pollution, which has affected water provisioning to local communities (Calvo-Alvarado et al. 2009). Many families have sold their land and have migrated to cities.

21.4.3 Cultural, Political, and Economic Dimensions

People of the two settlements analyzed in this case study recognized the fact that the protected area should provide more jobs (only 8% of park workers are locals). The results of the interviews with some of the local people indicate that "The Park occupies 36% of the whole 'Cantón' area, they have the best lands and the aspiration of the community would be that the Park will provide more job opportunities for locals and that the municipality can promote tourism development so local people can participate of the economic benefits generated by the park." Land tenure and land use are the subject of both debate and political conflict.

Another conflict relates to people's access to natural resources such as fish that provide income or food. According to the mayor of the town of La Cruz, which is located in one of the region's richest fishing areas, 110 families depend on fishing, and the municipality is trying to promote a fishing terminal. However, because of the importance of the marine life, the area was protected and became part of the Guanacaste National Park. Fishermen now require permits, and their fishing periods are regulated. Although the relevance of the protected area is understood and recognized by many in the local community, the people and industries that rely on the threatened marine life do not fully understand or agree with the ecological implications of overexploitation of the aquatic resources they are depleting, and object to regulations that protect these resources.

From colonial times in the Guanacaste region, the overexploitation of natural resources of *Hacendados* (owners of large areas) restricted the benefits of this natural capital to a few; this has continued through the nineteenth and twentieth centuries, when big landowners retained lands for cattle raising, commercial crops such as cotton, sugarcane, and other recent products for export such as watermelon. At present, tourism has become one of Costa Rica's main economic activities, with limited access to local people. Foreign and national investments in this sector also contribute to this situation.

It is also important to state that the Guanacaste National Park has a very active environmental education program on biological conservation. The Biological Education Program was created with the objective of developing understanding and sensitivity to nature (bio-alphabetization), and of promoting critical attitudes about the relationship between humans and nature. Students from 42 elementary schools as well as students from six high schools, all of whom were from areas surrounding the ACG, Area de Conservación Guanacaste participate in this program. More than 2200 children per year have the opportunity to observe analyses, handle tools, perform experiments, and understand the human/TDF relationships and natural history of the region. This program allows children to gain hands-on experience and instruction on basic biology, knowledge that will help them in the future when making decisions concerning the environment. Due to the bio-geographical characteristics of the ACG, the program has been structured into three areas of study: the dry forest, the rain forest, and the coastal region. In each of these areas, the teachers work with children from schools in that region, while also making visits to other areas to make comparative assessments of other ecosystems. Other projects of the Biological Education Program include workshops where elementary and high school teachers and parents accompany students on visits to other conservation areas. This program is conducted in conjunction with the Ministry of Education of Costa Rica in coordination with the directors of the *Direccion Regional de Ensenanza de Liberia y Upala,* the principals and teachers of schools, and local authorities.

21.5 Parque Estadual da Mata Seca, Minas Gerais, Brazil

Tropical dry forests cover 3% of all Brazilian territory and are mainly located in the northeast and center-west regions. In Brazil, TDFs extend over 27,367,815 ha, representing 3.21% of its territory (see Chapter 5), and are predominantly distributed in the semiarid northeastern region (Figure 3.1 of Chapter 3). This vegetation represents 10.21% of native flora in Minas Gerais state (covering 2,040,920 ha), and 23.56% of native flora in the northern areas of that state (occupying 1,594,519 ha). TDFs are present in 78 out of 89 municipalities in the northern region of Minas Gerais (IEF and UFLA 2005). These forests have been intensively logged and cleared for cash crop farms and cattle-ranching areas. The region is inhabited by a diversity of local people living in traditional communities, with a way of life that is considered primitive and poorly developed when evaluated using conventional economic indicators. The study focused on the Parque Estadual da Mata Seca (PEMS), which is one of the seven conservation areas created in Brazil to protect TDFs.

21.5.1 Methods and Analyses

In order to analyze the social, political, economic, and historical processes influencing changes in TDFs (commonly known as "Matas Secas"), the following steps were taken: (1) identification of stakeholders involved in environmental conflicts in the areas surrounding TDFs; (2) documentation of the position of each individual with regard to the disposition of the TDFs; (3) definition of the roles of participants in terms of their relationship to the TDFs, particularly regarding environmental appropriation; and (4) determination of the complexity and structure of the power relations among all relevant social actors. In addition to analyzing environmental policies, our procedures included (a) the investigation of historical documents referring to common regional occupations; (b) the interpretation of census-based data; and (c) participant observation, which involved the organization of workshops, talks to the public, and informal meetings with some of the stakeholders involved in the territorial dispute for TDF control. Semi-structured interviews were also conducted with key members of organizations that represent and support traditional people in the region, often acting as mediators in their struggles to retain their way of life. Some of these cultural groups have legal recognition for their unique cultural practices and history of environmental reciprocity, along with their associations to ancestral African culture. All interviewees signed an informed consent decree, in accordance with the ethical procedures described by Brazilian law.

21.5.2 Historical Dimension of Society

Territorial disputes in TDF regions at the study site generally involve a competition among traditional people, agribusiness enterprises, and the state for access to TDF resources. A case in point is the PEMS, which is one of the seven conservation areas created as environmental compensation for damage done during the Jaíba irrigation project, conceived in the 1970s, constituting one of the largest public investments for fruit production in Latin America (Figure 1.4 of Chapter 1). Being located in the extreme north of the state of Minas Gerais, the surroundings of PEMS comprise the municipalities of Itacarambi, Manga, Matias Cardoso, and São João das Missões. Rodrigues (2000) has identified the presence of three historically significant social groups in the region, namely, native indigenous people, European colonizers (mainly Portuguese), and African slave labor brought to the region by force in the mid-sixteenth century.

The first human settlements in the region date back nearly 12,000 years. The first indigenous societies were seminomads and subsisted by fishing, hunting, and gathering abundant wild fruits. Another important group to settle in this region was comprised of escaped African slaves and their descendants. During the late sixteenth century, these groups penetrated the north of Minas Gerais and established themselves as the first *quilombos*. Thousands of lakes and extensive areas of TDFs facilitated the establishment of African-descendent communities, as these forests served as natural barriers, shielding these communities and making them practically invisible to the indigenous populations, pioneers, and *Bandeirantes* (mineral prospectors fanning out from SãoPaulo through the Brazilian backcountry in the early 1500s) (Lessa 2007). Costa (2005) acknowledges the arrival of the black *aquilombado* people within Mata da Jaíba, who interacted with indigenous people at a time preceding the arrival of European colonizers in the backcountry. The quilombo people cultivated cassava, corn, rice, beans, and bananas, among others crops, and fished in many lakes. Free-range cattle ranching was also a productive activity. A multiethnic, multicultural society existed that was modified on the arrival of the *Bandeirantes* (Lessa 2007). The Bandeirantes instituted a farm-centered society, implementing extensive cattle raising and cultivation of cassava for flour and sugarcane for liquor and brown sugar. These agricultural products were both traded (primarily with Bahia) and consumed locally. Small peasant communities associated with these farms were called *Cultura Sertaneja* (Countryside Culture or Hinterland Culture), and they supported themselves by means of agriculture, natural resources extraction, and cattle raising in natural habitats (Costa 1997). From 1750 to 1947, the dwellers of the mid-São Francisco River traded in and subsisted on beef, cassava, cereals, brown sugar produced from sugarcane, vegetable oils, wild fruits, and honey; and also fished in the abundant waters of the region and hunted (Mata-Machado 1991). From the end of the 1940s onward and more intensively in the mid-1960s, government policies in the region were driven by a move toward industrialization. The following programs that promoted four major industries were implemented in rural areas: (1) production of irrigated cash crops, primarily fruit trees; (2) extensive pastureland cattle ranching; (3) eucalyptus plantations for lumber; and (4) cotton plantations for textile fiber. However, government implementation took place in an uneven way, benefiting selected individuals, groups, and regions and resulting in varying intensities of development from one municipality and/or micro region to the next. Substantiated by a market rationale, such policies promoted the linkage between the region and the external market, which changed longstanding social dynamics and productive practices.

21.5.3 Cultural, Political, and Economic Dimensions

The use of space in the north of Minas Gerais is characterized by three basic pillars: (1) extensive cattle raising; (2) conservation sites; and (3) small-scale agriculture for food production. The importance of cattle raising in the region is shown by the size of the cattle herd: 146,287 head of cattle spread over a pasture area of 118,836 ha (Census of Agriculture 2006). Agriculturally dominated areas included about 24,481 ha, as described in 2006, that were distributed among 2828 rural settlements. Numbers indicate that food production in the area surrounding PEMS, for the most part, occurs on small family farms; whereas the bovine livestock operations tend to be medium and large rural establishments. Conservation sites established as parks (with no people) arose as a policy response to environmental consequences, resulting from the expansion and maintenance of the Jaíba project (Anaya et al. 2006). On the one hand, food production on small rural farms has been restricted by the increase of livestock areas and the expansion of agricultural crops for biofuel (sugarcane for ethanol production); on the other hand, it has been restricted by the creation of parks (Unidades de Proteção Integral).

The dispute over what exactly defines a TDF presents an interesting and symbolic conundrum: Big landowners or *Ruralistas* use arability and ease of conversion to pasturage to discern biome type, and classify TDFs as part of the Caatinga biome; whereas environmentalists use plant physiognomy to discern biome type, and classify TDFs as part of the Atlantic Rain Forest biome. When seen as part of the Caatinga, the Ruralistas are allowed to cut up to 70% of the TDF stands on their properties. Conversely, deforestation is totally prohibited in advanced secondary and primary forests that are classified as a part of the Atlantic Forest biome. In 2008, a political alliance among Ruralistas in the parliament of the state of Minas Gerais ordered a technical study to enable the promulgation of a new law that regulated deforestation through the recognition of TDFs as a part of the Caatinga, revoking previous legislation that recognized it as a part of the Atlantic Forest. At present, an agricultural census is being conducted (IBGE 1996) in order to revoke the Ruralista law and to define TDF as a phytophisiognomy-type representative of the Atlantic Forest, rendering it a protected ecosystem. This dispute not only has a normative-legal dimension, but also addresses a key conservation issue in Brazilian society, which locates the Amazon and Atlantic Forests in prime condition for biodiversity protection and transfers the role of an agricultural frontier to the Cerrados and Caatinga.

A third classification system occurs among the indigenous people who live in TDF areas. These "traditional populations or communities" include the *quilombolas*, a population comprised of slave-descendants, and the self-identified *vazanteiros*, who dwell along river margins, and whose way of life is dictated by their relationship with TDF riparian areas. As opposed to tribal indigenous groups, a vazanteiro is a vazanteiro, because his or her way of life is associated with the ebb banks of the San Francisco River. For these people, TDFs represent food, shelter, and a way of life that has existed for generations.

The TDF conservation areas have caused some tension between park managers/inspectors and traditional people. As TDFs become expropriated by powerful farmers and development programs, the traditional people residing in them are displaced. Similarly, as TDFs are protected under conservation policies, traditional people are displaced, because environmental policies restrict the use of natural resources in protected areas and disallow residential use. As a result of their struggle to protect their claim to the lands that they have occupied for generations, traditional people have established social networks to promote their point of view. With the assistance of advice and support groups (Pastoral Land Commission and Center for Alternative Agriculture), and researchers in academic centers (State University of Montes Claros), they seek to emphasize the ethnic and identity factors associated with their territory, giving a different meaning to environmental disputes. Hence, the movement of the "enclosureds," as they call themselves, which focuses on their precarious economic and sociocultural position as they find access to their traditional lands increasingly restricted, severely limiting their established customs. Social segments within the state are also trying to establish policies for the recognition and protection of these people.

21.6 Human Dimension in TDFs: A Synthesis

Historical reconstruction shows that although differences exist among the three case studies, all of them developed via similar processes. The three sites shared the characteristic of being large areas of land that were claimed by European settlers during the colonial period. In all cases, the hacienda system was used to expropriate lands that had been traditionally occupied by indigenous people, though the Brazilian hacienda system was complicated by the presence of refugee African slave populations and their descendants. The TDF areas at all sites were initially used for subsistence agriculture, hunting, and forage. Cattle farming occurred at a small scale and then became more extensive over time. The majority of TDF transformation, however, took place during the twentieth century, particularly from the 1950s to 1980s, especially in Costa Rica and Mexico. Governmental policies contributed to deforestation when promoting colonization (Mexico) and subsidizing cattle ranching (all three sites). It should be noted that these policies were strongly supported by financial policies outlined by the World Bank, the International Monetary Fund, and the Inter-American Bank for Development. Such institutions explicitly endorsed the emergence of industrial agriculture and animal husbandry in Latin America to fulfill the growing demand for fruits and beef in the Americas and other regions (Stiglitz 2006).

One additional aspect emerged from our analysis with regard to the contradictions related to the ideology of "environmental conservation" that guides state actions in the three countries (Zhouri 2001). Political actions centered on biodiversity and ecosystems conservation as a way of sustaining "environmental ethics" tend to result in the creation of protected or conservation areas (Parks and Biosphere Reserves). In Costa Rica and Mexico, these protected areas are segregated parcels where human presence is prohibited (following the protected area idea of the 1970s) (Price 2002a). Several issues still concern people in the influence regions of TDFs: In Costa Rica, local people claimed that parks should provide more jobs. In Mexico, the more benign local opinion asserts that locals do not benefit from the Biosphere Reserve. They are more troubled by the Ecological Land-Use Policy of the Jalisco Coast, which they believe operates with a system of favoritism and is often in direct conflict with their interests as peasants. In Brazil, TDF conservation units have been created in the territory of traditional people as reparations for large-scale fruit irrigation projects (there is a political-legal sanctioning of deforestation when undertaken to advance the expansion of agribusiness). In this perspective, state conservation policies generate a paradox: In an attempt to compensate for the loss of TDF lands from traditional people to agribusiness, the government further restricts and expropriates the lands used by traditional people for conservation.

Tourism development presents another arena in which difficulties and conflicts may arise and where conservation efforts may result in continuing environmental injustice. In all three case studies, stakeholders with sufficient economic and political power have begun to convert TDFs to accommodate tourism. Traditional people's economic needs and access to natural resources continue to be critical factors that remain unaddressed. Although the long-term cumulative effects of tourism on TDF habitats still remain unknown, some of the potential effects include (1) damage to coastal and mangrove areas along the beach; (2) improper sewage and garbage disposal; (3) disturbance of nesting areas of endangered sea turtles; (4) destruction of natural forests for tourist developments; and (5) depletion of watersheds and aquifers for water provisioning. To ensure the continued well-being of TDF ecosystems, it will be necessary for the government to implement and enforce stricter regulations for the development of the tourism industry.

Approximately 550,000 sq km of TDFs covered Mesoamerica at the time the Spaniards arrived (Janzen 1988). Today, less than 20% of the mature forest remains, which is mostly found in Mexico (Portillo-Quintero and Sánchez-Azofeifa 2010), but less than 4% is protected and restricted to countries such as Costa Rica. We estimate that within Mesoamerica alone, more than 20 million people inhabit TDF regions and enjoy only limited access to basic services. Though so many lives depend on these TDFs, little effort has been made to preserve and maintain the mature forest that is capable of providing current and future environmental services for these and neighboring communities. In South America, the level of TDF protection is great (37%), but management practices and outcomes shaping TDF status are roughly the same. Therefore, an interdisciplinary analysis of socio-ecological processes is needed to develop new management strategies that secure TDFs' long-term maintenance and to ensure a continuity of ecosystem services. Some of these important environmental services are (1) pollination of crops by wild pollinators; (2) protection of watersheds and aquifers for water provisioning and for preventing natural disasters; and (3) carbon sequestration by mature and regenerating forests, which is crucial for mitigating the impact of global climate change.

22

Conflicts between Conservation Units and Traditional Communities in a Brazilian Tropical Dry Forest

Felisa C. Anaya, Rômulo S. Barbosa, and Andréa Zhouri

CONTENTS

22.1 Introduction

The environmental field as a contemporary social phenomenon can be considered a field of ecological praxis, a political and institutional space for the environmental debate (Zhouri 1998; Carneiro 2005). This notion was reinterpreted from the concept of "field" conferred by Bourdieu (2007) to the juridical and political fields, meaning a social space of differentiations, forces, and power disputes between the involved agents. In this perspective, the environmental field has its own political and juridical rules and a particular structure, being composed of actors who are organized around the issue of the ethical belief of *nature as an asset** (Carvalho 2001). Currently, the environmental field has the dominant ideology of "sustainable development," which is the *doxa*† that has been guiding the society–environment

* The conditions for the emergence of an environmental phenomenon and its ethos in the context of modernity were marked by the political–cultural challenge of the 1960s. The configuration of the counterculture movements, including the ecological movement, produced a broader critique of the forms of capitalist production and its institutions in a context of environmental crisis. This scenario enabled the understanding of nature as an asset (ethos) as well as the emergence of an environmental rationality and its counterpart, which is the ecological self (Carvalho 2001).

† "According to Bourdieu, every field develops a *doxa*, a common sense; and a *nomos*, which are general laws that govern it. All agents agree with the *doxa*." (Thiry-Cherques 2006)

relationship for the past 25 years. Such doxa, forged in the context of re-emergence of a market economy and neoliberal policies in the late 1970s, by the eco-development, served as a source and paradigm of the environmental policies. The eco-development puts forth the Radical Ecology proposals and combines them with developmental proposals, taking on more conciliatory features of environmentalism and combining them with the practices of capitalist exploitation of nature. In this perspective, the environmental dimension would integrate with successful economic planning that considers the conditions and potential of ecosystems and wise management of resources (Sachs 1993).

Formally institutionalized in the 1980s by the Brundtland Report or "Our Common Future" (World Commission on Environment and Development 1987) and consolidated in the 1990s with the United Nations Conference on Environment and Development (the Rio Earth Summit; UNCED 1992), sustainable development is presented as a political concept that promises a solution to global concerns related to production processes and environmental degradation in the context of ecological crisis. Sustainable development is settled on the comprehension of the environment as an external means to sociopolitical dynamics, and it achieves environmental management by a market that changes and adapts itself through technical efficiency (Guerra et al. 2007). In this sense, development is marked by an evolutionary and adaptive design of economic growth based on environmental resources, supporting the belief that the global ecological crisis and local conflicts are likely to be bypassed through the use of technical adjustments—promoters of progress and modernization—without changing the capitalist mode of production.

In this way, the incorporation of broader issues in the sustainable development discourse, such as equity, social justice, and governance, coupled with the notion of sustainability—used for almost all things considered desirable—provided its rapid assimilation by official national and international agendas and common sense. Thus, sustainable development was established as the dominant ecological thinking in political, business, and scientific environments, with focus on the issue of scarcity and waste as environmental variables external to the processes of capitalist production. From this perspective, there was a homogenization of the distribution of industrial risks among different actors, disregarding the criticism about what is produced, how it is produced, and to whom it is produced (Acselrad 2009). Therefore, sustainable development ideology performs a global-scale analysis of the environmental crisis, in a hegemonic project of economic reform defined by Zhouri et al. (2005) as *environmentalism of results*. In this way, the global market is established as the regulator of environmental and social policies limited to instruments that are compatible with the market's economic interests, such as certification of green products, environmental compensations, and the use of so-called "clean technologies" in order to reduce carbon emissions.

Based on the *Paradigm of Ecological Modernization** (Acselrad 2004; Martínez-Alier 2007), also referred to as the *Paradigm of Environmental Adequacy* by Zhouri et al. (2005) and *Paradigm of the Mitigation Game* by Carneiro (2005), the model of sustainable development culminated in the emergence of numerous environmental conflicts. In this context, there are several different major conflicts that are currently considered "obstacles to development" in Brazil (Zhouri and Laschefski 2010), such as (1) the continuous struggle and dispute over the demarcation of boundaries at the Indian reserve Raposa Serra do Sol, which collides with the interests of the major rice producers in the state of Roraima in Brazil; (2) the clashes between environmentalists and ruralists around the Forest Code reform; (3) the controversy over the construction of the Belo Monte dam on the Xingu River Basin in the state of Pará, Brazil, which is considered the largest project developed by the Growth Acceleration Program, created by the federal government; and (4) the controversial issue of the São Francisco River transposition project that has been discussed by social movements, local communities, governments, and private sector representatives. Also included here are the conflicts related to the creation of conservation units (CUs), which is one of the most important conservation strategies in Brazil and worldwide (Diegues 2002). Brazilian CUs can be broadly divided into two main categories: restricted use units (RUUs) and sustainable use units (SUUs) (MMA 2000). The distinction between them is based on the form of nature conservation, and the resulting restriction of access and use. In the case of the RUUs, human occupation and direct exploitation of nature are not allowed. Their purpose is the preservation of the natural ecosystems, allowing the development of scientific research, ecological tourism, and educational activities.

The CUs are created mostly as a government compensation to the environmental degradation caused by economic activities, aiming at accommodating spaces to be explored and spaces to be preserved. Thus, CU creation follows from the paradigm of ecological modernization to "fit" (Zhouri et al. 2005) nature into the purposes of capital or to "mitigate" (Carneiro 2005) the damage caused by this capital by creating "untouched" spaces. According to Cunha and Almeida (2002), conservation units and other protected areas of nature (areas of permanent preservation and statutory reserves) are related to multiple territorial projects that express their intentions with a certain degree of institutionalization and legitimization in

* According to Martínez-Alier (2007, p. 28), the term *ecological modernization* was created by Martin Jaenicke and Arthur Mol during the 1990s. It combines the economic sphere with eco-fees, markets of license emissions, and an ecological sphere, and is based on measures that are aimed at saving energy and raw materials. According to the author, the ecological modernization follows the gospel of eco-efficiency, which is one of the three lines of thought of environmentalism, referred to as (1) "cult of wilderness," (2) "eco-efficiency," and (3) "environmentalism of the poor." In this regard, *eco-efficiency* is considered a synonym of *ecological modernization*, and it presents its corporate relationship and connection with *sustainable development*.

order to appropriate an area. In this perspective, most implemented CUs were superimposed on areas that are historically occupied and used by more economically vulnerable populations, such as traditional communities using techniques with low impact on the environment. According to Schaik and Rijksen (2002), about 70% of tropical parks in the world have people living within them. Thus, the creation of RUUs has fueled a surge in environmental conflicts, as these units surround and expropriate local populations from their territories, affecting their economic opportunities and sociocultural practices.

This is the specific case of the CUs situated in the tropical dry forests (TDFs—locally called "Matas Secas") at the north of the Minas Gerais state, Brazil, which have been in conflict with the local traditional populations. Historically dispossessed by economic development policies for that region during "intersocietal contexts of conflict"* (Little 2002), these communities now find themselves "trapped" in their own territories due to the establishment of RUUs such as state parks. Created as an environmental compensation for the large agribusiness project of irrigated agriculture called "Jaíba," the state parks "Lagoa do Cajueiro," "Verde Grande," and "Mata Seca" were overlaid on the traditional territories of communities that historically inhabit the margins of the São Francisco River (called "vazanteiros"). Three of these communities, named "Quilombo da Lapinha," "Pau Preto," and "Pau de Légua," are the main subjects of this study. In this context, there is an emergence of environmental territorial conflicts due to a contradictory model of sustainable development that enables an intensive and extensive expansion of agribusiness in the region through the clearing of TDF areas in order to expand the Jaíba project, to the detriment of other forms of material appropriation in the environment. Thus, in this chapter, we explore the case study of the RUUs situated in the TDFs at the north of Minas Gerais, and aim to disclose the internal contradictions of an environmental policy that follows an economic rationale and maintains the same mechanisms of unequal distribution of access to the environment and division of costs, risks, and impacts resulting from the dominant practices of appropriation of the natural resources.

However, due to the current processes of territory expropriation carried out by the creation of parks, which are interpreted here as "environmental injustice"† (Acselrad 2009), the vazanteiros and also the quilombolas

* Intersocietal contexts of conflict are processes involving some social groups that have their land constantly invaded by economically dominant agents or are negatively affected by developmental public policies. These invasions transforms the environmental, cultural and production relations that once existed in these groups. Such contexts usually unify those groups internally, aiming at the defense and protection of their territories. (Little 2002, p. 3).

† "Environmental injustice is defined as the mechanism by which economically and socially unequal societies provide the greatest burden of environmental damage caused by development to low-income, racially discriminated groups, traditional ethnic groups, working-class neighborhoods, and marginalized vulnerable populations." (Acselrad 2009, p. 41)

(slave-descendants who live in quilombos, or maroons) are organizing social struggles and aiming for greater political power. Social struggles are renewed during the building of a new political order and a new paradigm. Using a strategy of appreciation of their ethnic identities and cultural values, they seek reappropriation and recognition of their ecological and cultural means of production, suggesting the establishment and use of an alternative model of sustainability. This implies the creation of Extractive Reserves and Sustainable Development Reserves (two categories of SUUs defined in the National System of Conservation Units; see MMA [2000]) associated with the appreciation of the traditional practices of the local populations, coupled with agroecological techniques that constitute an alternative conservation strategy.

22.2 Meanings and Social Projects in a Tropical Dry Forest: The Vision from the Actors in Conflict

Following an analytical tradition in the sociology of environmental conflicts, Zhouri and Laschefski (2010) performed a critical observation of the so-called *paradigm of the ecological modernization*, which guides the discourses and actions in the environmental field. The authors consider environmental conflicts as composed of a diversity and heterogeneity of actors with different ways of thinking about the world and of how to design their action plans and use them. Thus, the most significant meaning of the local designation "Matas Secas" (TDFs) is to guide the use and the fate of these various players who have distinct ways of perceiving and acting in the environment. Therefore, they fight for the symbolic power in order to impose their outlook on the legitimate definition of this spatial area, which is crossed by a multitude of classifications related to each particular interest. From this perspective, the meanings brought into dispute appear now as (1) an economic good for ruralists (large farmers who raise cattle for export and agrobusiness representatives); (2) a natural asset for environmentalists (NGO representatives, environmental managers, and researchers); and (3) a territorial asset for traditional populations (self-recognized, culturally distinct groups, with a particular social organization, who occupy and use natural resources as a condition for their cultural, social, religious, and economic perpetuation, using knowledge, innovations, and practices that are generated and transmitted by tradition [Brasil 2007]).

The appraisal of the true value of TDFs and the need to develop mechanisms that protect this forest are associated with the context of environmental crisis that has been emerging since the 1990s. After the political situation of the Rio Earth Summit in 1992, the federal government, through the

Decree 750, which was issued in February 1993, stated that TDFs were a part of the Atlantic Forest biome and were, therefore, under the same level of protection ("specially protected ecosystem"; see Espírito-Santo et al. [2009, 2011]). Thus, TDF use and exploitation are limited to early regenerating forests, and clear-cutting is prohibited in intermediate and late successional stages as well as in primary forests. After a couple of years, by means of resolutions, decisions, and regulatory laws at the federal and state level, a legal construction was started, establishing criteria for identifying TDF successional stages and terms of use and protection.*

In the north of the state of Minas Gerais, Brazil, about 50% of the region has TDF remnants. In 2005, the Atlas of Biodiversity of Minas Gerais (Fundação Biodiversitas 2005) considered these areas as priorities for conservation and scientific research based on their "extreme biological importance." From this perspective, TDFs should be preserved through the implementation of environmental strategies aimed at the creation of conservation and research areas, in opposition to TDF intensive use and suppression by large agricultural projects and charcoal production.

These legal regulations for environmental conservation in the north of the state of Minas Gerais put into sharp focus the dispute over the significance of TDFs as part of the Atlantic Forest biome (as defended by environmentalists) or as belonging to the Caatinga biome (as claimed by ruralists). The symbolic and political struggle in the environmental field put in the spotlight the different meanings, relations of power, and social projects for the region. On one side, there were the environmentalists who pointed out the need for full TDF protection, due to its ecological and aesthetic significance. In opposition, the ruralists emphasized the almost unconditional use of TDFs to expand the agribusiness boundaries. In the particular case of TDFs at Minas Gerais, rural associations pressed for deforestation levels up to 70% of each property (Espírito-Santo et al. 2009, 2011). In order to impose their vision on the field of social struggles, ruralists lined up in defense of their interests and formed the SOS to the north of the Minas Gerais movement.† This was a part of a political strategy that used the developmental speech in which the restriction of TDF deforestation would cause an economic recession, generate unemployment, and intensify social struggles (Figure 22.1).

* CONAMA's Resolution of May 4, 1994; State Law 14309 of February 19, 2002; COPAM's Normative Deliberation No. 72 of September 8, 2004; State Law 15972 of January 12, 2006; Federal Law 11428 of December 22, 2006; CONAMA's Resolution 392 of July 25, 2007; State Law 17353 of January 18, 2008. See the analysis of the effects of these regulations on the TDFs in Anaya (2012).

† The SOS North of Minas Gerais movement is articulated by the rural society of the north of the state of Minas Gerais, whose goal is to legally remove the local TDFs from the protective "umbrella" of the Atlantic Rain Forest biome, with the purpose of increasing livestock and agriculture for exportation. For people to be able to follow this movement, a blog was created (http://movimentososnortedeminas.blogspot.com/).

FIGURE 22.1
(**See color insert.**) Poster disseminated by rural associations in the north of Minas Gerais, inviting people for a debate about the legal dispute over the Tropical Dry Forests in the region. Preserving the forests means "poverty," whereas converting them to irrigated agriculture means "progress."

Despite the tensions between ruralists and environmentalists about the ways of using the local TDFs, the CUs issue puts both groups in a controversial situation. Driven by the ideal of nature as an asset that should be preserved, environmental institutions sanctioned a legal–political feature that allowed the expansion of agribusiness by the deforestation of large TDF areas, as in the case of the Jaíba project. In this case, civil and government environmental agencies complied with the preservation of TDF remnants as small islands inside compensatory CUs.

In the late 1990s, before the commitment of the Minas Gerais government to fund the expansion of the Jaíba project, the Environmental Policies Council (COPAM) demanded the creation of a system of protected TDF areas surrounding the irrigated lands. As a consequence, five RUUs and two legal reserves (preservation areas inside private properties), totaling more than 91,000 ha, were created in three counties that were affected by the Jaíba project (Matias Cardoso, Jaíba, and Manga). These protected areas were created without any previous public consultation, disregarding the different existing territorialities and overlapping with the traditional territories. This strategy aggravated the existing conflicts in the region by reproducing the historical expropriation of traditional populations by large farmlands in the second half of the twentieth century.

These expropriation processes are related to the capitalist appropriation of the land base of social groups. Referred to as "territorial environmental conflicts" by Zhouri and Laschefski (2010), these are generated, in most cases, by the boundary expansion of commodity production that clashes with the territoriality of groups which have, on their resources, the basic, fundamental elements to sustain their sociocultural reproduction and their presence in disputes about which type of development is the best for the country (Zhouri and Laschefski 2010, p. 17).

The anthropological aspects that mark and characterize the territoriality of the vazanteiros and quilombolas groups make this conflict different from the other land struggles, as these groups have different ways of producing and appropriating the territory under dispute, a territory that goes beyond what the biologists consider TDFs. Most environmentalists and ruralists have a dichotomous view of the environment (untouched versus exploited areas), whereas the traditional populations have different ways to consider, use, and appropriate their territories. The territory is perceived as an asset that is necessary for production and reproduction, and it ensures the survival of the population as a whole (Zhouri and Laschefski 2010, p. 25). Thus, the land is interpreted as a territorial asset that is not necessarily linked to legal ownership but to appropriation, which is related to the complex relationship between place, culture, and nature, and focuses on the meanings and symbolisms attributed to this place. The places where these groups of individuals live consist of a landscape of forests, highlands, lakes, islands, and rivers, linking material and symbolic elements together, which are experienced dialectically along with their identity and culture. Thus, in response to the territorial expropriation process carried out by conservation policies, the vazanteiros and quilombolas, assisted by supporting organizations and political mediation actors such as the Pastoral Land Commission (CPT)* and the Center for Alternative Agriculture (CAA),† are organized in the field of social struggle, searching for greater visibility and political power. In this way, these groups are emerging as subjects from what Almeida (2008) refers

---

* With its origins in the social movements of the 1970s, during the period of the Brazilian dictatorship, the CPT develops a pastoral work that consists of the understanding of the rights that rural groups have on land, labor supply, and survival. In this context, the advice and mediation on territorial environmental conflicts provided by the CPT allowed these communities to go through a subjective process, resulting in the emergence and rise of collective subjects from a process of cultural and identity appreciation, awareness of their land and territorial rights, and strengthening of their claims. In the north of Minas Gerais, the relationship of the CPT with the traditional populations initially resulted in the formation of the Movement of the Enclosed Populations, which later became the Vazanteiros in Movement.

† The CAA is another important adviser and mediator in this process, and it was created during the 1980s when the processes of modernization of hinterlands and the regional development model were receiving criticism. The approach of the CPT allowed the CAA to incorporate issues such as those related to land conflicts, inequality of ownership of natural resources, and traditional and agroecological knowledge of the rural and traditional communities from the north of Minas Gerais, in the discussion about alternative agrarian technologies and in the environmental debate.

to as *territorialization movements*. Recognized as *"Vazanteiros in Movement,"* they claim for specific identities in order to ensure their economic production, sociocultural features, and permanence on territory.

Environmentalists, ruralists, and the traditional populations are part of a classificatory fight for the legitimate representation of nature and the distribution of power over territorialized resources in the TDFs at the north of Minas Gerais. For the vazanteiros and quilombolas, the rivers and woodlands in this place do not have the same meaning as the enterprises of farming irrigators of the Jaíba project. The biodiversity that is cultivated by traditional communities does not have the same logic as the untouched biodiversity valued by environmentalists. From this perspective, the environmental conflicts and territorial disputes should be analyzed in the context of the political struggles of particular actors in specific spatiotemporal situations.

22.3 Territorialization Processes: The Construction of the "Vazanteiros in Movement"

The creation of new sociocultural units similar to those claimed by the vazanteiros and quilombolas from the Pau Preto, Pau de Légua, and Lapinha communities was based on the emergence of new ethnic identifications, connected to the field of territorial rights, and focused on the reworking of the culture and its link to the past. The context of intersocietal conflict, which was common to all three communities, and in which they gradually lost the symbolic and material control over their territorial resources, is evidenced in the report listed next. This report summarizes the territorial processes involving expropriations that were initiated through public policies that aimed at the colonization and economic development of the north of Minas Gerais, during the 1970s, to integrate it to the south-central region of this state.

> We are now in "Pau Preto", between the margins of two rivers: the Verde Grande river which is over here, and the São Francisco river which is over there. We've been all living along these river shores; in this farm here, where the "Catelda" farm is located. From the 1970s to now, the "RURALMINAS" was created; it is a national government organ similar to the IEF. This organ said at the time that they came to settle the land. We didn't know anything. So they started giving land to the big guys. That was also when the law that prohibited us from raising free-range livestock was adopted. Then farmers came and began to push the little ones. We used to raise goats and pigs; we had small herds. Then farmers started killing our animals. At that time, if we dare to complain to the big guys about anything we'd put ourselves into trouble; there were gunmen around so there was nothing that we could really do about it. Then something called SUDENE came from the North and funded the farmers

> allowing them to grow, destroy all the woodlands, and cut the "aroeira" trees (Brazilian pepper, rose pepper). Then we were dislodged and had to move to town. But in 1979 we returned and repossessed the area. Now, the State comes here once again, takes our lands, and legalize the area for parks, and trampled us. (Community leader from Pau Preto [Anaya 2012])

The occupation and domination of this space by farmers and agricultural companies of Montes Claros, São Paulo, and other regions, led to serious social and environmental consequences that affected the territoriality of these groups. In this context, conservation units created by public conservation policies reproduced the deterritorialization of these groups, which were subjected to a developmental logic that intensified the territorial conflicts in that particular spatial area, promoting economic development for some people (i.e., the *big guys* as the natives and locals used to say) and unsustainability for others (i.e., the *small* ones). So, the dispossession of the black communities of vazanteiros from Pau Preto, Pau de Légua, and Quilombo de Lapinha is, in this context, updated and revitalized by the creation of the Verde Grande, Lagoa do Cajueiro State Park (in 1998) and the Mata Seca State Parks (in 2000), which were superimposed on their traditional territories. There was a clash of contradictory territorialities that was mainly focused on the conflict between the traditional communities and the State Forestry Institute (IEF)*—the latter manages the state CUs in Minas Gerais. So, the deterritorialization of these communities, which is intertwined with incessant processes of frontier expansion of the financial capital by government institutions such as SUDENE and Ruralminas, is reproduced these days through compensatory environmental policies that are triggered by the agribusiness.

Rodrigues (2000) showed evidence of the first territorializations that happened in the north of Minas Gerais, originally from three basic groups: (1) the indigenous (native) occupants; (2) the European settlers—mainly Portuguese—who introduced cattle ranching in the seventeenth century; and (3) the Africans who were taken by force in order to join the slave labor force during the same period. It should be noted, however, that there is evidence of the establishment of black groups (the "aquilombados" fugitive slaves) within the Jayba Forest† even before the arrival of explorers (Costa 1998), setting up autonomous groups that established interactions and exchanges with indigenous groups. With the settlement of the pioneers from the Minas Gerais and São Paulo states, livestock was extensively introduced along the

* The IEF is an agency that belongs to the State Secretariat for the Environment and Sustainable Development, and it is in charge of environmental conservation, sustainable use of natural resources, research on biodiversity, elaboration of forest inventories, and vegetation cover mapping in the Minas Gerais state.

† According to Costa-Filho (2008), the persistence of non-white people in the valleys of the Gorutuba and Verde Grande rivers is frequently attributed to the occurrence of diseases such as malaria in this region. Costa (1998) explores the regional designation "Jaíba forest," explaining that it would mean "a weird place of difficult access," with "Jahyba" being a term from the indigenous nation Tupi's language, which means "bad waters."

margins of the São Francisco River through the establishment of large farms. Associated with these farms, nuclei of small farmers were also formed, constituting a peculiar way of social life that some scholars classify as "country culture" (Costa 2001). Mata-Machado (1991) describes the extensive livestock farming system as the economic matrix of large farms that were established with the use of indigenous and black slave manpower and whose production was destined to supply the market of the Bahia State and mining centers, setting up a self-sufficient production unit. In addition, small owners, squatters, and households scattered throughout the hinterland regions called *sertão* (semiarid region in northeastern Brazil) also constituted an economic entity through the collective production based on relations of kinship, neighborhood, and cronyism, which enabled a diversified extractive agriculture associated with raising free-range livestock.

In the nineteenth century, when slavery was abolished, a new wave of black people who were freed reached the valleys of the São Francisco, Verde Grande, and Gorutuba rivers, where they settled (Costa-Filho 2008). Although the economic government policies since the Brazilian Empire were to integrate the region into the national economy, thereby ensuring the unity of the country, it was only in the late 1940s that the new process of integration of the north of Minas Gerais with southeastern Brazil and the occupation of the region actually happened. This period coincides with the emerging industrial development of the Brazilian society. In the mid-1960s, federal and state government policies for this region started being guided by a perspective of industrialization, encouraging the invasion and occupation of this area by squatters and farmers, pushing the traditional populations to occupy isolated parts of large farms or moving them to remote areas. Underpinned by market logic, these policies promoted the linkage of the region to foreign markets, changing the social dynamics and the actual productive logic. The result was the territorial expropriation of small farmers, the degradation of natural resources, and the increase of land concentration throughout the region. Therefore, the traditional populations that historically occupied the region resisted and upgraded their practices and links with the territory. They produced specific knowledge regarding the use and management of the environment, and became traditionally known as vazanteiros because of their lifestyle at the ebb of the São Francisco River. Anthropological studies done by Luz de Oliveira (2005) and Araújo (2009) confirmed that the São Francisco River is an important reference of territorial organization for the vazanteiros and became a major constitutive element of their territories. From the river flooding to ebb, ways of material and symbolic appropriation of the lands are defined, associated with a production system that consists of low-water, upland and riverside planting, fishing, free-range livestock raising, plant extraction, and hunting. Their landscape can be divided into upland complex and island complex, interconnected by a third unit, the river complex (Figure 22.2), whose cycle defines the ecological dynamics and, therefore, the management strategies of each unit (Luz de Oliveira 2005).

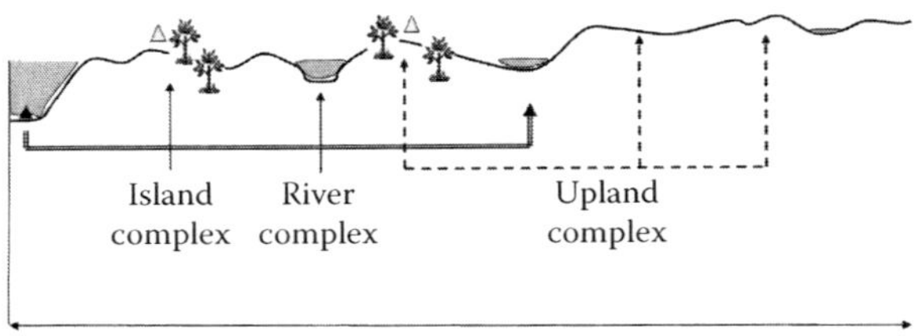

FIGURE 22.2
(See color insert.) Schematic definition of the territorial organization of the vazanteiros at the margins of the São Francisco River, whose cycle of floods and ebbs defines the ecological dynamics and, therefore, the management strategies of each complex. (Modified from Fernandes, L.A. et al., Proposta de ocupação e uso dos ambientes pelos vazanteiros de Pau Preto: Novos indicativos à proposta da RDS no context do diálogo dos Vazanteiros de Pau Preto com o Instituto Estadual de Florestas. Technical Report, Montes Claros, 2010.)

However, since the late 1990s, the Jaíba project and its CUs disregarded the presence of these communities and again changed the forms of use and ownership of TDFs in this region, imposing another dynamic there. The context in which these CUs were created, similar to many others in Brazil, shows that this process is always carried out in an authoritarian, vertical manner, without any previous consultation of the affected populations and disrespecting social issues. Until the creation of the CUs, the vazanteiros were going through a process of repossession of their territories. Currently, they have limited access to the communal natural resources that they had been using over the years (extraction, hunting, fishing, and agriculture), preventing their social, cultural, and economic production. Such traditional practices are not allowed inside CUs and have started being considered violations by environmental agencies such as the Brazilian Institute of Environment (IBAMA) and the IEF. In this sense, penalties such as fines, confiscation of fishing equipment and canoes, moral constraints, and physical violence have been used.

In response to the enclosure by CUs, the vazanteiros bonded together in 2005 through the political mediation of advisory and support entities such as the CPT and CAA, as well as scientific research centers (Montes Claros State University [UNIMONTES] and Federal University of Minas Gerais [UFMG]). Several meetings were held in settlements of traditional populations along the São Francisco River from 2007 to 2010, during which regular reports about the territorial conflicts in CUs as well as the issue of tradition—which was reaffirmed by historical origins and communal lifestyle—were presented. This process allowed the recovery of their identities, distinguishing them from the more general designation of squatters, and they had recognition as vazanteiros and quilombolas. Thus, land rights affected the environmental field by recognizing the existence of these people and their rights on their territories, leaving open the possibility to negotiate the rules of the National System of Conservation Units (MMA 2000).

In this context, several attempts were made to negotiate that the vazanteiros from Pau de Légua, Pau Preto, and Quilombo da Lapinha should be recognized by the state of Minas Gerais, and also by public organizations such as the IEF, IBAMA, the National Institute for Colonization and Agrarian Reform (INCRA), the Secretary of Union Assets (SPU), and the Public Ministry. Several attempts of negotiation were made by using letters, proposals, and manifestos that claimed the creation and establishment of sustainable reserves, and the request of an anthropological report so that the territory Quilombola da Lapinha could be delimited by the INCRA. Due to the numerous unsuccessful attempts to negotiate and because the territorial conflicts were ignored by the state, the "Movement of the Enclosed Populations" decided to articulate the involved groups of vazanteiros and quilombolas and to take over their land again. These actions started with the Quilombo da Lapinha in October 2006 when they took over the "Casa Grande" farm as part of their traditional territory. This invasion was considered a symbolic act, and it illustrated the conflicts between the remnants from the Quilombo da Lapinha and the large properties formed during the first expropriation process. Another important action was the self-delimitation of the "Pau de Légua" Sustainable Development Reserve in 2010, showing the traditional territory of the vazanteiros, that was communicated through a proposal letter sent to the state. In 2011, the vazanteiros from Pau Preto implemented the self-delimitation of their Sustainable Development Reserve and took over the "Catelda" farm. All these actions were coordinated with the participation of three communities that started calling themselves the *"Vazanteiros" in Movement: The People of the Rising Waters and Lands*. Claims from this movement would seek an alternative vision of conservation and environment where there is no dissociation between nature and culture, pointing out the fact that the use of the environment by these communities depends to a large degree on the pace of regeneration of the TDF ecosystem.

22.4 Final Considerations

There is currently a situation of tension between the actors who dispute the designations and social projects for the TDF territory, configuring the environmental conflicts in the north of Minas Gerais. This complex social–environmental scenario can be summarized in Figure 22.3. This scenario involves ruralists, environmentalists, and traditional populations with their different forms to relate themselves to this particular TDF area, either as an asset to be exploited or as a territorial asset to be preserved, using different strategies to impose their views in the field of social struggles. The ruralists exert a strong political pressure to change the environmental legislation at both the state and federal levels, whereas traditional populations

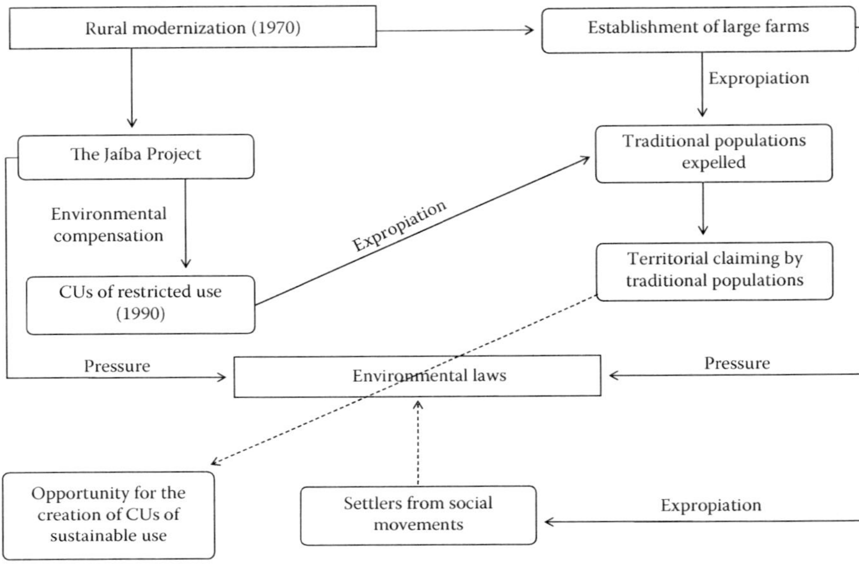

FIGURE 22.3
Diagram of past and present actors and processes that compose the social–environmental scenario in the north of Minas Gerais. Continuous lines represent processes that already occurred or are in course, whereas broken lines represent processes that are likely to occur in the near future.

seek a legal breach that would offer the possibility of changing the RUUs to SUUs. In this context, there is a dubious role of the state: On one hand, the environmental policies cause the expropriation of traditional populations from their ancestral territories through the creation of CUs, regulating the expansion of agribusiness via compensatory strategies; on the other hand, the state enables the inclusion of these populations in CUs via territorial policies (MMA 2000; Brasil 2007). These conflicts originate from different practices of technical, social, and cultural appropriation of the material world, in which the cognitive basis for the speech and actions of each actor involved is arranged according to their hegemonic visions on the use of space. This scenario highlights situations of environmental injustice, and it emphasizes the inequalities in the distribution of natural resources and the consequences of economic development, which are shared asymmetrically by social groups of workers, low-income populations, racially discriminated segments, and more vulnerable and marginalized citizens.

However, the construction of identities has become a key element of political confrontation in the social struggles of traditional populations in the environmental field. The effort of conceptual reconstruction of the territorial and identity elements by researchers and the awareness of these groups that started considering themselves as traditional populations allow for the divorce from the reductionist and naturalizing schemes of thought that homogenize them. Its relational character was proposed by Montero (1997), as a consequence of a symbolic process of self-designation of cultural traits. Therefore, the difference is progressively politicized as the phenomenon of identity is moved from the conceptual to the political field. As a result, new alternatives of development arise, expressing the symbolic dimensions of the relationships between these actors and natural resources, and showing the existence of new ways of access, use, and appropriation based on references other than the industrial and capitalist ones. In this context, the vazanteiros from Pau Preto, Pau de de Légua, and Quilombo da Lapinha emerge as a category that is connected with both their place and their ethnic identity, with the latter being interpreted as a contemporary phenomenon of reaffirmation of cultural differences. Thus, the defense of the place is linked to the practice and actions of a group of actors that include social movements, particularly those of the populations from TDFs, and it is based on the statement of four fundamental rights: (1) identity, (2) territory, (3) political autonomy, and (4) its own vision of development (Escobar 2005). Due to environmental issues, one can think of a new division of political labor by combining science, militant movements, and a repertoire of specific knowledge from each peculiar reality.

The political organization of the populations enclosed by CUs during their identity construction, along with the advisors and mediators, enabled these social groups to reimagine themselves as actors in social struggles; they gained momentum while moving from victims to active participants in the process of recognition of their identities and territorial struggle in environmental conflicts, redefining the Movement of the Enclosed Populations by the *Vazanteiros in Movement*.

23

Predicting Policy Impact on Tropical Dry Forests

Alexander Pfaff and Juan Robalino

CONTENTS

23.1 Introduction

Tropical dry forests (TDFs) provide important eco services, including carbon storage and species habitat. Although total forest is increasing in some developed countries, TDF has been lost at a significant rate over recent decades in many countries (Portillo and Sánchez-Azofeifa 2010). Policies for reducing the losses of dry forests are, therefore, one important part of global efforts to minimize biodiversity loss. Forest protection also is valuable for climate stability. Initiatives for "Reduced Emissions from Deforestation and Degradation" (REDD) have motivated critical consideration of a range of policies relevant for forest loss (Pfaff et al. 2011). Policies that would generate REDD would often, but not always, be a positive force in the protection of biodiversity.

In this chapter, we argue that public policy to protect TDFs should be based on a recognition that private incentives to conserve dry forest often do not reflect dry forests' full social value. Although a large city downstream from a dry forest might value that forest's hydrological services, private landowners upstream may ignore their impact on the city's welfare in deciding whether to clear. For biodiversity, this is even more striking, including in light of the sharp contrast between dry and tropical rainforest environments. Citizens around the world may value various species' existence, but such value is likely to have marginal influence on private land-use decisions to preserve dry forest habitats.

A central goal for those who would like to see additional and improved policy for TDFs is to find ways in which relevant private actors will take these societal values into consideration. This is a big challenge. At the national level, policy makers themselves should consider significant tradeoffs when making decisions whether to protect forests or whether to allow their conversion. Furthermore, national policies may not reflect the global value of the dry forests. Because many forest services are global public goods, optimal forest conservation is inevitably an international issue.

From a public perspective, a key policy issue is that private incentives to clear vary dramatically across TDF locations. In a similar manner, the impact from preventing deforestation varies. The rationale here is very direct: impact varies with incentives to clear because the effective impact of a conservation policy is equal to the private deforestation rate that is blocked. Policy impacts on development vary by location as well. Conservation and development needs can be sensibly balanced only if spatial variation in policy impacts is understood. If we assume that policies' impact on conservation and development are uniform across the entire landscape, society could adopt more conservation policies with a large economic loss for small species gains. It also might not reject development policies with large species losses for small economic gain.

We argue to integrate consideration of location into policy planning in the following three ways, to help predict policy impacts on loss of TDFs, thereby improving policy choices: (1) Policy impacts vary by location, because the rates of private deforestation that policies block vary with the characteristics (market distances, slopes, soils, etc.) of the locations in a landscape; (2) different mixes of political–economic pressures drive the final locations for various policies; and (3) policies trigger "second-order" or "spillover" effects, which are likely to differ by location. In this chapter, we review recent high-quality evaluations of policies' forest impacts—including protected areas, ecosystem-service payments, and development policies such as road investments. Forest impacts of well-enforced conservation rise with private clearing pressure, supporting (1). Protection types (e.g., federal vs. state) differ in locations and, thus, in impacts, supporting (2). Differences in the development process can explain the different signs for spillovers, supporting (3).

Section 23.2 of this chapter lays out a conceptual framework, with a standard landscape model that illustrates varied private deforestation pressure across all of the locations of TDFs. Pressure directly affects conservation impact and is linked to both political economy and spillovers. Evidence in Sections 23.3, 23.4, and 23.5 supports foci (1), (2), and (3), respectively. Section 23.6 is the conclusion.

23.2 Conceptual Framework: Three Issues for Assessing Policy Impact on Tropical Dry Forests

23.2.1 Private Pressure on Tropical Dry Forests, by Location

Private land-use decision making often implies varied deforestation pressure across a landscape. From von Thunen (1966) to the monocentric model of urban land use, many landscape analysts assert that clearing pressure falls as we move outward along a road leading from a market center (a city,

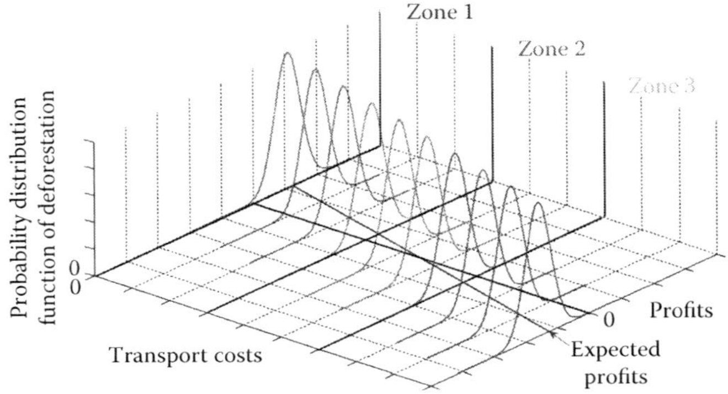

FIGURE 23.1
(**See color insert.**) Variation in private deforestation pressure across the landscape.

in Figure 23.1 where the "0" axis hits the left axis). Transport costs imply that, all else equal, while moving to the right, profits fall in agricultural production, whose output is to be sold in the city. If all land is originally forest and only transport costs matter, forests remain farther from markets (in Figure 23.1, the forest remains to the right of where the "Expected Profits" line crosses the "0" axis).

Of course, factors other than transport also affect relative profits from agriculture versus forests: High slopes near markets may stay forested, and good soils that are far from market may be cleared. From an analyst's point of view, some of these factors are observed; whereas others are unobserved, as there are limits on all datasets. The empirical analyses we cite include some observed factors, although Figure 23.1 focuses on representing the unobserved factors, in the form of a distribution.

23.2.1.1 Conservation's Short-Run Tropical-Dry-Forest Impact, by Location

Conservation policies aim to keep existing forests standing. This does not always work. Further, even if all forests in a policy's boundary remain standing, it may not imply policy impact. The impact of even perfect protection is equal to the baseline deforestation rate that is avoided. Thus, if private land use also would have featured standing forest, the policy would not have had an impact.

More generally, a conservation policy's impact equals the private or baseline deforestation rate that would have arisen without the policy minus the deforestation rate observed with the policy. Assuming that transport cost is a significant factor in the private (baseline) rate of clearing, a fully forested protected area that is far to the right in Figure 23.1 may not have much forest impact, the reason being that such locations very far from markets are mostly forests without protection.

23.2.1.2 Development's Short-Run Tropical-Dry-Forest Impact, by Location

Development policies, such as subsidies or road investments, increase profits from cleared land. Our model predicts that their impacts vary non-monotonically with private deforestation pressure (unlike conservation policy's impacts, which were just seen to rise and fall with private clearing). New roads may not generate significant deforestation if private clearing pressure is already high. Impacts of new development policy on deforestation are higher when past private pressures are intermediate and when many parcels remain at the margin of profitability (Pfaff et al. 2012a).

Perhaps non-intuitively, new roads within pristine forests beyond the frontier have a lower impact. Roads raise profits, but still, a large set of parcels do not reach the point of being worth clearing. However, we emphasize that this is a short-run result; holding all other relevant conditions constant, we will consider the dynamic development processes in which other conditions will change. Thus, even if a new road has little impact in earlier years, it may lead to additional investments that may imply higher long-run impacts. For instance, future roads may follow these new roads.

23.2.2 Political Pressures That Affect Tropical-Dry-Forest Policies' Locations

If a policy's forest impacts vary by location due to private deforestation pressure, as in our model, the distribution of conservation's locations across the landscape determines average impact. If most conservation is near low deforestation pressure, average impacts are relatively low. There are reasons to suspect that development tradeoffs push conservation toward low pressure. Land with a high opportunity cost in production, that is, where agricultural profits would be high, is expensive to buy for conservation. When public lands are being allocated, without a price, those lobbying for allocation for production may lobby more intensely when the profits are high.

Importantly, though, not all conservation decisions are made in the same manner. The incentives for national actors differ from those for global or local actors. The mix of pressures brought to bear on policy location can vary by decision maker. Incentives also may differ across types of policy. For instance, tradeoffs for creating strict reserves likely differ from those for multiple-use areas. Thus, the suite of political–economic influences on location also could vary by decision type.

23.2.3 Tropical Dry Forest Spillovers from Private and Public Responses, by Location

The total impact of a conservation policy is more than the impact of that policy within the boundary. Once a policy is established, private actors often respond, affecting forest and other outcomes in nearby areas. For instance, people and capital sometimes (although certainly not within all cases) pursue tourism-based opportunities that are created by a new public or private park or biological reserve.

Alternatively, migration and investment may shift away from areas previously thought attractive if an exclusive reserve signals that a government will make no further development investments. Then, public actors might not invest or assist, as they do not anticipate ongoing migrations. Instead, new roads and health clinics may follow on new roads, avoiding the areas with parks. Private actors may respond to new roads and to additional public development policies.

Critically, any such second-order or spillover effects are very likely to differ across forest locations. The level of tourism generated by conservation policy clearly varies significantly across settings. Development dynamics are likely to be different for new frontiers versus already developed sites.

23.3 Policy Impact Varies with Private Pressure on Tropical Dry Forests

This section, along with the two sections that follow, presents our review of recent high-quality evidence concerning conservation and development policies' short-run impacts on tropical deforestation. Before presenting this evidence, though, we highlight a critical feature of high-quality evidence: Outcomes for policy locations are compared with outcomes in similar locations without any policy. This approach is suggested by our land-use model and is incorporated in the evidence we review.

23.3.1 Improved Policy-Impact Estimates for Tropical Dry Forests, by Location

23.3.1.1 Controlling for Private Deforestation Baselines, by Location

In our model, expected deforestation without any policy varies across locations in the landscape. This variation has implications for empirically evaluating the impact of a forest-relevant policy, as impact estimates result from comparing the outcomes with policy to outcomes without policy. Once policy is established, no longer do we actually observe what would have happened without a policy. Thus, to estimate impact, we estimate what that baseline outcome would have been.

A simple method would be to use the average deforestation rate on all lands without any policies to estimate that unobserved baseline for lands that have a policy; for example, all of the protected lands. However, what if policy is far from private deforestation pressure, for example, to the right in Figure 23.1? Then, using the average unprotected land will overstate the deforestation that policy has avoided. Generally, high-quality evidence should compare parcels that are in the same zone in Figure 23.1. Since private deforestation pressure levels actually are multifactorial and do not appear on maps, one might measure characteristics such as the distances to roads and cities, slope, and soil quality, and then use those data to estimate the private deforestation rate had there been no conservation policy.

The evidence we cite uses matching for such apples-to-apples comparisons. For any parcel with policy, for example, in a protected area, one searches the unprotected lands for the parcels with the most similar characteristics. The average deforestation outcomes for most similar unprotected lands provide an estimate of the baseline private deforestation rate that policy avoided for that parcel. As there is no best definition of *similar*, one checks for robustness to the definitions of similarity.

23.3.1.2 Allowing for Varied Impacts, by Location

Once impacts are well estimated, it is also important to allow for varied impacts across locations. For instance, perfect protection may accomplish more to the left within Figure 23.1 than to the right. Since for integrated spatial policy planning we want to know each policy's impacts *by location*, we have focused our empirical review on the analyses that explicitly distinguish subsamples.

23.3.2 Policies' Short-Run Tropical-Dry-Forest Impacts Vary with Private Deforestation Pressure

23.3.2.1 Evidence on Conservation Impact, by Location

In a review of protection-impacts literature, Joppa and Pfaff (2010a) highlight a lack of explicit matching of baseline private deforestation pressure, based on measured location characteristics. Many researchers have compared clearing in protected areas with clearing at all unprotected sites or with clearing of the spatial buffer. Most of the evidence that we cite next has used matching, based on measured location characteristics, in order to control for baseline private pressure.

An early paper in terms of explicit focus on varying impacts is Pfaff et al. (2009), on effects of protected areas in Costa Rica on 1986–1997 deforestation. Areas farther from pressure avoid deforestation—as delineated using various proxies—which is often not statistically significant; whereas areas closer to private pressure (e.g., cities, roads) avoid deforestation well above average. In addition, protection on lands with relatively high slopes did not avoid any deforestation, statistically speaking; whereas protection on lower slopes is estimated to have avoided significant deforestation.

One might not expect Costa Rica to predict all other settings. In fact, the nature of these results, as well as the model in Figure 23.1, characterizes the global results for protection's varied impacts. Joppa and Pfaff (2010b) use global datasets for the years 2000 and 2005 to check the very same issues within each of more than 100 countries (every country listing more than 100 km^2 of protected area). Naturally, the experience of each country differs. However, the median and average results strongly support the results that conservation's forest impacts vary with private deforestation pressure.

Policy choice in light of such knowledge will have the most impact on future deforestation frontiers. Thus, it is worth checking whether such results hold where most deforestation is going to occur. For instance, while considering the entire Legal Brazilian Amazon region, Pfaff et al. (2012b) found that protection's impacts on deforestation during 2000–2004 and 2004–2008 are higher if nearer to pressure.

All of the same logic should apply just as well to an increasingly common conservation policy, payments for ecosystem services (PES). Costa Rican evidence for PES confirms that private pressure affects forest impact. Arriagada et al. (2011) found that when an NGO helped target deforestation pressure, the avoided deforestation was higher. Pfaff et al. (2012c) found that the PES impact varied considerably across agency offices. For Mexico's hydroservice-payment program, Alix-Garcia et al. (2012a) highlight that the PES impact was higher where poverty rates were lower.

23.3.2.2 Evidence on Development Impact, by Location

Our model predicted lower short-run forest impacts of new development policies both nearest to and farthest from private clearing pressure, with higher impacts for intermediate pressure levels. In the limited empirical work on spatially varied impact, our review finds support for our model. Again, though, we emphasize that this is a short-run result (see development dynamics in Section 23.5).

Empirical tests of such predictions are quite limited, but some do exist for new road investments. With regard to central Mexico, Nelson and Hellerstein (1997) consider the existence of prior roads, which we view as a proxy for location in a higher pressure area. They found that this influences new road impact. Andersen et al. (2002) use another proxy for private deforestation pressure: before forest clearing. For the Brazilian Amazon, they found that more prior forest clearing lowers a new road's impact. For an Amazonian region, Delgado et al. (2008) also found that the new road's impact was lower when prior development was higher. These results are consistent with our model's high-pressure prediction.

However, our model predicts a non-monotonic relationship between private pressure and the impact of new development policies such as new road investments; that is, the relationship changes in its sign. Pfaff et al. (2012a) test this using disaggregated data (census tracts) for the Brazilian Amazon.

While studying deforestation during 1976–1987, that is, for a relatively early decade in the development of this forest frontier, they confirmed the model's low-high-low prediction for impact given pressure. Thus, new road impact is lower both close to and far from baseline deforestation pressure.

As development unfolds, that is, while studying deforestation for 1986–1992, much of this result not only still holds but we also see changes, which suggest the importance of unobserved shifts in drivers over time. Defining lower or higher prior private pressure using prior deforestation, which results from the influences of all factors, Pfaff et al. (2012a) again confirm short-run predictions of all the models. However, defining lower private pressure using only the absence of any prior road investments, that is, a definition that can miss the evolution over time of unobservable deforestation drivers, new road impacts for defined *low private pressure* look more similar to those for intermediate pressure.

23.4 Policy Impact on Tropical Dry Forests Varies with Political–Economic Pressures

The evidence that deforestation impacts of conservation policies vary significantly, by location, implies that the distribution of conservation locations will determine the average policy impact. This section continues our review of evidence, which is focused on our second issue for assessing impact: the differing mixes of political–economic pressures that affect policies' impacts through location.

23.4.1 Evidence on Average Conservation Locations and Impacts

For Costa Rican protected areas, Andam et al. (2008) demonstrate the importance of biases in the average location of conservation policy toward locations with low private deforestation pressure. Protected lands are significantly farther from roads and cities, and are also are on higher slopes than is the median or average unprotected forest parcel. This fact is not reflected when the forest outcomes within protected areas are compared with the outcomes for average unprotected parcels; this comparison suggests that more than 40% of protected forest would have been cleared without a policy. However, if comparing with similar unprotected parcels, estimated impact is in the order of only 10%. This result makes it clear that the average location's influence on average impact can be enormous. This point applies to any type of forest, tropical or otherwise, including TDFs.

Considering whether this result is representative for other contexts, Joppa and Pfaff (2010b) found exactly the same pattern globally, for the median country and on average for more than 100 countries (location bias in protected networks for these countries is documented in Joppa and Pfaff [2009]). Matching estimate of protection's average impact by Joppa and Pfaff (2010b) is less than half the estimate generated when ignoring the bias in location. Sims (2010) found similar bias and results for Thailand; whereas for the Brazilian Amazon, Pfaff et al. (2012b) generate similar conclusions. This suite of similar results clearly has implications for optimal policy in TDFs.

The distribution of PES conservation policies also, on average, has been biased to low pressure. For Costa Rica's early PES, Robalino and Pfaff (2012a) found a countrywide bias. Other analyses found that PES participants differ from others in characteristics that affect land use (see, e.g., Miranda et al. [2003]; Ortiz et al. [2003]; and Zbinden and Lee [2005]). For Mexican PES, Munoz et al. (2008) suggest that early hydroservices payments were located in lower pressure areas and, as a result, had a low deforestation impact on average. It is worth mentioning, however, that the same agency shifted this PES policy to address these biases within the location by targeting a higher risk of deforestation.

Such analyses have direct implications for not only the evaluation of but also the establishment of new policies that affect TDFs. Conflicts between conservation and development clearly arise with regard to TDFs, and they can be expected to affect policy locations. Thus, looking backward at the impacts of past policies, the locations of protected areas within TDFs clearly need to be taken into account to evaluate past impact most accurately. Further, looking ahead, in making tradeoffs across candidate locations for new policies, decision makers should recognize that the impacts on deforestation rates will vary across the landscape.

23.4.2 Evidence on Political–Economic Variation, by Decision Maker and Decision Type

We ask whether different decision makers' incentives affect such biases in policy location; for instance, it seems eminently reasonable to expect federal and state actors' incentives to differ. We also ask whether for a given decision maker, location incentives differ across decision types; for instance, establishing a strict protected area can involve a different set of local tradeoffs with livelihoods than establishing a multiple-use protected area that permits some local production. These empirical examples consider only a few of the many possible examples of such differences, yet they provide important demonstrations of the relevance of the political economy to forest impact.

23.4.2.1 Conservation Policy Locations by Whom

A global actor giving funds to incent actions for biodiversity will, in the end, fund a local actor. Which actor is an important choice, for example, to the top or the bottom, is one key question: Should all the funding go through the federal actor(s) or, instead, go directly to some more local actors? This could affect impact if the actors have different goals and would choose different locations. For example, typical public-economic modeling distinguishes between federal and state motives. States put greater weight on local development opportunity costs, relative to benefits from forests. The reason is that some of the benefits of standing forests flow to people who are outside the state. Thus, we might predict that state conservation locations would be farther from private pressure.

To test this, Pfaff et al. (2012d) examined location choices for Brazilian Amazon protected areas to compare the locations of federal conservation policies with the locations of state conservation. They found that as is predicted by public-economics perspectives concerning important differences between these decision makers, the average federally protected area is located closer to private deforestation pressure, and is estimated to avoid more deforestation, than the average state area. Since development–conservation conflicts are eminently apparent around TDFs, and since the weights placed on dry-forest local actors also will vary across federal and state agencies, the impacts of typical policies for TDFs also are likely to vary by the state level.

23.4.2.2 Conservation Policy Locations for Whom

Conservation-development tradeoffs appear in another form if we consider types of protection. One globally relevant distinction is between the categories of strict, that is, excluding extraction, and multiple use, which in various ways allows some extraction inside the protected area. Clearly, more deforestation occurs inside multiple-use areas, suggesting that they have lower impacts. As far as we are aware, this point holds for not only tropical rainforests but also TDFs, as the distinctions across strict and multiple use certainly have been applied to the dry forest.

Reviewing the evidence on the deforestation impacts of different types of protection, with quite different land-use restrictions, suggests that location choice is critical and may reverse this ranking; that is, even if this is not intuitive, multiple-use protection can prevent more deforestation than does strict protection. Quite clearly, multiple-use protection goes to different locations, compared with strict protection. Specifically, on average, the strict protection is farther from private deforestation pressure, thus multiple-use areas can have higher impact.

For the whole Brazilian Amazon, Pfaff et al. (2012d) found that among federally protected areas (that have higher impacts—see above), multiple-use areas have greater impacts than strict areas. In the Amazonian state of Acre, multiple-use areas are closer to roads than is strict protection (see, e.g., Delgado and Pfaff (2008), for study of a specific case, Chico Mendes Extractive Reserve). Pfaff et al. (2012e) found this to be so significant that it outweighs the legal internal deforestation; that is, whatever political economy permits multiple-use areas to occupy higher-pressure locations, that difference in locations reverses the intuitive ranking of impact. The multiple-use areas avoid more deforestation. These deforestation results are consistent with global fire analysis by Nelson and Chomitz (2011); looking across many countries, they found that the multiple-use areas have greater average impact, which is again due to location. It appears that protection type generally affects location, and, thus, impact.

As far as we are aware, this point has not yet been tested specifically for TDFs. However, as noted earlier, all of the compelling logic concerning location should apply well, as the same tradeoffs of conservation and development certainly also arise for dry forests. That said, as TDFs differ, it is worth specifically checking these claims here.

23.5 Policy Impact on Tropical Dry Forests Varies with Policy Spillovers, by Location

23.5.1 Private Responses to Conservation Policy, by Location

23.5.1.1 Evidence on Forest Spillovers (Including Leakage), by Location

Protection may affect forests not only inside their boundaries but also on neighboring lands. Because the mechanisms are various, and can be simultaneous, the net spillover can vary across a landscape. Protection can promote neighboring forests by attracting tourism. Protection instead can reduce neighboring forests if crop markets are local, such that the reduction of output due to protection raises crop prices and profit on neighboring lands (Robalino 2007). For empirical consideration of forest spillovers, Robalino and Pfaff (2012b) examine the net effects of private private land use on neighboring private land use in Costa Rica. Spillovers mimic the initial land use: Clearing increases neighbor clearing, and conservation increases neighboring forest conservation.

Spatial spillovers to forests from protected areas may differ from spillover due to private land use. Public protected areas usually are larger, and public conservation of standing forest could be seen as longer lasting. For Costa Rica, an early rigorous estimate of spillovers found insignificance (Andam et al. 2008), yet that could reflect positive and negative effects. Robalino et al. (2012) considered mechanisms by which impact can vary across locations and examined nearby roads and nearby park entrances, where tourism may be a much more significant force than a nearby boundary where nobody enters. Around the roads, they found significant deforestation leakage, that is, higher deforestation in the adjacent areas than would be expected without protection. However, net effects are insignificant even around roads when near to entrances, which is consistent with private forest conservation in tourism.

This logic should apply to payments for environmental services as well, which raise another potential mechanism—effects on expectations. If neighbors learn that it is possible to get payments, their expectations of the future streams of revenues from forests could rise, discouraging clearing. Anecdotes from Costa Rica support this possibility. Even when little impact is found within PES boundaries, analyses suggest spillovers using coarse units that blend paid and neighboring lands (Sánchez-Azofeifa et al. [2007] and Arriagada et al. [2011], e.g., found slightly higher total impacts).

For Mexican PES, Alix-Garcia et al. (2012a) distinguish increased clearing on unpaid property owned by those who receive payments from increased deforestation in regions with high levels of program participation. Evidence of both was found, and the magnitude and direction of the spillovers varied across locations. Additional deforestation was higher within poor communities, which was consistent with credit constraints; whereas it was lowered in wealthier ones. Another case of spatially varying spillovers in Mexico is provided in Alix-Garcia et al. (2012b). They found that new income from *Oportunidades*, a randomized conditional-transfer program in poor communities, increased deforestation on average and more so in more isolated communities.

Such spillovers are relevant in developed countries, as seen in the debate about leakage in the U.S. Conservation Reserve Program (Wu 2000). Challenges for empirical comparisons are seen in the debate on identification (Roberts and Bucholtz 2005), although spillovers are also suggested by other analytical approaches (see Sohngen et al. [1999], Sedjo and Sohngen [2000], and Sedjo [2005]). Leakage could undermine deforestation reductions (Gan and McCarl 2007).

There is every reason to believe that all such interactions apply to TDFs as well, although again, as far as we are aware, there have been no specific tests for dry forest regions. Simply considering the logics, the price and tourism, and labor and income stories all may fit. This said, as noted earlier, the net spillover in any given location can be a blend of such effects, which is another reason that testing specifically in areas of particular interest seems worthwhile.

23.5.1.2 Evidence on Socioeconomic Spillovers, by Location

Private responses, such as those discussed earlier, surely could affect more than rates of deforestation. Theoretically, conservation policies could have important distributional effects, for example, through prices. Wages can fall if the demand for agricultural workers decreases (Robalino 2007). However, policies can also have positive economic effects on local tourism income (Sims 2010). Once again, the signs of net spillovers are ambiguous, and they can vary across locations.

In Costa Rica and Thailand, protected areas' locations were poorer than their national averages. Choosing comparisons from similar locations helps isolate policy's impact. By controlling for confounding influences in this way, it has been shown that protection can have positive effects on consumption and can lower poverty levels (Andam et al. 2010; Sims 2010). A similar study exists for Bolivia (Canavire and Hanauer 2012), showing that protected areas have significantly reduced poverty (a finding which is robust to different ways of measuring poverty). Gains can result from increased tourism around protected areas and are likely to vary by location. Sims (2010) found the largest net impacts at intermediate distances from major cities in Thailand.

We would like to understand the channels through which such effects may arise (Hanauer 2011). One approach is to focus on particular variables that are linked to poverty, such as employment or wages. For instance, parks have had positive effects on wages and employment in Costa Rica, implying that at least a part of spillover benefit is being channeled through labor markets (Villalobos 2009; Robalino and Villalobos 2010). Using another statistical method, Hanauer (2011) found similar results for Costa Rica; nearly half the poverty reduction in a previous study is due to tourism. We also want to look for variations across locations. The studies for Costa Rica found big wage effects that are only close to the parks' entrances, suggesting a link to tourism (Robalino and Villalobos 2010).

23.5.2 Public–Private Development-Conservation Dynamics Can Affect Tropical Dry Forests

Earlier, we consider varied private land, labor, and investment responses to conservation policy. Here, we consider development policies, which both drive private choices and respond to them. For example, similar to protected areas, new roads investments could attract or repel capital and labor. In general, when considering where to put a development investment, private responses are key. For instance, it would make no sense to build a road where one is sure nobody will ever venture. Further, if private actions reveal willingness to invest in the development of particular locations, the expected marginal development gains of putting public infrastructure there seem higher.

23.5.2.1 Evidence on Road Responses

If such dynamics were to occur, one natural empirical focus would be the private responses. For instance, following development policies, one could track flows of people and investments. Unfortunately, for the forested developing countries, the data requirements could be a challenge. However, if public policies follow, in turn, on the private response to initial shifts in public policies, one could instead examine the reduced-form relationship in which new policies follow old.

Pfaff et al. (2012f) study the reduced-form implication in examining road investments over time. Breaking Brazilian Amazon roads into initial investment as well as further investment over time, they study where road investments go as a function of the prior road investments up to that point. They found that paving investments tend to follow unpaved roads; that is, initial directions are continued. They also found that paving investments tend to go where there are prior neighboring unpaved roads. Within more pristine areas, unpaved investments follow prior neighboring-area paved roads, further suggesting dynamics. These results suggest that long-run road impacts are above short-run ones.

Some additional results concerning neighbor forest outcomes are consistent with such dynamics. Pfaff (1999), for instance, found at a decadal scale that neighboring deforestation is significantly higher next to counties with roads. Greatly improving on this evidence using more precise data, Pfaff et al. (2007) test for impacts of road investments on deforestation in neighboring countries that do not receive any road investment. Consistent with a model of development that spreads out from initial access roads, these new roads significantly increased the neighboring deforestation.

23.5.2.2 More Evidence on Forest Spillovers (Blockage), by Location

Should such public–private development dynamics be common in countries with tropical forests, the possibility exists of an additional longer-run impact of protected areas on deforestation. Protection on a frontier could signal to private actors that the state will not be investing further in that area to stimulate development (though if a park has tourism, that can go in the other direction). Such expectations could affect labor and capital movements, discouraging such private response. Private non-responses, in turn, discourage public investment (justifying expectations, *ex-post*).

Thus, we showed that protection had little impact, if far from roads; here, we add that if a new road would have been built but was not due to protection, it implied additional impact. Put another way, it would illustrate another mechanism for positive forest spillovers from policies. For the Brazilian Amazon, Pfaff et al. (2012g) provide the only evidence we know on this topic. For deforestation in 2000–2004 and in 2004–2008, they found that the land next to protected areas does not feature higher deforestation than the estimated baseline (leakage) but lower deforestation (blockage). Such a result could arise due to inadequate control for isolation of protected areas within the empirical analysis of impact. However, if it were due to poor control, the result should appear larger the farther the areas are from prior development. In fact, the locations closer to prior roads show the greatest blockage. This suggests a further impact of protection.

23.6 Conclusion

Tropical dry forest policies balance conservation's local costs with its benefits—local to global—in biodiversity, the mitigation of climate change, and other eco services, such as water quality. Both challenge and opportunity are implied by variations across locations in these core tradeoffs.

We argued for considering location in three ways to help predict policy impacts and to improve policy: (1) Policies' impacts vary by location with the rates of private deforestation that policies block; (2) different mixes of political–economic pressures drive the final location for different policies; and (3) policies trigger second-order or spillover effects that are likely to differ by location as well. Analyses exploring these points for TDF regions, in particular, would contribute to better policy planning.

Two additional considerations for policy impact, which are likely to vary with location within a landscape, are monitoring and enforcement. These were not featured in the studies we reviewed; however, they could have been responsible, in part, for some observed variations across space. For instance, Sims (2010) found impacts of protection in Thailand to be slightly lower near cities, in contrast to the results for Costa Rica and consistent with weaker enforcement in Thailand. To explain enforcement-based spatial variation in impact, Albers (2010), for instance, models a game between protected-area managers and neighboring villages. Monitoring by local stakeholders could also be part of the explanation for the relatively high impacts of multiple-use protection. This is consistent with work by Albers and Robinson (2011) on locals having a stake in protection. Exploring these additional dynamics could further contribute to an understanding of TDFs.

With biodiversity as a focus, the benefit from one unit of forest is another key spatial variation. Critical variations include species' estimated densities and the values people place on species. Where species are dense and valuable clearly should affect the targeting of parks or roads. There is considerable work reflecting the species differences across different types of forests and for TDF policy evaluation and optimization, these habitat issues clearly apply.

Biodiversity's need for effective habitat may also suggest another lens for measuring benefits, focused on not only total standing forest but also remaining forests' spatial patterns or fragmentation. For example, for the Mayan forest, Conde and Pfaff (2008) found lower impact of new roads farther from current threats. Thus, in the short run, a new road in an isolated area may clear fewer trees. However, such an intrusion in an otherwise-uninterrupted large area of forest could matter more for species outcomes. The work by Conde (2008) on effective jaguar habitat shows that a road within a pristine forest might not lead to much clearing but still can significantly affect species presence. Such concerns also matter for protected areas, which may have an impact on fragmentation as well. Sims (2011) finds evidence that protected areas reduced fragmentation of forests in Thailand, whereas Albers and Robinson (2012) provide a review of work on spatial pattern in forest extraction.

These additional considerations help emphasize the core theme of our review of the evidence, that TDF's policy impact will vary significantly across locations within a landscape. We provided empirical evidence demonstrating that all three of our considerations are important. Forest impacts of well-enforced conservation rise with private clearing pressure, supporting (1). Protection types (e.g., federal vs. state) differ in locations and, thus, in impacts, supporting (2). Differences in development processes explain the different signs for spillovers, supporting (3). This support for our conceptual framework suggests that an understanding of spatial variation in impact is required for TDF conservation and development to be sensibly balanced. Future research testing these claims specifically for TDFs would add perspective.

24

Traditional Ecological Knowledge of Rural Communities in Areas of a Seasonally Dry Tropical Forest in Serra do Cipó, Brazil

Emmanuel Duarte-Almada, Marcel Coelho, André Vieira Quitino, Geraldo Wilson Fernandes, and Arturo Sánchez-Azofeifa

CONTENTS

24.1 Introduction

Understanding the value of biodiversity through interdisciplinary scientific research is one tool for mitigating the looming environmental crisis (Leff 2006; Hissa 2008). In this sense, ethnobotany, the study of relationships between humankind and plants can substantially contribute to overcoming the environmental crisis (Alcorn 1995; Albuquerque 2005; Oliveira et al. 2009). Ethnobotany lies by definition within the ethnosciences field and seeks to build a fruitful dialog between empirical knowledge coming from local communities and connect it to traditional livelihoods and academic scientific knowledge (Blaikie et al. 1997; Berkes 2008). From this dialog, it is possible to build a new hybrid knowledge closer to sociobiodiversity (Sillitoe 2009). The knowledge of rural and traditional communities about their environment is not restricted to the utilitarian character of plant species but also encompasses their distribution in space and time as well as the interactions established with the local fauna (e.g., pollinators). This traditional ecological knowledge, in addition to being an important tool for conservation policies, is the basis for maintaining the livelihoods and cultural diversity of rural and traditional communities. Thus, communities and traditional knowledge have played an essential role in the conservation debate to achieve alternative development models and promote sociobiodiversity.

Brazil, as the most megadiverse country in the world, is in a privileged position to develop new ways to manage natural heritage that allow the creation of a sustainable society (Lewinsohn and Prado 2006). However, to achieve this goal it is necessary to deepen the scientific knowledge about Brazilian biodiversity without forgetting the traditional and ecological knowledge relating to species and ecosystem processes (Diegues 2000; Diegues and Arruda 2001; Mauro and Hardison 2002). This knowledge covers both the following: (1) knowledge related to the use of natural resources and their spatial and temporal distribution and (2) values, feelings, and mystical aspects related to biodiversity (Gadgil et al. 1993; Abele 1997; Marques 2001; Berkes 2008).

The ethnobotany field, as well as ethnobiology, ethnoecology, and ethnopedology, comes especially from an understanding of how traditional ecological knowledge has allowed human populations to adapt to and transform the environment (see Posey 1984; Balée 1998). In recent decades, ethnobotanical studies in Brazil have advanced significantly in terms of both number and methodology (Oliveira et al. 2009). However, considering the enormous cultural and biological diversity in Brazil, there still remains a vast research gap. Throughout the 1960s, 1970s, and 1980s, many studies have been conducted in the Amazon and more recently in the Atlantic forest and the caatinga (Posey 1984; Balée 1998; Hanazaki et al. 2000; Albuquerque and Andrade 2002; Begossi et al. 2002). However, phytogeographic domains such as the cerrado, despite representing 22% of the national territory, is one of the biomes with the fewest ethnobotanical studies, despite their high plant richness, high cultural diversity (Maroon, indigenous peoples, caboclos, geraizeiros, and vazanteiros), and vulnerable status (Myers et al. 2000; Klink and Machado 2005; Mazzetto 2009). With approximately 35% of the remaining area, the cerrado has annual deforestation rates higher than those recorded for the Amazon, suffering from the advance of cattle and large monocultures, besides mining and extensive urbanization, throughout its domains (Eiten 1994; Klink and Machado 2005).

The cerrado comprises a mosaic of landscapes, ranging from "rupestrian grasslands" with extremely poor, shallow, and acidic soils to forest formations characterized by a well-developed canopy (Oliveira and Marquis 2002). Among this landscape diversity, deciduous forests or simply tropical dry forests (TDFs) stand out for their uniqueness. Disjunct and scattered TDFs are mostly associated with limestone outcrops, extending from southern Minas Gerais southern Bahia, Brazil. TDFs contain a heterogeneous floristic composition strongly influenced by surrounding vegetation (Espírito-Santo et al. 2006, 2008; Meguro et al. 2007; Coelho et al. 2012).

Several species of economic and cultural interest such as *pequi* (*Caryocar brasiliensis*), *baru* (*Dipteryx alata*), *fava-danta* (*Dimorphandra mollis*), *barbatimão* (*Stryphnodendron adstringens*), *sempre-vivas* (Eriocaulaceae), *jatobá* (*Hymenaea* sp.), *pau d'óleo* (*Copaifera langsdorffii*), *sucupira* (*Bowdichia virgilioides*), and *araticum* (*Annona crassiflora*), comprise the plant diversity of cerrado. However, information related to the vegetation uses in cerrado by traditional populations is still lacking. Given the degree of threat that the cerrado is facing, as well as the lack of ethnobotanical studies, our objective was to evaluate and describe the perception of ecosystems and knowledge about the use of plant resources by rural communities in Serra do Cipó, Minas Gerais, focusing on TDFs. This work will contribute to the ethnobotanical knowledge of the cerrado, as well as demonstrate the rich relationship networks established by the communities with their environment, despite the whole process of social and environmental transformations that is underway.

24.2 Materials and Methods

The study was conducted in five rural communities located close to TDFs associated with limestone outcrops in Serra do Cipó, municipality of Santana do Riacho, Minas Gerais, southeastern Brazil. Serra do Cipó is located in the southern portion of Espinhaço, which extends from Serra de Ouro Branco, southern Belo Horizonte, to southern Bahia, Chapada Diamantina. The area has a global relevance for conservation policies, as part of the Espinhaço Biosphere Reserve. Serra do Cipó is a mosaic of landscapes, including the savanna formations at the lower altitudes and the rupestrian grasslands at altitudes above 900–1000 m a.s.l. However, there are enclaves of limestone that support a tree-sized vegetation with distinct floristic composition called TDFs (Meguro et al. 2007; Coelho et al. 2012). Serra do Cipo is known for its high species diversity, especially for its high plant richness with a high endemism level (Giulietti et al. 1987). Several species from popular medicine and those with considerable pharmacological research are also present, such as *pacari* (*Lafoensia pacari*), barbatimão (*S. adstringens*), and *jatobá* (*Hymenaea* sp.) (see the studies by Braga et al. [2000], Sólon et al. [2000], and Santos et al. [2006]). For example, a study conducted by Brandão et al. (1996) indicated the presence of 134 species of medicinal use in the rupestrian grassland from Minas Gerais, including Serra do Cipó.

The rural communities studied are located throughout the unpaved road that connects the district of Serra do Cipó to Santana do Riacho, besides the banks of limestone outcrops. Each community consisted of 14 houses on average, ranging from 7 to 19 houses, hosting approximately 71 families and 210 people in five communities (data from the Department of Tourism and Environment of Santana do Riacho). Many residents work in the county seat and do not practice agriculture. These communities were established in the region at least four generations ago, having practiced ranching and farming as main activities as well as the extraction of native species for different uses as medicine, ornament, and food. However, today they are going through intense sociocultural process transformations with the arrival of urban industrial elements. Moreover, Serra do Cipó is internationally recognized for its great tourism potential, which has led to changes in the economic activities and abandonment of agricultural practices. The predominant religion of Serra Do Cipó is Catholicism with rich manifestations such as *congadas, folias de reis, and benzeções*, the result of mixing European religiosity with indigenous practices going back to African origins. Our research was primarily exploratory to describe the communities' uses and knowledge in relation to local physiognomies.

A reconnaissance field trip for establishing the first contact with the communities was conducted between January 5 and 7, 2009. After the first contact, and the research proposal presentation to the communities, between January 21 and 26, 2009, we conducted 17 semistructured interviews with people from the communities *Melo, Mato Grande, Paraúnas, Usina, Picador, and Tenda* (43°36′42.3″W, 19°18′24.1″S). Our sample included 5 women and 12 men with ages ranging from 23 to 84 years. The number of respondents represented approximately 7% of the communities' population.

Interviews were conducted with five key informants, four men and one woman, selected by the technique called snowball (Biernacki and Waldorf 1981; Bernard 2005; Faria and Neto 2006). All the key informants interviewed were engaged in agricultural activities and were born in the region. Moreover, visits were attempted to other houses across communities; however, most of the houses were empty and in some cases residents refused to participate. In the houses visited, an adult was selected for the interview, totaling 12 more informants.

Prior consent of each informant was obtained for the interviews after explaining the research aims. Throughout the interviews conducted individually with each informant, three questions were asked: (1) Which native plants do you usually use or know? (2) What kind of use is intended for each species? (3) Where do the species occur? Furthermore, whenever possible, we went for walks with the informants, allowing species indications in the field, thus facilitating scientific identification. All other ecological information provided by the informants about the species cited (such as morphological characteristics and interactions with other species) were also recorded. Although our research was directed toward native species, the exotic species mentioned by the informants also were included in our species list as known or used by the communities. Some exotic species could be recognized as native since they were introduced in the region decades, or hundreds of years, ago and were thereby perceived by the communities as native. In addition to interviews, field and guided trips were conducted with key informants to check the species identification and collect botanical material regarding the plant species. The dried material was identified by comparing it to the scientific literature and with the assistance of technicians from the Systematics Plant Laboratory (Biological Sciences Institute, Federal University of Minas Gerais) and deposited in the collection of the Laboratory of Evolutionary Ecology and Biodiversity from the same institution. The field visit with the members of the communities possibilited the definition of the landscape used by them, as well as the use description of each environment. The methods applied in this study to collect qualitative data have been widely used in ethnoecological studies and are efficient for general characterizations of the relationship between human populations and the ecological systems in which they interact (Marques 2001; Albuquerque and Lucena 2004). Based on the methods used in other studies (e.g., Amoroso [2002], Lozada and Ladio [2006], Pinto et al. [2006]), it was decided to group the uses of species into the following categories: food, medicine, magic, ornament, fuel, and construction. The notes from the field diary also helped to describe the communities' perceptions about the local ecosystems as well as their understanding of temporal changes in plant uses as a consequence of the socioeconomic changes that took place in the past decades.

24.3 Results and Discussion

From the citations made by the informants, we recorded 218 species related to 65 botanical families. The most frequent plant families were Fabaceae (32), Bignoniaceae (13), Asteraceae (12), and Myrtaceae (11). Among all species, 9 were identified only to the family level and 15 species had no kind of scientific identification since they were cited by the informants without enough details that could allow the collection of biological material for later identification. Among the uses listed, medical use was the most important among the species mentioned, accounting for 45% of the uses listed (see Figure 24.1) and most species (185) had only one use indicated by the informants. Cerrado, including its forest and savanna formations, was the phytophysiognomy with more species mentioned, with only 34 species occurring in dry forest and 36 being exotic (Figure 24.2).

The heterogeneous landscape in Serra do Cipó is described by the classification system of the local communities, which defined four environment types: (1) "Campos" (savanna) or cerrado, (2) "serra" (corresponding to areas of higher altitudes, where rupestrian fields predominate), (3) "matas" (gallery forests and forests), and (4) limestone outcrops (which are associated with TDFs). The limestone outcrops are recognized by communities as *lapa* or *lapeiro* due to the presence of caves. In several interviews, there were references to the association of some species to TDFs, as in the case of a ficus popularly known as *gameleira* (*Ficus calyptroceras*) and the cactus *quiabo-do-capeta* (*Cereus jamacaru*). According to the informants, TDFs now occupied by rural communities were once used as shelter by primitive people. Despite the species richness mentioned by the informants, it was emphasized that the use of native species in recent years has been heavily regulated by government legislations and environmental

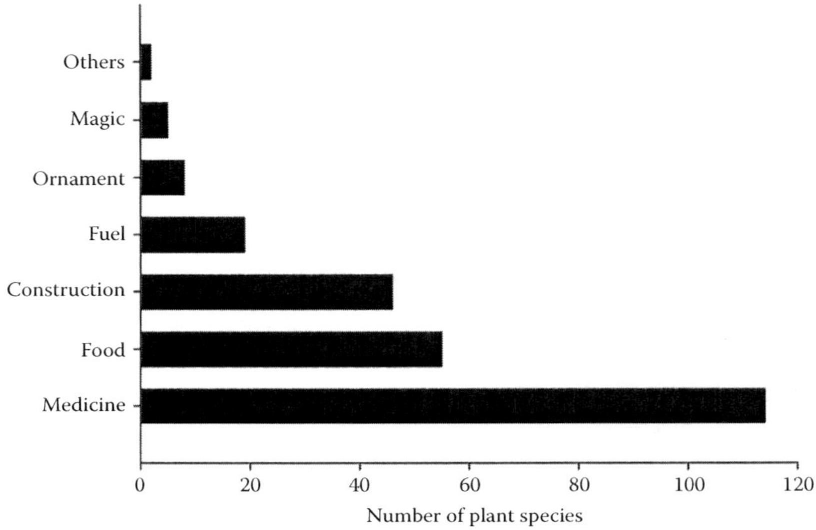

FIGURE 24.1
Number of plant species by categories used by the interviewed residents of Serra do Cipó.

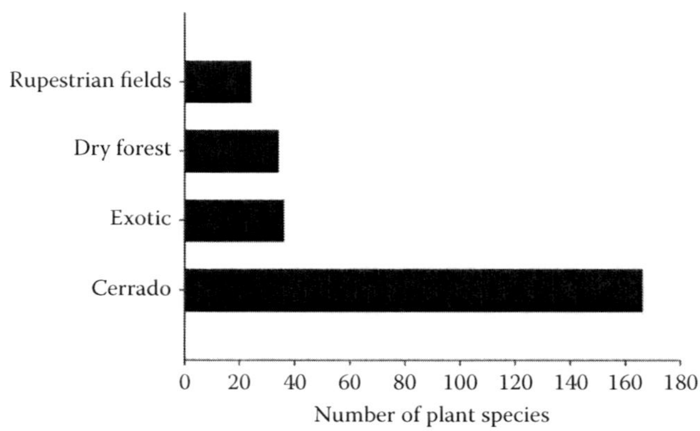

FIGURE 24.2
Number of plant species in each ecosystem of Serra do Cipó used by the interviewed residents Serra do Cipó.

management schemes. The natural resources of Serra do Cipó have been historically used for cattle grazing and extraction of firewoods and ornamental plants (especially evergreens, orchids, and bromeliads) (ICMBio 2009) (see Figure 24.3).

Major changes in the use of plant resources due to socioeconomic changes were reported in the interviews, such as the extinction of the use of candeia (*Eremanthus erythropappus*) for lighting homes and night walks as a

FIGURE 24.3
(See color insert.) The ecosystems used by the local communities in Serra do Cipó. At the top right the cerrado ecosystem (first plan), followed by rock outcrops lapeiros and mountain ecosystems. A significant number of species are used as firewood (photographs at the top left). The palm *Acromia aculeata* (at the bottom left) is one of the most abundant species and is used by the local communities in constructions and for oil extraction from its fruits (at the bottom right).

consequence of the arrival of electricity as well as other technologies. Also, in relation to the use of medicinal plants, according to the informants, a progressive replacement of medicinal plants by drugs from the pharmaceutical industry is taking place. Another interesting example is the past use of plants for dyeing outfits, such as *pacari* (*L. pacari*) and *caparosa* (*Neea theifera*) giving the colors red and pink, respectively, to clothes. Another case is that of *barriguda* (*Corisia speciosa*), which was once used for building canoes and feeding troughs for cattle. Another species highlighted was *mangaba* (*Hancornia speciosa*), which has latex that was used in the production of rubber and small balls (toys).

Informants showed well-defined criteria for each type of use for the local woody species. Those species with a well-developed core are generally not used as firewood, but are used in construction and the manufacture of furniture and fixtures, such as cedro (*Cedrela odorata*), *vinhático* (*Plathymenia* spp.), and barriguda (*C. speciosa*) (Table 24.1). The use of wood for fence construction was frequent in the local community; these should be resistant to attack by herbivores and detritivores.

The ideal time to extract the wood is also defined according to climatic variations and even lunar cycles. According to the informants, the most appropriate time to extract the wood from the forest is usually in the dry season (winter) and during the waning-moon phase. The *faveira* (*Peltophorum dubium*), for example, is cut only in the dry season.

For the communities investigated in the present study, the distinction between phylogenetically related species is accomplished by observing sensitive characteristics, such as taste. On the distinction between *pereira* (*Aspidosperma tomentosum*) and *canudo* (*Aspidosperma olivacium*), for example, an informant states that pereira is bitter while canudo is not. The odor is also a key feature for the reference to some species such as *tamboril-de-cheiro* (*Platymiscium* sp.), *bálsamo* (unidentified species), and *articum-cupim* (*Dughetia furfuracea*). The latter, according to the informants, has a similar smell to *cupim-caiana* (Termita), hence its name.

Communities also have traditional ecological knowledge about ecosystem processes and environmental changes. For example, one of our informants stated thus: "Armadillo, bird, and fish from the rivers disappeared recently. Now there is less forest than before. People have hunted a lot. We are close to the end times. Much disrespect. That attract sickness." Moreover, another resident presented some interesting information about food webs, "… is the small monkey (*Calytrix* sp.) that carries the bromeliad seed back and forth… " and "*pacu fish* (*Myleus micans*) who is eating the gameleira (*F. calyptroceros*) leaves."

The mystical aspects of various species were highlighted by some informants. The *pata-de-vaca* (*Bauhinia* sp.) cannot be burned, because according to local tradition it brings bad luck to the one who burns it. Other species have magical properties like *cipó*-são-joão (*Pyrostegia venu*sta), the *olho-de-boi* (*Dicloea* sp, and *assa-peixe* (*Piptocarpa rotundifolia*). There are also plants that can protect

TABLE 24.1

Species, Habitats, and Their Uses by Rural Communities from Serra do Cipó

Family	Species	Common name	Habitat	Med	Fuel	Cons	Food	Orn	Mag	Oth
				Uses						
Agavaceae	*Agave americana* L.	Piteira	Cer							x
Alismataceae	*Echinodorus grandiflorus* (Cham. & Schltdl.) Micheli	Chapéu-de-Couro	Cer	x						
Anacardiaceae	*Anacardium humili* A. St.-Hil.	Cajuzinho Caju-do-Mato	Cer				x			
Anacardiaceae	*Anacardium occidentale* L.	Caju	Ex				x			
	Astronium fraxinifolium Schott ex Spreng.	Gonçalves	Cer			x				
	Mangifera indica L.	Manga	Ex				x			
	Myracruodum urundeuva Allen	Aroeira	Cer, DF		x	x				
	Schinus terebinthifolia Beauverd	Aroeirinha	Cer	x	x	x				
Annonaceae	*Annona crassiflora* Mart.	Articum	Cer				x			
	Duguetia furfuracea (A. St.-Hil.) Saff.	Articum-Cupim	Cer				x			
	Rollinia sylvatica (A. St.-Hil.)	Articum cagão	Cer, DF				x			
	Xylopia aromatica (Lam.) Mart.	Pindaíba	Cer, DF			x				
	Annonaceae sp1	Articum do Mato	Cer			x	x			
Apocynaceae	*Aspidosperma ramiflorum* Müll. Arg.	Tambu	Cer			x				

(*Continued*)

TABLE 24.1 (*Continued*)

Species, Habitats, and Their Uses by Rural Communities from Serra do Cipó

				Uses						
Family	**Species**	**Common name**	**Habitat**	**Med**	**Fuel**	**Cons**	**Food**	**Orn**	**Mag**	**Oth**
	Aspidosperma cylindrocarpon Müll. Arg.	Peroba	Cer, DF			x				
	Aspidosperma macrocarpon Mart.	Peroba rosa	Cer, DF			x				
	Aspidosperma olivacium M. Arg.	Canudo	Cer, DF		x	x				
	Aspidosperma tomentosum Mart.	Pereira	Cer	x		x				
	Hancornia speciosa Gomes	Mangaba	Cer	x			x			
	Macrosiphonia velame (A. St.-Hil.) Müll. Arg.	Velame	Cer, DF	x						
Araceae	*Monstera sp.*	Imbé	Cer	x						
Arecaceae	*Acrocomia aculeata* (Jacq.) Lodd. ex Mart.	Macaúba Côco de espinho	Cer			x	x			
	Syagrus coronata (Martius) Beccari	Licuri	Cer				x			
	Arecaceae sp1	Palmito do Mato	Cer				x			
Aristolochiaceae	*Aristolochia* sp.	Milono Cipó Milomen	Cer	x						
Asteraceae	*Achyrocline satureoides* D.C.	Macela	Cer	x						
	Arctium lappa L.	Bardana	Ex	x						
	Artemisia absinthium (Mill.) DC.	Losna	E	x						

	Baccharis sp.	Alecrim	Cer, RF, DF	x				
	Baccharis trimera (Less.) DC.	Carqueja	Cer, DF, RF	x				
	Bidens pilosa L.	Picão	Cer, DF	x				
	Lichnophora sp.	Arnica	CR	x				
	Piptocarpha rotundifolia (Less.) Baker.	Assa-peixe-do-Mato	Cer	x				
	Eremanthus erythropappus (DC.) MacLeish	Candeia	Cer, DF		x	x		
	Taraxacum sp.	Dente-de-leão	Cer, DF	x				
	Vernonia polyanthes Less.	Assa-peixe-do-Reino	Cer, DF	x				x
	Asteraceae sp1	Mulata-na-sala	Ex				x	
Bignoniaceae	*Anemopaegma* sp1	Catuaba Fêmea	RF, Cer	x				
	Anemopaegma sp2	Catuaba Macho	RF, Cer	x				
	Jacaranda decurrens Cham.	Jacarandá	Cer		x	x		
	Jacaranda sp1	Caroba-da-Serra	RF	X				
	Jacaranda sp2	Carobinha	Cer	X				
	Jacaranda sp3	Carobinha-do-Campo	Cer	X				
	Pyrostegia venusta (Ker Gawl.) Miers	Cipó-de-São- João	Cer	X				
	Tabebuia impetiginosa (Mart. ex DC.) Standl.	Ipê-roxo Pau d'arco	Cer	X				
	Tabebuia sp1.	Ipê-amarelo	Cer	X				
	Tabebuia sp2.	Ipê-branco	Cer			x		
	Tynnanthus elegans (Cham.) Miers.	Cipó Cravo Trindade	RF	X				
	Zeyheria digitalis (Vell.) L.B. Sm. & Sandwith	Bolsa-de-pastor	Cer			x		

(*Continued*)

TABLE 24.1 (*Continued*)

Species, Habitats, and Their Uses by Rural Communities from Serra do Cipó

Family	Species	Common name	Habitat	Uses						
				Med	Fuel	Cons	Food	Orn	Mag	Oth
Bromeliaceae	*Ananas comosus* (L.) Merr.	Abacaxi	Ex				x			
	Ananas sp.	Abacaxizinho do cerrado Ananas	Cer				x			
	Bromeliaceae sp1	Bromélia	DF					x		
Burseraceae	*Protium heptaphyllum* (Aubl.) Marchand	Amescla Armesca	Cer		x					
Cactaceae	*Cereus jamacaru* subsp. *calcirupicola* DC.	Quiabo-do-Capeta	DF				x			
Cannabaceae	*Celtis* sp.	Grão-de-Galo	Cer, DF		x					
Caprifoliaceae	*Sambucus nigra* L.	Sabueiro	Cer	x						
Caricaceae	*Carica papaya* L.	Mamão	Ex				x			
Caryocaraceae	*Caryocar brasiliensis* Camb.	Pequi	Cer				x			
Celastraceae	*Maytenus* sp1	Espinheira-Santa	Cer	x						
	Maytenus sp2	Espinheira-Santa-do-Campo (folha larga)	Cer	x						
	Maytenus sp3	Espinheira-Santa-do-Mato	Cer	x						
	Salacia crassifólia (Mart. ex Schult.) G. Don	Bacupari Bago-de-pari	Cer	x			x			
Chenopodiaceae	*Chenopodium* sp.	Santa Maria	Ex	x						
Combretaceae	*Terminalia glabrescens* Mart.	Cambuí	RF			x				
	Terminalia sp.	Capitão	RF, DF		x	x				
Convolvulaceae	*Operculina macrocarpa* (Linn) Urb.	Jalapa	Cer	x						

Crassulaceae	Crassulaceae sp1					x			
Curcubitaceae	*Momordica* sp.	São-Caetano	Ex	x					
	Curcubitaceae sp1	Melão-branco	Cer				x		
Dilleniaceae	*Davilla elliptica* A. St.-Hil.	Bugre da Serra	RF	x					
Ebenaceae	*Diospyros obovata* Jacq.	Maria-preta	Cer		x	x			
Equisetaceae	*Equisetum* sp.	Cavalinha	Cer	x					
Euphorbiaceae	*Manihot utilíssima*	Mandioca	Ex				x		
	Pera sp.	Canela	Cer			x			
	Phyllanthus niruri L.	Quebra-pedra	Ex	x					
	Phyllanthus sp.	Quebra-pedra-do-campo	Cer	x					
	Sebastiana brasiliensis Spreng.	Cantarina	Cer			x			
Fabaceae	*Anadenanthera colubrina* (Vell.) Brenan	Angico-Branco	Cer, DF			x			
	Anadenanthera sp.	Angico	Cer, DF			x			
	Bauhinia sp1	Escada-de-Macaco	Cer	x					
	Bauhinia sp2	Pata-de-vaca	Cer	X					
	Bauhinia sp.	Unha-de-boi	Cer, DF						x
	Bowdichia virgilioides Kunth	Sucupira	Cer			x			
	Calliandra sp.	Candongueira	Cer						x
	Centrosema bracteosum Benth.	Cervejinha do campo	Cer	X					
	Chaesalpinea sp.	Pau-Ferro	Cer	x					
	Copaifera langsdorffii Desf.	Pau D'óleo Copaíba	Cer	x					
	Enterolobium gummiferum (Mart.)	Espirradeira	Cer			x			

(*Continued*)

TABLE 24.1 (*Continued*)

Species, Habitats, and Their Uses by Rural Communities from Serra do Cipó

Family	Species	Common name	Habitat	Uses: Med	Fuel	Cons	Food	Orn	Mag	Oth
	Eritrina sp.	Mulungu	Cer	x						
	Hymenaea courbaril L.	Jatobá-do-Mato	Cer	x						
	Hymenaea sp.	Jatobá-do-Campo	Cer	x						
	Inga sp.	Angá	Cer				x			
	Melanoxylon brauna Schot.	Braúna	Cer			x				
	Dicloea sp.	Olho-de-boi	Cer	x					x	
	Parapiptadenia sp.	Angico-Roxo	Cer, DF			x				
	Peltophorum dubium (Spreng.) Taub.	Faveira	Cer		x	x				
	Phaseolus vulgaris L.	Feijão	Ex				x			
	Plathymenia reticulata Benth	Vinhático	Cer, DF			x				
	Plathymenia sp.	Vinhático-do-campo	Cer			x				
	Platymiscium sp2.	Tamboril de cheiro	Cer			x				
	Pterodon sp1.	Pau Mijolo	Cer, DF	x						
	Pterodon sp2.	Sucupira Amarela	Cer, DF			x				
	Sclerolobium aureum var.	Pau –Bosta	Cer		x					
	Senna cathartica (L.)	Sene	Cer	x						
	Senna macranthera (DC. ex Collad.) H.S. Irwin & Barneby	Cabo verde	Cer	x						
	Stryphnodendron adstringens (Mart.) Coville	Barbatimão	Cer	x		x				

	Stylosanthes viscosa (L.) Sw.	Maria Babenta	Cer		x	
	Vatairea macrocarpa (Benth.) Ducke	Arcaçu Acaçu	Cer	x		
	Vatairea macrocarpa (Benth.) Ducke	Angelim	Ex	x	x	
Flacourtiaceae	*Casearia sylvestris Sw.*	Língua-de-Tiú	Cer	x		
Iridaceae	*Trimesia* sp.	Junco	Cer, RF	x		
Lamiaceae	*Hyptis carpinifolia* Benth.	Sulfato	Cer	x		
	Leonotis nepetaefolia (L.) R. Br.	Cordão-de-São-Francisco	Cer	x		
	Cunila microcephala Benth.	Poejinho	Ex	x		
	Mentha pulegium L.	Poejo	Ex	x		
	Mentha sp.	Hortelã	Ex	x		
	Plectranthus barbatus Andrews	Boldo	Ex	x		
Lauraceae	*Ocotea pretiosa* (Nees) Mez	Sassafrás	Cer	x		
	Persea americana var.	Abacate	Ex		x	
Lecythidaceae	*Cariniana estrellensis* (Raddi) Kuntze	Jequitibá Carimbó	Cer	x		
Loganiaceae	*Buddleja brasiliensis* Jacq. ex Spreng.	Barbasco	Cer	x		
	Strychnos sp1	Quina-de-Papagaio Quina-do-Cerrado	Cer	x		
	Strychnos sp2	Quina-de-Varinha	Cer, RF	x		
Lythraceae	*Cuphea carthagenensis* (Jacq.) J.F. Macbr.	Sete-Sangrias	Cer	x		
	Lafoensia pacari A. St.-Hil.	Pacari	Cer	x		x

(Continued)

TABLE 24.1 (*Continued*)

Species, Habitats, and Their Uses by Rural Communities from Serra do Cipó

Family	Species	Common name	Habitat	Uses						
				Med	Fuel	Cons	Food	Orn	Mag	Oth
Malpighiaceae	*Banisteriopsis* sp.	Cipó Prata	Cer, DF	x						
	Byrsonima sp1	Murici maior	Cer, RF				x			
	Byrsonima sp2	Murici menor	Cer, RF				x			
Malvaceae	*Bombocpsis* sp.	Amendoezeira	Ex				x			
	Chorisia speciosa A. St.-Hil.	Barriguda	Cer			x				
	Gossypium hirsutum L.	Algodão	Ex	x						
	Guazuma ulmifolia Lam.	Mutamba	Cer, DF		x	x	x			
Melastomataceae	*Clidemia* sp.	Cú-de-pinto	Cer				x			
	Tibouchina candolleana Cogn.	Quaresma	Cer					x		
Meliaceae	*Cedrela odorata* L.	Cedro	Cer	x		x				
	Guarea sp.	Canjerana	Cer			x				
Memecylaceae	*Mouriri pusa* Gardner	Manapuçá	Cer				x			
Moraceae	*Artocarpus heterophyllus* Lam.	Jaca	Ex				x			
	Brosimum gaudichaudii Trecul	Mama-cadela	Cer	x			x			
	Dorstenia brasiliensis Lam.	Carapiá	Cr, Cer	x						
	Ficus calyptroceras (Miq.) Miq.	Gameleira	DF			x				
	Moraceae sp1	Pau-Moreira	Cer	x						
Myrtaceae	*Blepharocalyx salicifolius* (Kunth) O. Berg	Maria-preta	Cer	x						

	Campomanesia pubescens (DC.) O. Berg	Gabiroba	Cer				x			
	Eugenia dysenterica DC.	Cagaitera	Cer	x			x			
	Eugenia involucrata DC.	Pitanga-do-Cerrado	Cer			x	x			
	Eugenia klotzschiana O. Berg	Pêra João-Boro	Cer				x			
	Eugenia uniflora L.	Pitanga	Ex				x			
	Myrcia splendens (Sw.) DC.	Folha-miúda Vassourinha	DF, Cer		x	x				
	Syzygium jambos (L.) Alston	Jambo branco	Ex				x			
	Eucalytptus sp.	Eucalipto	Ex	x		x				
	Psidium guajava L.	Goiaba	Cer				x			
	Psidium sp.	Goiabinha-do-Cerrado	Cer				x			
Nyctagynaceae	*Neea theifera* Oerst.	Capa rosa	Cer, RF	x						
Opiliaceae	*Agondra* sp.	Marmelinha	Cer	x						
Orchidaceae	*Orchidaceae* sp1	Florzinha da lapa	MS							x
	Orchidaceae sp2	Orquídeas	RF, DF					x		
Piperaceae	*Pothomorphe umbellata* (L.) Miq.	Capeba	RF	x			x			
Plantaginaceae	*Plantago* sp.	Transagem	Ex	x						
Poaceae	*Coix lacryma-jobi* L.	Lágrima de Nossa Senhora	Ex	x						
	Cymbopogon citratus	Erva Cidreira	Ex	x						
	Zea maiz Vell.	Milho	Ex				x			
Polygonaceae	*Polygonum hydropiperoides* Michx.	Erva-de-bicho	Cer	x						

(Continued)

TABLE 24.1 (*Continued*)

Species, Habitats, and Their Uses by Rural Communities from Serra do Cipó

Family	Species	Common name	Uses							
			Habitat	Med	Fuel	Cons	Food	Orn	Mag	Oth
Polypodiaceae	*Pteridium aquilinum* (L.) Kuhn	Samambaia do Mato	Cer				x			
Portulacaceae	*Portulaca oleracea* L.	Beldroega	Ex				x			
Punicaceae	*Punica granatum* L.	Romã	Ex	x						
Rosaceae	*Eriobotrya japonica* (Thunb.) Lindl.	Ameixa-amarela	Ex				x			
	Rubus brasiliensis Mart.	Amora do mato	Cer				x			
Rubiaceae	*Alibertia* sp.	Marmelada	Cer		x					
	Chiococca Alba (L.) Hitchc	Cainca	RF	x						
	Coffea arabica L.	Café	Ex				x			
	Palicourea rigida Kunth	Dom Bernardo	Cer	x						
	Palicuria sp.	Congonha de Bugre (Bugre)	RF	x						
	Randia sp.	Coroa-de-Santo-Antonio	Cer, MS, RF		x					
	Sabiceae sp.	Sangue-de-Cristo	Cer	x						
Rutaceae	*Citrus aurantium* L.	Laranja	Ex				x			
	Zanthoxylum riedelianum Engl.	Mama-de-Porca pequena	Cer	x						
	Zanthoxylum rhoifolium Lam.	Mama-de-porca	Cer	x						
Sapindaceae	*Cupania oblongifolia* Mart.	Pau-magro/Camboatá	Cer, MS	x		x				
	Magonia pubescens A. St.-Hil.	Tingui	Cer		x	x				

	Matayba juglandifolia Radlk.	Caxuá-Branco	Cer	x				
	Sapindus saponaria L.	Sabãozinho	Cer	x				
Sapotaceae	*Pouteria ramiflora* (Mart.) Radlk.	Figo-do-Cerrado	Cer				x	
	Pouteria torta (Mart.)	Acá	Cer				x	
	Pouteria sp.	Acá da Mata	Cer				x	
Schizaeaceae	*Anemia phyllitidis* (L.) Sw.	Avenca	Cer, DF	x				x
Smilacaceae	*Smilax* sp1	Jupicanga	Cer	x				
	Smilax sp2	Salsaparrilha	Cer	x				
	Smilax sp3	Salsaparrilha folha larga	Cer	x				
Solanaceae	*Solanum cernuum* Vell.	Panacéia	Cer	x				
	Solanum paniculatum L.	Jurubeba	Cer	x				
	Solanum sp.	Jurubeba do campo	Cer	x				
Urticaceae	*Cecropia* sp.	Embaúba branca	Cer	x				
Velloziaceae	*Vellozia* sp.	Canela-de-ema	RF		x			
Verbenaceae	*Aloysia virgata* (Ruiz & Pav.) Juss.	Cambará-branco	Cer, DF	x				
	Citharexylum myrianthum Cham.	Pau-pombo	Cer		x			
Violaceae	*Anchieta salutaris* St Hill.	Suma-roxa	Cer	x				
	Anchieta sp.	Suma-branca	Cer	x				
Vitaceae	*Cissus erosa* Rich.	Uva-do-mato	Cer	x				
Vochysiaceae	*Qualea grandiflora* Mart.	Pau-terra	Cer, RF		x	x		
	Salvertia convallariaeodora St. Hil.	Bananeira-do-campo	Cer			x		
Zingiberaceae	*Costus spirulis* (Jacq.) Roscoe	Cana-de-Macaco	Ex	x				
Not identified		Azeitona-do-mato	Cer	x				
Not identified		Cansanção	Cer				x	

(Continued)

TABLE 24.1 (*Continued*)

Species, Habitats, and Their Uses by Rural Communities from Serra do Cipó

				Uses						
Family	**Species**	**Common name**	**Habitat**	**Med**	**Fuel**	**Cons**	**Food**	**Orn**	**Mag**	**Oth**
Not identified		Caroquinha	Cer	x						
Not identified		Casadinha	Cer	x						
Not identified		Faísca	Cer						x	
Not identified		Flor da Noite	Cer	x						
Not identified		Landinho	Cer	x						
Not identified		Macieira	Cer				x			
Not identified		Madrugada	Ex					x		
Not identified		Maria-mole	Cer					x		
Not identified		Murta, aperta-o-cu	Cer				x			
Not identified		Murtinha	Cer				x			
Not identified		Piúna	Cer			x				
Not identified		Salsamburana	Cer	x						
Not identified		Saracura Canela-de-Saracura	Cer, RF	x						
TOTAL		**218**		**114**	**19**	**46**	**55**	**8**	**5**	**2**

Note: Medicinal = Med, fuel = Fuel, construction = Cons, food = Food, ornomental = Orn, magic = Mag, others = Oth, cerrado = Cer, dry forest = DF, exotic = Ex, rupestrian fields = RF.

against natural phenomena, such as the *faísca* (not identified), which has the power to keep lightning away from the natives during storms. Another interesting example is in relation to *candongueira* (*Calliandra* sp.). According to the belief, a child that breaks a twig of this species starts telling undue things in front of adults. There is also a common belief that people who use barbatimão (*S. adstringens*) in baths get the power to give women their virginity back.

The large number of species used cited, especially when compared with other studies in Brazilian biomes (see the studies by Rossato et al. [1999], Amoroso [2002], Pasa et al. [2005], and Hanazaki et al. [2006]), is quite high considering the low number of respondents in our survey. As found by several studies carried out in other biomes, such as the Atlantic forest and the caatinga (Silva and Andrade 2004; Hanazaki et al. 2006; Leitão et al. 2009), the main plant species use in the communities located in Serra do Cipó was for medicinal purposes. The use of plants in health care is prevalent in many low-income communities and communities from rural areas in developing countries (Hamilton 2004). Many species used by the communities have a great potential for the pharmaceutical industry; some of them were already a research subject as the extracts of barbatimão (*S. adstringens*), pacari (*L. pacari*), and *espinheira-santa* (*Maytenus* sp.) (Lima et al. 1998; Sólon et al. 2000; Gonzalez et al. 2001).

Morphological and sensory elements have proved important for species recognition and classification by the communities from Serra do Cipó. As demonstrated by Jernigan (2008) for Aguaruna Jivaro in Peru, odor is a key factor for tree species taxonomy. In Brazil, Ramos et al. (2008), when studying the communities in rural areas from caatinga, northeast Brazil, indicated that the physical species property used as a fuel source may interfere with the communities' preference. In the communities from Serra do Cipó, sensorial aspects are the base of system classification and species use, establishing connections even between animal and plant species such as in the case of articum-cupim (*D. furfuracea*), which is recognized by its characteristic odor, similar to a native termite. The use of timber species is also based on the evaluation of the quality of its core, and whitewood, which has a poorly developed core, is of lesser value for communities when they are looking for wood for building purposes.

There is also a difference in species use by different social groups. For example, many fruit species cited by the informants actually have their consumption practically limited to children, such as mangaba and *goiabinha-do-cerrado*, and are present only in the memory of adults. As demonstrated by the extensive literature, it reinforces the observation that environments are perceived and used differently according to age, besides gender influence; social class; and religion, all of which are aspects not investigated in this study (Martin 1995; Schultes and Von Reis 1995; Voeks 2007; Ayantunde et al. 2008; Dovie et al. 2008).

Although the sampling design does not allow an accurate quantitative analysis, it is possible to note that that some informants concentrate much of the information related to plant uses. One of the key informants, for example, is known locally as one of the main healers and has cited the use of 63 out of the 203 species recorded in our sample (excluding those that not identified). However, the absence of some species in the interviews given by some informants may simply indicate that these are not important in their daily lives and not a complete lack of knowledge about the uses of these species. In this study, only one interview per informant was conducted, which also limits the chances of recalling more species.

In the current environmental debate, there has been an increasing interest in correlations between biological and cultural diversities. Many authors have proposed since the 1990s the use of biocultural diversity indexes (Bennett et al. 1975; Bridgewater 2002; Loh and Harmon 2005; Maffi 2005; Cocks 2006), and terms such as sociobiodiversity are increasingly being used in policies and academic discussions. The sociocultural changes associated with the expansion of urban industrial lifestyles have transformed and generated knowledge losses associated with ecosystems used by traditional communities (Tsuji 1996; Escobar 1998; Reyes-García et al. 2005; Drew and Henne 2006; Furusawa 2009). In Serra do Cipó communities, some species such as quiabo-do-capeta (*C. jamacaru*), which according to the locals occurs only on lapeiros, is currently rarely consumed by the population. Another example is *samambaia-do-mato* (*Pteridium aquilinum*), formerly widely used as a typical meal accompanied by polenta and cracklings. The access to processed foods, compared with the difficulties in agricultural production, tends to lead to the abandonment of many food traditions besides the losses of native seed varieties. Moreover, the restriction on plant resource uses ruled by inspection agencies, can alter species usage patterns (see the study by Zuchiwschi et al. [2010]), also leading to changes in the ecological knowledge associated with them.

Our results indicate that most plant species used by the communities come from the cerrado ecosystem. However, due their floristic and phenological characteristics, TDFs provide important species that generally occur at lower density in other environments, such as *aroeira* (*Myracrodruon urundeuva*) and *angico-branco* (*Anadenanthera colubrina*).

The lapas or lapeiros, as TDFs are known locally, are also an important cultural inheritance, already guarding the archeological records of indigenous populations. Furthermore, it is common for these environments to become traditional sites for religious pilgrimages, as is the case with some Catholic feasts held in Serra do Cipó.

24.4 Conclusions and Final Remarks

Our results indicate a deep knowledge and use of native plants by Serra do Cipó communities, despite restrictions imposed by inspection agencies and also by socioeconomic changes that the communities faced in the recent decades. Serra do Cipó has been witnessing deep changes caused by the paving of the MG10 highway throughout the 1980s, and more keenly in the 1990s, promoting an acceleration of the tourist sector and the speculation and arrival of several mining ventures (ICMBio 2009). This situation has caused dramatic changes in social relations in rural communities. Many family members have abandoned farming and migrated to urban centers or have engaged in expanding tourism activities, which is now the main economic sector in Serra do Cipó. The widespread access to cultural and technological urban industrial elements discourages the new generation from learning traditional knowledge accumulated throughout history.

Despite this new reality, the large species richness mentioned and the ecological knowledge provided by the communities demonstrate that they still have a broad knowledge related to local biodiversity and maintain a sentimental relationships with these ecosystems that goes beyond the simple use of species as resources. This valuation has great importance in conservation, especially in the case of TDFs, which represent one of the most unique and threatened environments in the whole region. Any conservation policy must not only take into account the technical aspects of protection and ecosystem restoration but also bring to the table the motivations responsible for guiding individual and collective attitudes in relation to biodiversity conservation.

References

Abele, F. 1997. Traditional knowledge in practice. *Artic* 50:3–4.

Acselrad, H. 2004. Conflitos ambientais–a atualidade do objeto. In *Conflitos ambientais no Brazil*, ed. H. Acselrad, 8–11. Rio de Janeiro: Editora Relume Dumará.

Acselrad, H. 2009. Mapeamento, Identidades e Territórios. In *Anais do 33° Encontro anual da associação nacional de pós-graduação e pesquisa em ciências sociais*. ANPOCS: Caxambu.

Adams, W., M.D. Brockington, J. Dyson, and B. Vira. 2003. Managing tragedies: Understanding conflict over common pool resources. *Science* 302:1915–1916.

Aguiar, L.M.S. and J.S. Marinho-Filho. 2004. Activity patterns of nine phyllostomid bat species in a fragment of the Atlantic Forest in southeastern Brazil. Revista *Brasileira de Zoologia* 21:385–390.

Ahl, D.E., S.T. Gower, S.N. Burrows, N.V. Shabanov, R.B. Myneni, and Y. Knyazikhin. 2006. Monitoring spring canopy phenology of a deciduous broadleaf forest using MODIS. *Remote Sensing of Environment* 104:88–95.

Aide, T.M. 1992. Dry season leaf production: An escape from herbivory. *Biotropica* 24:532–537.

Aide, T.M., J.K. Zimmerman, J.B. Pascarella, L. Rivera, and H. Marcano-Vega. 2000. Forest regeneration in a chronosequence of tropical abandoned pastures: Implications for restoration ecology. *Restoration Ecology* 8:328–338.

Aizen, M. and P. Feisinger. 1994. Forest fragmentation, pollination and plant reproduction in a Chaco Dry Forest, Argentina. *Ecology* 75(2):330–351.

Albers, H.J. 2010. Spatial modeling of extraction and enforcement in developing country protected areas. *Resource and Energy Economics* 32:165–179.

Albers, H.J. and E.J.Z. Robinson. 2011. The trees and the bees: Using enforcement and income projects to protect forests and rural livelihoods through spatial joint production. *Agricultural and Resource Economics Review* 40(3):424–438.

Albers, H.J. and E.J.Z. Robinson. 2012. A review of the spatial economics of non-timber forest product extraction: Implications for policy. *Ecological Economics* (forthcoming).

Albert-Puentes, D., A. López-Almirall, and M. Roudná. 1993. Phenological observation in tropical trees. Methodological considerations. *Fontqueria* 36:257–26.

Albert-Puentes, D., A. Urquiola Cruz, I. Baró Oviedo, P. Herrera Oliver, L. González Oliva, and A. Urquiola Cabrera. 2008. Phenological behavior of 143 species in western Cuba. *Acta Botanica Cubana* 201:1–11.

Albuquerque, U.P. 2005. *Etnobiologia e Biodiversidade*. Recife: NUPEEA/Sociedade Brasileira de e EtnOecologia.

Albuquerque, U.P. and L.H.C. Andrade. 2002. Conhecimento botânico tradicional e conservação em uma área de caatinga no estado de Pernambuco, nordeste do Brazil. *Acta botanica brasilica* 16:273–285.

Albuquerque, U.P. and R.F.P. Lucena, 2004. *Métodos e técnicas na pesquisa etnobotânica*. Recife. Livro Rápido ed./NUPEEA.

Alcorn, J.B. 1995. The scope and aims of ethnobotany in a developing world. In *Ethnobotany, evolution of a discipline*, eds. R.E. Schultes and S. von Reis, 23–39. USA: Dioscorides Press.

Alfaro, E.A., A. Alvarado, A. Chaverri, 2001. Cambios edáficos asociados a tres etapas sucesionales de bosque tropical seco en Guanacaste, Costa Rica. *Agronomía Costarricense* 25:7–20.

Alix-Garcia, J., C. McIntosh, K.R.E. Sims, and J.R. Welch. 2012a. The ecological footprint of poverty alleviation: Evidence from mexico's oportunidades program. *The Review of Economics and Statistics* (forthcoming).

Alix-Garcia, J., E. Shapiro, and K. Sims. 2012b. Forest conservation and slippage: Evidence from mexico's national payments for ecosystem services program. *Land Economics* (forthcoming).

Allen, E.B., E. Rincon, M.F. Allen, A.P. Jimenez, and P. Huante, 1998. Disturbance and seasonal dynamics of mycorrhizae in a tropical deciduous forest in Mexico. *Biotropica* 30:261–274.

Allen, R.G., L.S. Pereira, D. Raes, and M. Smith. 1998. *Crop evapotranspiration, guidelines for computing crop water requirements*. Rome, Italy: United Nations Food and Agriculture Organisation.

Almeida, A.W.B. 2008. *Terras tradicionalmente ocupadas: Terras de quilombo, terras indígenas, babaçuais livres, castanhais do povo, faxinais e fundos de pasto*. Manaus: Editora da Universidade do Amazonas.

Almeida-Cortez, J.S. 2004. Dispersão e banco de sementes. In *Germinação: Do básico ao aplicado*, eds. A.G. Ferreira and F. Borghetti, 225–235. Porto Alegre: Artmed.

Almeida, H.S. and E.L.M. Machado. 2007. Relações Florísticas entre Remanescentes de Floresta Estacional Decídua no Brazil. *Revista Brasileira de Biociências* 5:648–650.

Álvares, V.V.H., L.E. Dias, C.A. Ribeiro, and R.B. Souza. 1999. Uso de gesso agrícola. In *Recomendação para o uso de corretivos e fertilizantes em Minas Gerais: 5. Aproximação*, eds. A.C. Ribeiro, P.T.G. Guimarães and V.H. Alvarez, 67–78. Viçosa: Comissão de Fertilidade do Solo do Estado de Minas Gerais.

Alvarez-Añorve, M.Y., M. Quesada, A. Sánchez-Azofeifa, L.D. Avila-Cabadilla, and J. Gamon. 2012. Functional regeneration and spectral reflectance of trees during succession in a highly diverse tropical dry forest ecosystem. *American Journal of Botany* 99:816–826.

Alvarez-Clare, S. and M.C. Mack. 2011. Influence of precipitation on soil and foliar nutrients across nine Costa Rican forests. *Biotropica* 43:433–441.

Alvarez, E., I. Mendoza, M. Pacheco, Cogollo Gutierrez Benitez, O.C. Ramirez, J.C. Dib, A. Roldan, E. Carbon, E. Zarza, L.A. Velasquez, M. Serna, C. Velasquez, Y. Alvarez, O. Jimenez, and M. Artinez (in press). Ten years of monitoring the Colombian Caribbean dry forest: preliminary results and recommendations. *Intropica Journal*. http://intropica.unimagdalena.edu.co/.

Alvarez-Yépiz, J.C., A. Martínez-Yrízar, A. Burquez, and C. Lindquist. 2008. Variation in vegetation and soil properties related to land use history of old-growth and secondary tropical forests in northwestern Mexico. *Forest Ecology and Management* 256:355–366.

Alves, D.S. and D.L. Skole. 1996. Characterizing land cover dynamics using multitemporal imagery. *International Journal of Remote Sensing* 17(4):835–839.

Alves, E.U., R.L.A. Bruno, A.P. Oliveira, A.U. Alves, and A.U. Alves. 2006. Ácido sulfúrico na superação da dormência de unidades de dispersão de juazeiro (Zizyphus joazeiro Mart.). *Revista Árvore* 30:187–195.

Alves, R. 2008. *Zonamento ambiental e os desafios da implementação do Parque Estadual Mata Seca, Municipio de Manga, Norte de Minas Gerais*. Belo Horizonte, Brazil: Instituto de Geociencias da Universidade Federal de Minas Gerais.

Alvim, P. 1960. Moisture stress as a requirement for flowering of coffee. *Science* 132:354.

Amoroso, M.C.M. 2002. Uso e diversidade de plantas medicinais em Santo Antônio do leverger, MT, Brazil. *Acta botanica brasilica* 16:189–203.

Anaya, C.A., F. García-Oliva, and V.J. Jaramillo. 2007. Rainfall and labile carbon availability control litter nitrogen dynamics in a tropical dry forest. *Oecologia* 150:602–610.

Anaya Merchant, C., J.L. Jaramillo, A. Martínez-Yrízar, and F. García-Oliva. 2012. Large rainfall pulses control litter decomposition in a tropical dry forest: Evidence from an 8-Year study. *Ecosystems* 14:652–663.

Anaya, F. 2012. De Encurralados pelos Parques a Vazanteiros em Movimento: As reivindicações territoriais das comunidades vazanteiras de Pau Preto, Pau de Légua e Quilombo da Lapinha no campo Ambiental. PhD diss., Universidade Federal de Minas Gerais, Brazil.

Anaya F., R. Barbosa, and C. Sampaio. 2006. Sociedade e biodiversidade na mata seca mineira. *Unimontes Científica* 8:35–41.

Andam, K., P. Ferraro, A. Pfaff, J. Robalino, and A. Sanchez. 2008. Measuring the effectiveness of protected-area networks in reducing deforestation. *Proceedings of the National Academy of Sciences of the United States of America* 105(42):16089–16094.

Andam, K.S., P. Ferraro, K. Sims, A. Healy, and M. Holland. 2010. Protected areas reduced poverty in Costa Rica and Thailand. *Proceedings of the National Academy of Sciences of the United States of America* 107(22):9996–10001.

Andersen, L.E., C.W.J. Granger, E.J. Reis, D. Weinhold, and S. Wunder 2002. *The dynamics of deforestation and economic growth in the Brazilian Amazon*. Cambridge, U.K.: Cambridge University Press.

Anderson, M.J. 2001. A new method for non-parametric multivariate analysis of variance. *Austral Ecology* 26:32–46.

Andrade, J.L., F.C. Meinzer, G. Goldstein, and S.A. Schnitzer. 2005. Water uptake and transport in lianas and co-occurring trees of a seasonally dry tropical forest. *Trees-Structure and Function* 19(3):282–289.

Andrade-Lima, D. 1981. The caatinga dominium. *Revista Brasileira de Botânica* 4:149–153.

André, F., M. Jonard, F. Jonard, and Q. Ponette. 2011. Spatial and temporal patterns of throughfall volume in a deciduous mixed-species stand. *Journal of Hydrology* 400(1–2):244–254.

Antunes, F.Z. 1994. Caracterização climática: Caatinga do Estado de Minas Gerais. *Informe Agropecuário* 17:15–19.

Antunes, N.B., J.F. Ribeiro, and A.N. Salomão. 1998. Caracterização de frutos e sementes de seis espécies vegetais em matas de galeria do distrito federal. *Revista Brasileira de Sementes* 20:112–119.

Aragão, L.E.O.C., Y.E. Shimabukuro, F.D.B. Espírito-Santo, and M. Williams. 2005. Spatial validation of the collection 4 MODIS LAI product in eastern amazonia. *IEEE Transactions on Geoscience and Remote Sensing* 43(11):2526–2534.

Aranguren, C.I. 2011. Contribution of birds and bats to seed dispersal in a dry forest of Venezuela's Llanos. Master's Thesis, Instituto Venezolano de Investigaciones Científicas, Altos de Pipe, Venezuela. 117 p.

Aranguren, C.I., J.A. González-Carcacía, H. Martínez, and J.M. Nassar. 2011. Noctilio albiventris (Noctilionidae), a potential seed disperser in disturbed tropical dry forest habitats. *Acta Chiropterologica* 13:189–194.

Araújo, E.C. 2009. Nas margens do São Francisco: Sociodinâmicas ambientais, expropriação territorial e afirmação étnica do Quilombo da Lapinha e dos Vazanteiros do Pau de Légua. MSc. Thesis, Universidade Estadual de Montes Claros, Brazil.

Araújo, F.P., C.A.F. Santos, N.B. Cavalcanti, and G.M. Resende. 2001. Influência do período de armazenamento das sementes de umbuzeiro na sua germinação e no desenvolvimento da plântula. *Revista Brasileira de Armazenamento* 26:36–39.

Araújo-Neto, J.C., I.B. Aguiar, V.M. Ferreira, and T.J.D. Rodrigues. 2005. Armazenamento e requerimento fotoblástico de sementes de Acacia polyphylla DC. *Revista Brasileira de Sementes* 27:115–124.

Arcova, F.C.S., V. de Cicco, and P.A.B. Rocha. 2003. Precipitação efetiva e interceptação das chuvas por floresta de Mata Atlântica em uma microbacia experimental em Cunha - São Paulo. *Revista Árvore* 27:257–262.

Arriagada, R., P. Ferraro, E.O.Sills, S.P. Pattanayak, and S. Cordero. 2012. Do payments for environmental services affect forest cover? A farm-level evaluation from Costa Rica. Land Economics 88:382–399.

Arroyo-Mora, J.P., G.A. Sánchez-Azofeifa, B. Rivard, and J.C. Calvo-Alvarado. 2005. Quantifying successional stages of tropical dry forests using Landsat ETM+. *Biotropica* 37:497–507.

Arroyo-Mora, J.P., G.A. Sánchez-Azofeifa, B. Rivard, J.C. Calvo-Alvarado, and D.H. Janzen 2005. Dynamics in landscape structure and composition for the Chorotega region, Costa Rica from 1960 to 2000. *Agriculture, Ecosystems and Environment* 106:27–39.

Arroyo Mora, J.P., G.A. Sánchez-Azofeifa, M.E.R. Kalácska, B. Rivard, J.C. Calvo, and D.H. Janzen. 2005. Secondary forest detection in a neotropical dry forest landscape using landsat 7 ETM+ and IKONOS Imagery1. *Biotropica* 37(4):497–507.

Arruda, D.M., D.O. Brandão, F.V. Costa, G.S. Tolentino, R. Duque-Brazil, S. D'Ângelo-Neto, and Y.R.F. Nunes. 2011. Structural aspects and floristic similarity among tropical dry forest fragments with different management histories in northern Minas Gerais, Brazil. *Revista Árvore* 35:133–144.

Ascanio E.D. and M. García. 2005. *Bird inventory and checklist of Hato Piñero*. Caracas: Fundación Hato Piñero. 11 p.

Asner, G.P., J.M.O. Scurlock, and J.A. Hicke. 2003. Global synthesis of leaf area index observations: Implications for ecological and remote sensing studies. *Global Ecology and Biogeography* 12:191–205.

Asner, G.P. and R.E. Martin. 2008a. Airborne spectranomics: Mapping canopy chemical and taxonomic diversity in tropical forests. *Frontiers in Ecology and the Environment* 7(9):269–276.

Asner, G.P. and R.E. Martin. 2008b. Spectral and chemical analysis of tropical forests: Scaling from leaf to canopy levels. *Remote Sensing of Environment* 112(10):3958–3970.

Asner, G.P., R.E. Martin, A.J. Ford, D.J. Metcalfe, and M.J. Liddell. 2009. Leaf chemical and spectral diversity in Australian tropical forests. *Ecological Applications* 19(1):236–253.

Asquith, N. and M. Mejia-Chang. 2005. Mammals, edge effects and the loss of tropical forest diversity. *Ecology* 86(2):379–390.

Attarod, P., M. Aoki, D. Komori, T. Ishida, K. Fukumura, S. Boonyawat, P. Tongdeenok, M. Yokoya, S. Punkngum, and T. Pakoktom, 2006. Estimation of crop coefficients and evapotranspiration by meteorological parameters in a rain-fed paddy rice field, Cassava and Teak Plantations in Thailand. *Journal of Agricultural Meteorology* 62:93–102.

Attiwill, P.M. and M.A. Adams. 1993. Nutrient cycling in forests. *New Phytologist* 124:561–582.
Austin, A.T., L. Yahdjian, J.M. Stark, J. Belnap, A. Porporato, U. Norton, D.A. Ravetta, and S.M. Schaeffer. 2004. Water pulses and biogeochemical cycles in arid and semiarid ecosystems. *Oecologia* 141:221–235.
Austin, A.T. and P.M. Vitousek. 1998. Nutrient dynamics on a precipitation gradient in Hawai'i. *Oecologia* 113:519–529.
Avila-Cabadilla L.D., G.A. Sánchez-Azofeifa, K.E. Stoner, M.Y. Alvarez-Añorve, M. Quesada, and C.A. Portillo-Quintero. 2012. Local and landscape factors determining occurrence of phyllostomid bats in tropical secondary forests. *PLoS ONE* 7:e35228.
Avila-Cabadilla, L.D., K.E. Stoner, M. Henry, and M.Y. Alvarez Añorve. 2009. Composition, structure and diversity of phyllostomid bat assemblages in different successional stages of a tropical dry forest. *Forest Ecology and Management* 258:986–996.
Ayantunde, A.A., M. Briejer, P. Hiernaux, H.M.J. Udo, and R. Tabo. 2008. Botanical knowledge and its differentiation by age, gender and ethnicity in South-western Niger. *Human Ecology* 36:881–889.
Baker, H.G. 1974. The evolution of weeds. *Annual Review of Ecology and Systematics* 5:1–24.
Baker, R.J., S.R. Hoofer, C.A. Porter, and R.A. van den Bussche. 2003. Diversification among New World leaf-nosed bats: An evolutionary hypothesis and classification inferred from digenomic congruence of DNA sequences. Occasional Papers, Museum of Texas Tech University 230:1–32.
Balée, W. 1998. Historical ecology: Premises and postulates. In *Advances in historical ecology*. New York: Columbia University Press.
Balieiro, F. de C., A.A. Franco, R.L.F. Fontes, L.E. Dias, E.F.C. Campello, and S.M. de. Faria. 2007. Evaluation of the throughfall and stemflow nutrient contents in mixed and pure plantations of Acacia mangium, Pseudosamenea guachapele and Eucalyptus grandis. *Revista Árvore* 31:339–346.
Balvanera, P., E. Lott, G. Segura, C. Siebe, and A. Islas. 2002. Patterns of β-diversity in a Mexican tropical dry forest. *Journal of Vegetation Science* 13:145–158.
Balvanera, P., S. Quijas, and A. Pérez-Jiménez. 2011. Distribution patterns of tropical dry forest trees along a mesoscale water availability gradient. *Biotropica* 43(4):414–422.
Barbosa, A. and L.C.M. Lima. 2003. Fenologia de espécies lenhosas da Caatinga. In *Ecologia e conservação da Caatinga*, eds. I. Leal, M. Tabarelli, and J. Silva, 657–693. Recife: Editora Universitária-UFPE.
Barbosa, D.C.A. 2003. Estratégias de germinação e crescimento de espécies lenhosas da Caatinga com germinação rápida. In *Ecologia e conservação da caatinga*, eds. I.R. Leal, M. Tabarelli, and J.M.C. Silva, 625–656. Recife: Editora Universitária da UFPE.
Barbosa, D.C.A., M.C.A. Barbosa and L.C.M. Lima. 2003. Fenologia de espécies lenhosas da caatinga. In *Ecologia e conservação da caatinga*, eds. I.R. Leal, M. Tabarelli, and J.M.C. Silva, 657–694. Recife: Editora Universitária da UFPE.
Barr, A. G., T.A. Black, E.H. Hogg, N. Kljun, K. Morgenstern, and Nesic, Z. 2004. Inter-annual variability in the leaf area index of a boreal aspen-hazelnut forest in relation to net ecosystem production. *Agricultural and Forest Meteorology* 126(3–4):237–255.
Barrance, A., K. Schreckenberg, and J. Gordon. 2009. *Conservación mediante el uso: Lecciones aprendidas en el bosque seco tropical mesoamericano*. Londres: Overseas Development Institute. 141 p.

Baruah, U. and P. S. Ramakrishnan. 1989. Leaf dynamics of early versus late successional shrubs of sub-tropical moist forests of north-eastern India. *Proceedings of the Indian Academy of Science* 99:431–436.

Baskin, C.C. and J.M. Baskin. 1998. Seeds: Ecology, biogeography and Evolution of dormancy and germination. San Diego: Academic Press.

Batalha, M.A., S. Aragaki and W. Mantovani. 1997. Variações fenológicas das espécies do cerrado em Emas- (Pirassununga, SP). *Acta Botanica Brasilica* 11:61–78.

Batalha, M.A. and W. Mantovani. 2000. Reproductive phenological patterns of cerrado plant species at the Pé-de-Gigante Reserve (Santa Rita do Passa Quatro, SP, Brazil): A comparison between the herbaceous and woody floras. *Revista Brasileira de Biologia* 60:129–145.

Batchelet, E. 1981. *Circular statistics in biology*. London: Academic Press.

Bawa, K.S. 1974. Breeding systems of tree species of a lowland tropical community. *Evolution* 28:85–92.

Bawa, K.S. 1990. Plant-pollinator interactions in tropical rain forests. *Annual Review of Ecology and Systematics* 21:399–422.

Bawa, K.S., H. Kang, and M.H. Grayum. 2003. Relationships among time, frequency, and duration of flowering in tropical rain forest trees. *American Journal of Botany* 90:877–887.

Bawa, K.S., and S. Dayanandan. 1998. Global climate change and tropical forest genetic resources. *Climatic Change* 39:473–485.

Bazzaz, F.A. 1979. The physiological ecology of plant succession. *Annual Review of Ecology and Systematics* 10:351–371.

Bazzaz, F.A. and S.T.A. Pickett. 1980. Physiological ecology of tropical succession: A comparative review. *Annual Review of Ecology and Systematics* 11: 287–310.

Becknell, J.M., L. Kissing, L.K. Kucek, J.S. Powers. 2012. Aboveground biomass in mature and secondary seasonally dry tropical forests: A literature review and global synthesis. *Forest Ecology and Management* 276:88–95.

Begon, M., J.L. Harper, and C.R. Townsend. 2006. *Ecology: Individuals, populations and communities*. Oxford: Blackwell.

Begossi, A., N. Hanazaki, and J.Y. Tamashiro. 2002. Medicinal plants in the Atlantic forest (Brazil): Knowledge, use, and conservation. *Human Ecology* 30:281–299.

Behera, S.K., P. Srivastava, U.V. Pathre, and R. Tuli. 2010. An indirect method of estimating leaf area index in Jatropha curcas L. using LAI-2000 Plant Canopy Analyzer. *Agricultural and Forest Meteorology* 150(2):307–311.

Bell, S.B., E.D. McCoy, and H.R. Mushinsky. 1991. *Habitat structure, the physical arrangement of objects in space*. London: Chapman & Hall.

Bennett, K.A., R.H. Osborne, and R.J. Miller, 1975. Biocultural ecology. *Annual Review of Anthropology* 4:163–181.

Berkes, F. 2008. *Sacred Ecology: Traditional knowledge and resource management*. USA. Taylor & Francis.

Berkes, F., and C. Folke. 2000. Linking social systems and sustainability. In *Linking social and ecological systems: Mangement practices and social mechanisms for building resilience*, eds. F. Berkes and C. Folke.Cambridge: Cambridge University Press.

Bernard, E., A.L.K.M. Albernaz and W.E. Magnusson. 2001. Bat species composition in three localities in the Amazon Basin. *Studies on Neotropical Fauna and Environments* 36:177–184.

Bernard, H.R. 2005. Research methods in anthropology: Qualitative and quantitative approaches. California: Sage.

Bernate, J.F. and Fernández, F. 2002. Evaluación de la diversidad ecológica de las zonas áridas y semiáridas del área de influencia de la ecorregión estratégica de la Tatacoa en el departamento del Tolima [Evaluation of the ecological diversity of the arid and semi-arid areas of influence of strategic ecoregion of Tatacoa in the department of Tolima]. BSc Forestry Engineering thesis. Universidad del Tolima.

Bertsch C., and G. Barreto. 2008. Diet of the Yellow-knobbed curassow in the central Venezuelan Llanos. *The Wilson Journal of Ornithology* 120(4):767–777.

Bever, J.D., P.A. Schultz, A. Pringle, and J.B. Morton. 2001. Arbuscular mycorrhizal fungi: More diverse than meets the eye, and the ecological tale of why. *Bioscience* 51:923.

Bianconi, G.V., S.B. Mikich, S.D. Teixeira, and B.H.L.N.S. Maia. 2007. Attraction of fruit-eating bats with essential oils of fruits: A potential tool for forest restoration. *Biotropica* 39:136–140.

Biernacki, P. and Waldorf, D. 1981. Snowball sampling: Problems and techniques of chain referral sampling. *Sociological Methods Research* 2:141–163.

Bierregard R., T. Lovejoy, V. Kapos, A. Dos Santos, and R. Hutchings. 1992. The biological dynamics of tropical rain forest fragments. *Bioscience* 42(11):859–866.

Billings, W.D., and R.J. Morris. 1951. Reflection of visible and infrared radiation from leaves of different ecological groups. *American Journal of Botany* 38(5):327–331.

Birky, A.K. 2001. NDVI and a simple model of deciduous forest seasonal dynamics. *Ecological Modelling* 143(1):43–58.

Blackwell, M. 2011. The Fungi: 1,2,3…5.1 million species? *American Journal of Botany* 98(3):426–438.

Blaikie, P., K. Brown, M. Stocking, L. Tanga, P. Dixon, and P. Sillitoe. 1997 Knowledge in action: Local knowledge as a development resource and barriers to its incorporation in natural resource research and development. *Agricultural Systems* 55:217–237.

Blake, J.G. and B.A. Loiselle. 2001. Bird assemblages in second-growth and old-growth forests, Costa Rica: Perspectives from mist nets and point counts. *The Auk* 118:304–326.

Blake, J.G. and W.G. Hoppes. 1986. Influence of resource abundance on use of tree-fall gaps by birds in an isolated woodlot. *Auk* 103:328–340.

Blasco, F., T.C. Whitmore, and C. Gers. 2000. A framework for the worldwide comparison of tropical woody vegetation types. *Biological Conservation* 95:175–189.

Bobbink, R., K. Hicks, J. Galloway, T. Spranger, R. Alkemade, M. Ashmore, M. Bustamante, et al. 2010. Global assessment of nitrogen deposition effects on terrestrial plant diversity: A synthesis. *Ecological Applications* 20:30–59.

Bocchese, R.A., A.K.M. Oliveira, and E.C. Vicente. 2007. Taxa e velocidade de germinação de sementes de Cecropia pachystachya Trécul (Cecropiaceae) ingeridas por Artibeus lituratus (Olfers, 1818) (Chiroptera: Phyllostomidae). *Acta Scientiarum Biological Sciences* 29:395–399.

Boege-Pare, K., C. Castillo-Alvares, A. Garcia Aguayo, J.H. Vega-Rivera, A. Miranda-Garcia, A. Ruiz-Sanchez, and R. Rueda-Hernandez. 2010. Dictamen técnico de la manifestación de impacto ambiental del Project de desarrollo turístico Zafiro (Clave: 14JA2009T0017): Identificación de posibles impactos a las áreas naturales protegidas de la región. Universidad Nacional Autonoma de Mexico and Fundacion Ecologica de Cuixmala, A.C.

Bohlman, S.A. 2010. Landscape patterns and environmental controls of deciduousness in forests of central Panama. *Global Ecology and Biogeogaphy* 19:376–385.

Borchert, R. 1994a. Induction of rehydration and bud break by irrigation or rain in deciduous trees of a tropical dry forest in Costa Rica. *Trees* 8:198–204.

Borchert, R. 1994b. Soil and stem water storage determine phenology and distribution of tropical dry forest trees. *Ecology* 75:1437–1449.

Borchert, R. 1996. Phenology and flowering periodicity of Neotropical dry forest species: Evidence from herbarium collections. *Journal of Tropical Ecology* 12:65–80.

Borchert, R. and G. Rivera. 2001. Photoperiodic control of seasonal development and dormancy in tropical stem-succulent trees. *Tree Physiol* 21:213–221.

Borchert, R., G. Rivera, and W. Hagnauer. 2002. Modification of vegetative phenology in a tropical semi-deciduous forest by abnormal drought and rain. *Biotropica* 34:27–39.

Borghetti, F. and A.G. Ferreira. 2004. Interpretação de resultados de germinação. In *Germinação: Do básico ao aplicado*, eds. A.G. Ferreira and F. Borghetti, 209–222. Porto Alegre: Artmed.

Bourdieu, P. 2007. *O poder simbólico*. Rio de Janeiro: Bertrand Brazil.

Boza, M. 1993. Conservation in action: Past, present, and future of the national parks system in Costa Rica. *Conservation Biology* 7(2):239–247.

Braga, F.C., Wagner, F.C., Wagner, H., Lombardi, J.A. and De Oliveira A.B. 2000. Screening Brazilian plant species for in vitro inhibition of 5-lipoxygenase. *Phytomedicin* 6:447–452.

Brandão, M. 2000. Caatinga. In *Lista vermelha das espécies ameaçadas de extinção da flora de Minas Gerais*, eds. M.P. Mendonça and L.V. Lins, 75–85. Belo Horizonte: Fundação Biodiversitas e Fundação Zôo-Botânica de Belo Horizonte.

Brandão, M., M.L. Gavinales, and J.P. Laca-Buendia. 1996. Plantas medicamentosas de uso popular dos Campos Rupestres de Minas Gerais. *Daphne* 6:7–9.

Brazil, República Federativa do. 2007. Política Nacional de Desenvolvimento Sustentável dos Povos e Comunidades Tradicionais. Decreto 6.040, Brasília.

Bridgewater, P.B. 2002. Biosphere reserves: Special places for people and nature. *Environmental Science & Policy* 5:9–12.

Britton, N.L. and J.N. Rose. 1963. *The Cactaceae*. New York: General Publishing Company.

Brodribb, T.J. and N.M. Holbrook. 2004. Leaf physiology does not predict leaf habit; examples from tropical dry forest. *Trees* 19:290–95.

Brooks, S., S. Cordell, and L. Perry, 2009. Broadcast seeding as a potential tool to reestablish native species in degraded dry forest ecosystems in Hawaii. *Ecological Restoration* 27(3):300–305.

Brown, K.S. and A. Ab`Sáber. 1979. Ice-age refuges and evolution in the Neotropics: Correlation of paleoclimatological, geomorphological and pedological data with modern biological endemism. *Paleoclimas* 5:1–30.

Brown, S. and A.E. Lugo. 1990. Tropical secondary forests. *Journal of Tropical Ecology* 6:l–32.

Bruijnzeel, L.A. 2004. Hydrological functions of tropical forests: Not seeing the soil for the trees? *Agriculture, Ecosystems & Environment* 104(1):185–228.

Bruno, R.L.A., E.U. Alves, A.P. Oliveira, and R.C. Paula. 2001. Tratamentos pré-germinativos para superar a dormência de sementes de Mimosa caesalpiniaefolia Benth. *Revista Brasileira de Sementes* 23:136–143.

Buckman, H.O. and N.C. Brady, 1976. *Natureza e propriedade dos solos*. 4th ed. Rio de Janeiro: Freitas Bastos.

Buermann, W., S. Saatchi, B.R.Zutta, J. Chaves, B. Mila C.H. Graham, and T.B. Smith. 2008. Application of remote sensing data in predictive models of species' distribution. *Journal of Biogeography* 35:1160–1176. doi:10.1111/j.1365–2699.2007.01858.x

Bullock, S.H. 1986. Climate of Chamela, Jalisco, and trends in the south coastal region of Mexico. *Meteorology and Atmospheric Physics* 36:297–316.

Bullock, S.H. 1995. Plant reproduction in neotropical dry forest. In *Seasonally dry tropical forests*, eds. S.H. Bullock, H.A. Mooney, and E. Medina, 277–303. Cambridge, New York: Cambridge University Press.

Bullock, S.H., H.A. Mooney, and E. Medina (eds.). 1995. *Seasonally dry tropical forests*. Cambridge: Cambridge University Press. 450 p.

Bullock, S.H. and J.A. Solis-Magallanes. 1990. Phenology of canopy trees of a tropical deciduous forest in Mexico. *Biotropica* 22:22–35.

Burnham, K.P. and W.S. Overton. 1979. Robust estimation of population size when capture probabilities vary among animals. *Ecology* 60:927–936.

Burnham, R.J. 1997. Stand characteristics and leaf litter composition of a dry forest hectare in Santa Rosa National Park, Costa Rica. *Biotropica* 29:387–395.

Burquez, A. and A. Martínez-Yrízar. 2010. Límites geográficos entre las selvas bajas caducifolias y los matorrales espinosos y xerófilos: ¿qué conservar? In *Diversidad, amenazas y áreas prioritarias para la conservación de las selvas secas del Pacífico de México*, eds. G. Carlquist, S. 1975. *Ecological Strategies of Xylem Evolution*. Berkeley: University of California Press.

Bye, R. 2002. Etnobotánica en la región de Chamela, Jalisco, México. In *Historia natural de Chamela*, eds. F.A. Noguera, J.H. Vega, A. N. García-Aldrete and M. Quesada, 545–559. Mexico DF: Instituto de Biología, Universidad Nacional Autónoma de México.

Cabrera, E. and Galindo, G.A. 2006. *Methodological approach for the delimitationdry enclaves ecosystem: Pilot Case, river canyons and river Dagua, Tulua, Valle del Cauca, Colombia* [Methodological approach for the delimitationdry enclaves ecosystem: Pilot Case, river canyons and river Dagua, Tulua, Valle del Cauca. Colombia]. Bogotá, DC: Instituto de Investigación de Recursos Biológicos Alexander von Humboldt. 65 p.

Cabrera, R. 2007. *Tierra y Ganadería en Guanacaste*. Cartago: Editorial Tecnológica de Costa Rica.

Caetano, S., D. Prado, R.T. Pennington, S. Beck, A. Oliveira, R. Spichiger, and Y. Naciri. 2008. The history of Seasonally Dry Tropical Forests in eastern South America: inferences from the generic structure of the tree Astronium urundeuva (Anacardiaceae). *Molecular Ecology* 17:3147–3159.

Calder, I.R., I.R. Wright, and D. Murdiyarso. 1986. A study of evaporation from tropical rain forest—West Java. *Journal of Hydrology* 89(1–2):13–31.

Calle, Z., B. Schlumpberger, L. Piedrahita, A. Leftin, S. Hammer, A. Tye, and R. Borchert. 2010. Seasonal variation in daily insolation induces synchronous bud break and flowering in the tropics. *Trees* 24:865–877.

Calvo-Alvarado, J., A. Sánchez-Azofeifa, and C. Portillo-Quintero. 2012a. Neotropical seasonally dry forests. In *Encyclopedia of biodiversity*. San Diego, CA: Academic Press (In press).

Calvo-Alvarado, J., B. McLennan, A. Sánchez-Azofeifa, and T. Garvin. 2009a. Deforestation and forest restoration in Guanacaste, Costa Rica: Putting conservation policies in context. *Forest Ecology and Management* 258(6):931–940.

Calvo-Alvarado, J., C. Jiménez-Rodríguez, D. Carvajal-Vanegas, and D. Arias-Aguilar. 2009b. Rainfall Interception in Tropical Forest Ecosystems: Tree Plantations and Secondary Forest. In 2009 Virginia Water Research Conference. Presented at the Water Resources in Changing Climates, Virginia US: Virginia Tech, 74–83.

Calvo-Alvarado, J.C., A. Sánchez-Azofeifa, and M. Kalácska. 2008. Deforestation and Restoration of tropical dry forest: The case of Chorotega Region-Costa Rica. In *Applying ecological knowledge to landuse decisions. (SCOPE) Scientific Committee on Problems of the Environment*, eds. H. Tiessen and J. Stewart, 123–133. IAI, the Inter-American Institute for Global Change Research, and IICA, the Inter-American Institute for Cooperation on Agriculture.

Calvo-Alvarado, J.C., C. Jiménez-Rodríguez, and M. de Saá-Quintana 2012b. Intercepción de precipitación en tres estadios de sucesión de un Bosque húmedo Tropical, Parque Nacional Guanacaste, Costa Rica. *Revista Forestal Mesoamericana Kurú* (Costa Rica), 9 (22):1–9.

Campbell, C.J., F. Aureli, C.A. Chapman, G. Ramos-Fernandez, K. Matthews, S.E. Russo, A. Suarez, and L. Vick. 2005. Terrestrial behavior of Ateles spp. *International Journal of Primatology* 26:1039–1061.

Campo, J. and C. Vázquez-Yanes. 2004. Effects of nutrient limitation on aboveground carbon dynamics during tropical dry forest regeneration in Yucatán, Mexico. *Ecosystems* 7:311–319.

Campo, J., E. Solís, and M.G. Valencia. 2007. Litter N and P dynamics in two secondary tropical dry forests after relaxation of nutrient availability constraints. *Forest Ecology and Management* 252:33–40.

Campo, J., M. Maass, V.J. Jaramillo, A. Martinez-Yrizar, and J. Sarukhan. 2001. Phosphorus cycling in a Mexican tropical dry forest ecosystem. *Biogeochemistry* 53:161–179.

Campo, J., V.J. Jaramillo, and J.M. Maass. 1998. Pulses of soil phosphorus availability in a Mexican tropical dry forest: Effects of seasonality and level of wetting. *Oecologia* 115:167–172.

Canavire, G. and M. Hanauer. 2012. Estimating the Impacts of Bolivia's Protected Areas on Poverty. Discussion Paper Series. IZA

Cardoso da Silva, J.M., C. Uhl, and G. Murray. 1996. Plant succession, landscape management, and the ecology of frugivorous birds in abandoned Amazonian pastures. *Conservation Biology* 10:491–503.

Carlyle-Moses, D. and J.C. Gash. 2011. Rainfall interception loss by forest canopies. In *Forest Hydrology and Biogeochemistry*, eds. D.F. Levia, D. Carlyle-Moses, and T. Tanaka, 407–423. Netherlands: Springer.

Carneiro, E.J. 2005. Política ambiental e a ideologia do desenvolvimento sustentável. In *A insustentável leveza da política ambiental: Desenvolvimento e conflitos sócio-ambientais*, eds. A. Zhouri, C. Laschefski and D. Pereira, 27–48. Belo Horizonte: Autêntica.

Carneiro, J.G.A. and I.B. Aguiar. 1993. Armazenamento de sementes. In *Sementes florestais tropicais*, eds. I.B. Aguiar, F.C.M. Piña-Rodrigues and M.B. Figliolia, 333–350. Brasília: Abrates.

Carpenter, G. 1997. ART neural networks for remote sensing: Vegetation classification from Landsat TM and terrain data. *IEEE Transactions on Geoscience and Remote Sensing* 35(2):308–325.

Carrenho, R., S.F.B. Trufem, and V.L.R. Bononi. 2001. Fungos micorrízicos arbusculares em rizosferas de tres espécies de fitobiontes instaladas em área de mata ciliar revegetada. *Acta Botanica Brasilica* 15(1):115–124.

Carrillo, E.J., R.A. Morera, G.R. Wong. 1994. Depredación de tortuga lora (Lepidochelys olivacea) y de tortuga vrde (Chelonia mydas) por el jaguar (Panthera onca). *Vida Silvestre Neotropical* 3:48–49.

Carter J.M., W.K. Gardner, and A.H. Gibson. 1994. Improved growth and yield of Faba beans (Vicia faba cv Fiord) by inoculation with strains of Rhizobium-leguminosarum biovar viciae in acid soils in south-west Victoria. *Australian Journal of Agricultural Research* 45:613–623.

Carvajal-Vanegas, D. and J.C. Calvo-Alvarado. 2012. Intercepción de precipitación en dos especies forestales nativas: Vochysia guatemalensis Donn. Sm. y Vochysia ferruginea Mart. *Revista Forestal Mesoamericana Kurú* (Costa Rica) 9(22):32–39.

Carvalho, I.C.M. 2001. *A invenção ecológica*. Porto Alegre: Editora da UFRGS.

Carvalho, N.M. and J. Nakagawa. 2000. *Sementes: Ciência, tecnologia e produção*. Campinas: Fundação Cargill.

Carvalho, P.E.R. 1994. *Espécies florestais brasileiras: Recomendações silviculturais, potencialidades e uso da madeira*. Brasília: EMBRAPA/CNPF.

Casas A., A. Valiente-Banuet, J.L. Viveros, P. Dávila, R. Lira, J. Caballero, L. Cortés, I. Rodríguez. 2001. Plant resources of the Tehuacán Valley, Mexico. *Economic Botany* 55:129–166.

Cascante, A., M. Quesada, J.A. Lobo, and E.J. Fuchs. 2002. Effects of dry tropical forest fragmentation on the reproductive success and genetic structure of the tree, Samanea saman. *Conservation Biology* 16:137–147.

Castellanos, A.E. 1991. Photosynthesis and gas exchange of vines. In *The biology of vines*, eds. F.E. Putz and H.A. Mooney, 181–202. Cambridge: Cambridge University Press.

Castillo, A., C. Godinez, N. Schroeder, C. Galicia, A. Pujadas-Botey, and L. Martínez. 2009. Los bosques tropicales secos en riesgo: Conflictos entre el desarrollo turístico, el uso agropecuario y la provisión de servicios ecosistémicos en la costa de Jalisco, México. *Interciencia* 34:844–850.

Castillo, A., M.A. Magaña, A. Pujadas, L. Martínez, and C. Godinez. 2005. Understanding rural people interaction with ecosystems: A case study in a tropical dry forest of Mexico. *Ecosystems* 8:630–643.

Castillo, C.C.M. 1991. El proyecto de colonización de la costa de Jalisco. Primera etapa, 1944-1947. *Estudios Sociales* 11:86–115.

Castillo, M., B. Rivard, A. Sánchez-Azofeifa, J. Calvo-Alvarado, and R. Dubayah. 2012. LIDAR remote sensing for secondary tropical dry forest identification. *Remote Sensing of Environment* 121:132–143.

Castillo-Núñez, M., G.A. Sánchez-Azofeifa, A. Croitoru, B. Rivard, J. Calvo-Alvarado, and R.O. Dubayah. 2011. Delineation of secondary succession mechanisms for tropical dry forests using LiDAR. *Remote Sensing of Environment* 115(9):2217–2231.

Castrillo Marín, N. 2009: El cambio climático en Argentina. http://www.ambiente.gov.ar/archivos/web/UCC/File/09ccargentina.pdf

Castro-Esau, K., G. Sánchez-Azofeifa, and B. Rivard. 2003. Monitoring secondary tropical forests using space-borne data: Implications for Central America. *International Journal of Remote Sensing* 24(9):1853–1894.

Castro-Esau, K.L., G.A. Sánchez-Azofeifa, B. Rivard, S.J. Wright, and M. Quesada. 2006. Variability in leaf optical properties of Mesoamerican trees and the potential for species classification 1. *American Journal of Botany* 93(4):517–530.

Castro-Esau, K.L., G.A. Sánchez-Azofeifa, and T. Caelli. 2004. Discrimination of lianas and trees with leaf-level hyperspectral data. *Remote Sensing of Environment* 90:353–372.

Castro-Luna, A.A., V.J. Sosa, and G. Castillo-Campos. 2007. Quantifying phyllostomid bats at different taxonomic levels as ecological indicators in a disturbed tropical forest. *Acta Chiropterologica* 9:219–228.

Cavalcanti, N.B., G.M. Resende, and M.A. Drumond. 2006. Período de dormência de sementes de imbuzeiro. *Revista Caatinga* 19:135–139.

Cavelier, J., A. Ruiz, M. Santos, M. Quiñones, and P. Soriano. 1996. El proceso de degradación y sabanización del Valle Alto del Magdalena [The process of degradation and savannah of the Alto Magdalena Valley]. Unedited report, Fundación del Alto Magdalena. Bogotá.

Cavelier, J. and G. Vargas. 2002. Procesos hidrológicos. In *Ecología y Conservación de Bosques Neotropicales*, eds. M.R. Guariguata, G.H. Kattan, 145–165. Cartago, Costa Rica: Libro Universitario Regional.

Ceballos, G. 1995. Vertebrate diversity, ecology, and conservation in Neotropical deciduous forests. In *Seasonally dry tropical forests*, eds. S.H. Bullock, H.A. Mooney and E. Medina, 195–220. New York: Cambridge University Press.

Ceccon, E., I. Omstead, C. Vázquez-Yanes, and J. Campo-Alves. 2002. Vegetation and soil properties in two tropical dry forests of differing regeneration status in Yucatán. *Agrociencia* 36:621–631.

Ceccon, E., P. Huante, and J. Campo. 2003. Effects of nitrogen and phosphorus fertilization on the survival and recruitment of seedlings of dominant tree species in two abandoned tropical dry forests in Yucatán, Mexico. *Forest Ecology and Management* 182:387–402.

Chambers, J.Q., G.P. Asner, D.C. Morton, L.O. Anderson, S.S. Saatchi, F.D.B. Espirito-Santo, M. Palace, and C. Souza Jr. 2007. Regional ecosystem structure and function: Ecological insights from remote sensing of tropical forests. *Trends in Ecology & Evolution* 22(8):414–423.

Champlin, T.B., J.C. Kilgo, and C.E. Moorman. 2009. Food abundance does not determine bird use of early-successional habitat. *Ecology* 90:1586–1594.

Chapin, F.S. 1989. The cost of tundra plant structures: Evaluation of concepts and currencies. *American Naturalist* 133:1–19.

Chapotin, S.M., N.M. Holbrook, S.R. Morse and M.V. Gutierrez. 2003. Water relations of tropical dry forest flowers: Pathways for water entry and the role of extracellular polysaccharides. *Plant Cell Environment* 26:623–630.

Charles-Dominique, P. 1986. Inter-relations between frugivorous vertebrates and pioneer plants: Cecropia, birds and bats in French Guyana. In *Frugivores and seed dispersal*, eds. A. Estrada and T.H. Fleming, 119–135. Dordrecht: W. Junk Publishers.

Chason, J.W., D.D. Baldocchi, and M.A. Huston. 1991. A comparison of direct and indirect methods for estimating forest canopy leaf area. *Agricultural and Forest Meteorology* 57(1):107–128.

Chaturvedi, R.K., A.S. Raghubanshi, and J.S. Singh. 2011. Plant functional traits with particular reference to tropical deciduous forests: A review. *Journal of Biosciences* 36:1–19.

Chave, J., D. Coomes, S. Jansen, S.L. Lewis, N.G. Swenson, A.E. Zanne. 2009. Towards a worldwide wood economics spectrum. *Ecology Letters* 12:351–66.

Chave, J., D. Navarrete, S. Almeida, E. Álvarez, L.E.O.C. Aragão, J.S. Espejo, P. von Hildebrand, et al. 2009. Regional and temporal patterns of litterfall in tropical South America. *Biogeosciences Discuss* 6:7565–7597.

Chaves, O. and G. Avalos. 2008. Do seasonal changes in light availability influence the inverse lefing phenology of the neotropical dry forest understory shrub Bonellia nervosa? *Revista de Biología Tropical* 56(1):257–268.

Chazdon, R. and N. Fetcher. 1984. Photosynthetic light environments in a lowland tropical rainforest in Costa Rica. *Journal of Ecology* 72:553–564.

Chazdon, R., S. Letcher, M. van Breugel, M. Martinez-Ramos, F. Bongers, and B. Finegan. 2007. Rates of change in tree communities of secondary Neotropical forests following major disturbances. *Philosophical transactions of the Royal Society of London. Series B, Biological Sciences* 362:273–289.

Cheke, A.S., W. Nanakorn, and C. Yankoses. 1979. Dormancy and dispersal of seeds of secondary forest species under the canopy of a primary Tropical Rain Forest in Northern Thailand. *Biotropica* 11:88–95.

Chen, J.M. 1996. Evaluation of vegetation indices and a modified simple ratio for boreal applications. *Canadian Journal of Remote Sensing* 22(3):229–242.

Chen, J.M., G. Pavlic, L. Brown, J. Cihlar, S.G. Leblanc, H.P. White, and R.J. Hall, et al. 2002. Derivation and validation of canada-wide coarse-resolution leaf area index maps using high-resolution satellite imagery and ground measurements. *Remote Sensing of Environment* 80(1): 165–184.

Chen, J.M. and J. Cihlar. 1996. Retrieving leaf area index of boreal conifer forests using Landsat TM images. *Remote Sensing of Environment* 55(2):153–162.

Chen, J.M., P.D. Blanken, T.A. Black, M. Guilbeault, and S. Chen. 1997. Radiation regime and canopy architecture in a boreal aspen forest. *Agricultural and Forest Meteorology* 86:107–125.

Chen, J.M., T.A. Black, and R.S. Adams. 1991. Evaluation of hemispherical photography for determining plant area index and geometry of a forest stand. *Agricultural and Forest Meteorology* 56:129–143.

Chen, Y., J.T. Randerson, G.R. Van Der Werf, D.C. Morton, M.Q. Mu, and P.S. Kasibhatla. 2010. Nitrogen deposition in tropical forests from savanna and deforestation fires. *Global Change Biology* 16:2024–2038.

Chong, M.M., K. Dutchak, J. Gamon, Y. Huang, M. Kalácska, D. Lawrence, C. Portillo, J.P. Rodríguezy, A. SánchezAzofeifa. 2008. Remote Sensing. In *Manual of Methods: Human, ecological and biophysical dimensions of tropical dry forests*, eds. J.M. Nassar, J.P. Rodríguez, A. Sánchez-Azofeifa, T. Garvin, M. Quesad, 47–80. Caracas, Venezuela: Ediciones IVIC, Instituto Venezolano de Investigaciones Científicas (IVIC).

Clark, D.A. and Clark, D.B. 1992. Life history diversity of canopy and emergent trees in a neotropical rain forest. *Ecological Monographs* 62:315–344.

Clark, D.B., P.C. Olivas, S.F. Oberbauer, D.A. Clark, and M.G. Ryan, 2008. First direct landscape-scale measurement of tropical rain forest Leaf Area Index, a key driver of global primary productivity. *Ecology Letters* 11(2):163–172.

Clark, M.L., D.A. Roberts, and D.B. Clark. 2005. Hyperspectral discrimination of tropical rain forest tree species at leaf to crown scales. *Remote Sensing of Environment* 96(3–4):375–398.

Cleveland, C.C., A.R. Townsend, P. Taylor, S. Alvarez-Clare, M.M.C. Bustamante, G. Chuyong, S.Z. Dobrowski, et al. 2011. Relationships among net primary productivity, nutrients and climate in tropical rain forest: A pan-tropical analysis. *Ecology Letters* 14:939–947.

Cochrane, M.A. 2000. Using vegetation reflectance variability for species level classification of hyperspectral data. *International Journal of Remote Sensing* 21(10):2075–2087.

Cocks, P.M. 2006. Biocultural diversity: Moving beyond the realm of 'indigenous' and 'local' people. *Human Ecology* 34:185–200.

Coelho, M.S., E.D. Almada, A.V. Quintino, G.W. Fernandes, M.M.D. Espirito-Santo, and G.A. Sanchez-Azofeifa. 2013. Phenology and reproductive patterns of a tropical dry tropical forest on limestone outcrops at different successional stages in the Espinhaço Mountains, southeastern Brazil (submitted).

Coelho, M.S., E.D. Almada, A.V. Quintino, G.W. Fernandes, R.M.S. Santos, G.A. Sánchez-Azofeifa, and M.M.D. Espirito-Santo. 2012. Floristic composition and structure of a seasonally dry tropical forest at different successional stages in the Espinhaco Mountains, southeastern Brazil. *Interciencia* 37:190–196.

Coelho, M.S., E.D. Almada, G.W. Fernandes, M.A.C. Carneiro, R.M. Santos, A.V. Quintino, and A. Sanchez-Azofeifa. 2009. Gall inducing arthropods from a seasonally dry tropical forest in Serra do Cipó, Brazil. *Revista Brasileira de Entomologia* 53:404–414.

Cohen, W.B. and S.N. Goward. 2004. Landsat's role in ecological applications of remote sensing. *Bioscience* 54(6):535–545.

Cohen, W.B., T.K. Maiersperger, Z. Yang, S.T. Gower, D.P. Turner, W.D. Ritts, M. Berterretche, and S.W. Running. 2003. Comparisons of land cover and LAI estimates derived from ETM+ and MODIS for four sites in North America: A quality assessment of 2000/2001 provisional MODIS products. *Remote Sensing of Environment* 88(3):233–255.

Colon, S.M. and A.E. Lugo. 2006. Recovery of a subtropical dry forest after abandonment of different land uses1. *Biotropica* 38(3):354–364.

Colwell, R.K. 2005. EstimateS: Statistical estimation of species richness and shared species from samples. Version 7.5.

Colwell, R.K. and J.A. Coddington. 1994. Estimating terrestrial biodiversity through extrapolation. *Philosophical Transactions: Biological Sciences* 345:101–118.

Conde, D.A. 2008. Road impact on deforestation and jaguar habitat loss in the Mayan Forest. Doctor of Philosophy, Duke University

Conde, D.A. and A. Pfaff. 2008. Sequenced road investments and clearing of the Mayan Forest. Presentation (from Conde's PhD thesis) at North Carolina State University.

Condit, R., P.S. Ashton, N. Manokaran, J.V. LaFrankie, S.P. Hubbell, R.B. Foster. 1999. Dynamics of the forest communities at Pasoh and Barro Colorado: Comparing two 50-ha plots. *Philosophical Transactions of the Royal Society of London. Series B, Biological Sciences* 354:1739–1748.

Connell, J.H. 1978. Diversity in tropical rain forests and coral reefs. *Science* 199:1302–1310.

Connell, J.H., J.G. Tracey, and L.J. Webb. 1984. Compensatory recruitment, growth, and mortality as factors maintaining rain forest tree diversity. *Ecological Monographs* 54:141–164.

Connell, J.H. and R.O. Slatyer. 1977. Mechanisms of succession in natural communities and their role in community stability and organization. *American Naturalist* 111:1119–1144.

Corby, H.D.L. 1988. Types of rhizobial nodules and their distribution among the Leguminosae. *Kirkia* 13:53–123.

Córcega Pita, E. and Silva Escobar, O. 2011. Evaluación de la intercepción de la lluvia en plantaciones de cacao (Theobroma cacao L.), bosque tropical semideciduo y conuco en laderas de montaña. *Revista de la Facultad de Agronomía* (UCV) 37(2):47–54.

Corlett, R.T. 1995. Tropical secondary forests. *Progress in Physical Geography* 19:159–172.

Corlett, R.T. and R.B. Primack. 2011. *Tropical rain forests: An ecological and biogeographical comparison*. 2nd ed. West Sussex: Wiley-Blackwell. 336 p.

Cornwell, W.K., J.H.C. Cornelissen, K. Amatangelo, E. Dorrepaal, V.T. Eviner, O. Godoy, S.E. Hobbie, et al. 2008. Plant species traits are the predominant control on litter decomposition rates within biomes worldwide. *Ecology Letters* 11:1065–1071.

Costa-Filho, A. 2008. Os Gurutubanos: Territorialização, produção e sociabilidade em um quilombo do centro norte-mineiro. PhD diss., Universidade de Brasília, Brasília.

Costa, J. 2005. Cultura, Natureza e Populações Tradicionais: O norte de Minas como síntese da nação brasileira. *Revista Verde Grande* 1:8–45.

Costa, J.B.A. 1998. Do tempo da fartura dos crioulos ao tempo da penúria dos morenos: A identidade através de um rito em Brejo dos Crioulos. MSc. thesis, Universidade de Brasília, Brasília.

Costa, J.B.A. 2001. Brejo dos Crioulos e a sociedade negra da Jaíba: Novas categorias sociais e a visibilização do invisível na sociedade brasileira. *Revista brasileira de pósgraduação em Ciências Sociais* 5:99–122.

Costa, N.P., R.L.A. Bruno, F.X. Souza, and E.D.P.A. Lima. 2001. Efeito do estádio de maturação do fruto e do tempo de pré-embebição de endocarpos na germinação de sementes de umbuzeiro (Spondias tuberosa Arr. Cam.). *Revista Brasileira de Fruticultura* 23:738–741.

Costanza, R. 1992. Toward an operational definition of ecosystem health. In *Ecosystem health: New goals for environmental management*, eds. R. Costanza, B.G. Norton and B.D. Haskell, 239–256. Washington: Island Press.

Covacevich, F. and Echeverría, H.E. 2009. Mycorrhizal occurrence and responsiveness in tall fescue and wheatgrass are affected by the source of phosphorus fertilizer and fungal inoculation. *Journal of Plant Interactions* 4:101–112.

Covacevich, F., H.E. Echeverria, and L.A.N. Aguirrezabal. 2007. Soil available phosphorus status determines indigenous mycorrhizal colonization into field and glasshouse-grown spring wheat in Argentina. *Applied Soil Ecology* 35:1–9.

Covacevich, F., M.A. Marino, and H.E. Echeverria. 2006. The phosphorus source determines the arbuscular mycorrhizal potential and the native mycorrhizal colonization of tall fescue and wheatgrass in a moderately acidic Argentinean soil. *European Journal of Soil Biology* 42:127–138.

Crawley, M. 2002.*Statistical computing: An introduction to data analysis using S-Plus*. London: Wiley.

Creswell, J.W. 2003. *Research design. Qualitative, quantitative, and mixed methods approaches*. Thousand Oaks: Sage Publications.

Crockford, R.H. and Richardson, D.P. 2000. Partitioning of rainfall into throughfall, stemflow and interception: Effect of forest type, ground cover and climate. *Hydrological Processes*, 14(16–17):2903–2920.

Cuartas, L.A., J. Tomasella, A.D. Nobre, M.G. Hodnett, M.J. Waterloo, and J.C. Múnera. 2007. Interception water-partitioning dynamics for a pristine rainforest in Central Amazonia: Marked differences between normal and dry years. *Agricultural and Forest Meteorology* 145(1–2):69–83.

Cunha, M.C. and M.W.B. Almeida. 2002. Populações tradicionais e conservação ambiental. In *Biodiversidade na Amazônia Brasileira: Avaliação e ações prioritárias para a conservação, uso sustentável e repartição de benefícios*, eds. J.P.R. Capobianco, A. Veríssimo, A. Moreira, D. Sawer, S. Ikeda and L.P. Pinto, 184–193. São Paulo: ISA/Editora Estação Liberdade.

Curran, L.M. and M. Leighton. 2000. Vertebrate responses to spatiotemporal variation in seed production of mast-fruiting dipterocarpaceae. *Ecological Monographs* 70:101–128.

Curran, P.J., J.L. Dungan, and H.L. Gholz. 1992. Seasonal LAI in slash pine estimated with Landsat TM. *Remote Sensing of Environment* 39(1):3–13.

Curtis, J.T. and R.P. McIntosh. 1951. An upland forest continnum in the prairie-forest border region of Wisconsin. *Ecology* 32:476–496.

Curtis, J.T. 1959. *The Vegetation of Wisconsin. An ordination of plant communities*, University Wisconsin press, Madison Wisconsin, 657 pp.

Curtis, P.S., P.J. Hanson, P. Bolstad, C. Barford, J.C. Randolph, H.P. Schmid, and K.B. Wilson. 2002. Biometric and eddy covariance based estimates of ecosystem carbon storage in five eastern North American deciduous forests. *Agricultural and Forest Meteorology* 113:3–19.

Dalponte, M., L. Bruzzone, and D. Gianelle. 2008. Fusion of hyperspectral and LIDAR remote sensing data for classification of complex forest areas. *Geoscience and Remote Sensing, IEEE Transactions on* 46(5):1416–1427.

Daniels, B.A. and H.D. Skipper. 1982. Methods for the recovery and quantitative estimation of propagules from soil. In *Methods and principles of mycorrhizal research*, ed. N.C. Schenck, 9–53. New York: American Phytopathological Society.

Danson, F.M. and P.J. Curran. 1993. Factors affecting the remotely sensed response of coniferous forest plantations. *Remote Sensing of Environment* 43(1):55–65.

Davidson, E.A., C.J. Reis de Carvalho, A.F. Figueira, F.Y. Ishida, J.P. Ometto, G.B. Nardoto, R.T. Sabá, et al. 2007. Recuperation of nitrogen cycling in Amazonian forests following agricultural abandonment. *Nature* 447:995–998.

Davidson, E.A., C.J. Reis de Carvalho, I.C.G. Vieira, R. de O Figueiredo, P. Moutinho, F.Y. Ishida, M.T.P. dos Santos, J.B. Guerrero, K.Kalif, and R.T. Sabá. 2004. Nitrogen and phosphorus limitation of biomass growth in a tropical secondary forest. *Ecological Applications* 14:S150–S163.

Daws M.I., C.E. Mullins, D.F.R.P. Burslem, S. Paton, and J.W. Dalling. 2002. Topographic position affects the water regime in a semi-deciduous tropical forest in Panama. *Plant Soil* 238:79–90.

Delascio, F. and M. Ramia. 2001. Notas sobre la distribución y aspectos fenológicos de la botonera: *Borreria scabisoides* Cham. & schltdl. var. *anderssonii* (standl.) Steyerm. (Rubiaceae) en el Estado Cojedes, Venezuela. *Acta Botanica Venezuelica* 24:59–61.

De Souza, F.A., S.L. Stürmer, R. Carrenho, and S.F.B. Trufem. 2010. Classificação e Taxonomia de fungos micorrízicos arbusculares e sua diversidade e ocorrência no Brazil. In *Micorrizas: 30 anos de pesquisa no Brazil*, eds. J.O. Siqueira, F.A. de Souza, E.J.B.N. Cardoso, S.M. Tsai, 716. Lavras MG: Editora UFLA.

Del Hoyo, J., A. Elliott, J. Sargatal, and D. Christie. 1992. *Handbook of the birds of the World*. Barcelona: Lynx Ediciones.

Delalieux, S., B. Somers, W.W. Verstraeten, J.A.N. van Aardt, W. Keulemans, and P. Coppin. 2009. Hyperspectral indices to diagnose leaf biotic stress of apple plants, considering leaf phenology. *International Journal of Remote Sensing* 30:1887–1912.

Delalieux, S., J. van Aardt, W. Keulemans, E. Schrevens, and P. Coppin. 2007. Detection of biotic stress (Venturia inaequalis) in apple trees using hyperspectral data: Non-parametric statistical approaches and physiological implications. *European Journal of Agronomy* 27:130–143.

Delgado, C. and A. Pfaff, 2008. Will Nearby Protected Areas Constrain Road Impacts On Deforestation? Presentation at NASA LBA conference Amazon. In Perspective, Manaus. (Based upon C. Delgado's Duke MEM project at the Nicholas School of the Environment.)

Delgado, C., D.A. Conde, J.O. Sexton, F. Colchero, J.J. Swensen, and A. Pfaff 2008. Deforestation dynamics in response to the Evolution of the Western Amazonian Inter-Oceanic Highway. Duke University (a working paper based upon C. Delgado's final MEM project).

DeLonge, M., P. D'Odorico, and D. Lawrence. 2008. Feedbacks between phosphorus deposition and canopy cover: The emergence of multiple stable states in tropical dry forests. *Global Change Biology* 14:154–160.

Demarez,V., S. Duthoit, F. Baret, M. Weiss, and G. Dedieu. 2008. Estimation of leaf area and clumping indexes of crops with hemispherical photographs. *Agricultural and Forest Meteorology* 148(4):644–655.

Diaz, S. 1995. Elevated CO_2 responsiveness, interactions at the community level and plant functional types. *Journal of Biogeography* 22:289–295.

Diaz S., J.G. Hodgson, K. Thompson, M. Cabido, J.H.C. Cornelissen, A. Jalili, G. Montserrat-Martí, et al. 2004. The plant traits that drive ecosystems: Evidence from three continents. *Journal of Vegetation Science* 15:295–304.

Díaz, M. and E. Granadillo. 2005. The significance of episodic rains for reproductive phenology and productivity of trees in semiarid regions of northwestern Venezuela. *Trees* 19:336–348.

Didham, R. and J. Lawton. 1999. Edge structure determines the magnitude of changes in microclimate and vegetation structure in tropical forest fragments. *Biotropica* 31(1):17–30.

Diegues, A.C. 2000. *Etnoconservação: Novos rumos para a proteção da natureza nos trópicos*. São Paulo: Editora Hucitec NUPAUB-USP.

Diegues, A.C. and R.S. Arruda. 2001. *Saberes tradicionais e biodiversidade no Brazil*. São Paulo: Ministério do Meio Ambiente.

Diegues, A.C.A. 2002. Etnoconservação da natureza. In *Etnoconservação: Novos rumos para a proteção da natureza nos trópicos*, ed. A.C.A. Diegues, 1–46. São Paulo: Hucitec.

Doley, D. 1981. Tropical and subtropical forests and woodlands. In *Water Deficits and Plant Growth* Vol. VI, ed. T.T. Kozlowski, 209–323. New York: Academic Press.

Doligez, A. and H. Joly. 1997. Genetic diversity and spatial structure within a natural stand of a tropical forest tree species, Carapa procera (Meliaceae), in French Guiana. *Heredity* 79:72–82.

Dornelas, A.A.F., D.C. Paula, M.M. Espírito-Santo, G.A. Sánchez-Azofeifa, and L.O. Leite. 2012. Avifauna do Parque Estadual da Mata Seca, norte de Minas Gerais. *Rev Bras Ornitol* 20:378–391.

Dos Reis, N.R., A.L. Peracchi, W.A. Pedro, and I.P. de Lima (eds). 2007. *Bats of Brazil*. Londrina: Universidad Estadual de Londrina.

Douglas, E.A., T.G. Stith, N.B. Sean, V.S. Nikolay, B.M. Ranga, and K. Yuri. 2006. Monitoring spring canopy phenology of a deciduous broadleaf forest using MODIS. *Remote Sensing of Environment* 104:88–95.

Dovie, D.B.K., E.T.F. Witkowski, and C.M. Shackleton. 2008. Knowledge of plant resource use based on location, gender and generation. *Applied Geography* 28:311–322.

Drew, J.A. and A.P. Henne. 2006. Conservation biology and traditional ecological knowledge: Integrating academic disciplines for better conservation practice. *Ecology and Society* 11:34.

Droogers, P. and R.G. Allen. 2002. Estimating reference evapotranspiration under inaccurate data conditions. *Irrigation and Drainage Systems* 16(1):33–45.

Drummond, G.M., C.S. Martins, A.B.M. Machado, F.A. Saibo, and I. Antonini. 2005. *Biodiversidade em Minas Gerais: Um Atlas para sua Conservação*. Belo Horizonte: Fundação Biodiversitas.

Dufrene, E. and Breda, N. 1995. Estimation of deciduous forest leaf area index using direct and indirect methods. *Oecologia* 104:156–162.

Duno de Stefano, R., G. Aymard, and O. Hubber (eds.). 2007. *Vascular Flora of the Venezuelan Llanos*. Caracas: Fundación Empresas Polar (FUDENA). 738 pp.

Dupuy, J.M., J.L. Hernández-Stefanoni, R.A. Hernández-Juárez, E. Tetetla-Rangel, J.O. López-Martínez, E. Leyequién-Abarca, F. Tun-Dzul, and F. May Pat. 2012. Patterns and correlates of tropical dry forest structure and composition in a highly replicated chronosequence in Yucatan, Mexico. *Biotropica* 44:151–162.

Dupuy, S., E. Barbe, and M. Balestrant. 2012. An object-based image analysis method for monitoring land conversion by artificial sprawl use of rapideye and IRS data. *Remote Sensing* 4:404–423.

Durán, F.1980. Especies Ns promisorias para la reforestación en el bosque seco tropical, hacienda los guayabos Coyaima Tolima. BSc Forestry Engineering thesis, Universidad del Tolima.

Dytham, C. 2011. Choosing and using statistics: A biologist's guide. Oxford: Blackwell.

Eamus, D. 1991. The interaction of rising CO_2 and temperatures with water use efficiency. *Plant Cell Environment* 14:843–852.

Eamus, D. 1999. Ecophysiological traits of deciduous and evergreen woody species in the seasonally dry tropics. *Trends in Ecology and Evolution* 14:11–16.

Edelman, M. 1998. *La Lógica del Latifundio*. San José: Editorial Universidad de Costa Rica.

Egler, F.E. 1954. Vegetation science concepts I. Initial floristiccomposition: A factor in old field vegetation development. *Vegetation* 4:412–417.

Egley, G.H. and J.M. Chandler. 1983. Longevity of weed seeds after 5.5 years in the Stoneville 50-year buried-seed study. *Weed Science* 31:264–270.

Eiten, G. 1994. Vegetação do Cerrado. In *Cerrado: Caracterização, ocupação e perspectivas*, ed. M.N. Pinto, 17–73. Brasília: Editora da UNB.

Eklundh, L., K. Hall, H. Eriksson, J. Ardo, and P. Pilesjo, 2003. Investigating the use of Landsat thematic mapper data for estimation of forest leaf area index in southern Sweden. *Canadian Journal of Remote Sensing* 29:349–362.

Elliott, S., P.J. Baker, and R. Borchert. 2006. Leaf flushing during the dry season: The paradox of Asian monsoon forests. *Global Ecology and Biogeography* 15:248–257.

Elzinga, J.A., A. Atlan, A. Biere, L. Gigord, A.E. Weis, G. Bernasconi. 2007. Time after time: flowering phenology and biotic interactions. *Trends in Ecology and Evolution* 22:432–439.

EMBRAPA-Empresa Brasileira de Pesquisa Agropecuária. 1979. *Manual de métodos de análise de solo*. Rio de Janeiro: Serviço Nacional de Levantamento e Conservação de Solos.

Enquist, B.J. and A.J. Leffler. 2001. Long-term tree ring chronologies from sympatric tropical dry-forest trees: Individualistic responses to climatic variation. *Journal of Tropical Ecology* 17:41–60.

Enquist, B.J. and C.A.F. Enquist. 2011. Long-term change within a Neotropical forest: Assessing differential functional and floristic responses to disturbance and drought. *Global Change Biology* 17:1408–1424.

Enquist, B.J., G.B. West, E.L. Charnov, and J.H. Brown. 1999. Allometric scaling of production and life-history variation in vascular plants. *Nature* 401:907–911.

Erice, G., J.J. Irigoyen, P. Perez, R. Martinez-Carrasco, and M. Sanchez-Diaz. 2006. Effect of elevated CO2, temperature and drought on dry matter partitioning and photosynthesis before and after cutting of nodulated alfalfa. *Plant Science* 170:1059–1067.

Escobar, A. 1998. Whose knowledge, Whose nature? Biodiversity, conservation, and the political ecology of social movements. *Journal of Political Ecology* 5:53–82.

Escobar, A. 2005. O lugar da natureza e a natureza do lugar: Globalização ou pós-desenvolvimento? In *A colonialidade do saber: Eurocentrismo e ciências sociais. Perspectivas latino-americanas*, ed. E. Lander, 133–168. Buenos Aires: Colección Sur.

Espinal, S. 1977. Zonas de vida o formaciones vegetales de Colombia. In Memoria explicativa sobre el mapa ecológico [Life zones or vegetation formations Colombia. in Memory Guidance on the ecological map]. Bogotá: Instituto Geográfico Agustín Codazzi, http://library.wur.nl/isric/index2.html?url = http://library.wur.nl /WebQuery/isric/6600.

Espírito-Santo, M.M. 2007. Secondary seed dispersal of Ricinus communis Linnaeus (Euphorbiaceae) by ants in secondary growth vegetation in Minas Gerais. *Revista Árvore* 31:1013–1018.

Espírito-Santo, M.M., A.C. Sevilha, F.C. Anaya, R.S. Barbosa, G.W. Fernandes, A.G. Sanchez-Azofeifa, A.O. Scariot, S.E. Noronha, C. Sampaio. 2009. Sustainability of tropical dry forests: Two case studies in southeastern and central Brazil. *Forest Ecology and Management* 258:922–930.

Espírito-Santo, M.M., G.W. Fernandes, R.S. Barbosa, and F.C. Anaya. 2011. Mata seca é mata atlântica? *Ciência Hoje* 288:74–76.

Espírito-Santo, M.M., M. Fagundes, A.C. Sevilha, A.O. Scariot, G.A. Sánchez-Azofeifa, S.E. Noronha, and G.W. Fernandes. 2008. Florestas estacionais deciduais brasileiras: Distribuição e estado de conservação. *MG-Biota* 1:5–13.

Espírito-Santo, M.M., M. Fagundes, Y.R.F. Nunes, G.W. Fernandes, G.A. Sánchez-Azofeifa, and M. Quesada. 2006. Bases para a conservação e uso sustentável das florestas estacionais deciduais brasileiras: A necessidade de estudos multidisciplinares. *UNIMONTES Científica* 8:13–22.

Evans, C.E. and J.R. Etherington. 1990. The effect of soil water potential on seed germination of some British plants. *New Phytologist* 115:539–548.

Ewel, J. 1977. Differences between wet and dry successional tropical ecosystems. *Geo-Eco-Trop* 1:103–117.

Falcão, L.A.D. 2010. Variação espaço-temporal da assembléia de morcegos em uma floresta estacional decidual no norte de Minas Gerais. MSc. thesis, Universidade Estadual de Montes Claros, Montes Claros, Brazil.

FAO, 1988. FAO/UNESCO Soil Map of the Word, Revised Legend, with corrections and updates. Word Soil Resources Report 60, FAO, Roma. Reprinted with updates as Technical Paper 20, ISRIC, Wageningen, 1997.

Faria, A.A.C. and P.S.F. Neto. 2006. Ferramentas de diálogo: Qualificando o uso das técnicas de diagnóstico rural participativo. Brazil: MMA/IEB.

Farmer, W., K. Strzepek, C.A. Schlosser, P. Droogers, and X. Gao. 2011. *A method for calculating reference evapotranspiration on daily time scales*. No. 195. Massachusetts Institute of Technology.

Favreto, R. and R.B. Medeiros. 2006. Banco de sementes do solo em área agrícola sob diferentes sistemas de manejo estabelecida sobre campo natural. *Revista Brasileira de Sementes* 28:34–44.

Fedigan, L.M., L. Fedigan, and C.A. Chapman. 1985. A census of Alouatta palliata and Cebus capucinus monkeys in Santa Rosa National Park, Costa Rica. *Brenesia* 23:309–322.

Feinsinger, P. 2001. Design field studies for biodiversity conservation: The nature conservancy. London: Island Press. 236 p.

Feldpausch, T.R., M.A. Rondon, E.C.M. Fernandes, S.J. Riha, and E. Wandelli. 2004. Carbon and nutrient accumulation in secondary forests regenerating on pastures in central Amazonia. *Ecological Applications* 14(sp4):164–176.

Felfili, J.M., A.R.T. Nascimento, C.W. Fagg, and E.M. Meirelles. 2007. Floristic composition and community structure of a seasonally deciduous forest on limestone outcrops in central Brazil. *Revista Brasileira de Botânica* 30:611–621.

Felfili, J.M. 2003. Fragmentos florestais estacionais do Brazil central: diagnóstico e proposta de corredores ecológicos. In *Fragmentação florestal e alternativas de desenvolvimento rural na região Centro-Oeste*, ed. R.B. Costa, 139–160. Campo Grande: UCDB.

Fenner, M. 1985. *Seed ecology*. New York: Chapman and Hall.

Fensholt, R., I. Sandholt, and M.S. Rasmussen, 2004. Evaluation of MODIS LAI, fAPAR and the relation between fAPAR and NDVI in a semi-arid environment using in situ measurements. *Remote Sensing of Environment* 91:490–507.

Fernandes, A. and P. Bezerra. 1990. *Estudo fitogeográfico do Brazil*. Fortaleza: Stylus Comunicacões.

Fernandes, G.W. and P.W. Price. 1991. Comparison of tropical and temperate gall-inducing species richness: The roles of environmental harshness and plant nutrient status. In *Plant-animal interactions: Evolutionary ecology in tropical and temperate regions*, eds. P.W. Price, T.M. Lewinsohn, G.W. Fernandes, and W.W. Benson, 91–115. New York: John Wiley & Sons.

Fernandes G.W., Y. Oki, A. Sánchez-Azofeifa, G. Faccion, and H.C. Amaro-Arruda. 2011. Hail impact on leaves and endophytes of the endemic threatened Coccoloba cereifera (Polygonaceae) *Plant Ecology* 212:1687–1697.

Fernandes, L.A., C. Dayrell, C. Luz de Oliveira, and F. Anaya. 2010. Proposta de ocupação e uso dos ambientes pelos vazanteiros de Pau Preto: Novos indicativos à proposta da RDS no context do diálogo dos Vazanteiros de Pau Preto com o Instituto Estadual de Florestas. Technical Report, Montes Claros.

Ferreira Fadini, R. and P. de Marco Jr. 2004. Interações entre aves frugívoras e plantas em um fragmento de mata atlântica de Minas Gerais. *Ararajuba* 12:97–103.

Ferreira, S.J.F., F.J. Luizão, and R.L.G. Dallarosa, 2005. Precipitação interna e interceptação da chuva em floresta de terra firme submetida à extração seletiva de madeira na Amazônia Central. *Acta Amazonica* 35:55–62.

Finegan, B. and M. Camacho. 1999. Stand dynamics in a logged and silviculturally treated Costa Rican rain forest, 1988–1996. *Forest Ecology and Management* 121:177–189.

Finegan, B., M. Camacho, and N. Zamora. 1999. Diameter increment patterns among 106 tree species in a logged and silviculturally treated Costa Rican rain forest. *Forest Ecology and Management* 121:159–176.

Fitter, A.H., A. Heinemeyer, R. Husband, E. Olsen, K.P. Ridgway, and P.L. Staddon. 2004. Global environmental change and the biology of arbuscular mycorrhizas: Gaps and challenges. *Canadian Journal of Botany* 82:1133–1139.

Fleischbein, K., W. Wilcke, R. Goller, J. Boy, C. Valarezo, W. Zech, and K. Knoblich. 2005. Rainfall interception in a lower montane forest in Ecuador: Effects of canopy properties. *Hydrological Processes* 19(7):1355–1371.

Fleming, T.H. and E.R. Heithaus. 1981. Frugivorous bats, seed shadows and the structure of tropical forest. *Biotropica* 13:45–53.

Fleming, T.H., R. Breitwisch, and G.H. Whitesides. 1987. Patterns of tropical vertebrate frugivore diversity. *Annual Review of Ecology and Systematics* 18:91–109.

Fleming, T.H., R.A. Nuñez, and L. S.L. Sternberg. 1993. Seasonal changes in the diets of migrant and non-migrant nectarivorous bats as revealed by carbon stable isotope analysis. *Oecologia* 94:72–75.

Fleming, T.H. and W.J. Kress. 2011. A brief history of fruits and frugivores. *Acta Oecologica* 37:521–530.

Floriano, E.P. 2004. Germinação e dormência de sementes florestais. *Caderno Didático* 2:1–19.

Foody, G.M. 2003. Remote sensing of tropical forest environments: Towards the monitoring of environmental resources for sustainable development. *International Journal of Remote Sensing* 24(20):4035–4046.

Foody, G.M., G. Palubinskas, R.M. Lucas, P.J. Curran, and M. Honzak. 1996. Identifying terrestrial carbon sinks: Classification of successional stages in regenerating tropical forest from Landsat TM data. *Remote Sensing of Environment* 55(3):205–216.

Foody, G.M., M.E. Cutler, J. McMorrow, D. Pelz, H. Tangki, D.S. Boyd, and I. Douglas. 2001. Mapping the biomass of Bornean tropical rain forest from remotely sensed data. *Global Ecology and Biogeography* 10(4):379–387.

Foody, G.M. and P.M. Atkinson. 2003. *Uncertainty in Remote Sensing and GIS*. Chichester, U.K.: John Wiley & Sons.

Forzza, R.C., P.M. Leitman, A. F. Costa, Jr. Carvalho, A.L. Peixoto, B.M.T. Walter, C. Bicudo, D. Zappi, D.P. Costa, E. Lleras, G. Martinelli, H.C. Lima, J. Prado, J.R. Stehmann, J.F.A. Baumgratz, J.R. Pirani, L. Sylvestre, L.C. Maia, L.G. Lohmann, L.P. Queiroz, M. Silveira, M.N. Coelho, M.C. Mamede, M.N.C. Bastos, M.P. Morim, M.R. Barbosa, M. Menezes, M. Hopkins, R. Secco, T.B. Cavalcanti, and V.C. Souza. 2010. Lista de Espécies da Flora do Brazil. *Jardim Botânico do Rio de Janeiro*. http://floradobrasil.jbrj.gov.br/2012 (accessed October 1, 2012).

Fournier, L.A. 1974. Un método cuantitativo para la medición de características fenológicas en árboles. *Turrialba* 24:422–423.

Fowler, A.J.P. and A. Bianchetti. 2000. *Dormência em sementes florestais*. Colombo: Embrapa Florestas (Documentos, 40).

Frankie, G.W., H.G. Baker, and P.A. Opler. 1974. Comparative phenological studies of trees in tropical wet and dry forests in the lowlands of Costa Rica. *Journal of Ecology* 62:881–911.

Frankie, G.W., H.G. Baker, and P.A. Opler. 1980. Comparative phonological studies of trees in tropical wet and dry forests in the lowlands of Costa Rica. *Journal of Ecology* 68:167–188.

Frankie, G.W., S.B. Vinson, M.A. Rizzardi, T.L. Griswold, S. O'Keefe, and R.R. Snelling. 1997. Diversity and abundance of bees visiting a mass flowering tree species in disturbed seasonal dry forest, Costa Rica. *Journal of the Kansas Entomological Society* 70:281–296.

Frazer, G.W., Trofymow, J.A., and Lertzman, K.P. (2000). Canopy openness and leaf area in chronosequences of coastal temperate rainforests. *Canadian Journal of Forest Research* 30:239–256.

Freitas, A.D.S., E.V.S.B. Sampaio, C.E.R.S. Santos, and A.R. Fernandes. 2010. Biological nitrogen fixation in tree legumes of the Brazilian semi-arid caatinga. *Journal of Arid Environments* 74:344–349.

Fuchs, E.J., Lobo, J.A., Quesada, M. 2003. Effects of forest fragmentation and flowering phenology on the reproductive success and mating patterns of the tropical dry forest tree pachira quinata. *Conservation Biology* 17(1):149–157.

Fuentes, D.A., J.A. Gamon, H.L. Qiu, D.A. Sims, and D.A. Roberts. 2001. Mapping Canadian boreal forest vegetation using pigment and water absorption features derived from the AVIRIS sensor. *Journal of Geophysical Research-Atmospheres* 106:33,565–33,577.

Fundação Biodiversitas. 2005. *Biodiversidade em Minas Gerais: Um atlas para sua conservação*. Belo Horizonte: Fundação Biodiversitas.

Furley, P.A., J. Proctor, and J.A. Ratter (Eds). 1992. *Nature and dynamics of forest–savanna boundaries*. London: Chapman and Hall.

Furusawa, T. 2009. Changing ethnobotanical knowledge of the roviana Peolple, Solomon Islands: Quantitative approaches to its correlation with modernization. *Human Ecology* 37:147–159.

Gabler, R., J. Petersen, L. Trapasso, and D. Sack. 2008. *Physical Geography*. California: Brooks Cole 9th edition. 672 pp.

Gadgil, M., F. Berkes, and C. Folke. 1993. Indigenous knowledge for biodiversity conservation. *Ambio* 22:151–157.

Galindo-González, J., S. Guevara, and V.J. Sosa. 2000. Bat- and bird-generated seed rains at isolated trees in pastures in a tropical rainforest. *Conservation Biology* 14:1693–1703.

Gamon, J.A., A.F. Rahman, J.L. Dungan, M. Schildhauer, and K. F Huemmrich. 2006. Spectral Network (SpecNet), What is it and why do we need it? *Remote Sensing of Environment* 103(3):227–235.

Gamon, J.A., J. Penuelas, and C.B. Field. 1992. A narrow-waveband spectral index that tracks diurnal changes in photosynthetic efficiency. *Remote Sensing of Environment* 41(1):35–44.

Gamon, J.A., K. Kitajima, S.S. Mulkey, L. Serrano, and S.J. Wright. 2005. Diverse optical and photosynthetic properties in a neotropical dry forest during the dry season: Implications for remote estimation of photosynthesis. *Biotropica* 37(4):547–560.

Gamon, J.A., L. Serrano, and J.S. Surfus. 1997. The photochemical reflectance index: An optical indicator of photosynthetic radiation use efficiency across species, functional types, and nutrient levels. *Oecologia* 112(4):492–501.

Gan, J. and B. McCarl. 2007. Measuring transnational leakage of forest conservation. *Ecological Economics* 64:423–432.

García Oliva, F., J.M. Maass, and L. Galicia. 1995. Rainstorm analysis and rainfall erosivity of a seasonal tropical region with a strong cyclonic influence in the Pacific Coast of Mexico. *Journal of Applied Meteorology* 34:2491–2498.

Garwood, N.C. 1983. Seed germination in a seasonal tropical forest in Panama: A community study. *Ecological Monographs* 53:159–181.

Gash, J.H.C. 1979. An analytical model of rainfall interception by forests. *Quarterly Journal of the Royal Meteorological Society* 105(443):43–55.

Gates, D.M., H.J. Keegan, J.C. Schleter, and V.R. Weidner. 1965. Spectral properties of plants. *Applied Optics* 4(1):11–20.

Geiselman, C.K., S.A. Mori, and F. Blanchard. 2007. Database of Neotropical Bat/ Plant Interactions. http://www.nybg.org/botany/tlobova/mori/batsplants /database/dbase_frameset.htm.

Gelhausen S., M.W. Schwartz, and C.K. Augspurger. 2000. Vegetation and microclimatic edge effects in two mixed mesophytic forest fragments. *Plant Ecology* 147:21–35.

Gentry, A.H. 1982a. Neotropical floristic diversity: Phytogeographical connections between Central and South America, Pleistocene climatic fluctuations, or an accident of the Andean orogeny? *Annals of the Missouri Botanical Garden* 69(3):557–593.

Gentry, A.H. 1982b. Patterns of neotropical plant species diversity. *Evolution Biology* 15:1–84.

Gentry, A.H. 1983. Dispersal ecology and diversity in neotropical forest communities. *Sonderband Naturwissenschaftlicher Verein Hamburg* 7:303–314.

Gentry, A.H. 1995. Diversity and floristic composition of neotropical dry forests. In *Seasonally dry tropical forests*, eds. S.H. Bullock, H.A. Mooney and E. Medina, 146–194. Cambridge, New York: Cambridge University Press.

Gentry, A.H. 1996. *A Field Guide to the Families and Genere of Woody plants of Northwest South America*. Washington D.C.: Conservation Biology. 895 pp.

Genty, B., J.M. Briantais, and N.R. Baker. 1989. The relationship between the quantum yield of photosynthetic electron transport and quenching of chlorophyll fluorescence. *Biochimica et Biophysica Acta (BBA)-General Subjects* 990(1):87–92.

Germer, S., H. Elsenbeer, and J.M. Moraes, 2006. Throughfall and temporal trends of rainfall redistribution in an open tropical rainforest, south-western Amazonia (Rondônia, Brazil). *Hydrology and Earth System Sciences* 10(3):383–393.

Gerwing, J. 2002. Degradation of forests through logging and fire in the eastern brazilian amazon. *Forest Ecology and Management* 157:131–141

Ghazoul, J., and D. Sheil. 2010. *Tropical rain forest ecology, diversity, and conservation*. Oxford, UK: Oxford University Press.

Gianello, C., C.A. Bissani, and M.J. Tedesco. 1995. *Princípios de fertilidade do solo*. Porto Alegre: Departamento de Solos - UFRGS.

Giannini, N.P. 1999. Selection of diets and elevation by sympatric species of Sturnira in an Andean rainforest. *Journal of Mammalogy* 80:1186–1195.

Gillespie, T.W., A. Grijalva, and C.N. Farris. 2000. Diversity, composition, and structure of tropical dry forests in Central America. *Plant Ecology* 147(1):37–47.

Giovannetti, M., L. Avio, P. Fortuna E. Pellegrino, C. Sbrana, and P. Strani. 2006. At the root of the wood wide web self recognition and non-self incompatibility in mycorrhizal networks. *Plant Signaling & Behavior* 1:1–5.

Giraldo, J.P. and N.M. Holbrook. 2011. Physiological mechanisms underlying the seasonality of leaf senescence and renewal in seasonally dry tropical forests trees. In *Seasonally dry tropical forests ecology and conservation*, eds. R. Dirzo, H.S. Young, H.A. Mooney, and G. Ceballos, 129–140. Washington, DC: Island Press.

Gitay, H. and I.R. Noble. 1997. What are plant functional types and how should we seek them? In *Plant functional types*, eds. T.M. Smith, H.H. Shugart and F.I. Woodward, 3–19. Cambridge: Cambridge University Press.

Giulietti, A.M., N.L. Menezes, J.R. Pirani, M. Meguro, and M.G.L. Wanderley. 1987. Flora da Serra do Cipó, Minas Gerais: Caracterização e lista de espécies. *Boletim de Botânica* 9:1–151.

Gleason, S.M., D.W. Butler, K. Zieminska, P. Waryszak, and M. Westoby. 2012. Stem xylem conductivity is key to plant water balance across Australian angiosperm species. *Functional Ecology* 26:343–352.

Godfray, H.C.J. and J.H. Lawton. 2001. Scale and species numbers. *Trends in Ecology & Evolution* 16:400–404.

Gómez-Sapiens, M. 2001. Características foliares de especies arbóreas y arbustivas en tres tipos de vegetación de clima seco en Sonora, México. BSc. Thesis, Centro de Estudios Superiores del Estado de Sonora, Unidad Académica Hermosillo.

Gong, P., R. Pu, G.S. Biging, and M.R. Larrieu. 2003. Estimation of forest leaf area index using vegetation indices derived from Hyperion hyperspectral data. *Geoscience and Remote Sensing, IEEE Transactions* 41(6 Part 1):1355–1362.

Gonzalez, F.G., T.Y. Portela, E.J. Stipp, and L.C. Di Stasi. 2001. Antiulcerogenic and analgesic effects of Maytenus aquifolium, Sorocea bomplandii and Zolernia ilicifolia. *Journal of Ethnopharmacology* 77:41–47.

González, Y. 1985. *Continuidad y cambio en la historia agraria de Costa Rica*. San José: Editorial Costa Rica.

Google Inc. 2012. Google Earth [Software]. http://earth.google.com.

Goto, B.T., G.A. da Silva, D.M.A. Assis, D.K.A. Silva, R.G. Souza, K. Jobim, C.M.A. Mello, H.E.E. Vieira, L.C. Maia, and F. Oehl. 2012a. Intraornatosporaceae (Gigasporales), a new family with two genera and two new species. *Mycotaxon* 119:117–132.

Goto, B.T., G.A. Silva, L.C. Maia, and F. Oehl. 2010a. Dentiscutata colliculosa, a new species in the Glomeromycetes from Northeastern Brazil with colliculate spore ornamentation. *Nova Hedwigia* 90(3–4):385.

Goto, B.T., G.A. Silva, A.M. Yano-Melo, and L.C. Maia. 2010b. Checklist of the arbuscular mycorrhizal fungi (Glomeromycota) in the Brazilian semiarid. *Mycotaxon* 113:251–254.

Goto, B.T., J.G. Jardim, G.A. da Silva, E. Furrazola, Y. Torres-Arias, and F. Oehl. 2012b. Glomus trufemii (Glomeromycetes), a new sporocarpic species from Brazilian sand dunes. *Mycotaxon* 120:1–9.

Gotsch, S.G., J.S. Powers, and M.T. Lerdau. 2010. Leaf traits and water relations of 12 evergreen species in Costa Rican wet and dry forests: Patterns of intra-specific variation across forests and seasons. *Plant Ecology* 211:133–146.

Gough, C.M., C.S. Vogel, H.P. Schmid, H.B. Su, and P.S. Curtis. 2008. Multi-year convergence of biometric and meteorological estimates of forest carbon storage. *Agricultural and Forest Meteorology* 148:158–170.

Gower, S.T. and J.M. Norman. 1991. Rapid estimation of leaf area index in conifer and broad-leaf plantations. *Ecology* 72(5):1896–1900.

Greene, D.F., M. Quesada, and C. Calogeropoulos. 2008. Dispersal of seeds by the tropical sea breeze. *Ecology* 89:118–125.

Gregorin, R. and V.A. Taddei. 2002. Chave artificial para a identificação de molossídeos brasileiros (mammalia, chiroptera). *Mastozoología Neotropical* 9:13–32.

Griffith A.B., H. Alpert, and M.E. Loik. 2010. Predicting shrub ecophysiology in the Great Basin Desert using spectral indices. *Journal of Arid Environments* 74:315–326.

Grime, J.P. 1989. Seed bank in ecological perspective. In *Ecology of soil seed banks*, eds. M.A. Leck, V.T. Parker and R.L. Simpson, 150–204. San Diego: Academic press.

Griscom, H.P., E.K.V. Kalko and M.S. Ashton. 2007. Frugivory by small vertebrates within a deforested, dry tropical region of Central America. *Biotropica* 39:278–282.

Griscom, H.P. and M.S. Ashton. 2011. Restoration of dry tropical forests in Central America: A review of pattern and process. *Forest Ecology and Management* 261:1564–1579.

Griz, L.M.S. and I.C.S. Machado. 2001. Fruiting phenology and seed dispersal syndromes in caatinga, a tropical dry forest in the northeast of Brazil. *Journal of Tropical Ecology* 17:303–321.

Gruber, N. and J.N. Galloway. 2008. An Earth-system perspective of the global nitrogen cycle. *Nature* 451:293–296.

Guariguata, M.R. and R. Ostertag. 2001.Neotropical secondary succession: Changes instructural and functional characteristics. *Forest Ecology and Management* 148:185–206.

Guariguata, M.R., R.L. Chazdon, J.S. Denslow, J.M. Dupuy, and L. Anderson. 1997. Structure and floristics of secondary and old-growth forest stands in lowland Costa Rica. *Plant Ecology* 132:107–120.

Guerra, L.D., D.S. Ramalho, C.R.P. Vasconcelos. 2007. Ecologia política da construção da crise ambiental global e do modelo do desenvolvimento sustentável. *Rev Int Desenvol Local* 8:9–25.

Guevara, S. and J. Laborde. 1993. Monitoring seed dispersal at isolated standing trees in tropical pastures: Consequences for local species availability. In *Frugivory and seed dispersal: Ecological and evolutionary aspects*, eds. T.H. Fleming and A. Estrada, 319–338. Dordrecht: Kluwer Academic.

Guevara, S., S. Purata, and E. van der Maarel. 1986. The role of remnant forest trees in tropical secondary succession. *Vegetation* 66:77–84.

Gutiérrez-Soto, M.V. and J.J. Ewel. 2008. Water use in four model tropical plant associations established in the lowlands of Costa Rica. *Revista de Biología Tropical* 56:1947–57.

Guzmán, I. and J.C. Calvo-Alvarado. 2012. Water resources of the Upper Tempisque River Waershed, Costa Rica. Tecnología en Marcha (Accepted) 25:24 (4).

Haber, W.A. and G.W. Frankie. 1989. A tropical hawkmoth community: Costa Rican dry forest Sphingidae. *Biotropica* 21:155–172.

Haber, W.A. and R. Stevenson. 2002. Diversity, migration, and conservation of butterflies in northern Costa Rica. In *Biodiversity conservation in Costa Rica: Learning the lessons in a seasonal dry forest*, eds. G.W. Frankie, A. Mata and S.B. Vinson, 99–114. Berkeley: University of California Press.

Hacke, U.G., J.S. Sperry, W.T. Pockman, S.D. Davis, and K.A. McCulloh. 2001a. Trends in wood density and structure are linked to prevention of xylem implosion by negative pressure. *Oecologia* 126:457–461.

Hacke, U.G., V. Stiller, J.S. Sperry, J. Pittermann, and K.A. McCulloh. 2001b. Cavitation fatigue. Embolism and refilling cycles can weaken the cavitation resistance of xylem. *Plant Physiology* 125:779–786.

Hall, C. 1985. Costa Rica, a geographical interpretation in historical perspective. Boulder, Colorado: Westview Press.

Hall, F.G., J.R. Townshend, and E.T. Engman. 1995. Status of remote sensing algorithms for estimation of land surface state parameters. *Remote Sensing of Environment* 51(1):138–156.

Hall, S.J. and P.A. Matson. 1999. Nitrogen oxide emissions after nitrogen additions in tropical forests. *Nature* 400:152–155.

Hamilton, A.C. 2004. Medicinal plants, conservation and livelihoods. *Biodiversity and Conservation* 13:1477–1517.

Hamrick, J.L. and D.A. Murawski. 1990. The breeding structure of tropical tree populations. *Plant Species Biology* 5:157–165.

Hanauer, M. 2011. Causal mechanisms of protected area impacts. Working Paper.

Hanazaki, N., J.Y. Tamashiro, H.F. Leitão-Filho, and A. Begossi. 2000. Diversity of plant uses in two caiçara communities (Atlantic Forest coast, Brazil). *Biodiversity and Conservation* 9:597–615.

Hanazaki, N., V.C. Souza, and R.R. Rodrigues. 2006. Ethnobotany of rural people from the boundaries of Carlos Botelho State Park, São Paulo State, Brazil. *Acta botanica brasilica* 20:899–909.

Hansen, M.C., R.S. DeFries, J.R. Towndsend, M. Carroll, C. Dimiceli, and R. Sohlberg, 2003a. Global percent tree cover at a spatial resolution of 500 meters: First results of the MODIS vegetation continuous algorithm. *Earth Interactions* 7:1–15.

Hansen, M.C., R.S. DeFries, J.R. Townshend, M. Carroll, C. Dimiceli, and R. Sohlberg. 2003b. MOD44B: Vegetation continuous fields collection 3, version 3.0.0. Earth Interactions 1–20.

Hanson, P.E. 2011. Insect diversity in seasonally dry tropical forests. In *Seasonally dry tropical forests ecology and conservation*, eds. R. Dirzo, H.S. Young, H.A. Mooney and G. Ceballos, 71–84. Washington, DC: Island Press.

Hardwick, K., J.R. Healey, S. Elliott, and D. Blakesley. 2004. Research needs for restoring seasonal tropical forests in Thailand: Accelerated natural regeneration. *New Forest* 27:285–302.

Hargreaves, G.L., A.M. Asce, G.H. Hargreaves, F. Asce, J.P. Riley, and M. Asce. 1985. Irrigation water requirements for senegal river basin. *Journal of Irrigation and Drainage Engineering* 111:265–275.

Harmon, M.E., D.F. Whigham, J. Sexton, and I. Olmsted. 1995. Decomposition and mass of woody detritus in the dry tropical forests of the Northeastern Yucatan Peninsula, Mexico. *Biotropica* 27:305–316.

Harper, K., S.E. MacDonald, P.J. Burton, J. Chen, K.D. Brosofske, S.C. Saunders, E.S. Euskirche, D. Roberts, M.S. Jaiteh, P.A. Esseen. 2005. Edge influence on forest structure and composition in fragmented landscape. *Conservation Biology* 19(3):768–782.

Hartter, J., C. Lucas, A.E. Gaughan, and L.L. Aranda. 2008. Detecting tropical dry forest succession in a shifting cultivation mosaic of the Yucatán Peninsula, Mexico. *Applied Geography* 28(2):134–149.

Hayden, B., D.F. Greene, and M. Quesada. 2010. A field experiment to determine the effect of dry-season precipitation on annual ring formation and leaf phenology in a seasonally dry tropical forest. *Journal of Tropical Ecology* 26:237–242.

Heijden, M.G.A., R.D. Bardgett, N.M. van Straalen. 2008. The unseen majority: Soil microbes as drivers of plant diversity and productivity in terrestrial ecosystems. *Ecology Letters* 11:296–310.

Held, A., C. Ticehurst, L. Lymburner, and N. Williams. 2003. High resolution mapping of tropical mangrove ecosystems using hyperspectral and radar remote sensing. *International Journal of Remote Sensing* 24(13):2739–2759.

Heltshe, J.F. and N.E. Forrester. 1983. Estimating species richness using the jackknife procedure. *Biometrics* 39:1–11.

Henao, E.I. and B.E. Moreno. 2001. Selección fenotípica y manejo del germoplasma de la especie Cordia alliodora para el establecimiento de un jardín clonal en áreas del bosque seco tropical en el norte del Tolima [Phenotypic selection and germplasm management for Cordia alliodora for the establishment of a clonal garden areas tropical dry forest in northern Tolima]. BSc Forestry Engineering thesis, Universidad Del Tolima.

Herbst, M., P.T.W. Rosier, D.D. McNeil, R.J. Harding, and D.J. Gowing. 2008. Seasonal variability of interception evaporation from the canopy of a mixed deciduous forest. *Agricultural and Forest Meteorology* 148(11):1655–1667.

Herrerías-Diego, Y., M. Quesada, K.E. Stoner, and J.A. Lobo. 2006. Effect of forest fragmentation on phenological patterns and reproductive success of the tropical dry forest tree Ceiba aesculifolia. *Conservation Biology* 20:1111–1120.

Hesketh, M. and G.A. Sánchez-Azofeifa. 2012. The effect of seasonal spectral variation on species classification in the Panamanian tropical forest. *Remote Sensing of Environment* 118:73–82.

Hietz, H., B.L. Turner, W. Wanek, A. Richter, C.A. Nock, and S.J. Wright. 2011. Long-term change in the nitrogen cycle of tropical forests. *Science* 334:664–666.

Hijmans, R.J., S.E. Cameron, J.L. Parra, P.G. Jones, and A. Jarvis. 2005. Very high resolution interpolated climate surfaces for global land areas. *International Journal of Climatology* 25:1965–1978.

Hilty, S.L. 2003. *Birds of venezuela*. Princeton: Princeton University Press. 878 p.

Hissa, C.E.V. 2008. *Saberes ambientais: Desafios para o conhecimento disciplinar*. Belo Horizonte: Editora UFMG.

Hoekstra, J.M., T.M. Boucher, T.H. Ricketts, and C. Roberts. 2005. Confronting a biome crisis: Global disparities of habitat loss and protection. *Ecology Letters* 8(1):23–29.

Holbrook, N.M., J. Whitbeck, and H.A. Mooney. 1995. Drought responses of neotropical dry forest trees. In *Seasonally Dry Tropical Forests*, eds. S.H. Bullock, H.A. Mooney, and E. Medina, 243–276. New York: Cambridge University Press.

Holdridge, L.R. 1947. Determination of world plant formations from simple climatic data. Science 105(2727):367–368.

Holdridge, L.R. 1967. *Life zone ecology*. rev. ed. San José, CR: Tropical Science Center. 197p.

Holdridge, L.R., W.C. Grenke, W.H. Hatheway, T. Liang, and J.A. Tosi Jr. 1971. *Forest environments in tropical life zones: A pilot study*. New York: Pergamon Press.

Holl, K.D. 1999. Factors limiting tropical rain forest regeneration in abandoned pasture: Seed rain, seed germination, microclimate and soil. *Biotropica* 31:229–242.

Holwerda, F., L.A. Bruijnzeel, L.E. Muñoz-Villers, M. Equihua, and H. Asbjornsen. 2010. Rainfall and cloud water interception in mature and secondary lower montane cloud forests of central Veracruz, Mexico. *Journal of Hydrology* 384(1–2):84–96.

Hoshino, D., D. Nishimura, S. Yamamoto. 2002. Dynamics of major conifer and deciduous broad-leaved tree species in an oldgrowth Chamaecypans obtusa forest, central Japan. *Forest Ecology and Management* 159:133–144.

Houghton, R.A., K.T. Lawrence, J.L. Hackler, and S. Brown. 2001. The spatial distribution of forest biomass in the Brazilian Amazon: A comparison of estimates. *Global Change Biology* 7(7):731–746.

Housman, D.C., E. Naumburg, T.E. Huxman, T.N. Charlet, R.S. Nowak, and S.D. Smith. 2006. Increases in desert shrub productivity under elevated carbon dioxide vary with water availability. *Ecosystem* 9:374–385.

Howe, H.F. 1990. Seed dispersal by birds and mammals: Implications for seedling demography. In *Reproductive ecology of tropical forest plants*, eds. K.S. Bawa and M. Hadley, 191–218, 422. Paris: UNESCO and Parthenon Publishing Group.

Howe, H.F. and J. Smallwood. 1982. Ecology of seed dispersal. *Annual Review of Ecology and Systematics* 13:201–228.

Huber, A. and A. Iroumé. 2001. Variability of annual rainfall partitioning for different sites and forest covers in Chile. *Journal of Hydrology* 248(1–4):78–92.

Huber, O. and C. Alarcón. 1988. *Mapa de Vegetación de Venezuela*. Caracas: MARNR-The Nature Conservancy.

Huemmrich, K.F., J.L. Privette, M. Mukelabai, R.B. Myneni, and Y. Knyazikhin. 2005. Time-series validation of MODIS land biophysical products in a Kalahari woodland, Africa. *International Journal of Remote Sensing* 26(19):4381–4398.

Huertas, F.A. 2001. Flora arbórea de las zonas áridas de los departamentos de Huila y Tolima. Facultad De Ingeniería Forestal, Universidad del Tolima.

Hulshof, C.M., C. Violle, M. Spasojevic, B.J. McGill, E. Damschen, S. Harrison, and B.J. Enquist. 2013. Functional variation reveals the importance of abiotic and biotic drivers of species diversity across elevational and latitudinal gradients. *Journal of Vegetation Science: Online Early*.

Humphrey, S.R. and F.J. Bonaccorso. 1979. Population and community ecology. In *Biology of bats of the New World family Phyllostomidae, Part III*, eds. R.J. Baker, J.K. Jones, and D.C. Carter, vol. 16, 1–441. Austin: Special Publications, The Museum, Texas Tech University.

IAvH. Instituto Alexander Von Humboldt. 1998. El bosque seco tropical (bs-t) en Colombia programa de inventario de la biodiversidad. Grupo de exploraciones y monitoreo ambiental gema [The tropical dry forest (bs-t) in Colombia program inventory of biodiversity. Exploration group and environmental monitoring (GEMA)]. pp 1–24.

IAvH. Instituto de Investigación de Recursos Biológicos Alexander von Humboldt. 2002. Análisis preliminar de representatividad ecosistémica e identificación de vacios de conservación y alternativas para el SIRAP del Departamento del Valle del Cauca utilizando sistemas de información geográfica. Informe interno [Preliminary analysis of ecosystem representation and identification of conservation gaps and alternatives for the Department SIRAPCauca Valley using geographic information systems. Internal report]. Bogotá. Colombia. 48 p.

IAvH. Instituto de Investigación de Recursos Biológicos Alexander von Humboldt. 1997. *Informe Nacional sobre el estado de la Biodiversidad Colombia Tomo I* [National Report on the Status of Biodiversity Colombia Volume I]. Bogotá D.C. Colombia.

Ibarra-Manríquez, G. and K. Oyama. 1992. Ecological correlates of reproductive traits of Mexican rain forest trees. *American Journal of Botany* 79:383–394.

IBGE. 1992. *Manual Técnico da Vegetação Brasileira*. Rio de Janeiro: Fundação Instituto Brasileiro de Geografia e Estatística.

IBGE. 1996. *Censo Agropecuário*. Rio de Janeiro: Fundação Instituto Brasileiro de Geografia e Estatística.
IBGE. 2006. *Censo Agropecuário-Resultados Preliminares*. Rio de Janeiro: Fundação Instituto Brasileiro de Geografia e Estatística.
IBGE. 2008. *Mapa da área de aplicação da Lei no 11,428 de 2006, Decreto no 6.660, de 21 de novembro de 2008*. Brasília: Instituto Brasileiro de Geografia e Estatística, Diretoria de Geociências.
IDEAM, IGAC, IAvH, Invemar, I. Sinchi. and e IIAP. 2007. Ecosistemas continentales, costeros y marinos de Colombia. Instituto de Hidrología, Meteorología y Estudios Ambientales, Instituto Geográfico Agustín Codazzi, Instituto de Investigación de Recursos Biológicos Alexander von Humboldt, Instituto de Investigaciones Ambientales del Pacífico Jhon von Neumann, Instituto de Investigaciones Marinas y Costeras José Benito Vives De Andréis e Instituto Amazónico de Investigaciones Científicas Sinchi. Bogotá, D. C, 276 p. + 37 hojas cartográficas.
Idol, T.W., P.J. Baker, and D. Meason. 2007. Indicators of forest ecosystem productivity and nutrient status across precipitation and temperature gradients in Hawaii. *Journal of Tropical Ecology* 23:693–704.
Ijdo, M., N. Schtickzelle, S. Cranenbrouck, and S. Declerck. 2010. Do arbuscular mycorrhizal fungi with contrasting life-history strategies differ in their responses to repeated defoliation? *FEMS Microbiology Ecology* 72(1):114–122.
Instituto Chico Mendes de Conservação da biodeiversidade (ICMBio). 2009. Plano de Manejo do Parque Nacional da Serra do Cipó. Serra do Cipó e da Área de Proteção Ambiental Morro da Pedreira, MG. Minas Gerais. Ministério do Meio Ambiente.
IPCC. 2007a. Climate Change 2007: Synthesis Report. In *Contribution of Working Groups I, II and III to the Fourth Assessment Report of the Intergovernmental Panel on Climate Change*, eds. Core Writing Team, R.K. Pachauri, and A. Reisinger, 104. Geneva, Switzerland: IPCC.
IPCC. 2007b. Climate Change 2007: The physical science basis. In *Contribution of Working Group I to the fourth assessment report of the Intergovernmental Panel on Climate Change*, eds. S. Solomon, D. Qin, M. Manning, Z. Chen, M. Marquis, K.B. Averyt, M. Tignor, and H.L. Miller. Cambridge: Cambridge University Press.
ISA. 2005. Estudio etnobotánico del área del cabildo anacarco Natagaima–departamento del Tolima [Ethnobotanical study Indigenus area Anacarco Natagaima, Department of Tolima]. Unedited report.
Ishida, A., S. Diloksumpun, P. Ladpala, D. Staporn, S. Panuthai, M. Gamo, K. Yazaki, M. Ishizuka, L. Puangchit. 2006. Contrasting seasonal leaf habits of canopy trees between tropical dry-deciduous and evergreen forests in Thailand. *Tree Physiology* 26:643–656.
Ivanauskas, N.M. and R.R. Rodrigues. 2000. Florística e fitossociologia de remanescentes de floresta estacional decidual em Piracicaba, São Paulo, Brazil. *Revista Brasileira de Botânica* 23:291–304.
Jackson, P.C., F.C. Meinzer, M. Bustamante, G. Goldstein, A. Franco, P.W. Rundel, L. Caldas, E. Igler, and F. Causin. 1999. Partitioning of soil water among tree species in a Brazilian Cerrado ecosystem. *Tree Physiology* 19:717–724.
Jacomassa, F.A.F. and M.A. Pizo. 2010. Birds and bats diverge in the qualitative and quantitative components of seed dispersal of a pioneer tree. *Acta Oecologica* 36:493–496.

Jacquemoud, S., S.L. Ustin, J. Verdebout, G. Schmuck, G. Andreoli, and B. Hosgood. 1996. Estimating leaf biochemistry using the PROSPECT leaf optical properties model. *Remote Sensing of Environment* 56(3):194–202.

Jaimes, I. and N. Ramirez. 1999. Breeding systems in a secondary deciduous forest in Venezuela: The importance of life form, habitat, and pollination specificity. *Plant Systematics and Evolution* 215:23–36.

Janzen, D.H. 1967. Synchronization of sexual reproduction of tree within the dry season in Central America. *Evolution* 21:620–637.

Janzen, D.H. 1983. *Costa Rican natural history*. Chicago:The University of Chicago Press. 823 p.

Janzen, D.H. 1986a. *Guanacaste National park: Tropical ecological and cultural restoration*. San José, CR: Editorial Universidad Estatal a Distancia.

Janzen, D.H. 1987. Insect diversity of a Costa Rican dry forest: Why keep it, and how? *Biological Journal of the Linnean Society* 30:343–356.

Janzen, D.H. 1988a. Chapter 14 Tropical dry forests: The most endangered major tropical ecosystem. In *Biodiversity*, ed. E.O. Wilson, 130–137, 521. Washington, DC: National Academy Press.

Janzen, D.H. 1988b. Management of habitat fragments in a tropical dry forest: Growth. *Annals of Missouri Botanical Garden* 75:105–116.

Janzen, D.H. 2000. Costa Rica's area de conservación guanacaste: A long march to survival through non-damaging biodevelopment. *Biodiversity* 1:7–20.

Jaramillo, V.J., A. Martínez-Yrízar, and R.L. Sanford Jr. 2011. Primary productivity and biogeochemistry of primary tropical dry forests. In *Seasonally dry tropical forests. Ecology and Conservation*, eds. R. Dirzo, H.S. Young, H.A. Mooney, and G. Ceballos, 109–128. Washington, DC: Island Press.

Jasper, D.A. 1994. Management of mycorrhizas in revegetation. In *Management of mycorrhizas in agriculture, horticulture and forestry*, ed. A.D. Robson, L.K. Abbott and N. Malajczuk, 211–219. Boston: Kluwer Academic.

Jastrow, J.D. and R.M. Miller. 1998. Soil aggregate stabilization and carbon sequestration: Feedbacks through organomineral associations, pp. 207–223. In R. Lal, J.M. Kimble, R.F. Follett, and B.A. Stewart (eds.), *Soil Processes and the Carbon Cycle*. CRC Press LLC, Boca Raton, FL.

Jensen, J.R. 2000. *Remote sensing of the environment: An earth resource perspective*. Upper Saddle River, NJ: Prentice Hall.

Jernigan, K.A. 2008. The importance of chemosensory clues in Aguaruna tree classification and identification. *Journal of Ethnobiology and Ethnomedicine* 4:12.

Jetten, V.G. 1996. Interception of tropical rain forest: Performance of a canopy water balance model. *Hydrological Processes* 10(5):671–685.

Jiang, H., C. Peng, M.J. Apps, Y. Zhang, P.M. Woodard, and Z. Wang. 1999. Modelling the net primary productivity of temperate forest ecosystems in China with a GAP model. *Ecological Modelling* 122:225–238.

Jiménez, A. 2006. Gobernanza en áreas protegidas de Centroamérica: Discusión conceptual y avances regionales. San José, CEMEDE, Cuadernos de Estudios Mesoamericanos 5: 60.

Jiménez, E., F. Fernández, T.M. Arias, F.H. Lozano-Zambrano (eds.). 2008. *Sistemática, biogeografía y conservación de las hormigas cazadoras de Colombia*. Bogotá, DC, Colombia: Instituto de Investigación de Recursos Biológicos Alexander von Humboldt. xiv + 609 p.

Jiménez-Rodríguez, C., 2010. *Intercepción de lluvia en tres estadios sucesionales del Bosque seco Tropical*. Santa Rosa, CR: Parque Nacional.

Johansson, J.F., L.R. Paul, and R.D. Finlay. 2004. Microbial interactions in the mycorrhizosphere and their significance for sustainable agriculture. *FEMS Microbiology Ecology* 48:1–13.

Johns, A.D. 1991. Responses of Amazonian rain forest birds to habitat modification. *Journal of Tropical Ecology* 7:417–437.

Johnson, N.C., and D.A.Wedin. 1997. Soil carbon, nutrients, and mycorrhizae Turing conversion of dry tropical forest to grassland. *Ecological Applications* 7(1):171–182.

Jonckheere, I., S. Fleck, K. Nackaerts, B. Muys, P. Coppin, M. Weiss, and F. Baret. 2004. Review of methods for in situ leaf area index determination Part I. Theories, sensors and hemispherical photography. *Agricultural and Forest Meteorology* 121:19–35.

Joppa, L. and A. Pfaff. 2009. High & Far: Biases in the location of protected areas. *PLoS ONE* 4(12):e8273. doi: 10.1371/journal.pone.0008273

Joppa, L. and A. Pfaff. 2010a. Re-assessing the forest impacts of protection: The challenge of non-random protection & a corrective method. *Annals of the New York Academy of Sciences* 1185:135–149.

Joppa, L. and A. Pfaff. 2010b. Global Park Impacts: How could protected areas avoid more deforestation? *Proceedings of the Royal Society B*. doi:10.1098/rspb.2010.1713

Jose, S., A. Gillespie, S.J. George, and B.M. Kumar. 1996. Vegetation responses along edge-to-interior gradients in a high altitude tropical forest in peninsular India. *Forest Ecology and Management* 87(1–3): 51–62.

Kahmen, S. and P. Poschlod. 2004. Plant functional trait responses to grassland succession over 25 years. *Journal of Vegetation Science* 15:21–32.

Kalácska, M. 2005. Use of remotely sensed data to assess Neotropical dry forest structure and diversity. A thesis submitted to teh Faculty of Graduate Studies and Research in partial fulfillment of the requirements for the degree of Doctor of Philosophy., Department of Earth and Atmospheric Science, University of Alberta, Edmonton.

Kalácska, M., G.A. Sánchez-Azofeifa, B. Rivard, J.C. Calvo-Alvarado, A.R.P. Journet, J.P. Arroyo-Mora, and D. Oritz-Oritz. 2004. Leaf area index measurements in a tropical moist forest: A case study from Costa Rica. *Remote Sensing of Environment* 91:134–152.

Kalácska, M., G.A. Sánchez-Azofeifa, B. Rivard, T. Caelli, H.P. White, and J.C. Calvo-Alvarado. 2007. Ecological fingerprinting of ecosystem succession: Estimating secondary tropical dry forest structure and diversity using imaging spectroscopy. *Remote Sensing of Environment* 108(1):82–96.

Kalácska, M., G.A. Sánchez-Azofeifa, J.C. Calvo-Alvarado, M. Quesada, B. Rivard, and D.H. Janzen. 2004. Species composition, similarity and diversity in three sucessional stages of a seasonally dry tropical forest. *Forest Ecology and Management* 200(1–3):227–247.

Kalácska, M., G.A. Sánchez-Azofeifa, T. Caelli, B. Rivard, and B. Boerlage. 2005. Estimating leaf area index from satellite imagery using Bayesian networks. *IEEE Transactions on Geoscience and Remote Sensing* 43(8):1866–1873.

Kalácska, M., J.C. Calvo-Alvarado, and G.A. Sánchez-Azofeifa. 2005. Calibration and assessment of seasonal changes in leaf area index of a tropical dry forest in different stages of succession. *Tree Physiology* 25:733–744.

Kalácska, M., S. Bohlman, G. Sánchez-Azofeifa, K. Castroesau, and T. Caelli. 2007. Hyperspectral discrimination of tropical dry forest lianas and trees: Comparative data reduction approaches at the leaf and canopy levels. *Remote Sensing of Environment* 109(4):406–415.

Kalácska, M., G.A. Sánchez-Azofeifa, J.C. Calvo-Alvarado, B. Rivard, and M. Quesada. 2005. Effects of season and successional stage on leaf area index and spectral vegetation indices in three mesoamerican tropical dry forests. *Biotropica* 37(4):486–496.

Kang, H. and K.S. Bawa. 2003. Effects of successional status, habit, sexual systems, and pollinators on flowering patterns in tropical rain forest trees. *American Journal of Botany* 90:865–876.

Kapos, V. 1989. Effects of isolation on the water status of forest patches in the Brazilian Amazon. *Journal of Tropical Ecology* 5:173–185.

Kapos, V., E. Wandelli, J.L. Camargo, and G. Ganade. 1997. Edge-related changes in environment and plant responses due to forest fragmentation in Central Amazonia. In *Tropical Forest Remnants: Ecology, management, and conservation of fragmented communities*, eds. W.F. Lawrence and R.O. Bierregaard Jr., 33–44, 612. Chicago, USA: The University of Chicago Press.

Karr, J.R., D.W. Schemske, and N. Brokaw. 1982. Temporal variation in the undergrowth bird community of a tropical forest. In *The ecology of a tropical forest: Seasonal rhythms and long-term changes*, eds. E.G. Leigh Jr., A.S. Rand, and D.M. Windsor, 441–453, 480. Washington, DC: Smithsonian Institution Press.

Kellman, M. and N. Roulet. 1990. Stemflow and throughfall in a tropical dry forest. *Earth Surface Processes and Landforms* 15(1):55–61.

Kennard, D.K. 2002. Secondary forest succession in a tropical dry forest: Patterns of development across a 50-year chronosequence in lowland Bolivia. *Journal of Tropical Ecology* 18:53–66.

Kennard, D.K., K. Gould, F.E. Putz, T.S. Fredericksen, and F. Morales. 2002. Effect of disturbance intensity on regeneration mechanisms in a tropical dry forest. *Forest Ecology and Management* 162:197–208.

Kernaghan G. 2005. Mycorrhizal diversity: Cause and effect? *Pedobiologia* 49:511–520.

Kerr, J.T. and M. Ostrovsky. 2003. From space to species: Ecological applications for remote sensing. *Trends in Ecology and Evolution* 18(6):299–305.

Khurana, E. and J.S. Singh. 2001. Ecology of seed and seedling growth for conservation and restoration of tropical dry forest: A review. *Environmental Conservation* 28:39–52.

Kimes, D.S., R.F. Nelson, W.A. Salas, and D.L. Skole. 1999. Mapping secondary tropical forest and forest age from SPOT HRV data. *International Journal of Remote Sensing* 20(18):3625–3640.

Kikuzawa, K. 1995. Leaf phenology as an optimal strategy for carbon gain in plants. *Canadian Journal of Botany* 73:158–163.

Kingston, T. 2009. Analysis of species diversity of bat assemblages. Pages 195–215. In *Ecological and behavioral methods for the study of bats*, eds. T.H. Kunz and S. Parsons, Baltimore: Johns Hopkins University Press. 920 p.

Kirk, P.M., P.F. Cannon, D.W. Minter, and J.A. Stalpers. 2008. *Dictionary of the fungi*. 10th ed. Cromwell Press, Trowbridge, U.K. 784 p.

Kissling, W.D., K.B. Gaese, and W. Jetz. 2009. The global distribution of frugivory in birds. *Global Ecology and Biogeography* 18:150–162.

Klaassen, W., F. Bosveld, and E. de Water. 1998. Water storage and evaporation as constituents of rainfall interception. *Journal of Hydrology* 212–213:36–50.

Klein, R.M. 1980. Ecologia da flora e vegetação do vale do Itajaí. *Sellowia*. 32:165–389.

Klingbeil, B.T. and M.R. Willig. 2010. Seasonal differences in population-, ensemble- and community-level responses of bats to landscape structure in Amazonia. *Oikos* 119:1654–1664.

Klink, C.A. and Machado, R.B. 2005. A conservação do cerrado brasileiro. *Megadiversidade* 1:147–155.

Knipling E.B. 1970. Physical and physiological basis for the reflectance of visible and near-infrared radiation from vegetation. *Remote Sensing of Environment* 1:155–159.

Koerselman, W. and A.F.M. Meuleman. 1996. The vegetation N:P ratio: A new tool to detect the nature of nutrient limitation. *Journal of Applied Ecology* 33:1441–1450.

Kohlmann, B. 1991. Dung beetles in subtropical North America. In *Dung beetle ecology*, eds. I. Hanski and Y. Camberfort, 166–132. New Jersey: Princenton University Press.

Koning, F., R. Olschewski, E. Veldkamp, P. Benitez, M. López-Ulloa, T. Schlichter, and M. Urquiza. 2005. The ecological and economic potential of carbon sequestration in forests: Examples from South America. *AMBIO: A Journal of the Human Environment* 34(3):224–229.

Körner C. 2003. Ecological impacts of atmospheric CO_2 enrichment on terrestrial ecosystems. *Philosophical Transactions of the Royal Society* 361:2023–2041.

Körner C. 2004. Through enhanced tree dynamics carbon dioxide enrichment may cause tropical forests to lose carbon. *Philosophical Transactions of the Royal Society* 359:493–498.

Krebs, C.J. 1989. *Ecological methodology*. New York: Harper & Row.

Kruskal, W.H. and W.A. Wallis. 1952. Use of Ranks in One-Criterion Variance Analysis. *Journal of the American Statistical Association* 47:583–621.

Kucharik, C.J., J.M. Norman, and S.T. Gower. 1998. Measurements of branch area and adjusting leaf area index indirect measurements. *Agricultural and Forest Meteorology* 91:69–88.

Kudo, G., Y. Nishikawa, T. Kasagi, and S. Kosuge. 2004. Does seed production of spring ephemerals decrease when spring comes early? *Ecological Research* 19:255–259.

Kushwaha, C.P. and K.P. Singh. 2005. Diversity of leaf phenology in a tropical deciduous forest in India. *Journal of Tropical Ecology* 21:47–56.

Kushwaha, C.P., S.K. Tripathi, and K.P. Singh. 2011. Tree specific traits affect flowering time in Indian dry tropical forest. *Plant Ecology* 212:985–98.

Kuusk, A. 1995. A Markov chain model of canopy reflectance. *Agricultural and Forest Meteorology* 76(3):221–236.

Kuusk, A. and T. Nilson. 2000. A directional multispectral forest reflectance model. *Remote Sensing of Environment* 72(2):244–252.

Labouriau, L.G. 1983. *Germinação das sementes*. Washington: OEA.

Lal, C.B., C. Annapurna, A.S. Raghubanshi, J. S. Singh. 2001a. Effect of leaf habit and soil type on nutrient resorption and conservation in woody species of a dry tropical environment. *Canadian Journal of Botany* 79:1066–1075.

Lal, C.B., C. Annapurna, A.S. Raghubanshi, J.S. Singh. 2001b. Foliar demand and resource economy of nutrients in dry tropical forest species. *Journal of Vegetation Science* 12:5–14.

Laliberté, E. and P. Legendre. 2010. A distance-based framework for measuring functional diversity from multiple traits. *Ecology* 91:299–305.

Lambert, J.D.H., J.T. Arnason, and J.L. Gale. 1980. Leaf-litter and changing nutrient levels in a seasonally dry tropical hardwood forest, Belize, C. A. *Plant and Soil* 55:429–443.

Lambin, E. F. 1999. Monitoring forest degradation in tropical regions by remote sensing: Some methodological issues. *Global Ecology and Biogeography* 8(3):191–198.

Lampe, M.G., Y. Bergeron, R. Mcneil, and A. Leduc. 1992. Seasonal flowering and fruiting patterns in tropical semi-arid vegetation of Northeastern Venezuela. *Biotropica* 24:64–76.

Laurance, S. 2004. Responses of understory rain forest birds to road edges in Central Amazonia. *Ecological Applications* 14(5):1344–1357.

Laurance, W. 1997. Hyperdisturbed parks: Edge effects and the ecology of isolated rainforest reserves in tropical Australia, In *Tropical Forest remnants: Ecology, management and Conservation of Fragmented communities*, eds. W. Laurance and R. Bierregaard, 71–83, 612. USA: The University of Chicago Press.

Laurance, W., L. Ferreira, R. De-Merona, and S. Laurence. 1998. Rain forest fragmentation and the dynamics of Amazonian tree communities. *Ecology* 79(6): 2032–2040.

Laurance, W. and T. Curran. 2008. Impacts of wind disturbance on fragmented tropical forests: A review and synthesis. *Austral Ecology* 33:399–408.

Laurance, W., T.E. Lovejoy, H.L. Vasconcelos, E.M. Bruna, R.K. Didham, P.C.Stouffer, C. Gascon, R.O. Bierregaard, S.G. Laurance, and E. Sampiao. 2002. Ecosystem decay of Amazonian forest fragments: A 22-year investigation. *Conservation Biology* 16(3):605–618.

Lawrence, D. 2005. Regional-scale variation in litter production and seasonality in the tropical dry forests of Southern Mexico. *Biotropica* 37:561–570.

Lawton, J.H. 1999. Are there general laws in ecology? *Oikos* 84:177–192.

Le Maire, G., C. Francois, and E. Dufrene. 2004. Towards universal broad leaf chlorophyll indices using PROSPECT simulated database and hyperspectral reflectance measurements. *Remote Sensing of Environment* 89(1):1–28.

Leblanc, S.G. and J.M. Chen. 2001. A practical scheme for correcting multiple scattering effects on optical LAI measurements. *Agricultural and Forest Meteorology* 110:125–139.

Lebrija-Trejos, E. 2009. Tropical dry forest recovery processes and causes of change. Doctoral Dissertation, Wageningen University, Wageningen, The Netherlands. 189 p.

Lebrija-Trejos, E., E.A. Pérez-García, J.A. Meave, F. Bongers, and L. Poorter. 2010. Functional traits and environmental filtering drive community assembly in a species-rich tropical system. *Ecology* 91:386–398.

Lebrija-Trejos, E.E., J. Meave, L. Poorter, E.A. Pérez-García, and F. Bongers. 2010. Pathways, mechanisms and predictability of vegetation change during tropical dry forest succession. *Perspectives in Plant Ecology, Evolution, and Systematics* 12:267–275.

Leff, E. 2006. *Racionalidade ambiental: A reapropriação social da natureza*. Rio de Janeiro: Ed. Civilização Brasiliera.

Lefsky, M.A., W.B. Cohen, D.J. Harding, G.G. Parker, S.A. Acker, and S.T. Gower. 2002. Lidar remote sensing of above-ground biomass in three biomes. *Global Ecology and Biogeography* 11(5):393–399.

Lehmann, M.P. 1992. Deforestation and changing land-use patterns in Costa Rica. In *Changing tropical forests: Historical perspectives on today's challenges in Central and South America*, ed. H.K. Steen and R.P. Tucker, 59–76. USA: Forest History Society.

Leishman, M.R. and B.R. Murray. 2001. The relationaship between seed size and abundance in plant communities: Model predictions and observed patterns. *Oikos* 93:151–161.

Leitão, F., V.S. Fonseca-Kruel, I.M. Silva, and F. Reinert. 2009. Urban ethnobotany in Petrópolis and Nova Friburgo (Rio de Janeiro, Brazil). *Revista Brasileira de Farmacognosia* 19:333–342.

Leite, L.O., M.A.Z. Borges, C.A. Lima, R.M.M. Gonçalves, and P.R. Siqueira. 2008. Variação espaço-temporal do uso de recursos pela avifauna do Parque Estadual da Mata Seca. *MG-Biota* 1:54–60.

Leite, M.R., J.L.S. Brito, M.E. Leite, M.M. Espírito-Santo, C.M.S. Clemente, and J.W.L. Almeida. 2011. Sensoriamento remoto como suporte para quantificação do desmatamento de floresta estacional decidual no Norte de Minas Gerais. In *Anais do XV Simpósio Brasileiro de Sensoriamento Remoto*, 8583–8590. São José dos Campos: Instituto Nacional de Pesquisas Espaciais. http://www.dsr.inpe.br/sbsr2011/files/p1196.pdf.

Leiva, J.A., R. Mata, O.J. Rocha, and M.V. Gutierrez. 2009. Cronología de la regeneración del bosque tropical seco en Santa Rosa, Guanacaste, Costa Rica. I. Características edáficas. *Revista Biología Tropical* 57:801–815.

Leonard, H.J. 1987. *Natural resources and economic development in Central America*. Rutgers, New Jersey: International Institute for Environment and Development.

Lessa, S.N. 2007. Mesonorte: *Diagnóstico para a agenda de desenvolvimento integrado e sustentável da Mesorregião do norte de Minas*. Montes Claros, MG: UNIMONTES.

Levésque, M., K.P. Mclaren, and M.A. Mcdonald. 2011.Recovery and dynamics of a primary tropical dry forest in Jamaica, 10 years after human disturbance. *Forest Ecology and Management* 262:817–826.

Levey, D.J. 1988. Tropical wet forest treefall gaps and distributions of understory birds and plants. *Ecology* 69:1076–1089.

Lewinsohn, T.M. and P.I. Prado. 2002. *Biodiversidade brasileira: Síntese do estado atual do conhecimento*. São Paulo: Editora Contexto.

Lewinsohn, T.M., and P.I. Prado. 2005. How many species are there in Brazil? *Conservation Biology* 19:619–624.

Lewinsohn, T.M., V. Novotny, and Y. Basset. 2005. Insects on plants: Diversity of herbivore assemblages revisited. *Annual Review of Ecology, Evolution, and Systematics* 36:597–620.

Lewis, S.L., O.L. Phillips, D. Sheil, B. Vinceti, T.R. Baker, S. Brown, A.W. Graham, et al. 2004. Tropical forest tree mortality, recruitment and turnover rates: Calculation, interpretation and comparison when census intervals vary. *Journal of Ecology* 92:929–944.

Li, F., S. Kang, and F. Zhang. 2003. Effects of CO_2 enrichment, nitrogen and water on photosynthesis, evapotranspiration and water use efficiency of spring wheat. *Ying Yong Sheng Tai Xue Bao* 14:387–393.

Li, X. and A.H. Strahler. 1986. Geometric-optical bidirectional reflectance modeling of a conifer forest canopy. *IEEE Transactions on Geoscience and Remote Sensing* 30:276–291.

Li, X. and A.H. Strahler. 1992. Geometric-optical bidirectional reflectance modeling of the discrete crown vegetation canopy: Effect of crown shape and mutualshadowing. *Geoscience and Remote Sensing, IEEE Transactions* 30(2):276–292.

Licor Inc. 2010 Li-191SA Line Quantum Sensor Brochure. Licor Environmental Division. Lincoln, NE. USA. http://www.licor.com/env/PDF/191sa.pdf

Lieberman, D. 1982. Seasonality and phenology in a dry tropical forest in Ghana. *Journal of Ecology* 70:791–806.

Lieberman, D. and M. Lieberman. 1987. Forest tree growth and dynamics at La Selva, Costa Rica (1969–1982). *Journal of Tropical Ecology* 3:347–358.

Lilliefors, H.W. 1967. On the Kolmogorov-Smirnov test for normality with mean and variance unknown. *Journal of the American Statistical Association* 62:399–402.

Lima, J.C.S., D.T.O. Martins, and P.T. De Souza Jr. 1998. Experimental evaluation of stem bark of Stryphnodendron adstringens (Mart.) Coville for antiinflammatory activity. *Phytotherapy Research* 12:218–220.

Lima, V.V.F., D.L.M. Vieira, A.C. Sevilha and A.N. Salomão. 2008. Germinação de espécies arbóreas de floresta estacional decidual do vale do rio Paranã em Goiás após três tipos de armazenamento por até 15 meses. *Biota Neotropica* 8:89–97.

Linares, E.L. and E.A. Moreno-Mosquera. 2010. Morphology of Crecopia (Crecopiaceae) fruitlets of the Colombian Pacific and its taxonomic value in the bats diets study. *Caldasia* 32:275–287.

Linares, O. 1986. Bats of Venezuela. Caracas: Cuadernos Lagoven. 119 p.

Linares, O. 1998. Mammals of Venezuela. Caracas: Sociedad Conservacionista Audubon de Venezuela.

Linares, R.J. and M.C. Fandiño. 2009. Estado del bosque seco tropical e importancia relativa de su flora leñosa, islas de la Vieja Providencia y Santa Catalina, Colombia, Caribe suroccidental. *Revista de la Academia Colombiana de Ciencias* 33(126):1–12.

Link, T.E., M. Unsworth, and D. Marks. 2004. The dynamics of rainfall interception by a seasonal temperate rainforest. *Agricultural and Forest Meteorology* 124(3–4):171–191.

Lisboa, F.J.G., E. Jesus, G. Chaer, S.M. Faria, F.S. Goncalves, F.M. Santos, A.F. Castilho, R.L.L. Berbara. 2012. The influence of the litter quality on the direct relationship between vegetation and the variation of below-ground compartments: A Procrustean approach. *Plant and Soil*. 360:412–418.

Little, P. 2002. *Territórios sociais e povos tradicionais no Brazil: Por uma antropologia da territorialidade*. Gramado, Brazil: Anais da XXIII Reunião Brasileira de Antropologia.

Lloyd, C.R. and F.A.de.O. Marques. 1988. Spatial variability of throughfall and stemflow measurements in Amazonian rainforest. *Agricultural and Forest Meteorology* 42(1):63–73.

Lloyd, C.R., J.H.C. Gash, W.J. Shuttleworth, and F.A.de.O. Marques. 1988. The measurement and modelling of rainfall interception by Amazonian rain forest. *Agricultural and Forest Meteorology* 43(3–4):277–294.

Loh, J. and D. Harmon. 2005. A global index of biocultural diversity. *Ecological Indicators* 5:231–241.

Loisele, B. and J.G. Blake. 1994. Annual variation in birds and plants of a second-growth woodland. *Condor* 96:368–380.

Longhi, S.J., E.J. Brun, D.M. Oliveira, L.E.B. Fialho, J.C. Wojciechowski and S. Vaccaro. 2005. Banco de sementes do solo em três fases sucessionais de uma floresta estacional decidual em Santa Tereza, RS. Ciência Florestal 15:359–370.

Longman, K.A. and J. Jenik. 1974. *Tropical forest and its environment*. London: Longman.

Loomis, W.E. 1965. Absorption of radiant energy by leaves. *Ecology* 14–17.

Lorenzi, H. 1992. *Árvores brasileiras: Manual de identificação e cultivo de plantas arbóreas nativas do Brazil*. Nova Odessa: Plantarum.

Lorenzi, H. 1998. *Árvores brasileiras: Manual de identificação e cultivo de plantas arbóreas nativas do Brazil*. Nova Odessa: Plantarum.

Losada-Prado, S. and Y.G. Molina-Martínez. 2011 Tropical dry forest's birds in Tolima Department (Colombia): Community analysis *Caldasia* 33(1):271–294.

Lott, E. and H. Atkinson. 2002. Biodiversidad y fitogeografía de Chamela Cuixmala, Jalisco. In *Historia natural de chamela*, eds. F.A. Noguera, M. Quesada, J. Vega, and A. Garcia-Aldrete, 83–97. México: Instituto de Biología, Universidad Nacional Autónoma de México.

Lott, E.J., S.H. Bullock, and J.A. Solis-Magallanes. 1987. Floristic diversity and structure of upland and arroyo forests of coastal Jalisco. *Biotropica* 228–235.

Louman, B., J. Valerio, and W. Jiménez. 2001. Bases ecológicas. In *Sivicultura de bosques latifoliados húmedos con énfasis en América Central*, eds. B. Louman, D. Quirós, and M. Nilsson, 19–78. Turrialba, CR: CATIE.

Lozada, M. and A. Ladio 2006. Cultural transmission of ethnobotanical knowledge in a rural community of northwestern Patagonia, Argentina. *Economic Botany* 60:374–385.

Lozano, L.A. 2005. Patrones ecológicos de un relicto de bosque seco tropical ribereño, en el C.U.R.N.e la Universidad del Tolima, Armero-Guayabal-Colombia [Ecological patterns of a relict coastal tropical dry forest in the CURN the University of Tolima in Armero, Colombia]. MSc thesis Biological Sciences, Universidad del Tolima.

Lozano, L.A., F.A. Gomez, and S. Valderrama. 2011. Estado de fragmentación de los bosques naturales en el Norte del Tolima. *Revista Tumbaga* [State fragmentation natural forests in northern of Tolima. *Tumbaja Journal*] 6:125–140.

Lucas, R.M., M. Honzak, P.J. Curran, G.M. Foody, R. Milne, T. Brown, and S. Amaral. 2000. Mapping the regional extent of tropical forest regeneration stages in the Brazilian Legal Amazon using NOAA AVHRR data. *International Journal of Remote Sensing* 21(15):2855–2881.

Lucena, E.A.R.M. 2007. Fenologia, biologia da polinização e da reprodução de Pilosocereus Byles & Rowley (Cactaceae) no Nordeste do Brazil. Ph Thesis. Univ. Federal de Pernambuco, Recife.

Lugo, A.E., and E. Helmer. 2004. Emerging forests on abandoned land: Puerto Rico's new forests. *Forest Ecology and Management* 190:145–161.

Lugo, M.A. and M.N. Cabello. 2002. Native arbuscular mycorrhizal fungi (AMF) from mountain grassland (Córdoba, Argentina) I. Seasonal variation of fungal spore diversity. *Mycologia* 94:579–586.

Luna, M.Y., A. Morata, C. Almarza, and M.L. Martín. 2006. The use of GIS to evaluate and map extreme maximum and minimum temperatures in Spain. *Meteorological Applications* 13:385–392.

Luz de Oliveira, C. 2005. Vazanteiros do Rio São Francisco: Um estudo sobre populações tradicionais e territorialidade no Norte de Minas Gerais. MSc. thesis, Universidade Federal de Minas Gerais, Belo Horizonte.

Maass, J., J.M. Vose, W.T. Swank, and A. Martínez-Yrízar. 1995. Seasonal changes of leaf area index (LAI) in a tropical deciduous forest in west Mexico. *Forest Ecology and Management* 74(1–3):171–180.

Maass, M. and A. Burgos. 2011. Water dynamics at the ecosystem level in seasonally dry tropical forests. In *Seasonally dry tropical forests: Ecology and conservation*, eds. R. Dirzo, H.S. Young, H.A. Mooney, and G. Ceballos, 141–156. Washington, DC: Island Press.

Macfarlane, C., M. Hoffman, D. Eamus, N. Kerp, S. Higginson, R. McMurtrie, and M. Adams. 2007. Estimation of leaf area index in eucalypt forest using digital photography. *Agricultural and Forest Meteorology* 143(3–4):176–188.

Machado, I.C.M. and A.V. Lopes. 2004. Floral traits and pollination systems in the Caatinga. *Annals of Botany* 94:365–376.

Machado, I.C.S. and L.M. Barros. 1997. Phenology of Caatinga species at Serra Talhada, PE, Northeastern Brazil. *Biotropica* 29:57–68.

Madeira, B.G. 2008. Diversidade de borboletas frugívoras no norte de Minas Gerais. PhD. dissertation, Universidade Federal de Viçosa, Viçosa, Brazil.

Madeira, B.G., M.M. Espírito-Santo, S. D'Ângelo-Neto, Y.R.F. Nunes, G.A. Sánchez-Azofeifa, G.W. Fernandes, and M. Quesada. 2009. Changes in tree and liana communities along a successional gradient in a tropical dry forest in south-eastern Brazil. *Plant Ecology* 201:291–304.

Madeira, J.A. and G.W.Fernandes. 1999. Reproductive phenology of sympatric taxa of *Chamaecrista* (Leguminosae) in Serra do Cipó, Brazil. *Journal of Tropical Ecology* 15:463–479.

Maffi, L. 2005. Linguistic, cultural, and biological diversity. *Reviews in Anthropology* 29:599–617.

Maguire, J.D. 1962. Speed of germination aid in selection and evaluation for seedling emergence and vigor. *Crop Science* 2:176–177.

Mahecha, E.E. 2002. Árboles de las zonas boscosas pertenecientes a la eco-región estratégica De la Tatacoa y sus áreas de influencia en el Tolima. BSc Forestry Engineering thesis, Universidad del Tolima.

Maia, G.N. 2004. *Caatinga: Árvores, arbustos e suas utilidades*. São Paulo: D&Z.

Maia, L.C., G.A. Silva, A.M. Yano-Melo, and B.T. Goto. 2010. Fungos micorrízicos arbusculares no bioma Caatinga. In Micorrizas: 30 anos de pesquisas no Brazil, eds. J.O. Siqueira, F.A. de Souza, E.J.B.N. Cardoso, and S.M. Tsai, v.1, 311–340.

Maia, N.G. 2004. *Caatinga: Árvores e arbustos e suas utilidades*. São Paulo: Editora Livro e Arte.

Malhi, Y., J.T. Roberts, R.A. Betts, T.J. Killeen, W. Li, and C.A. Nobre. 2008. Climate change, deforestation, and the fate of the Amazon. *Science* 319(5860):169–172.

Malhi, Y., L.E.O.C. Aragão, D. Galbraith, C. Huntingford, R. Fisher, P. Zelazowski, S. Sitch, C. McSweeney, and P. Meir. 2009. Exploring the likelihood and mechanism of a climate-change-induced dieback of the Amazon rainforest. *Proceedings of the National Academy of the United States of America* 106:20610–20615.

Maluf de Souza, F. and J.L. Ferreria Batista. 2004. Restoration of seasonal semideciduous forests in Brazil: Influence of age and restoration design on forest structure. *Forest Ecology and Management* 191:185–200.

Manu, S., W. Peach, and W. Cresswell. 2007. The effects of edge, fragment size and degree of isolation on avian species richness in highly fragmented forest in West Africa. *Ibis* 149:287–297.

Marcos Filho, J. 2005. *Fisiologia de sementes de espécies cultivadas*. Piracicaba: FEALQ.

Marin, C.T., W. Bouten, and J. Sevink. 2000. Gross rainfall and its partitioning into throughfall, stemflow and evaporation of intercepted water in four forest ecosystems in western Amazonia. *Journal of Hydrology* 237(1–2):40–57.

Marin, G.S., R. Nygard, B.G. Rivas, and P.C. Oden. 2005. Stand dynamics and basal area change in a tropical dry forest reserve in Nicaragua. *Forest Ecology and Management* 208:63–75.

Markesteijn, L., J. Iraipi, F. Bongers, and L. Poorter. 2010. Seasonal variation in soil and plant water potentials in a Bolivian tropical moist and dry forest. *Journal of Ecology* 26:497–508.

Markesteijn, L., L. Poorter, H. Paz, F. Bongers, and L. Sack. 2011. Hydraulics and life history of tropical dry forest tree species: Coordination of species' drought and shade tolerance. *New Phytologist* 191:480–495.

Marques, J.G. 2001. *Pescando Pescadores*. São Paulo: NUPAUB-USP.

Marques, M.C.M., J.J. Roper, and A.P. Baggio Salvalaggio. 2004. Phenological patterns among plant life-forms in a subtropical forest in southern Brazil. *Plant Ecology* 203–213.

Marques, T. 2011a. Diversidade de Formicidae em Florestas Estacionais Decíduas em diferentes escalas temporais e espaciais. PhD diss., Universidade Federal de Viçosa, Brazil.

Marquis, R.J. 1988. Phenological variation in the neotropical understory shrub Piper arieianum: Causes and consequences. *Ecology* 69:552–1565.

Martin, G.J. 1995. *Ethnobotany, a methods manual*. London: Chapman & Hall.

Martin, P.H., R.E. Sherman, and T.J. Fahey. 2004. Forty years of tropical forest recovery from agriculture: Structure and floristics of secondary and old-growth riparian forests in the Dominican Republic. *Biotropica* 36:297–317.

Martin, R.E., G.P. Asner, and L. Sack. 2007. Genetic variation in leaf pigment, optical and photosynthetic function among diverse phenotypes of Metrosideros polymorpha grown in a common garden. *Oecologia* 151(3):387–400.

Martínez-Alier, J. 2007. *O ecologismo dos pobres: Conflitos ambientais e linguagens de valoração*. São Paulo: Contexto.

Martínez Yrízar, A., M. Burquez, and M. Maass. 2000. Structure and functioning of tropical deciduous forest in Western Mexico. In *The tropical deciduous forest of Alamos: Biodiversity of a threatened ecosystem in Mexico*, eds. R.H. Robichaux and D. Yetman, 19–35. Arizona: University of Arizona Press.

Martínez-Yrízar, A., S. Núñez, H. Miranda, and A. Burquez. 1999. Temporal and spatial variation of litter production in Sonoran Desert communities. *Plant Ecology* 145:37–48.

Mata-Machado, B.N. 1991. *História do Sertão Noroeste de Minas Gerais*. Belo Horizonte: Imprensa Oficial de Minas Gerais.

Matson, P.A., K.A. Lohse, and S.J. Hall. 2002. The globalization of nitrogen deposition: Consequences for terrestrial ecosystems. *Ambio* 31:113–119.

Mauro, F. and D. Hardison. 2002. Traditional knowledge of indigenous and local communities: International debate and policy initiatives. *Ecological Applications* 10:1263–1269.

Maxwell, J.F. and S. Elliott. 2001. *Vegetation and vascular flora of doi Sutep–Pui National Park, Chiang Mai Province, Thailand. Thai Studies in Biodiversity 5*. Bangkok: Biodiversity Research and Training Programme. 205 p.

Mazzetto, C.E. 2009. *O Cerrado em Disputa: Apropriação global e resistências locais*. Brasília: CONFEA.

McArdle, B.H. and M.J. Anderson. 2001. Fitting multivariate models to community data: A comment on distance based redundancy analysis. *Ecology* 82:290–297.

McDonald, A.J., F.M. Gemmell, and P.E. Lewis. 1998. Investigation of the utility of Spectral Vegetation Indices for determining information on coniferous forests-an overview. *Remote Sensing of Environment* 66(3):250–272.

Mearns, L.O., I. Bogardi, F. Giorgi, I. Matyasovszky, and M. Palecki. 1999. Comparison of climate change scenarios generated from regional climate model experiments and statistical downscaling. *Journal of Geophysical Research* 104(6):6603–6623.

Medellín, R., M. Equihua, and M.A. Amín. 2000. Bat diversity and abundance as indicators of disturbance in Neotropical rainforest. *Conservation Biology* 14:1666–1675.

Medellín, R. and O. Gaona. 1999. Seed dispersal by bats and birds in forest and disturbed habitats of Chiapas, Mexico. *Biotropica* 31:478–485.

Medellín, R.A., H.T. Arita, and O. Sánchez-Herrera. 1997. Identification of Bats of Mexico. A.C. Distrito Federal, México: Asociación Mexicana de Mastozoología. 83 p.

Medina, A., C.A. Harvey, D. Sánchez Merlo, S. Vílchez, and B. Hernández. 2007. Bat diversity and movement in an agricultural landscape in Matiguás, Nicaragua. *Biotropica* 39:120–128.

Medina, E. 1966. Producción de hojarasca, respiración edáfica y productividad vegetal en bosques deciduos de Los llanos centrales de Venezuela. pp 97–108. UNESCO para América Latina. Montevideo, Uruguay.

Medina, E. 1967. Intercambio gaseoso de árboles de la sabana de *Trachypogonen* Venezuela. *Boletín de la Sociedad Venezolana de Ciencias Naturales* 111:56–69.

Medina, E., J. Silva, and E. Castellanos. 1969. Variaciones estacionales del crecimiento y la respiración foliar de plantas leñosas de las sabanas de *Trachypogon*. *Boletín de la Sociedad Venezolana de Ciencias Naturales* 115–116:62–82.

Medina, E., D. Marín, and E. Olivares. 1985. Ecophysiological adaptations in the use of water and nutrients by woody plants of arid and semi-arid tropical regions. *Medio Ambiente* 7:91–102.

Medina, E., V. García, and E. Cueva. 1990. Sclerophylly and oligotrophic environments: relationship between leaf structure, mineral nutrients content and drought resistance in tropical rain forest of the upper Rio Negro region. *Biotropica* 22:51–64.

Medina, E. 1995. Diversity of life forms of higher plants in Neotropical dry forests. In *Seasonally Tropical Forests*, eds. S.H. Bullock, H.A. Mooney, and E. Madina. UK: Cambridge University Press.

Meguro, M., J.R. Pirani, R. Mello-Silva, and I. Cordeiro. 2007. Composição florística e estrutura das florestas estacionais decíduas sobre calcário a oeste da Cadeia do Espinhaço, Minas Gerais, Brazil. *Boletim de Botânica da Universidade de São Paulo* 25:147–171.

Meir, P. and T.B. Pennington. 2011. Climatic change and seasonally dry tropical forests. In *Seasonally dry tropical forests: Ecology and conservation*, eds. R. Dirzo, H.S. Young, H.A. Mooney and G. Ceballos, 279–300. Washington, DC: Island Press.

Mello, C.M.A., G.A. Silva, H.E.E. Vieira, I.R. Silva, L.C. Maia, and F. Oehl. 2012. Fuscutata aurea, a new species in the Glomeromycetes from cassava and maize fields in the Atlantic rainforest zone in Northeastern Brazil. *Nova Hedwigia* 95(1–2):267–275.

Mello, M.A.R. 2009. Temporal variation in the organization of a Neotropical assemblage of leaf-nosed bats (Chiroptera: Phyllostomidae). *Acta Oecologica* 35:280–286.

Mello, M.A.R., E.K.V. Kalko, and W.R. Silva. 2009. Ambient temperature is more important than food availability in explaining reproductive timing of the bat Sturnira lilium (Mammalia: Chiroptera) in a montane Atlantic Forest. *Canadian Journal of Zoology* 87:239–245.

Mello, M.A.R., F. Marquitti, P. Guimarães, E. Kalko, P. Jordano, and M. de Aguiar. 2011. The modularity of seed dispersal: Differences in structure and robustness between bat–and bird–fruit networks. *Oecologia* 167:131–140.

Mello, M.A.R., F.M.D. Marquitti, P.R. Guimarães Jr., E.K.V. Kalko, P. Jordano, and M.A. Martinez de Aguiar. 2011. The missing part of seed dispersal networks: Structure and robustness of bat-fruit interactions. *PLoS ONE* 6:e17395.

Melo, O. 2002. Estudios de caracterización biofísica y socioeconómica de la ecorregión estratégica del valle del alto Magdalena. Componente Flora Ministerio de Ambiente, CORTOLIMA, CAM, Universidad del Tolima and Universidad Surcolombiana. Ibagué – Tolima. 600p.

Melo, O. and R. Vargas. 2003. *Evaluación ecológica y silvicultural de Ecosistemas Boscosos*. Ibagué, Colombia: Impresiones Conde. 222p.

Menaut, J.C., M. Lepage, and L. Abbadie. 1995. Savannas, woodlands and dry forests in Africa. In *Seasonally Dry Tropical Forests*, eds. S.H. Bullock, H.A. Mooney, and E. Medina, 64–88. New York: Cambridge University Press.

Mendoza, C.H. 1999. Estructura y riqueza florística del bosque seco tropical en la región Caribe y el valle del río Magdalena, Colombia. *Caldasia* 21(1):70–94.

Meroni, M., R. Colombo, and C. Panigada. 2004. Inversion of a radiative transfer model with hyperspectral observations for LAI mapping in poplar plantations. *Remote Sensing of Environment* 9:195–206.

Meurer, E.J. 2007. Fatores que influenciam o crescimento e o desenvolvimento das plantas. In *Fertilidade do solo*, eds. R.F. Novais, V.H.V. Alvarez, N.F. Barros, R.L.F. Fontes, R.B. Cantarutti, and J.C.L. Neves, 65–90. Viçosa: Sociedade Brasileira de Ciência do Solo.

Meza, V.H. and F. Mora. 2002. Crecimiento del diámetro y del área basal en tres parcelas permanentes en el bosque seco tropical, parque nacional Guanacaste, Costa Rica. In *Ecosistemas forestales de bosque seco tropical: Investigaciones y resultados en mesoamérica*, 198–209. Heredia, CR: Universidad Nacional, INISEFOR.

Mikich, S.B. and S.M. Silva. 2001. Composição florística e fenologia das espécies zoocóricas de remanescentes de Floresta Estacional Semidecidual no centro-oeste do Paraná, Brazil. *Acta Botanica Brasilica* 15:89–113.

Miles, L., A. Newton, R. DeFries, C. Ravilioui s, I. May, S. Blyth, V. Kapos, J. Gordon. 2006. A global overview of the conservation status of tropical dry forests. *Journal of Biogeography* 33(3):491–505.

Miller-Rushing, A.J. and R.B. Primack. 2008. Global warming and flowering times in Thoreau's concord: A community perspective. *Ecology* 89:332–341.

Miralles, D.G., R.A.M. De Jeu, J.H. Gash, T.R.H. Holmes, and A.J. Dolman, 2011. Magnitude and variability of land evaporation and its components at the global scale. *Hydrology and Earth System Sciences* 15(3):967–981.

Miranda, M., I. Porras, and M. Moreno. 2003. *The social impacts of payments for environmental services in Costa Rica. A quantitative field survey and analysis of the Virilla watershed*. London: IIED (processed).

Mistry, J. 1998. Fire in the Cerrado (savannas) of Brazil: an ecological review. *Progress in Physical Geography* 22:425–448.

MMA - Ministério do Meio Ambiente. 2000. *Sistema Nacional de Unidades de Conservação - SNUC*. Brasília, Brazil: Ministério do Meio Ambiente.

MMA - Ministério do Meio Ambiente. Instrução Normativa nº 6, de 23 de setembro de 2008. http://www.mma.gov.br/estruturas/179/_arquivos/179_05122008033615.pdf (accessed April 12, 2012).

Mohan, J.E., L. Ziska, W.H. Schlesinger, R.B. Thomas, R.C. Sicher, K. George, and J.S. Clark. 2006. Biomass and toxicity responses of poison ivy (Toxicodendron radicans) to elevated atmospheric CO_2. *Proceedings of the National Academy of Sciences of the United States of America* 103:9086–9089.

Molina, M.A.2002. Inducción del proceso de restauración del bosque seco tropical en el Área de Conservación Guanacaste (ACG), Costa Rica. In *Ecosistemas forestales de bosque seco tropical: Investigaciones y resultados en mesoamérica*, 41–47. Heredia, CR: Universidad Nacional, INISEFOR.

Molino, J.F. and D. Sabatier. 2000. Tree diversity in tropical rain forest: A validation of the intermediate disturbance hypothesis. *Science* 294:1702–1704.

Monasterio, M. and G. Sarmiento. 1976. Phenological strategies of plant species in the tropical savanna and the semidecidous forest of the Venezuelan Llanos. *Journal of Biogeography* 3:325–356.

Monge, A., R. Quesada, and E. González. 2002. Estudio de la dinámica del bosque seco tropical a partir de parcelas permanentes de muestreo en el parque nacional Palo Verde, Bagaces, Costa Rica. In *Ecosistemas forestales de bosque seco tropical: Investigaciones y resultados en mesoamérica*, 175–184. Heredia, CR: Universidad Nacional, INISEFOR.

Montagnini, F. and C.F. Jordan. 2005. *Tropical forest ecology: The basis for conservation and management*. Berlin: Springer Verlag. 295 p.

Mooney H.A., S.H. Bullock, and E. Medina. 1995. Introduction. In *Seasonally dry tropical forests*, eds. S.H. Bullock, H.A. Mooney, and A. Medina, 1–8. Cambridge, NY: Cambridge University Press.

Morellato, L.P.C. 1992. Sazonalidade e dinâmica de ecossistemas florestais de uma área florestal no sudeste do Brazil. In *História natural da Serra do Japi: Ecologia e preservação de uma área florestal no sudeste do Brazil*, ed. L.P.C. Morellato, 98–110. Campinas: Editora da UNICAMP/FAPESP.

Morellato, L.P.C. and H.F. Leitão-Filho. 1992. Padrões de frutificação e dispersão na Serra do Japi. In *História natural da Serra do Japi: Ecologia e preservação de uma área florestal no sudeste do Brazil*, ed. L.P.C. Morellato, 112–141. Campinas: Editora da UNICAMP/FAPESP.

Morellato, L.P.C., R.R. Rodrigues, H.F. Leitão-Filho, and C.A. Joly. 1989. Estudo comparativo da fenologia de espécies arbóreas de floresta de altitude e floresta mesófila semidecídua na Serra do Japi, Jundiaí, São Paulo. *Revista Brasileira de Botânica* 12:85–98.

Morellato, P., D. Talora, A. Takahasi, C. Bencke, E. Romera, and V. Zipparro. 2000. Phenology of Atlantic Rain Forest Trees: A Comparative Study. *Biotropica* 32:811–823.

Moreno, C.E. and G. Halffter. 2001. Assessing the completeness of bat biodiversity inventories using accumulation curves. *Journal of Applied Ecology* 37:149–158.

Morim, M.P. and G.M. Barroso. 2007. Leguminosae arbustivas e arbóreas da Floresta Atlântica do Parque Nacional Itatiaia, Sudeste do Brazil: Subfamílias Caesalpinioideae e Mimosoideae. *Rodriguesia* 58:423–468.

Morisita, M. 1959. Measuring of interspecific association and similarity between communities. *Memoirs of the Faculty of Science of Kyushu University, Series E, Biology* 3:65–80.

Morrison, D.W. 1978. Lunar phobia in a Neotropical fruit bat, Artibeus jamaicensis (Chiroptera: Phyllostomidae). *Animal Behavior* 26:852–855.

Mountjoy J.B. 2008a. Arqueología de la zona costera de Jalisco y del municipio de Villa Purificación. En P. A. Regalado. Miscelánea histórica de Villa Furificación. Testimonios del 475 aniversario de su fundación. Ayuntamiento Constitucional de Villa Purificación, Jalisco. pp 21–39.

Moura, A.E.S.S. de, Correa, M.M., Silva, E.R. da, Ferreira, R.L.C., Figueiredo, A. de C., and Possas, J.A.M.C. 2009. Interceptação das chuvas em um fragmento de floresta da Mata Atlântica na Bacia do Prata, Recife, PE. *Revista Árvore* 33:461–469.

MPPA [Ministerio del Poder Popular para el Ambiente]. 2009. Climatología del estado Cojedes. Informe del plan operativo para el manejo integral de los desechos sólidos en el estado Cojedes 30 p.

MPPA [Ministerio del Poder Popular para el Ambiente]. 2010. Informe del regristro de precipitación del Estado Cojedes-Venezuela. 10 p.

Mueller-Dombois, D. and H. Ellenberg. 1974. *Aims and methods of vegetation ecology*. New York: John Wiley & Sons.

Mulkey, S.S., R.L. Chazdon, and A.P. Smith. 1996. *Tropical forest plant ecophysiology*. New York: Chapman & Hall.

Munoz, C., A. Guevara, J.M. Torres, and J. Brana. 2008. Paying for the hydrological services of Mexico's forests: Analysis, negotiations and results. *Ecological Economics* 65:725–736.

Murali, K.S. 1997. Patterns of seed size, germination and seed viability of tropical tree species in southern Índia. *Biotropica* 29:271–279.

Murcia, C. 1995. Edge effects in fragmented forests: Implications for conservation. *Trends in Ecology and Evolution* 10:58–62.

Murphy, P.G. and A.E. Lugo. 1995. Dry forests of Central America and the Caribbean islands. In *Seasonally dry tropical forests*, eds. S.H. Bullock, H.A. Mooney, and E. Medina, 9–34. Cambridge, NY: Cambridge University Press.

Murphy, P.G. and A.E. Lugo, 1986a. Ecology of tropical dry forest. *Annual Review of Ecology and Systematics* 17:67–88.

Murphy, P.G. and A.E. Lugo. 1986b. Structure and biomass of a subtropical dry forest in Puerto Rico. *Biotropica* 18:89–96.

Murray, K.G. 1988. Avian seed dispersal of three Neotropical gap-dependent plants. *Ecological Monographs* 58:271–298.

Muscarella, R. and T.H. Fleming. 2007. The role of frugivorous bats in tropical forest succession. *Biological Reviews* 82:573–590.

Mużyło, A., P. Llorens, and F. Domingo. 2011. Rainfall partitioning in a deciduous forest plot in leafed and leafless periods. *Ecohydrology* 5(6):759–767.

Muzylo, A., P. Llorens, F. Valente, J.J. Keizer, F. Domingo, and J.H.C. Gash. 2009. A review of rainfall interception modelling. *Journal of Hydrology* 370(1–4):191–206.

Myers, N., R.A. Mittermeier, C.G. Mittermeier, G.A.B. Fonseca, and J. Kent. 2000. Biodiversity hotspots for conservation priorities. *Nature* 403:853–858.

Myneni, R.B., F.G. Hall, P.J. Sellers, and A.L. Marshak. 1995. The interpretation of spectral vegetation indexes. *Geoscience and Remote Sensing, IEEE Transactions* 33(2):481–486.

Myneni, R.B., R. Ramakrishna, R. Nemani, and S.W. Running. 1997. Estimation of global leaf area index and absorbed par usingradiative transfer models. *IEEE Transactions on Geoscience and Remote Sensing* 35(6):1380–1393.

Myneni, R.B., S. Hoffman, Y. Knyazikhin, J. Privette, J. Glassy, Y. Tian, Y. Wang, et al. 2002. Global products of vegetation leaf area and fraction absorbed PAR from year one of MODIS data. *Remote Sensing of Environment* 83:214–231.

Myneni, R.B., S. Maggion, J. Iaquinta, J.L. Privette, N. Gobron, B. Pinty, M.M. Verstaete, et al. 1995. Optical remote sensing of vegetation: Modeling, caveats, and algorithms. *Remote Sensing of Environment* 51:169–188.

Nagel, J.M., X. Wang, J.D. Lewis, H.A. Fung, D.T. Tissue, and K.L. Griffin. 2005. Atmospheric CO_2 enrichment alters energy assimilation, investment and allocation in Xanthium strumarium. *New Phytologist* 166:513–523.

Nagendra, H. 2001. Using remote sensing to asses biodiversity. *International Journal of Remote Sensing* 12:2377–2400.

Nagendra, H. and M. Gadgil. 1999. Satellite imagery as a tool for monitoring species diversity: An assessment. *Journal of Applied Ecology* 36(3):388–397.

Nardoto, G.B., J.P.H.B. Ometto, J.R. Ehleringer, N. Higuchi, M.M.C. Bustamante, and L.A. Martinelli. 2008. Understanding the influences of spatial patterns on N availability within the Brazilian Amazon forest. *Ecosystems* 11:1234–1246.

Nascimento, A.R.T., J.M. Felfili, and E.M. Meirelles. 2004. Florística e estrutura dacomunidade arbórea de um remanescente de Floresta Estacional Decidual de Encosta, Monte Alegre, GO, Brazil. *Acta Botanica Brasilica* 18:650–669.

Nassar, J.M., J.P. Rodríguez, A. Sánchez-Azofeifa, T. Garvin, and M. Quesada (eds.). 2008. *Human, Ecological and Biophysical dimensions of Tropical dry forests. Manual of Methods.* Caracas: Ediciones IVIC. 135 p.

Navas, M.L., C. Roumet, A. Bellmann, G. Laurent, and E. Garnier. 2010. Suites of plant traits in species from different stages of a Mediterranean secondary succession. *Plant Biology* 12:183–196.

Neelin, J.D., M. Munnich, H. Su, J.E. Meyerson, and C. Holloway. 2006. Tropical drying trends in global warming models and observations. *Proceedings of the National Academy of Sciences of the United States of America* 103:6110–6115.

Nelson, A. and K. Chomitz. 2011. Effectiveness of strict vs. multiple use protected areas in reducing tropical forest fires: A global analysis using matching methods. *PLoS ONE* 6:e22722.

Nelson, G.C. and D. Hellerstein. 1997. Do roads cause deforestation? Using satellite images in econometric analysis of land use. *American Journal of Agricultural Economics* 79:80.

Neves, F.S. 2009. Dinâmica espaço-temporal de insetos associados a uma floresta estacional decidual. PhD. dissertation, Universidade Federal de Minas Gerais, Belo Horizonte, Brazil.

Neves, F.S., B.G. Madeira, V.H.F. Oliveira, and M. Fagundes. 2008. Insetos como bioindicadores dos processos de regeneração em matas secas. *MG Biota* 1:46–53.

Neves, F.S., L.S. Araújo, M. Fagundes, M.M. Espirito-Santo, M. Fagundes, G.W. Fernandes, G.A. Sanchez-Azofeifa, and M. Quesada. 2010a. Canopy herbivory and insect herbivore diversity in a dry forest-savanna transition in Brazil. *Biotropica* 42:112–118.

Neves, F.S., R.F. Braga, M.M. Espírito-Santo, J.H.C. Delabie, G.W. Fernandes, and G.A. Sánchez-Azofeifa. 2010b. Diversity of arboreal ants in a Brazilian tropical dry forest: Effects of seasonality and successional stage. *Sociobiology* 56:177–194.

Neves, F.S., V.H. Oliveira, M.M Espírito-Santo, F.Z. Vaz-De-Mello, J. Louzada, and G.W. Fernandes. 2010c. Successional and seasonal changes in a community of dung beetles (Coleoptera: Scarabaeinae) in a Brazilian tropical dry forest. *Nature Conservancy* 8:160–164.

Newstrom, L.E., G.W. Frankie, H.G. Baker, and R.K. Colwell. 1994. Diversity of long-term flowering patterns. In *La Selva: Ecology and natural history of a neotropical rain forest*, ed. L.A. Lucinda, K.S. Bawa, H. Hespenheide, G.S. Hartshorn, 142–160. Chicago: Chicago University Press.

Nichol, C.J., K.F. Huemmrich, T.A. Black, P.G. Jarvis, C.L. Walthall, and J. Grace. 2000. Remote sensing of photosynthetic-light-use efficiency of boreal forest. *Agricultural and Forest Meteorology* 101(2–3):131–142.

Nieschulze, J., S. Erasmi, J. Dietz, and D. Hölscher. 2009. Satellite-based prediction of rainfall interception by tropical forest stands of a human-dominated landscape in Central Sulawesi, Indonesia. *Journal of Hydrology* 364(3–4):227–235.

Niklas, K.J. 1995. Size-dependent allometry of tree height, diameter and trunk taper. *Annals of Botany* 75:217–227.

Nimer, E. and A.M.P.M. Brandão. 1989. *Balanço hídrico e clima da região dos cerrados*. Rio de Janeiro: IBGE.

Noguera, F.A., J.H. Vega Rivera, A.N. García Aldrete, and M. Quesada Avendaño (eds.). 2002. *Natural history of Chamela*. UNAM, Mexico: Instituto de Biología. 568 p.

North, G.B., L. Moore, and P.S. Nobel. 1995. Cladode Cunder current and doubled CO_2 concentrations. *American Journal of Botany* 82:159–166.

Nunes, Y.R.F., A.V.R. Mendonça, L. Bottezelli, E.L.M. Machado, and A.T. Oliveira-Filho. 2003. Variações da fisionomia, diversidade e composição de guildas da comunidade arbórea em um fragmento de Floresta semidecidual em lavras, MG. *Acta Botanica Brasilica* 17:213–229.

Nunes, Y.R.F., M. Fagundes, H.S. Almeida, and M.D.M. Veloso. 2008. Aspectos ecológicos da aroeira (Myracrodruon urundeuva Allemão – Anacardiaceae): Fenologia e germinação de sementes. *Revista Árvore* 32:233–243.

Nunes, Y.R.F., M. Fagundes, M.R. Santos, R.F. Braga, and A.P.D. Gonzaga. 2006. Germinação de sementes de Guazuma ulmifolia Lam. (Malvaceae) e Heteropterys byrsonimifolia A. Juss (Malpighiaceae) sob diferentes tratamentos de escarificação tegumentar. *Unimontes Científica* 8:43–52.

Nunes, Y.R.F., M. Fagundes, R.M. Santos, E.B.S. Domingues, H.S. Almeida, and A.P.D. Gonzaga. 2005. Atividades fenológicas de Guazuma ulmifolia Lam. (Malvaceae) em uma Floresta Estacional Decidual no norte de Minas Gerais. *Lundiana* 6:99–105.

Nunes, Y.R.F. and M. Petrere Jr. 2012. Structure and dynamics of a Cariana estrellensis (Lecythidaceae) population in a fragment of Atlantic Forest in Minas Gerais, Brazil. *Rodriguésia* 63:257–267.

O'Dea, N. and R.J. Whittaker. 2007. How resilient are Andean montane forest bird communities to habitat degradation? *Biodiversity and Conservation* 16:1131–1159.

Oehl, F., Z. Sýkorováz, J. Blaszkowsi, I. Sánchez-Castro, D. Coynes, A. Tchabi, L. Lawouin, F.C.C. Hountondji, and G.A. Silva, 2011. Acaulospora sieverdingii, an ecologically diverse new fungus in the Glomeromycota, described from lowland temperate Europe and tropical West Africa. *Journal of Applied Botany and Food Quality* 84(1):47–53.

Olivares, E. and E. Medina. 1992. Water and nutrient relations of woody perennials from tropical dry forests. *Journal of Vegetation Science* 3:383–392.

Oliveira, F.C., U.P. Albuquerque, V.S. Fonseca-Kruel, and N. Hanazaki. 2009. Avanços nas pesquisas etnobotânicas no Brazil. *Acta botanica brasilica* 23:590–605.

Oliveira, P.E.A.M. and A.G. Moreira. 1992. Anemocoria em espécies do cerrado e mata de galeria de Brasília, DF. *Revista Brasileira de Botânica* 15:163–174.

Oliveira, P.E. and P.E. Gibbs. 2000. Reproductive biology of woody plants in a Cerrado community of Central Brazil. *Flora* 195:311–329.

Oliveira, P.E., H. Behling, M.P. Ledru, M. Barbieri, M. Bush, M.L. Salgado-Laboriau, M.J. Garcia, S. Medeanic, O.M. Barth, M.A. Barros, and R. Scheelybert. 2005. Paleovegetação e paleoclimas do quaternário do Brazil. In *Quaternário do Brazil* eds. C.R.D. Souza, K. Suguio, A.M.S. Oliveira, and P.E. Oliveira, 52–74. Ribeirão Preto: Holos Editora.

Oliveira, P.S. and R.J. Marquis. 2002. *The Cerrados of Brazil: Ecology and natural history of a neotropical savanna*. New York: Columbia University Press.

Oliveira-Filho, A.T. 2006. *Catálogo das árvores nativas de Minas Gerais: mapeamento e inventário da flora nativa e dos reflorestamentos de Minas Gerais*. Lavras: Editora da UFLA.

Oliveira-Filho, A.T., J.A. Jarenkow, and M.J.N. Rodal. 2006. Floristic relationships of seasonally dry forests of eastern South America based on tree species distribution patterns. In *Neotropical savannas and dry forests: Plant diversity, biogeography, and conservation*, eds. R.T. Pennington, G.P. Lewis, and J.A. Ratter, 59–192. Oxford: Taylor & Francis CRC Press.

Oliveira-Filho, A.T., J.M. Mello, and J.R.S. Scolforo. 1997. Effects of past disturbance and edges on tree community structure and dynamics within a fragment of tropical semideciduous forest in Southeastern Brazil over a five-year period (1987–1992). *Plant Ecology* 131:45–66.

Oliveira-Filho, A.T., N. Curi, E.A.Vilela, and D.A. Carvalho. 1998. Effects of canopy gaps, topography and soils on the distribution of woody species in a central Brazilian deciduous dry forest. *Biotropica* 30:362–375.

Oliveira Júnior, J.C. and H.C.T. Dias. 2005. Precipitação efetiva em fragmento secundário da Mata Atlântica. *Revista Árvore* 29:9–15.

Olofsson, P., L. Eklundh, F. Lagergren, P. Jönsson, and A. Lindroth. 2007. Estimating net primary production for Scandinavian forests using data from Terra/MODIS. *Advances in Space Research* 39:125–130.

Olson, D.M. 2000. *The Global 200: A representation approach to conserving the Earth's distinctive ecoregions*. World Wildlife Fund-US: Conservation Science Program.

Olson, D.M., E. Dinerstein, E.D. Wikramanayake, N.D. Burgess, G.V.N. Powell, E.C. Underwood, J.A. D'Amico, et al. 2001. Terrestrial ecoregions of the world: A new map of life on Earth. *Bioscience* 51:933–938.

Opler, P.A., G.W. Frankie, and H.G. Baker. 1976. Rainfall as a factor in the release, timing, and sychronization of anthesis by tropical tress and shrubs. *Journal of Biogeography* 3:231–236.

Opler, P.A., H.G. Baker, and G.W. Frankie. 1980. Plant reproductive characteristics during secondary succession in neotropical lowland forest ecosystems. *Biotropica* 12:40–46.

Ormond, W.T., M.C.B. Pinheiro, H.A. Lima, M.C.R. Correia, and M.L. Pimenta. 1993. Estudo das recompensas florais das plantas da restinga de Marica-Itaipuacu, RJ. I - Nectariferas. *Bradea* 6:179–195.

Ormrod, D.P., V.M. Lesser, D.M. Olszyk, and D.T. Tingey. 1999. Elevated temperature and carbon dioxide affect chlorophylls and carotenoids in douglas-fir seedlings. *International Journal of Plant Sciences* 160:529–534.

Ortiz, E. and F. Carrera. 2002. Estadística básica para inventarios forestales. In *Inventarios forestales para bosques latifoliados en América Central*, Orozco, L., C. Brumér, eds. Turrialba, CR: CATIE. 71–99.

Ortiz, E., L. Sage, and C. Borge. 2003. *Impacto del programa de pago de servicios ambientales en Costa Rica como medio de reducción de la pobreza en los medios rurales*. Serie de Publicaciones RUTA. San José: Unidad Regional de Asistencia Técnica.

Ortiz-Pulido, R., J. Laborde, and S. Guevara. 2000. Frugivory by birds in a fragmented landscape: consequences on seed dispersal. *Biotropica* 32(3):473–488.

Owensby, C.E., J.M. Ham, A.K. Knapp, and L.M. Auen. 1999. Biomass production and species composition change in a tallgrass prairie ecosystem after long-term exposure to elevated atmospheric CO_2. *Global Change Biology* 5:497–506.

Palmeirim, J.M., D.L. Gorchov, and S. Stoleson. 1989. Trophic structure of a neotropical frugivore community: Is there competition between birds and bats? *Oecologia* 79:403–411.

Paré, L. 1995. Transformación de los sistemas productivos y deterioro del medio ambiente en una región étnica del trópico veracruzano. In *Globalización, deterioro ambiental y reorganización social en el campo*, ed. H. Carton de Grammont, 122–158. Mexico DF: Juan Pablos Editor/UNAM.

Parga, P. and A. Quevedo, 2002. Estudios de caracterización biofísica y socioeconómica de la ecorregión estratégica del valle del alto Magdalena. Componente Fauna. Ministerio de Ambiente, CORTOLIMA, CAM, Universidad del Tolima and Universidad Surcolombiana. Ibagué – Tolima. 600 p.

Pasa, M.C., J.J. Soares, and G. Guarim-Neto. 2005. Estudo etnobotânico na comunidade de Conceição-Açu (alto da bacia do rio Aricá Açu, MT, Brazil). *Acta Botanica Brasilica* 19:195–207.

Pascarella, J., T. Aide, and J. Zimmerman. 2004. Short-term response of secondary forests to hurricane disturbance in Puerto Rico, USA. *Forest Ecology and Management* 199:379–393.

Passos, M.A., K.M.P. Tavares, and A.R. Alves. 2007. Germinação de sementes de sabiá (Mimosa caesalpiniaefolia Benth.). *Revista Brasileira de Ciências Agrárias* 2:51–56.

Patton, M.Q. 2002. *Qualitative research and evaluation methods*. Thousand Oaks: Sage Publications.

Pearson, H.L. and P.M. Vitousek. 2001. Stand dynamics, nitrogen accumulation, and symbiotic nitrogen fixation in regenerating stands of Acacia koa. *Ecological Applications* 11:1381–1394.

Pech-Canche, J.M., C.E. Moreno, and G. Halffter. 2011. Additive partitioning of phyllostomid bat richness at fine and coarse spatial and temporal scales in Yucatan, Mexico. *Ecoscience* 18:42–51.

Peddle, D.R., F.G. Hall, and E.F. Ledrew. 1999. Spectral mixture analysis and geometric-optical reflectance modeling of boreal forest biophysical structure. *Remote Sensing of Environment* 67(3):288–297.

Pedesoli, J.L, and J.L.A. Martins. 1972. A vegetação dos afloramentos de calcário. *Oreades* 5:27–29.

Pedralli, G. 1997. Florestas secas sobre afloramentos de calcário em Minas Gerais: florística e ficionomia. *Bios* 5:81–88.

Peña, E.P. and Peña, M.C. 2003. Caracterización fenológica de 10 especies forestales del bosque seco tropical ubicadas en el Centro Universitario Regional del Norte CURN en el municipio de Armero Guayabal [Characterization phenological of 10 forest species of tropical dry forest located in the University Center North Regional CURN in the town of Armero Guayabal]. BSc Forestry Engineering thesis. Universidad Del Tolima.

Pennington, R.T., D.E. Prado, and C.A. Pendry. 2000. Neotropical seasonally dry forests and Quaternary vegetation changes. *Journal of Biogeography* 27:261–273.

Pennington, R.T., G.P. Lewis, and J.A. Ratter. 2006a. An overview of the plant diversity, biogeography and conservation of neotropical savannas and seasonally dry forests. In *Neotropical savannas and seasonally dry forests plant diversity biogeography and conservation*, eds. R.T. Pennington, G.P. Lewis, and J.A. Ratter, 1–29. Boca Raton, FL: CRC Press.

Pennington, R.T., M. Lavin, and A. Oliveira-Filho. 2009. Woody plant diversity, evolution, and ecology in the tropics: Perspectives from seasonally dry tropical forests. *Annual Review of Ecology, Evolution, and Systematics* 40:437–457.

Peñuelas, J. and I. Filella. 1998. Visible and near-infrared reflectance techniques for diagnosing plant physiological status. *Trends Plant Science* 3:151–156.

Penuelas, J., I. Filella, and J.A. Gamon. 1995. Assessment of photosynthetic radiation-use efficiency with spectral reflectance. *New Phytologist* 131(3):291–296.

Perea, J.J. 2001. Estado de los suelos de ecorregión estratégica dela Tatacoa. In *Caracterización biofísica y socioeconómica de la ecorregión estratégica de la Tatacoa y su área de influencia*, Tomo III, Componente Suelos, 180. Universidad del Tolima [State soils Tatacoa dela strategic ecoregion. in Biophysical and socioeconomic characterization of strategic ecoregion Tatacoa and its area of influence, Volume III, Component Soil] 180. Universidad del Tolima.

Pereira, T.S., M.L.M.N. Costa, L.F.D. Moraes, and C. Luchiari. 2008. Fenologia de espécies arbóreas em floresta atlântica da Reserva Biológica de Poço das Antas, Rio de Janeiro, Brazil. *Iheringia* 63:329–339.

Pérez-García, E.A., A.C. Sevilha, J.A. Meave, and A. Scariot. 2009. Floristic differentiation in limestone outcrops of southern Mexico and Central Brazil: a beta diversity approach. *Boletin de la Sociedad Botanica de Mexico* 84:45–58.

Peters, V.E., R. Mordecai, C.R. Carroll, R.J. Cooper, and R. Greenberg. 2010. Bird community response to fruit energy. *Journal of Animal Ecology* 79:824–835.

Peterson, D.L., J.D. Aber, P.A. Matson, D.H. Card, N. Swanberg, C. Wessman, and M. Spanner. 1988. Remote sensing of forest canopy and leaf biochemical contents. *Remote Sensing of Environment* 24(1):85–108.

Peterson, D.L., M.A. Spanner, S.W. Running, and K.B. Teuber. 1987. Relationship of thematic mapper simulator data to leaf area index of temperate coniferous forests. *Remote Sensing of Environment* 22(3):323–341.

Pezzini, F.F., D.O. Brandão, B.D. Ranieri, M.M. Espirito-Santo, C.M. Jacobi, and G.W. Fernandes. 2008. Polinização, dispersão de sementes e fenologia das espécies arbóreas no Parque Estadual da Mata Seca. *MG Biota* 1:37–45.

Pfaff, A. 1999. What drives deforestation in the brazilian Amazon? Evidence from satellite and socioeconomic data. *Journal of Environmental Economics and Management* 37(1):26–43.

Pfaff, A., G.S. Amacher, and E.O. Sills. 2013a. Realistic REDD: Improving the forest impacts of domestic policies in different settings. *Review of Environmental Economics and Policy* 7(1):114–135.

Pfaff, A. and J. Robalino. 2012. Protecting forests, biodiversity and the climate: Predicting policy impact to improve policy choice. *Oxford Review of Economic Policy* 28(1):164–179.

Pfaff, A., J. Robalino, D. Herrera, and C. Sandoval. 2012. Protected areas' forest spillovers in the Brazilian Amazon: What might explain 'blockage'? AERE Summer Conference, June 4, Asheville, NC.

Pfaff, A., J. Robalino, and R. Walker. 2010. Spatial Transport Dynamics on the Brazilian Amazon Frontier: Roads follow roads, making long-run impacts higher than short run. Duke University mimeo.

Pfaff, A., J. Robalino, R. Walker, S. Perz, W. Laurance, C. Bohrer, J.A. Robalino, S. Aldrich, E. Arima, M. Caldas, and K. Kirby. 2011. 'Clean Development' from infrastructure location: spatially varied road impacts on Brazilian Amazon deforestation. ASSA/AERE conference, January 7, Denver, CO.

Pfaff, A., J.A. Robalino, E. Walker, S. Reis, C. Perz, S. Bohrer, E. Aldrich, M. Arima, Caldas, W Laurance and K. Kirby. 2007. Road investments, spatial intensification and deforestation in the Brazilian Amazon. *Journal of Regional Science* 47:109–123.

Pfaff, A., J.A. Robalino, and G.A. Sánchez-Azofeifa. 2007. Changing the deforestation impacts of ecopayments: Evolutions (2000–2005) within Costa Rica's PSA. NBER Summer Institute, July 24, Cambridge, MA.

Pfaff, A., J.A. Robalino, G.A. Sánchez-Azofeifa, K. Andam and P. Ferraro 2009. Park location affects forest protection: Land characteristics cause differences in park impacts across Costa Rica, The B.E. *Journal of Economic Analysis & Policy* 9(2) (Contributions), Article 5. http://www.bepress.com/bejeap/vol9/iss2/art5.

Pfaff, A. and J.A. Robalino. 2011. Decentralization given environment-development tradeoffs: Federal versus state conservation and impacts on Amazonian deforestation. AERE Summer Conference, June 9–10, Seattle, WA.

Pfaff, A., J.A. Robalino, E. Lima, C. Sandoval, and L.D. Herrera. 2013b. Governance, location & avoided deforestation from protected areas: Greater restrictions can have lower impact, due to differences in location. *World Development* (forthcoming).

Phillips, O. and A.H. Gentry. 1993. The useful plants of Tambopata, Peru: I. Statistical hypothesis tested with a new quantitative technique. *Economic Botany* 47(1):15–32.

Phillips, O.L., R.V. Martínez, L. Arroyo, T.R. Baker, T. Killeen, S.L. Lewis, Y. Malhi, et al. 2002. Increasing dominance of large lianas in Amazonian forests. *Nature* 418:770–774.

Piña-Rodrigues, F.C.M. and R.M. Jesus. 1992. Comportamento das sementes de cedro-rosa (Cedrela angustifolia S. ET. MOC.) durante o armazenamento. *Revista Brasileira de Sementes* 14:31–36.

Pinheiro, C.E.G. and J.V.C. Ortiz. 1992. Communities of fruit-feeding butterflies along a vegetation gradient in central Brazil. *Journal of Biogeography* 19:505–511.

Pinto, E.P.P., M.C.M. Amorozo, and A. Furlan. 2006. Conhecimento popular sobre plantas medicinais em comunidades rurais de Mata Atlântica – Itacaré, BA, Brazil. *Acta Botanica Brasilica* 20:751–762.

Pinto, J.R.R., A.T. Oliveira-Filho, and J.D.V. Hay. 2005. Influence of soil and topography on the composition of a tree community in a central brazilian valley forest. *Edinburgh Journal of Botany* 62:69–90.

Piotto, D., D. Craven, F. Montagnini, and F. Alice. 2010. Silvicultural and economic aspects of pure and mixed native tree species plantations on degraded pasturelands in humid Costa Rica. *New Forests* 39:369–385.

Pohlman, C.L., S.M. Turton, and M. Goosem. 2007. Edge effects of linear canopy openings on tropical rain forest understory microclimate. *Biotropica* 39(1): 62–71.

Pohlman, C.L., S.M. Turton, and M. Goosem. 2009. Temporal variation in microclimatic edge effects near powerlines, highways and streams in Australian tropical rainforest. *Agricultural and Forest Meteorology* 149:84–95.

Poorter, L. 2005. Resource capture and use by tropical forest tree seedlings and their consequences for competition. In *Biotic interactions in the Tropics; their role in the maintenance of species diversity*, eds. D. Burslem, M. Pinard, and S. Hartley, 35–64. Cambridge: Cambridge University Press.

Poorter, L. 2009. Leaf traits show different relationships with shade tolerance in moist versus dry tropical forests. *New Phytologist* 181:890–900.

Popinigis, F. 1977. Dormência. In *Fisiologia da semente*. ed. F. Popinigis, 75–93. Brasília: AGIPLAN.

Portigal, F., R. Holasek, G. Mooradian, P. Owensby, M. Dicksion, and M. Fene. 1997. Vegetation classification using red edge first derivative and green peak statistical moment indices with the Advanced Airborne Hyperspectral Imaging System (AAHIS).

Portillo-Quintero, C.A. and G.A. Sánchez-Azofeifa. 2010. Extent and conservation of tropical dry forests in the Americas. *Biological Conservation* 143:144–155.

Posey, D. 1984. A preliminary report on diversified management of tropical forest by the Kayapó Indians of the Brazilian Amazon. In *Ethnobotany in the Neotropics*, eds. G. Prance and J. Kallunki, 112–126. New York: The New York Botanical Garden.

Poulin, B., G. Lefebvre, and R. McNeil. 1994. Characteristics of feeding guilds and variation in diets of bird species of three adjacent tropical sites. *Biotropica* 26:187–197.

Powers, J.S. and P. Tiffin. 2010. Plant functional type classifications in tropical dry forests in Costa Rica: Leaf habit versus taxonomic approaches. *Functional Ecology* 24:927–936.

Powers, J.S. and S. Salute. 2011. Macro- and micronutrient effects on decomposition of leaf litter from two tropical tree species: Inferences from a short-term laboratory incubation. *Plant and Soil* 346:245–257.

Powers, J.S., J.M. Becknell, J. Irving, and D. Pèrez-Aviles. 2009a. Diversity and structure of regenerating tropical dry forests in Costa Rica: Geographic patterns and environmental drivers. *Forest Ecology and Management* 258:959–970.

Powers, J.S., R.A. Montgomery, E.C. Adair, F.Q. Brearley, S.J. DeWalt, C.T. Castanho, J. Chave. 2009b. Decomposition in tropical forests: A pan-tropical study of the effects of litter type, litter placement and mesofaunal exclusion. *Journal of Ecology* 97:801–811.

Prado, D.E. and P.E. Gibbs. 1993. Patterns of species distributions in the dry seasonal forest South America. *Annals of the Missouri Botanic Garden* 80:902–927.

Prance, G.T. 1973. Phytogeographical support for the theory of Pleistocene forest refuges in the Amazon basin, based on evidence from distribution patterns in Caryocaraceae, Chrysobalanaceae, Dichapetalaceae and Lecythidaceae. *Acta Amazonica* 3:5–28.

Prance, W. 2006. Tropical savannas and seasonally dry forests: An introduction. *Journal of Biogeography* 33:385–386.

Price, M.F. 2002. The periodic review of biosphere reserves: A mechanism to foster sites of excellence for conservation and sustainable development. *Environmental Science and Policy* 5:13–18.

Price, P.W. 2002. Resource-driven terrestrial interaction webs. *Ecological Research* 17:241–247.

Pringle, E.G., R.I. Adams, E. Broadbent, P.E. Busby, C.I. Donatti, E.L.Kurten, K. Renton, and R. Dirzo. 2011. Distinct leaf-trait syndromes of evergreen and deciduous trees in a seasonally dry tropical forest. *Biotropica* 43:299–308.

Prior, L.D., D. Eamus, and D.M.J.S. Bowman. 2003. Leaf attributes in the seasonally dry tropics: A comparison of four habitats in northern Australia. *Functional Ecology* 17:504–515.

Pritchard, S.G., H.H. Rogers, S.A. Prior, and C.M. Peterson. 1999. Elevated CO_2 and plant structure: A review. *Global Change Biology* 5:807–837.

Privette, J.L., W.J. Emery, and D.S. Schimel. 1996. Inversion of a vegetation reflectance model with NOAA AVHRR data. *Remote Sensing of Environment* 58:187–200.

Pujadas, A. and A. Castillo. 2007. Social participation in conservation efforts: A case study of a biosphere reserve on private lands in Mexico. *Society and Natural Resources* 20:57–72.

Pujadas-Botey, A. 2003. Comunicación y participación social en el programa de Ordenamiento Ecológico Territorial de la Costa de Jalisco y la Reserva de la Biosfera Chamela-Cuixmala. Tesis de Maestría. Institución: Instituto de Ecología, UNAM.

Pypker, T.G., B.J. Bond, T.E. Link, D. Marks, and M.H. Unsworth. 2005. The importance of canopy structure in controlling the interception loss of rainfall: Examples from a young and an old-growth Douglas-fir forest. *Agricultural and Forest Meteorology* 130(1–2):113–129.

Qi, J., A. Chehbouni, A.R. Huete, Y.H. Kerr, and S. Sorooshian. 1994. A modified soil adjusted vegetation index. *Remote Sensing of Environment* 48(2):119–126.

Quesada, M., E.J. Fuchs, and J.A. Lobo. 2001. Pollen load size, reproductive success, and progeny kinship of naturally pollinated flowers of the tropical dry forest tree Pachira quinata, Bombacaceae. *American Journal of Botany* 88:2113–2118.

Quesada, M. and K.E. Stoner. 2004. Threats to the conservation of the tropical dry forest in Costa Rica. In *Biodiversity conservation in Costa Rica: Learning the lessons in a seasonal dry forest*, eds. G.W. Frankie, A. Mata, and S.B. Vinson, 266–280. Berkeley: University of California Press.

Quesada, M., K.E. Stoner, J.A. Lobo, Y. Herrerias-Diego, C. Palacios-Guevara, M.A. Munguía-Rosas, K.A.O. Salazar, and V. Rosas-Guerrero. 2004. Effects of forest fragmentation on pollinator activity and consequences for plant reproductive success and mating patterns in bat-pollinated bombacaceous trees. *Biotropica* 36:131–138.

Quesada, M., K.E. Stoner, V. Rosas-Guerrero, C. Palacios-Guevara, and J.A. Lobo. 2003. Effects of habitat disruption on the activity of nectarivorous bats (Chiroptera: Phyllostomidae) in a dry tropical forest: Implications for the reproductive success of the neotropical tree Ceiba grandiflora. *Oecologia* 135:400–406.

Quesada, M., R. Aguilar, F. Rosas, L. Ashworth, V.M. Rosas-Guerrero, R. Sayago, J.A. Lobo, Y. Herrerías-Diego, and G. Sánchez-Montoya. 2011. Human impacts on pollination, reproduction and breeding systems in tropical forest plants. In *Seasonally dry tropical forests*, ed. R. Dirzo, H. Mooney, and G. Ceballos, 173–194. Washington, DC: Island Press.

Quesada, M.G.A., M. Sánchez-Azofeifa, K. Alvarez-Anorve, L. Stoner, J. Avila-Cabadilla, A. Calvo-Alvarado, M.M. Castillo, et al. 2009. Succession and management of tropical dry forests in the Americas: Review and new perspectives. *Forest Ecology and Management* 258:1014–1024.

Quesada, R. 1980. *Costa Rica: La frontera sur de Mesoamérica*. 2nd ed. San José, Costa Rica: Instituto Costaricense de Turismo. Instituto de la Caza Fotográfica (INCAFO).

Quimbayo, L.C. 2009. Determinación del Estado de Fragmentación del Bosque Seco Tropical en las Zonas Secas del Centro del Tolima, con el fin de Identificar Áreas de Interés Para La Conservación [Determination of the State of the Dry Forest Fragmentation Tropical Dry Area Tolima Center, in order to identify areas of Interest for Conservation]. BSc Bilogy thesis, Universidad del Tolima, Ibague, Colombia.

Quiroga, J.A. and H.Y. Roa. 2002. Estructura de los fragmentos boscosos de la ecorregión de la Tatacoa y su área de influencia, en el Tolima [Structure of the forest fragments Tatacoa ecoregion and its area of influence in Tolima]. BSc Forestry Engineering thesis, Universidad del Tolima, Ibague, Colombia.

R Development Core Team. 2009. *R: A language and environment for statistical computing*. R foundation for statistical computing. http://www.r-project.org.

R Development Core Team. 2011. *R: A language and environment for statistical computing*. R Foundation for Statistical Computing.

Raghubanshi, A.S. 1992. Effect of topography on selected soil properties and nitrogen mineralization in a dry tropical forest. *Soil Biology and Biochemistry* 24:145–150.

Ragusa-Netto, J. and R.R. Silva. 2007. Canopy phenology of a dry forest in western Brazil. *Brazilian Journal of Biology* 67:569–575.

Rahbek, C. and G.R. Graves. 2001. Multiscale assessment of patterns of avian species richness. *Proceedings of the National Academy of Science* 98:4534–4539.

Rahman, A., J. Gamon, D. Fuentes, D. Roberts, and D. Prentiss. 2001. Modeling spatially distributed ecosystem flux of boreal forest using hyperspectral indices from AVIRIS imagery. *Journal of Geophysical Research. D. Atmospheres* 106:33.

Raich, J.W., R.H. Riley, and P.M. Vitousek. 1994. Use of root-ingrowth cores to assess nutrient limitations in forest ecosystems. *Canadian Journal of Forest Research* 24:2135–2138.

Ramia, M. 1977. Observaciones fenológicas en las sabanas del medio Apure. *Acta Botanica Venezuelica* 12:171–206.

Ramos, M.A., P.M. Medeiros, A.L.S. Almeida, A.L.P. Feliciano, and U.P. Albuquerque. 2008. Can wood quality justify local preferences for firewood in an area of caatinga (dryland) vegetation. *Biomass & Bioenergy* 32:503–509.

Rangel-Salazar, J.L., P.L. Enríquez, and E.C. Sántiz-López. 2009. Diversity variation of understory birds in Lagos National Park, Montebello, Chiapas, México. *Acta Zoológica Mexicana* 25:479–495.

Ratter, J.A., G.P. Askew, R.F. Montgomery, and D.R. Gifford. 1978. Observations on the vegetation of Northeastern Mato Grosso. II. Forests and soils of the rio Suia-Missu area. *Proceedings of the Royal Society of London. Series B, Biological Sciences* 203:191–208.

Read, L. and D. Lawrence. 2003. Litter nutrient dynamics during succession in dry tropical forests of the Yucatán: Regional and seasonal effects. *Ecosystems* 6:747–761.

Reatto, A., J. Correia, and S. Spera. 1998. Solos do bioma Cerrado: Aspectos pedológicos. In *Cerrado: Ambiente e flora*, eds. S. Sano and S. Almeida, 47–86. Planaltina: Embrapa-CPAC.

Reed, R.A., M.E. Finley, W.H. Romme, and M.G. Turner. 1999. Aboveground net primary production and leaf-area index in early postfire vegetation in Yellowstone National Park. *Ecosystems* 2(1):88–94.

Reich, P.B. 1995. Phenology of tropical forests: Patterns, causes, and consequences. *Canadian Journal of Botany* 73:164–174.

Reich, P.B. and R. Borchert. 1982. Phenology and ecophysiology of the tropical tree, *Tabebuia neochrysantha* (Bignoniaceae). *Ecology* 63:294–299.

Reich, P.B. and R. Borchert. 1984. Water stress and tree phenology in a tropical dry forest in the lowlands of Costa Rica. *Journal of Ecology* 72:61–74.

Reich, P.B., M.B. Walters, and D.S. Ellsworth. 1997. From tropics to tundra: Global convergence in plant functioning. *Proceedings of the National Academy of Sciences* 94:13730–13734.

Rentería, L.Y. and V.J. Jaramillo. 2011. Rainfall drives leaf traits and leaf nutrient resorption in a tropical dry forest in Mexico. *Oecologia* 165:201–211.

Rentería, L.Y., V.J. Jaramillo, A. Martínez-Yrízar, and A. Pérez-Jiménez. 2005. Nitrogen and phosphorus resorption in trees of a Mexican tropical dry forest. *Trees–Structure and Function* 19:431–441.

Repizo, A.A. and C.A. Devia. 2008. Árboles y arbustos del valle seco del río Magdalena y e la región Caribe colombiana: su ecología y usos. Bogotá, D.C. Facultad de Estudios Ambientales y Rurales, Pontificia Universidad Javeriana. 150 p.

Restall, R., C. Rodner, and M. Lentino. 2007. *Birds of nothern South America. An identification guide*. New Heaven: Yale University Press. 880 p.

Rex, K., D.H. Kelm, K. Wiesner, T.H. Kunz, and C.C. Voigt. 2008. Species richness and structure of three Neotropical bat assemblages. *Biological Journal of the Linnean Society* 94:617–629.

Rey, A.E. 2008. Efecto de la explotación agrícola y ganadera sobre las especies vegetales de la sucesión temprana en áreas de bosque seco tropical en el municipio de Armero Guayabal – Tolima [Effect of farming and ranching on the plant species Early succession in tropical dry forest areas in the municipality Guayabal Armero – Tolima]. MSc Biological Sciences thesis, Universidad Del Tolima, Ibague, Colombia.

Reyes-García, V., V. Vadez, T. Huanca, W. Leonard, and D. Wilkie. 2005. Knowledge and uses of wild plants: A comparative study in two Tsimane' villages in the Bolivian lowlands. *Ethnobotany Research Applications* 3:201–207.

Ribeiro, J.F. and B.M.T. Walter. 2008. As principais fitofisionomias do bioma Cerrado. In *Cerrado: Ecologia e flora*. eds. S.M. Sano, S.P. Almeida, and J.F. Ribeiro, 151–212. Planaltina: Embrapa Informação Tecnológica, Brasília: Embrapa Cerrados.

Ribeiro, M.L.R.C. and C.F. Barros. 2006. Variação intraespecífica do lenho de Pseudopiptadenia contorta (DC.) G.P. Lewis & M.P. Lima (Leguminosae – Mimosoideae) de populações ocorrentes em dois remanescentes de Floresta Atlântica. *Acta Botanica Brasilica* 20:839–844.

Ricklefs, R.E. and D. Schluter. 1993. *Species diversity in ecological communities: Historical and geographical perspectives*. Chicago: University of Chicago Press.

Ridgely, R.S. and G. Tudor. 1989. *The birds of South America*. Vol I. The Oscine Passerines. Austin: University of Texas Press. 596 p.

Ridgely, R.S. and G. Tudor. 1994. The birds of South Amercia. Vol II. The Suboscine Passerines. Austin: University of Texas Press. 940 p.

Ries, L., R.J. Fletcher, J. Battin, and T.D. Sisk. 2004. Ecological responses to habitat edges: Mechanisms, Models, and Variability Explained. *Annual Review of Ecology, Evolution, and Systematics* 35:491–522.

Rilling, M. 2004. Arbuscular mycorrhizae and terrestrial ecosystem processes. *Ecology Letters* 7:740–754.

Rillig, M.C. and D.L. Mummey. 2006. Mycorrhizas and soil structure. *New Phytologist*, 171:41–53.

Rivard, B., G.A. Sánchez-Azofeifa, S. Foley, and J.C. Calvo-Alvarado. 2008. Species classification of tropical tree leaf reflectance and dependence on selection of spectral bands. In *Hyperspectral remote sensing of tropical and sub-tropical forests*, eds. M. Kalácska and G.A. Sánchez-Azofeifa. Boca Raton, FL: CRC Press.

Rivera, G., S. Elliott, L. Caldas, G. Nicolossi, V. Coradin, and R. Borchert. 2002. Increasing day-length induces spring flushing of tropical dry forest trees in the absence of rain. *Trees* 16:445–456.

Rizzini, C.T. 1997. *Tratado de fitogeografia do Brazil*. São Paulo: HUCITEC.

Robalino, J. 2007. Land conservation policies and income distribution: Who bears the burden of our environmental efforts? *Environment and Development Economics* 12(4):521–533.

Robalino, J. and L. Villalobos. 2010. Evaluating the Effects of Land Conservation Policies on Labor Markets using Matching Techniques. Mimeo, CATIE.

Robalino, J.A. and A. Pfaff. 2012a. Ecopayments and deforestation in Costa Rica: Countrywide analysis of PSA's initial years (1997–2000). Mimeo, CATIE.

Robalino J.A. and A. Pfaff. 2012b. Contagious development: Neighbor interactions in deforestation. *Journal of Development Economics* 97:427–436.

Robalino, J.A., A. Pfaff, G.A. Sánchez-Azofeifa, and L. Villalobos. 2012. Evaluating Spillover Effects of Land Conservation Policies in Costa Rica. Mimeo, CATIE.

Roberts, D.A., B.W. Nelson, J.B. Adams, and F. Palmer. 1998. Spectral changes with leaf aging in Amazon caatinga. *Trees-Structure and Function* 12(6):315–325.

Roberts, M.J. and S. Bucholtz. 2005. Slippage in the conservation reserve program or spurious correlation? A comment. *American Journal of Agricultural Economics, Agricultural and Applied Economics Association* 87:244–250.

Robson, C. 1994. *Real World Research: A resource for social scientists and practitoner-researchers*. Oxford: Blackwell Science.

Rodríguez, B.M. 1991. La integración de la costa de Jalisco. *Estudios Sociales* 11:116–124.

Rodríguez, J.P., A.B. Taber, P. Daszak, R. Sukumar, C. Valladares-Padua, S. Padua, L.F. Aguirre, R.A. Medellin, M. Acosta, and A.A. Aguirre. 2007. Globalization of conservation: A view from the south. *Science* 317(5839):755.

Rodríguez, J.P., J.K. Balch, and K.M. Rodriguez-Clark. 2007. Assessing extinction risk in the absence of species-level data: Quantitative criteria for terrestrial ecosystems. *Biodiversity and Conservation* 16(1):183–209.

Rodríguez, J.P. and J.M. Nassar. 2011. TROPI-DRY: Human and Biophysical Dimensions of Tropical Dry Forests in the Americas, CRN II-021, Technical Report Year 5 - Venezuela. Centro de Ecología, IVIC, Venezuela. p. 20.

Rodrígues, L. 2000. Formação econômica do Norte de Minas e o período recente. In *Formação Social e Econômica do Norte de Minas Gerais*, eds. M.F.M. Oliveira, L. Rodrigues, J.M.A. Machado, and T.R. Botelho, 105–170. Montes Claros: Editora Unimontes.

Rodríguez, M. 1989. Población y poblamiento de la costa de Jalisco. *Estudios Sociales* 6:5–22.

Rodrigues, L.A. and G.M. Araújo. 1997. Levantamento florístico de uma mata decídua em Uberlândia, Minas Gerais, Brazil. *Acta Botânica Brasilica* 11:229–236.

Rodrigues, R.R. and S. Gandolfi. 1996. Recomposição de florestas nativas: Princípios gerais e subsídios para uma definição metodológica. *Revista Brasileira de Horticultura Ornamental* 2:4–15.

Rojas-Arechiga, M. and C. Vasquez-Yanes. 2000. Cactus seed germination: A review. *Journal of Arid Environments* 44:85–104.

Rojas-Sandoval, J., J.A. Lobo, and M. Quesada. 2008. Phenological patterns and reproductive success of Ceiba pentandra (Bombacaceae) in tropical dry and wet forests of Costa Rica. *Revista Chilena De Historia Natural* 81:443–454.

Rojas, A.M., L.A. Lozano, and M.G. Yaya. 2011. Evaluación ecológica y estructural de los bosques del departamento del Tolima [*Ecological and structral assessment of forest in the Tolima Departament*]. Universidad del Tolima. 237p.

Rojas, J. 2001. Caracterización biofísica y socioeconómica de la ecorregión estratégica de la Tatacoa y su área de influencia. Tomo IV, Componente Fauna. 250 p.

Rojas, Z.P. 1995. Uso popular de las plantas medicinales en la zona plana del municipio de Armero-Guayabal. Biology Bachelor Tesis. Universidad del Tolima. 82 p.

Rolstad, J. 1991. Consequences of forest fragmentation for the dynamics of bird populations: Conceptual issues and the evidence. *Biological Journal of the Linnean Society* 42:149–163.

Rolston, M.P. 1978. Water impermeable seed dormancy. *The Botanical Review* 44:365–396.

Ronce, O., S. Brachet, I. Olivieri, P.H. Gouyon, and J. Clobert. 2005. Plastic changes in seed dispersal along ecological succession: Theoretical predictions from an evolutionary model. *Journal of Ecology* 93:431–440.

Rossato, S., H.F. Leitão-Filho, and A. Begossi. 1999. Ethnobotany of Caiçaras of the Atlantic Forest Coast (Brazil). *Economic Botany* 53:387–395.

Roth, L.C. 1999. Anthropogenic change in subtropical dry forest during a century os settlements in Jaiquí Picado, Santiago Province, Dominican Republic. *Journal of Biogeography* 26:739–759.

Roughgarden, J., S.W. Running, and P.A. Matson. 1991. What does remote sensing do for ecology? *Ecology* 72(6):1918–1922.

Roy, S. and J.S. Singh. 1994. Consequences of habitat heterogeneity for availability of nutrients in a dry tropical forest. *Journal of Ecology* 82:503–509.

Rozendaal, D.M.A. 2010. *Looking backwards: Using tree rings to evaluate long-term growth patterns of Bolivian forest trees*. Riberalta, Bolivia: PROMAB Scientific Series 12. 151 p.

Ruiz-Suescún, O.A., J.J. Acosta-Jaramillo, and J.D. León-Perez, 2005. Escorrentía superficial en bosques montanos naturales y plantados de Piedras Blancas, Antioquía (Colombia). *Revista Facultad NAcional de Agronomía Medellín* 58(1).

Ruiz, J., M.C. Fandino, and R.L. Chazdon. 2005. Vegetation structure, composition, and species richness across a 56-year chronosequence of dry tropical forest on Providencia Island, Colombia. *Biotropica* 37:520–530.

Rulequest Research. 2008. See5 (Classification Software), Version 2. www.rulequest.com.

Rundel, P. and K. Boonpragob. 1995. Dry forest ecosystems of Thailand. In *Seasonally Dry Tropical Forests*, eds. S.H. Bullock, H.A. Mooney, and E. Medina, 93–119. New York: Cambridge University Press.

Runkle, J. 1992. Guidelines and sample protocol for sampling forest gaps. USDA Forest Service General Technical Report. PNW-GTR-283.

Running, S.W., D.L. Pederson, M.A. Spanner, and K.B. Teuber. 1986. Remote sensing of coniferous forest leaf area. *Ecology* 67(1):273–276.

Russo, S.E., K.L. Jenkins, S.K. Wiser, M. Uriarte, R.P. Duncan, D.A. Coomes. 2010. Interspecific relationships among growth, mortality and xylem traits of woody species from New Zealand. *Functional Ecology* 24:253–262.

Rutter, A.J., A.J. Morton, and P.C. Robins. 1975. A predictive model of rainfall interception in forests. II. Generalization of the model and comparison with observations in some coniferous and hardwood stands. *Journal of Applied Ecology* 12(1):367–380.

Sabogal, C., and L. Valerio. 1998. Forest composition, structure and regeneration in a dry foresto of the Nicarguan Pacific COSAT. In *Forest biodiversity in North Central and South America, and the Caribbean: Research and monitoring. Man and the Biospere Series* eds. F. Dallmeier and J.A. Comiskey, vol. 21, 187–212. New York, USA: UNESCO.

Sachs, I. 1993. *Estratégias de Transição para o século XXI: Desenvolvimento e Meio Ambiente*. São Paulo: Studio Nobel.

Sader, S.A., R.B. Waide, W.T. Lawrence, and A.T. Joyce. 1989. Tropical forest biomass and successional age class relationships to a vegetation index derived from Landsat TM data. *Remote Sensing of Environment* 28:143–198.

Sader, S.A. and A.T. Joyce. 1988. Deforestation rates and trends in Costa Rica 1940 to 1983. Biotropica 20: 11–14.

Sakai, S., K. Momose, T. Yumoto, T. Nagamitsu, H. Nagamasu, A.A. Hamid, and T. Nakasiiizuka. 1999. Plant reproductive phenology over four years including an episode of general flowering in a lowland Dipterocarp forest, Sarwak, Malaysia. *American Journal of Botany* 86:1414–1436.

Saldarriaga, J.G., D.C. West, M.L. Tharp, and C. Uhl. 1988. Long-term chronosequence of forest succession in the Upper Rio Negro of Colombia and Venezuela. *Journal of Ecology* 76:938–958.

Salis, S.M., M.P. Silva, P.P. Mattos, J.V. Silva, V.J. Pott, and A. Pott. 2004. Fitossociologia de remanescentes de floresta estacional decidual em Corumbá, Estado do Mato Grosso do Sul, Brazil. *Revista Brasileira de Botânica* 27:671–684.

Sampaio, A.B. and A. Scariot. 2011. Edge effect on tree diversity, composition and structure in a deciduous dry forest in Central Brazil. *Revista Árvore* 35:1121–1134.

Sampaio, A.B., K.D. Holl, and A. Scariot, 2007. Regeneration of seasonal deciduous forest tree species in long-used pastures in Central Brazil. *Biotropica* 39:655–659.

Sampaio, E.V.S.B. 1994. Overview of the Brasilian caatinga. In *Seasonally dry tropical forests*, eds. S.H. Bullock, H.A. Mooney, and E. Medina, 35–63. Cambridge: Cambridge University Press. 450 p.

Sampaio, E.V.S.B. 1995. Overview of Brazilian Caatinga. In *Seasonal dry tropical forests*, eds. S.H. Bullock, A.M. Harold, and E. Medina, 35–63. Cambridge: Cambridge University Press.

Sanches, L., C.M.A. Valentini, O.B.P. Júnior, J. de Souza Nogueira, G.L. Vourlitis, M.S. Biudes, C.J. da Silva, P. Bambi, F. de Almeida Lobo. 2008. Seasonal and interannual litter dynamics of a tropical semideciduous forest of the southern Amazon Basin, Brazil. *Journal of Geophysical Research* 113:1–9.

Sánchez-Azofeifa, A., M. Kalácska, M. Quesada, J. Calvo-Alvarado, J. Nassar, and J. Rodríguez. 2005a. Need for integrated research for a sustainable future in tropical dry forests. *Conservation Biology* 19(2):1–2.

Sánchez-Azofeifa, A., M. Quesada, P. Cuevas-Reyes, A. Castillo, and G. Sánchez. 2009. Land cover and conservation in the area of influence of the Chamela-Cuixmala Biosphere Reserve, Mexico. *Forest Ecology and Management* 258:907–912.

Sánchez-Azofeifa, A., M.E. Kalácska, M. Quesada, K.E. Stoner, J.A. Lobo, and P. Arroyo-Mora. 2003. *Tropical Dry Climates. Phenology: An integrative environmental science*. The Netherlands: Kluwer Academic Publisher. 121 p.

Sánchez-Azofeifa, A., Y. Oki, G.W. Fernandes, R.A. Ball, and J. Gamon. 2012. Relationships between endophyte diversity and leaf optical properties. *Trees (Berl)* 26:291–299.

Sánchez-Azofeifa, G., A. Pfaff, J. Robalino, and J. Boomhower. 2007. Costa Rica's payment for environmental services program: Intention, implementation and impact. *Conservation Biology* 21(5):1165–1173.

Sánchez-Azofeifa, G.A. 2000. Land use change in Costa Rica. In *Quantifying sustainable development*, 473–501.

Sánchez-Azofeifa, G.A., A.Pfaff, A. Robalino, and J. Boomhower. 2007. Payments for Ecosystems Services in Costa Rica: Examining their intention, implementation and impact. *Conservation Biology* 21(5):1165–1173.

Sánchez-Azofeifa, G.A., K.L. Castro, B. Rivard, M.R. Kalascka, and R.C. Harriss. 2003. Remote sensing research priorities in tropical dry forest environments. *Biotropica* 35(2):134.

Sánchez-Azofeifa, G.A., K.L. Castro-Esau, S. Joseph Wright, J. Gamon, M. Kalácska, B Rivard, S.A. Schnitzer, and J.L. Feng. 2009. Differences in leaf traits, leaf internal structure, and spectral reflectance between two communities of lanias and trees: Implications for remote sensing in tropical environments. *Remote Sensing of Environment* 113:2076–2088.

Sánchez-Azofeifa, G.A., M. Káacska, B. Rivard, P. Arroyo-Mora, R. Hall, and J. Zhang. 2001. Observations of phenological changes on Mesoamerican tropical dry forests and implications for conservation strategies. Paper read at International Conference on Tropical Ecosystems, at New Dehli.

Sánchez-Azofeifa, G.A., M. Kalácska, M. Quesada, J.C. Calvo-Alvarado, J.M. Nassar, and J.P. Rodriguez. 2005b. Need for integrated research for a sustainable future in tropical dry forests. *Conservation Biology* 19(2):285–286.

Sánchez-Azofeifa, G.A., M. Kalácska, M.M. Espirito-Santo, G.W. Fernandes, and S. Schnitzer. 2009. Tropical dry forest succession and the contribution of lianas to Wood Area Index (WAI). *Forest Ecology and Management* 258:941–948.

Sánchez-Azofeifa, G.A., M. Quesada, J.P. Rodríguez, J.M. Nassar, K.E. Stoner, A. Castillo, T. Garvin, et al. 2005c. Research priorities for Neotropical dry forests. *Biotropica* 37(4):477–485.

Sánchez-Azofeifa, G.A., M. Quesada, P. Cuevas-Reyes, A. Castillo, and G. Sánchez-Montoya. 2009. Land cover and conservation in the area of influence of the Chamela-Cuixmala Biosphere Reserve, Mexico. *Forest Ecology and Management* 258(6):907–912.

Sánchez, J., C.A. Alvarez, and A. Cadena. 2007. Bat assemblage structure in two dry forests of Colombia: Composition, species richness, and relative abundance. *Mammalian Biology* 72:82–92.

Sandquist, D.R. and S. Cordell. 2007. Functional diversity of carbon-gain, water-use, and leaf-allocation traits in trees of a threatened lowland dry forest in Hawaii. *American Journal of Botany* 94:1459–1469.

Santana, D.G. and M.A. Ranal. 2004. Análise estatística. In *Germinação: Do básico ao aplicado*, eds. A.G. Ferreira and F. Borghetti, 197–208. Porto Alegre: Artmed.

Santiago, L.S., E.A.G. Schuur, and K. Silvera. 2005. Nutrient cycling and plant-soil feedbacks along a precipitation gradient in lowland Panama. *Journal of Tropical Ecology* 21:461–470.

Santoro, H. 2002. Estudios de caracterización biofísica y socioeconómica de la ecorregión estratégica del valle del Alto Magdalena. Componente Aguas. Ministerio de Ambiente, CORTOLIMA, CAM, Universidad del Tolima and Universidad Surcolombiana, Ibagué, Tolima. 600 p.

Santos, C.F. 2011. Mosquitos (Diptera: Culicidae) do Parque Estadual da Mata Seca, MG: sazonalidade e impacto da pecuária bovina. MSc. Thesis, Universidade Estadual de Montes Claros, Montes Claros, Brazil.

Santos, J.C., I.R. Leal, J.S. Almeida-Cortez, G.W. Fernandes, and M. Tabarelli. 2011. Caatinga: The scientific negligence experienced by a dry tropical forest. *Tropical Conservation Science* 4:276–286.

Santos, R.M., A.C.M.C. Barbosa, H.S. Almeida, F.A. Vieira, P.F. Santos, D.A. Carvalho, and A.T. Oliveira-Filho. 2011. Estrutura e florística de um remanescente de caatinga em Juvenília, norte de Minas Gerais, Brazil. *Cerne* 17:247–258.

Santos, R.M., A.T. Oliveira-Filho, P.V. Eisenlohr, L.P. Queiroz, D.B.O.S. Cardoso, and M.J.N. Rodal. 2012. Identity and relationships of the Arboreal Caatinga among other floristic units of seasonally dry tropical forests (SDTFs) of northeastern and Central Brazil. *Ecology and Evolution*. 2:409–428.

Santos, R.M., F.A. Vieira, M. Fagundes, Y.R.F. Nunes, and E. Gusmão. 2007. Riqueza e similaridade florística de oito remanescentes florestais no norte de Minas Gerais, Brazil. *Revista Árvore* 31:135–144.

Santos, S.C., W.F. Costa, F. Batista, L.R. Santos, P.H. Ferri, H.D. Ferreira, and J.C. Seraphin. 2006. Seasonal variation in the content of tannins in barks of barbatimão species. *Revista Brasileira de Farmacognosia* 16:552–556.

Santos, V.L.S. 2010. Fungos micorrízicos arbusculares em ecossistema de Mata seca no Norte de Minas Gerais. MSc. Thesis, Universidade Federal Rural do Rio de Janeiro, Rio de Janeiro, Brazil.

Sarmiento, G. and M. Monasterio. 1983. Life forms and phenology. In *Tropical Savannas*, ed. F. Bourlière, 79–108. Amsterdam: Elsevier.

Saunders, P.A. Jr., R.J. Hobbs, and C.R. Margules. 1991. Biological consequences of ecosystem fragmentation: A review. *Conservation Biology* 5:18–32.

Sáyago, R., M. Lopezaraiza-Mikel, M. Quesada, M.Y. Álvarez-Añorve, A. Cascante-Marín, and J.M. Bastida. 2013. Evaluating factors that predict the structure of a commensalistic epiphyte-phorophyte networks. *Proceedings of the Royal Society B* 280(1756):20122821.

Scariot, A. and A.C. Sevilha. 2005. Biodiversidade, estrutura e conservação de florestas estacionais deciduais no Cerrado. In *Cerrado: Ecologia, biodiversidade e conservação*, eds. A. Scariot, J.C. Sousa-Silva, and J.M. Felfili, 121–139. Brasília: Ministério do Meio Ambiente.

Schaik, C.V. and H.D. Rijksen. 2002. Projetos integrados de conservação e desenvolvimento: Problemas e potenciais. In *Tornando os parques eficientes: Estratégias para conservação da natureza nos trópicos*, eds. C.V. Schaik, L. Davenport, M. Rao, and J. Terborgh, 37–51. Curitiba: Editora da Universidade Federal do Paraná.

Schalamuk, S., S. Velázquez, H. Chidichimo, and M. Cabello, 2006. Fungal spore diversity of arbuscular mycorrhizal fungi associated with spring wheat: Effect of tillage. *Mycologia* 98:16–22.

Schenck, N.C. and Y. Pérez. 1990. *Manual for the identification of VA mycorrhizal fungi*. 3rd ed. Gainesville: Synergistic.

Schimper, A.F.W. 1898. *Pjanzengeographie auf Physiologischer Grundlage*. Fischer, Germany: Jena.

Schnitzer, S.A. 2005. A mechanistic explanation for global patterns of liana abundance and distribution. *American Naturalist* 166(2):262–276.

Schnitzer, S.A. and F. Bongers. 2002. The ecology of lianas and their role in forests. *Trends in Ecology & Evolution* 17:223–230.

Schnitzer, S.A. and F. Bongers. 2011. Increasing liana abundance and biomass in tropical forests: Emerging patterns and putative mechanisms. *Ecology Letters* 14:397–406.

Schultes, R.E. and S. Von Reis. 1995. *Ethnobotany: Evolution of a discipline*. New York: Chapman & Hall.

Schüssler, A. 2002. Molecular phylogeny, taxonomy, and evolution of Geosiphon pyriformis and arbuscular mycorrhizal fungi. *Plant and Soil* 244:75–83.

Schüßler, A. and C. Walker. 2010. The Glomeromycota: A species list with new families and new genera. Gloucester. Published in libraries at The Royal Botanic Garden Edinburgh, The Royal Botanic Garden Kew, Botanische Staatssammlung Munich, and Oregon State University; available at www.amf-phylogeny.com.

Schuur, E.A.G. 2001. The effect of water on decomposition dynamics in mesic to wet Hawaiian montane forests. *Ecosystems* 4:259–273.

Schuur, E.A.G. 2003. Productivity and global climate revisited: The sensitivity of tropical forest growth to precipitation. *Ecology* 84:1165–1170.

Schuur, E.A.G. and P.A. Matson. 2001. Net primary productivity and nutrient cycling across a mesic to wet precipitation gradient in Hawaiian montane forest. *Oecologia* 128:431–442.

Schwartz, M.D. and B.C. Reed. 1999. Surface phenology and satellite sensor-derived onset of greenness: An initial comparison. *International Journal of Remote Sensing* 20(17):3451–3457.

Scolforo, J.R. and L.M.T. Carvalho. 2006. *Mapeamento e Inventário da Flora Nativa e dos Reflorestamentos de Minas Gerais*. Lavras: Editora da Universidade Federal de Lavras.

Sedjo, R. and B. Sohngen. 2000. Forestry Sequestration of CO_2 and Markets for Timber. Discussion Papers. Resources For the Future.

Sedjo, R. 2005. Global agreements and U.S. forestry: Genetically modified trees. *Journal of Forestry* 103:109–113.

Segura, G., P. Balvanera, E. Durán, and A. Pérez-Jiménez. 2003. Tree community structure and stem mortality along a water availability gradient in a Mexican tropical dry forest. *Plant Ecology* 169:259–271.

Sellers, P.J., R.E. Dickinson, D.A. Randall, A.K. Betts, F.G. Hall, J.A. Berry, G.J. Collatz, et al. 1997. Modeling the exchanges of energy, water, and carbon between continents and the atmosphere. *Science* 275:502–509.

Serbin, S., S. Gower, and D. Ahl. 2009. Canopy dynamics and phenology of a boreal spruce wildfire chronosequence. *Agricultural and Forest Meteorology* 149:187–204.

Sevilha, A.C., A. Scariot, and S.E. Noronha. 2004. Estado atual da representatividade de unidades de conservação em Florestas Estacionais Deciduais no Brazil. In *Biomas florestais*, ed. Sociedade Brasileira de Botânica, 1–63. Viçosa: Editora da Universidade Federal de Viçosa.

Sheil, D. 2001. Long-term observations of rain forest succession, tree diversity and responses to disturbance. *Plant Ecology* 155:183–199.

Sheil, D., D.F.R.P. Burslem, and D. Alder. 1995. The interpretation and misinterpretation of mortality rate measures. *Journal of Ecology* 83:331–333.

Shull, C.A. 1929. A spectrophotometric study of reflection of light from leaf surfaces. *Botanical Gazette* 583–607.

Silbergauer-Gottsberger, I. and G. Gottsberger. 1988. A polinizacao de plantas do Cerrado. *Revista Brasileira de Biologia* 48:651–663.

Siles, P., P. Vaast, E. Dreyer, and J.-M. Harmand. 2010. Rainfall partitioning into throughfall, stemflow and interception loss in a coffee (Coffea arabica L.) monoculture compared to an agroforestry system with Inga densiflora. *Journal of Hydrology* 395(1–2):39–48.

Sillitoe, P. 2009. *Local science vs global science: Approaches to indigenous knowledge in international development*. Oxford: Berghahn.

Silva, A.J.R. and L.H.C. Andrade. 2004. Etnobotânica nordestina: Estudo comparativo da relação entre comunidades e vegetação na Zona do Litoral – Mata do Estado de Pernambuco, Brazil. *Acta botanica brasilica* 19:45–60.

Silva, C.R. and Z.M. Souza. 1998. *Eficiência do uso de nutrientes em solos ácidos: Manejo de nutrientes e uso pelas plantas*. Ilha Solteira: Faculdade de Engenharia de Ilha Solteira-FEIS.

Silva, J.O., M.M. Espírito-Santo, and G.A. Melo. 2012. Herbivory on Handroanthus ochraceus (Bignoniaceae) along a successional gradient in a tropical dry forest. *Arthropod-Plant Interactions* 6:45–57.

Silva, L.A. and A. Scariot. 2004a. Composição e estrutura da comunidade arbórea de uma floresta estacional decidual sobre afloramento calcário no Brazil central. *Revista Árvore* 28:69–75.

Silva, L.A. and A. Scariot. 2004b. Comunidade arbórea de uma floresta estacional decidual sobre afloramento calcário na bacia do rio Paranã. *Revista Árvore* 28:61–67.

Silva, L.A. and A. Scariot. 2003. Composição florística e estrutura da comunidade arbórea em uma floresta estacional decidual em afloramento calcário (Fazenda São José, São Domingos, GO, Bacia do Rio Paranã). *Acta Botanica Brasilica* 17:305–313.

Silver, W.L., F.N. Scatena, A.H. Johnson, T.G. Siccama, and F. Watt. 1996. At what temporal scales does disturbance affect below-ground nutrient pools? *Biotropica* 28:441–457.

Sims, D.A. and J.A. Gamon. 2002. Relationships between leaf pigment content and spectral reflectance across a wide range of species, leaf structures, and developmental stages. *Remote Sensing of Environment* 81:337–354.

Sims, K.R.E. 2010. Conservation and development: Evidence from Thai protected areas. *Journal of Environmental Economics and Management* 60:94–114.

Sims, K.R.E. 2011. Do protected areas reduce forest fragmentation? A microlandscapes approach. Working paper, Amherst College, December 2011.

Singh, J.S., A.S. Raghubanshi, R.S. Singh, and S.C. Srivastava. 1989. Microbial biomass acts as a source of plant nutrients in dry tropical forest and savanna. *Nature* 338:499–500.

Singh, K.P. and C.P. Kushwaha. 2005. Paradox of leaf phenology: *Shorea robusta* is a semi-evergreen species in tropical dry deciduous forests in india. *Current Science* 88(11):1820–1824.

Singh, K.P. and C.P. Kushwaha. 2006. Diversity of flowering and fruiting phenology of trees in a tropical deciduous forest in India. *Annals of Botany* 97:265–276.

Sistema Nacional de Áreas de Conservación (SINAC). 2008. Sistemas Nacional de Áreas de Conservacion. http://www.minae.go.cr/dependencias!desconcentras!sistema-nacional-areas-conservación.html

Sizer, N. and E. Tanner. 1999. Responses of woody plant seedlings to edge formation in a lowland tropical rainforest, Amazonia. *Biological Conservation* 91:135–142.

Skole, D. and C. Tucker. 1993. Tropical deforestation and habitat fragmentation in the Amazon: Satellite Data from 1978 to 1988. *Science* 260(5116):1905–1910.

Skowronski, N., K. Clark, R. Nelson, J. Hom, and M. Patterson. 2007. Remotely sensed measurements of forest structure and fuel loads in the Pinelands of New Jersey. *Remote Sensing of Environment* 108(2):123–129.

Slaton M.R., E.R. Hunt, and W.K. Smith. 2001. Estimating near-infrared leaf reflectance from leaf structural characteristics. *American Journal of Botany* 88:278–284.

Smiderle, O.J. and R.C.P. Souza. 2003. Dormência em sementes de paricarana (Bowdichia virgilioides Kunth - Fabaceae Papilionoideae). *Revista Brasileira de Sementes* 25:48–52.

Smith, S.E. and D.J. Read. 2008. *Mycorrhizal symbiosis*. New York: Elsevier.

Sobrado, M.A. and Cuenca, G. 1979. Aspectos de uso de agua en especies deciduas y siempreverdes en un Bosque Tropical Seco de Venezuela. *Acta Científica Venezolana* 30:302–308.

Sobrado, M.A. 1991. Cost-benefit relationships in deciduous and evergreen leaves of tropical dry forest species. *Functional Ecology* 5:608–616.

Sohngen, B., R. Mendelsohn, and R. Sedjo. 1999. Forest management, conservation, and global timber markets. *American Journal of Agricultural Economics* 81:1–13.

Soil Survey Staff. 1975. Soil taxonomy: a basic system of soil classification for making and interpreting soil surveys. Washington: Department of Agriculture.

Solbrig, O.T. 2005. The Dilemma of Biodiversity Conservation. *Revista, Harvard Review of Latin America: Flora and Fauna*. David Rockefeller Center for Latin American Studies, Harvard University.

Solís, E. and J. Campo. 2004. Soil N and P dynamics in two secondary tropical dry forests after fertilization. *Forest Ecology and Management* 195:409–418.

Sólon, S., L. Lopes, J.P.T. Sousa, and G. Schmeda-Hirschmann. 2000. Free radical scavenging activity of Lafoensia pacari. *Journal of Ethnopharmacology* 72:173–178.

Song, C., C. Woodcock, K.C. Seto, M.P. Lenney, and S.A. Macomber. 2001. Classification and change detection using Landsat TM Data: When and how to correct atmospheric effects? *Remote Sensing of Environment* 75(2):230–244.

Sonnentag, O., J. Talbot, J.M. Chen, and N.T. Roulet. 2007. Using direct and indirect measurements of leaf area index to characterize the shrub canopy in an ombrotrophic peatland. *Agricultural and Forest Meteorology* 144(3–4):200–212.

Sørensen, T. 1948. A method for establishing groups of equal amplitude in plant sociology based on similarity of species content and its application to analyses of the vegetation on Danish commons. *Biologiske Skrifter* 5:1–34.

Soto, O. 2002. Estudios de caracterización biofísica y socioeconómica de la ecorregión estratégica del valle del Alto Magdalena. Componente Fisiografía y Suelos. Ministerio de Ambiente, CORTOLIMA, CAM, Universidad del Tolima and Universidad Surcolombiana. Ibagué, Tolima. 600 p.

Soulé P.T. and P.A. Knapp. 2006. Radial growth rate increases in naturally occurring ponderosa pine trees: A late-20th century CO_2 fertilization effect? *New Phytologist* 171:379–390.

Southworth, J. 2004. An assessment of Landsat TM band 6 thermal data for analysing land cover in tropical dry forest regions. *International Journal of Remote Sensing* 25(4):689–706.

Souza, Jr. C., D.A. Roberts, and M. Cochrane. 2005. Combining spectral and spatial information to map canopy damage from selective logging of forests fires. *Remote sensing of Environment* 98:329–343.

Souza, R.G., L.C. Maia, M.F. Sales, and S.F.B. Trufem. 2003. Diversidade e potencial de infectividade de fungos micorrízicos arbusculares em áreas de Caatinga, na região de Xingó, Estado de Alagoas, Brazil. *R Bras Bot* 26:49–60.

Souza, S.C.A., G.R.A. Borges, D.O. Brandão, A.M.M. Matos, M.D.M. Veloso, and Y.R.F. Nunes. 2007. Conservação de sementes de Myracrodruon urundeuva Freire Allemão (Anacardiaceae) em diferentes condições de armazenamento. *Revista Brasileira de Biociências* 5:265–267.

Spanner, M.A., L.L. Pierce, D.L. Peterson, and S.W. Running. 1990. Remote sensing of temperate coniferous forest leaf area index. The influence of canopy closure, understory vegetation and background reflectance. *International Journal of Remote Sensing* 11:95–111.

Spittler, P. 2002. Dinámica de los bosques secundarios secos en la región Chorotega, Costa Rica. In *Ecosistemas forestales de bosque seco tropical: Investigaciones y resultados en mesoamérica*, 198–209. Heredia, CR: Universidad Nacional, INISEFOR.

Šraj, M., M. Brilly, and M. Mikoš. 2008. Rainfall interception by two deciduous Mediterranean forests of contrasting stature in Slovenia. *Agricultural and Forest Meteorology* 148(1):121–134.

Staddon, P.L., K. Thompson, I. Jakobsen, J.P. Grime, A.P. Askew, and A.H. Fitter. 2003. Mycorrhizal fungal abundance is affected by longterm climatic manipulations in the field. *Global Change Biology* 9:186–194.

Staelens, J., A. De Schrijver, K. Verheyen, and N.E.C. Verhoest. 2008. Rainfall partitioning into throughfall, stemflow, and interception within a single beech (Fagus sylvatica L.) canopy: Influence of foliation, rain event characteristics, and meteorology. *Hydrological Processes*, 22(1):33–45.

StatSoft Inc. 2003. STATISTICA 6.0 (data analysis software system), version 6. www.statsoft.com.

Stearns F.W. 1974. Phenology and environmental education. In *Phenology and seasonality modeling*, ed. H. Lieth, 425–429. New York: Springer-Verlag.

Sterck, F., L. Markesteijn, F. Schieving, and L. Poorter. 2011. Functional traits determine trade offs and niches in a tropical forest community. *Proceedings of the National Academy of Sciences of the United States of America* 108:20627–20632.

Sterck, F.J., D.B. Clark, D.A. Clark, and F. Bongers. 1999. Light fluctuations, crown traits, and response delays for tree saplings in a Costa Rican lowland rain forest. *Journal of Tropical Ecology* 15:83–95.

Stevens, R.D., M.R. Willig, and R.E. Strauss. 2006. Latitudinal gradients in the phenetic diversity of New World bat communities. *Oikos* 112:41–50.

Stiglitz, J.E. 2006. *Cómo hacer que funcione la globalización*. Buenos Aires: Taurus.

Stoner, K.E. 2005. Phyllostomid bat community structure and abundance in two contrasting tropical dry forests. *Biotropica* 37:591–599.

Stoner, K.E. and G. Arturo Sánchez-Azofeifa. 2009. Ecology and regeneration of tropical dry forests in the Americas: Implications for management. *Forest Ecology and Management* 258(6):903–906.

Strong. D.R., J.H. Lawton, and R. Southwood. 1984. *Insects on plants: Community patterns and mechanisms*. England: Blackwell Scientific.

Stylinski, C., J. Gamon, and W. Oechel. 2002. Seasonal patterns of reflectance indices, carotenoid pigments and photosynthesis of evergreen chaparral species. *Oecologia* 131(3):366–374.

Sundarapandian, S.M. and P.S. Swamy. 1999. Litter production and leaf-litter decomposition of selected tree species in tropical forests at Kodayar in the Western Ghats, India. *Forest Ecology and Management* 123:231–244.

Sutherland, W.J., I. Newton, and R.E. Green. 2004. *Bird ecology and conservation: A handbook of techniques*. New York: Oxford University Press. 408 p.

Swaine, M.D. and T.C. Whitmore. 1988. On the definition of ecological species groups in tropical rain forest. *Vegetatio* 75:81–86.

Swenson, N.G. and B.J. Enquist. 2007. Ecological and evolutionary determinants of a key plant functional trait: Wood density and its community-wide variation across latitude and elevation. *American Journal of Botany* 94:451–459.

Tabarelli, M. and W.A. Mantovani. 1999. Regeneração de uma floresta tropical Montana após corte e queima (São Paulo-Brazil). *Revista Brasileira de Biologia* 59:239–250.

Tabarelli, M., M.J.C. Silva, and C. Gascon. 2004. Forest fragmentation, synergisms ans the impoverishment of neotropical forests. *Biodiversity and Conservation* 13:1419–1425.

Tan, Z., Y. Zhang, G. Yu, L. Sha, J. Tang, X. Deng, and Q. Song. 2010. Carbon balance of a primary tropical seasonal rain forest. *Journal of Geophysical Research* 115:D00H26, 17 PP.

Tanaka, N., T. Kume, N. Yoshifuji, K. Tanaka, H. Takizawa, K. Shiraki, C. Tantasirin, N. Tangtham, and M. Suzuki. 2008. A review of evapotranspiration estimates from tropical forests in Thailand and adjacent regions. *Agricultural and Forest Meteorology* 148(5):807–819.

Taylor, N. and D. Zappi. 2004. *Cacti of eastern Brazil*. Kew: Royal Botanic Gardens.

Tchabi, A., F. Houtondji, L. Laouwin, D. Coyne, and F. Oehl. 2009. Racocetra beninensis from sub-Saharan savannas: A new species in the Glomeromycetes with ornamented spores. *Mycotaxon* 110:199–209.

Teketay, D. and A. Granstrom. 1997. Seed viability of afromontane tree species in forest soil. *Journal of Tropical Ecology* 13:81–95.

Teófilo, E.M., S.O. Silva, A.M.E. Bezerra, S. Medeiros-Filho, and F.D.B. Silva. 2004. Qualidade fisiológica de sementes de aroeira (Myracrodruon urundeuva Allemão) em função do tipo de embalagem, ambiente e tempo de armazenamento. *Revista Ciência Agronômica* 35:371–376.

Ter Braak, C.J.F. 1987. The analysis of vegetation environment relationship by Canonical Correspondence Analysis. *Vegetatio* 69:69–77.

Ter Braak, C.J.F. and P. Smilauer. 1998. *CANOCO reference manual and user's guide to Canoco for Windows: Software for Canonical Community Ordination (version 4).* New York: Microcomputer Power.

Teramura, A.H., W.G. Gold, and I.N. Forseth. 1991. Physiological ecology of mesic, temperate woody lianas. In *The biology of vines*, eds. F.E. Putz and H.A. Mooney, 245–285. Cambridge: Cambridge University Press.

Terborgh, J. 1986. Community aspects of frugivory in tropical forests. In *Frugivores and seed dispersal*, eds. A. Estrada and T.H. Fleming, 371–384, 392. Dordrecht: Dr. W. Junk Publishers.

Thiry-Cherques, H.R. 2006. Pierre Bourdieu: a teoria na prática. *RAP* 40:27–55.

Thomas, C.D., A. Cameron, R.E. Green, M. Bakkenes, L.J. Beaumont, Y.C. Collingham, E.F. Erasmus, et al. 2004. Extinction risk from climate change. *Nature* 427: 145–148.

Thomas, S.C. 2005. Increased leaf reflectance in tropical trees under elevated CO_2. *Global Change Biology* 11:197–202.

Thompson, K. and J.P. Grime. 1979. Seasonal variation in the seed banks of herbaceous species in ten contrasting habitats. *Journal of Ecology* 67:893–921.

Thornthwaite, C.W. 1948. An approach toward a rational classification of climate. *Geographical Review* 38:55–94.

Thürig, E., E. Kaufmann, R. Frisullo, and H. Bugmannc. 2005. Evaluation of the growth function of an empirical forest scenario model. *Forest Ecology and Management* 204:53–68.

Tian, Y., C.E. Woodcock, Y. Wang, J.L. Privette, N.V. Shabanov, L. Zhou, Y. Zhang. 2002. Multiscale analysis and validation of the MODIS LAI product I. uncertainty assessment. *Remote Sensing of Environment* 83(3):414–430.

Tian, Y., Y. Zhang, Y. Knyazikhin, R.B. Myneni, J.M. Glassy, G. Dedieu, and S.W. Running. (2000). Prototyping of MODIS LAI and FPAR algorithm with LASUR and LANDSAT data. *IEEE Transactions on Geoscience and Remote Sensing* 38(5 II):2387–2401.

Tilman, D. 1988. *Plant strategies and the dynamics and structure of plant communities.* Princeton: Princeton University Press.

Timm, R.M. and R.K. Laval. 1998. *A field key to the bats of Costa Rica.* Occasional Publication Series, 22:1–32. The University of Kansas Center of Latin American Studies.

Tobon-Marin, C., W. Bouten, and J. Sevink. 2000. Gross rainfall and its partitioning into throughfall, stemflow and evaporation of intercepted water in four forest ecosystems in western Amazonia. *Journal of Hydrology* 237(1–2):40–57.

Toda, M., K. Nishida, N. Ohte, M. Tani, and K. Mushiake. 2002. Observations of energy fluxes and evapotranspirationover terrestrial complex land covers in the tropical monsoon environment. *Journal of the Meteorological Society of Japan* 80:465–484.

Toledo, M. and J. Salick. 2006. Secondary succession and indigenous management in semideciduous forest fallows of the Amazon Basin. *Biotropica* 38:161–170.

Townsend, A.R., C.C. Cleveland, G.P. Asner, and M.M. Bustamante. 2007. Controls over foliar N:P ratios in tropical rain forests. *Ecology* 88(1):107–118.

Townsend, A.R., C.C. Cleveland, B.Z. Houlton, C.B. Alden, and J.W.C. White. 2011. Multi-element regulation of the tropical forest carbon cycle. *Frontiers in Ecology and the Environment* 9:9–17.

Treitz, P.M. and P.J. Howarth. 1999. Hyperspectral remote sensing for estimating biophysical parameters of forest ecosystems. *Progress in Physical Geography* 23(3):359.

Trejo, I. and R. Dirzo. 2000. Deforestation of seasonally dry tropical forest: A national and local analysis in Mexico. *Biological Conservation* 94:133–142.

Trejo, I. and R. Dirzo. 2002. Floristic diversity of seasonally tropical dry forests. *Biodiversity and Conservation* 11:2063–2048.

Treseder, K.K. and K.M. Turner. 2007. Glomalin in ecosystems. *Soil Science Society of American Journal* 71:1257–1266.

Tsuji, L.J.S. 1996. Loss of cree traditional ecological knowledge in the western james bay region of northern Ontario, Canada: A case study of the sharp-tailed grouse, Tympanuchus phasianellus phasianellus Tsuji. *The Canadian Journal of Native Studies* 16:283–292.

Turner, D.P., W.B. Cohen, R.E. Kennedy, K.S. Fassnacht, and J.M. Briggs. 1999. Relationships between Leaf Area Index and Landsat TM Spectral Vegetation Indices across three temperate zone sites-steps toward validating global map products. *Remote Sensing of Environment* 70(1):52–68.

Turton, S. and H.J. Freiburguer. 1997. Edge and aspect effects on the microclimate of a small tropical forest remnant on the Atherton Tableland, Northeastern Australia. In *Tropical Forest remnants: Ecology, management and conservation of fragmented communities*, eds. W. Laurance and R. Bierregaard, 45–54, 612. USA: The University of Chicago Press.

Uhl, C. and C.F. Jordan. 1984. Succession and nutrient dynamics following forest cutting and burning in Amazonia. *Ecology* 65:1476–1490.

Uhl, C., R. Buschbacher, and E.A.S. Serrao. 1988. Abandoned pastures in eastern Amazonia. I. Patterns of plant succession. *The Journal of Ecology* 76:663–681.

Universidad del Tolima, Universidad Surcolombiana, CAM, CORTOLIMA, Minambiente. 2002. Estudios de caracterización biofísica y socioeconómica de la ecorregión estratégica del valle del Alto Magdalena. Technicalreport. 600 p.

Ustin, S.L., D.A. Roberts, J.A. Gamon, G.P. Asner, and R.O. Green. 2004. Using imaging spectroscopy to study ecosystem processes and properties. *Bioscience* 54(9):523–534.

Vaccaro, S. 1997. Caracterização fitosociológica de três fases sucessionais de uma Floresta Estacional Decidual no município de Santa Tereza-RS. Msc. Dissertation. Santa Maria, Programa de Pós-graduação em Engenharia Florestal, Universidade Federal de Santa Maria.

Valdespino, P., R. Romualdo, L. Cadenazzi, and J. Campo. 2009. Phosphorus cycling in primary and secondary seasonally dry tropical forests in Mexico. *Annals of Forest Science* 66:107(1–8).

Valerio, J. and C. Salas. 1998. *Selección de prácticas silviculturales para bosques tropicales. Proyecto BOLFOR*. Santa Cruz, Bolivia: Editora El País.

Van Bloem, S.J., P.G. Murphy, and A.E. Lugo. 2004. Tropical dry forests. In *Encyclopedia of Foest Sciences*, eds. J. Burley, J. Evans, and J. Youngquist. Oxford: Elsevier.

Van de Pijl, D.L. 1982. *Principles of dispersal in higher plants*. New York: Springer Verlag.

Van der Putten, W.H., R.D. Bardgett, P.C. de Ruiter, W.H.G. Hol, K.M. Meyer, T.M. Bezemer, M.A. Bradford, et al. 2009. Empirical and theoretical challenges in aboveground–belowground ecology. *Oecologia* 161:1–14.

Van Schaik, C.P. 1986. Phenological changes in a Sumatran rain forest. *Journal of Tropical Ecology* 2:327–347.

Van Schaik, C.P., J.W. Terborgh, and S.J. Wright. 1993. The phenology of tropical forests: adaptive significance and consequences for primary producers. *Annual Review of Ecology and Systematics* 24:353–377.

Van Schaik, C.P., M. Ancrenaz, G. Borgen, B. Galdikas, C.D. Knott, I. Singleton, A. Suzuki, S.S. Utami, and M. Merrill. 2003. Orangutan cultures and the evolution of material culture. *Science* 299:102–105.

Vega, J.L. 1985. *Hacia una interpretación del desarrollo costarricense: ensayo sociológico*. San José, Editorial Porvenir.

Vega, M. 2002. Los incendios forestales en el bosque seco, un problema regional de soluciones locales. In *Ecosistemas forestales de bosque seco tropical: Investigaciones y resultados en mesoamérica*, 41–47. Heredia, CR: Universidad Nacional, INISEFOR.

Vela-Vargas, I.M. and J. Pérez-Torres. 2012. Bats associated with remnants of tropical dry forest in an extensive cattle breeding system (Colombia). *Chiroptera Neotropical* 18:1089–1100.

Vermote, E.F., D. Tanre, J.L. Deuze, M. Herman, and J.J. Morcette. 1997. Second simulation of the satellite signal in the solar spectrum, 6S: An overview. *Geoscience and Remote Sensing, IEEE Transactions* 35(3):675–686.

Vicente, A., A.M.M. Santos, and M. Tabarelli. 2003. Variações no modo de dispersão de espécies lenhosas em um gradiente de precipitação entre floresta seca e úmida no nordeste do Brazil. In *Ecologia e conservação da caatinga*, eds. I.R. Leal, M. Tabarelli, and J.M.C. Silva, 657–694. Recife: Editora Universitária da UFPE.

Vieira, D.L.M. and A. Scariot. 2006. Principles of natural regeneration of tropical dry forests for restoration. *Restoration Ecology* 14:11–20.

Vieira, D.L.M. and A. Scariot. 2008. Environmental variables and tree population structures in deciduous forests of Central Brazil with different levels of logging. *Brazilian Archieves of Biology and Technology* 51:419–431.

Vieira, D.L.M., A. Scariot, and K.D. Holl. 2006. Effects of habitat, cattle grazing and selective logging on seedling survival and growth in dry forests of central Brazil. *Biotropica* 39:269–274.

Vieira, D.L.M., V.V. Lima, A.C. Sevilha, and A. Scariot. 2008. Consequences of dry-season seed dispersal on seedling establishment of dry Forest trees: Should we store seed until the rains? *Forest Ecology and Management* 256:471–481.

Vieira, S., S. Trumbore, P.B. Camargo, D. Selhorst, J.Q. Chambers, N. Higuchi, and L.A. Martinelli. 2005. Slow growth rates of Amazonian trees: Consequences for carbon cycling. *Proceedings of the National Academy of Sciences of the United States of America* 102:18502–18507.

Viglizzo, E.F., A.J. Pordomingo, M.G. Castro, and F.A. Lertora. 2003: Environmental assessment of agriculture at a regional scale in the pampas of Argentina. *Environmental Monitoring and Assessment* 87:169–195.

Vile, D., B. Shipley, and E. Garnier. 2006. A structural equation model to integrate changes in functional strategies during old-field succession. *Ecology* 87:504–517.

Villalobos, L. 2009. The effects of national parks on local communities' wages and employment in Costa Rica. CATIE.

Villalobos, S. 2010. Fenología foliar y reproductiva en tres estadios sucesionales de un bosque seco de los llanos venezolanos. Tesis de Maestría. Instituto Venezolano de Investigaciones Científicas IVIC. 135 p.

Villar, R. and J. Merino. 2001. Comparison of leaf construction costs in woody species with differing leaf life-spans in contrasting ecosystems. *New Phytologist* 151:213–226.

Vitousek, P.M. 1984. Litterfall, nutrient cycling, and nutrient limitation in tropical forests. *Ecology* 65:285–298.

Vitousek, P.M. and R.L. Sanford. 1986. Nutrient cycling in moist tropical forest. *Annual Review of Ecology and Systematics* 17:137–167.

Voeks, R.A. 2007. Are women reservoirs of traditional plant knowledge? Gender, ethnobotany and globalization in northeast Brazil. *Singapore Journal of Tropical Geography* 28:7–20.

von Thunen, J.H. 1966. Der isolierte Staat in Beziehung der Landwirtschaft und Nationalokomie. In *The isolated state*, ed. D.P. Hall. Oxford, UK: Pergamon Press.

Walker, T.W. and J.K. Syers. 1976. The fate of phosphorus during pedogenesis. *Geoderma* 15:1–19.

Waltner-Toews, D., J.J. Kay, C. Neudoerffer, and T. Gitau. 2003. Perspective changes everything: managing ecosystems from the inside out. *Frontiers in Ecology and the Environment* 1:23–30.

Wang, D., G. Wang, and E.N. Anagnostou. 2007. Evaluation of canopy interception schemes in land surface models. *Journal of Hydrology* 347(3–4):308–318.

Wang, Q., J. Tenhunen, A. Granier, M. Reichstein, O. Bouriaud, D. Nguyen, and N. Breda. 2004. Long-term variations in leaf area index and light extinction in a Fagus sylvatica stand as estimated from global radiation profiles. *Theoretical and Applied Climatology* 79(3–4):225–238.

Wang, Q., J. Tenhunen, N.Q. Dinh, M. Reichstein, D. Otieno, A. Granier, and K. Pilegarrd. 2005. Evaluation of seasonal variation of MODIS derived leaf area index at two European deciduous broadleaf forest sites. *Remote Sensing of Environment* 96:475–484.

Warming, E. and M.G. Ferri. 1973. *Lagoa Santa e a vegetação de cerrados brasileiros*. São Paulo: EDUSP.

Watson, D.J. 1947. Comparative physiological studies in the growth of field crops. I. Variation in net assimilation rate and leaf area between species and varieties, and within and between years. *Annals of botany* 11, 41–76.

Weiher, E. and P.A. Keddy. 1995. Assembly rules, null models, and trait dispersion: new question from old patterns. *Oikos* 74:159–164.

Welles, J.M. and W.S Cohen. 1996. Canopy structure measurement by gap fraction analysis using commercial instrumentation. *Journal of Experimental Botany* 47:1335–1342.

Werneck, M.S., E.V. Franceschinelli, and E. Tameirão-Neto. 2000. Mudanças na florística e estrutura de uma floresta decídua durante um período de quatro anos (1994–1998), na região do Triângulo Mineiro, MG. *Revista Brasileira de Botânica* 23:401–413.

Westoby, M., D.S. Falster, A.T. Moles, P.A. Vesk, and I.J. Wright. 2002. Plant ecological strategies: Some leading dimensions of variation between species. *Annual Review of Ecology and Systematics* 33:125–159.

White, J.D., S.W. Running, R.R. Nemani, R.E. Keane, and K.C. Ryan. 1997. Measurement and remote sensing of LAI in Rocky Mountain montane ecosystems. *Canadian Journal of Forest Research* 27:1714–1727.

Whiting, G.J., D.S. Bartlett, M. Fan, P.S. Bakwin, and S.C. Wofsy. 1992. Biosphere atmosphere CO_2 exchange in tundra ecosystems - community characteristics and relationships with multispectral surface reflectance. *Journal of Geophysical Research Atmospheres* 97:16671–16680.

Wijdeven, S.M.J. and M.E. Kuzee. 2000. Seed availability as a limiting factor in forest recovery processes in Costa Rica. *Restoration Ecology* 8:414–424.

Wikander, T. 1984. Mecanismos de dispersión de diasporas de una selva decidua en Venezuela. *Biotropica* 16:276–283.

Williams-Linera, G. 1997. Phenology of deciduous and broad leaf evergreen tree species in a Mexican tropical lower montane forest. *Global Ecology and Biogeography* 6:115–127.

Williams-Linera, G. 2000. Leaf demography and leaf traits of temperate-deciduous and tropical evergreen-broadleaved trees in a Mexican montane cloud forest. *Plant Ecology* 149:233–244.

Williams, L.J., S. Bunyavejchewin, and P.J. Baker. 2008. Deciduousness in a seasonal tropical forest in western Thailand: interannual and intraspecific variation in timing, duration and environmental cues. *Oecologia* 155:571–582.

Williams, R.G. 1986. *Export agriculture and the crisis in Central America*. Chapel Hill: University of North Carolina Press.

Willson, M.F., A.K. Irvine, and N.G. Walsh. 1989. Vertebrate dispersal syndromes in some Australian and New Zealand plant communities, with geographic comparison. *Biotropica* 21:133–147.

Worbes, M. 1999. Annual growth rings, rainfall-dependent growth and long-term growth patterns of tropical trees from the Caparo Forest Reserve in Venezuela. *Journal of Ecology* 87:391–403.

World Commission on Environment Development. 1987. *Our common future*. London: Oxford University Press.

Wright, J. and F. Cornejo. 1991. Seasonal drought and leaf fall in a tropical forest. *Ecology* 71:1165–1175.

Wright, J. and O. Calderón. 1995. Phylogenetic patterns among tropical flowering phenologies. *Journal of Ecology* 83:937–948.

Wright, S.J. 1996. Phenological responses to seasonality in tropical forest plants. In *Tropical forest plant ecophysiology*, eds. S.S. Mulkey, R.L. Chazdon, and A.P. Smith, 444–460. New York: Chapman and Hall.

Wright, S.J. and C.P. van Schaik. 1994. Light and the phenology of tropical trees. *The American Naturalist* 143:192–199.

Wright, S.J., C. Carrasco, O. Calderón, and S. Paton. 1999. The El Niño Southern Oscillation, variable fruit production, and famine in a tropical forest. *Ecology* 80:1632–1647.

Wu, J. 2000. Slippage effects of the Conservation Reserve Program. *American Journal of Agricultural Economics* 82:979–992.

Wunderlee, J.M. 1997. The role of animal seed dispersal in accelerating native forest regeneration on degraded tropical lands. *Forest Ecology and Management* 99:223–235.

Xiao, Q., E.G. McPherson, S.L. Ustin, and M.E. Grismer. 2000. A new approach to modeling tree rainfall interception. *Journal of Geophysical Research* 105(D23):29173–29188.

Xu, Z.Z., G.S. Zhou, and Y.H. Wang. 2007. Combined effects of elevated CO_2 and soil drought on carbon and nitrogen allocation of the desert shrub *Caragana* intermedia. *Plant and Soil* 301:87–97.

Xuluc-Tolosa, F. 2003. Leaf litter decomposition of tree species in three successional phases of tropical dry secondary forest in Campeche, Mexico. *Forest Ecology and Management* 174:401–412.

Yamashita, N., S. Ohta, H. Sase, J. Luangjame, T. Visaratana, B. Kievuttinon, H. Garivait, and M. Kanzaki. 2010. Seasonal and spatial variation of nitrogen dynamics in the litter and surface soil layers on a tropical dry evergreen forest slope. *Forest Ecology and Management* 259:1502–1512.

Yang, C., C. Hamel, M.P. Schellenberg, J.C. Perez, and R.L. Berbara. 2010. Diversity and functionality of arbuscular mycorrhizal fungi in three plant communities in semiarid Grasslands National Park, Canada. *Microbial Ecology* 59:724–733.

Yin, R.K. 2003. *Case study research. Design and methods*. Thousand Oaks: Sage Publications (Applied social research methods series Volume 5).

Yodzis, P. 1986. Competition, mortality and community structure. In *Community ecology*, eds. J. Diamond and T.J. Case, 480–491. New York: Harper and Row.

Yoshifuji, N., T. Kumagai, K. Tanaka, N. Tanaka, H. Komatsu, M. Suzuki, and C. Tantasirin. 2006. Inter-annual variation in growing season length of a tropical seasonal forest in northern Thailand. *Forest Ecology and Management* 229(1–3):333–339.

Zahl, S. 1977. Jackknifing an index of diversity. *Ecology* 58:907–913.

Zaidan, L.B.P. and C.J. Barbedo. 2004. Quebra de dormência em sementes. In *Germinação: do básico ao aplicado*, eds. A.G. Ferreira and F. Borghetti, 135–146. Porto Alegre: Artmed.

Zalamea, M. and G. González. 2008. Leaf fall phenology in a subtropical wet forest in Puerto Rico: from species to community patterns. *Biotropica* 40:295–304.

Zangaro, W., R.A. Alves, L.E. Lescano, A.P. Ansanelo, and M.A. Nogueira. 2012. Investment in fine roots and arbuscular mycorrhizal fungi decrease during succession in three Brazilian ecosystems. *Biotropica* 44:141–150.

Zappi, D. 2008. Fitofisionomia da Caatinga associada a Cadeia do Espinhaço. *Megadiversidade* 4:34–38.

Zar, J.H. 1996. *Biostatistical analysis*. New Jersey: Prentice Hall.

Zbinden, S. and D.R. Lee. 2005. Paying for environmental services: An analysis of participation in Costa Rica's PSA Program. *World Development* 33(2):255–272.

Zhang, J., B. Rivard, A. Sánchez-Azofeifa, and K. Castro-Esau. 2006. Intra-and inter-class spectral variability of tropical tree species at La Selva, Costa Rica: Implications for species identification using HYDICE imagery. *Remote Sensing of Environment* 105(2):129–141.

Zhang, X., M. Friedl, C. Schaaf, A.H. Strahler, J.C. Hodges, F. Gao, B.C. Reed, and A. Huete. 2003. Monitoring vegetation phenology using MODIS. *Remote Sensing of Environment* 84:471–475.

Zhouri, A. 1998. Trees and people: An anthropology of British campaigners for the Amazon Rainforest. PhD diss., University of Essex, Essex, UK.

Zhouri, A. 2001. Ambientalismo e Antropologia. Descentrando a categoria de Movimentos Sociais. Teoria & Sociedade, No. 8.

Zhouri, A. and K. Laschefski. 2010. Desenvolvimento e conflitos ambientais: um novo campo de investigação. In *Desenvolvimento e conflitos ambientais*, eds. A. Zhouri and K. Laschefski, 11–31. Belo Horizonte: Editora UFMG.

Zhouri, A., K. Laschefski, and D. Pereira. 2005. *A Insustentável leveza da política ambiental: desenvolvimento e conflitos sócio-ambientais*. Belo Horizonte: Autêntica.

Zhu, S.D. and K.F. Cao. 2010. Contrasting cost-benefit strategy between lianas and trees in a tropical seasonal rain forest in southwestern China. *Oecologia* 163:591–599.

Zotz, G., N. Cueni, and C. Korner. 2006. in situ growth stimulation of a temperate zone liana (Hedera helix) in elevated CO_2. *Functional Ecology* 20:763–769.

Zuchiwschi, E., A.C. Fantini, A.C. Alves, and N. Peroni. 2010. Limitações ao uso de espécies florestais nativas pode contribuir com a erosão do conhecimento ecológico tradicional e local de agricultores familiares. *Acta botanica brasilica* 24:270–278.

Index

C

D

E

W

Y

Z